THE FUNDAMENTALS OF IMAGING PHYSICS AND RADIOBIOLOGY

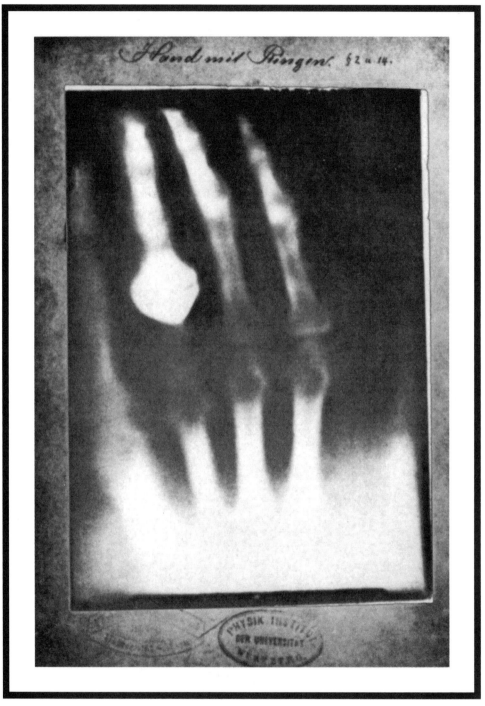

Copy of Roentgen's radiograph of his wife's "hand with rings," made soon after his discovery of x rays in November 1895 in Wurzburg, Germany. (See Glasser O. *Dr. W. C. Roentgen,* p. 39, 2nd ed., Springfield, Charles C Thomas, 1958.)

NINTH EDITION

THE FUNDAMENTALS
OF
IMAGING PHYSICS
AND
RADIOBIOLOGY

By

JOSEPH SELMAN, M.D., FACR, FACP

Medical Advisor
Radiation Therapy Center
East Texas Medical Center Cancer Institute

Charles C Thomas
PUBLISHER • LTD.
SPRINGFIELD • ILLINOIS • U.S.A.

Published and Distributed Throughout the World by

CHARLES C THOMAS • PUBLISHER
2600 South First Street
Springfield, Illinois 62794-9265

©*2000, 1994, 1985, 1977, 1972, 1965, 1961, 1957, and 1954*
by CHARLES C THOMAS • PUBLISHER
First Edition, 1954
Second Edition, 1957
Third Edition, 1961
Fourth Edition, 1965
Fifth Edition, 1972
Sixth Edition, 1977
Seventh Edition, 1985
Eighth Edition, 1994
Ninth Edition, 2000

ISBN 0-398-06987-5

Library of Congress Catalog Card Number: 99-035813

Printed in the United States of America
SM-HP-3

Library of Congress Cataloging-in-Publication Data

Selman, Joseph.
    The fundamentals of imaging physics and radiobiology / Joseph Selman --
9th ed.
       p.   cm.
    Prev. ed.: The fundamentals of x-ray and radium physics.
    Includes bibliographical references and index.
    ISBN 0-398-06987-5
    1. X-rays. 2. Radium. 3. Radiography. I. Selman, Joseph.
Fundamentals of x-ray and radium physics. II. Title.
QC481 .S456 2000
537.5'35--dc21
                                                99-035813
                                                   RW

Dedicated to my wife

PREFACE TO THE NINTH EDITION

A number of important changes have been made in this edition, the first being the obvious new title reflecting the demise of radium in medicine and its replacement by artificial radioisotopes (radionuclides).

To improve readability, a double-column format has been adopted, serving to bring more illustrations closer to their explanation in the text. A helpful addition should be the outline of the contents of each chapter below its heading.

Greater attention is paid to high-frequency generation of x rays. Because this involves microprocessors (computers), which also find wide application in the department generally, I have added a chapter on basic computer science. A typical modern control panel for comparison with the primitive type indicates how much progress has occurred in x-ray production and control.

The rapid increase in the number of approved mammography centers demands a corresponding need for specialized technologists. I have therefore updated the requirements for mammographic equipment and included a section outlining quality assurance, especially as it applies to the responsibilities of the designated mammographer. A separate chapter is now devoted to mammography.

Digital imaging has been assigned a new chapter to include more information about subtraction angiography.

A new chapter takes up the basic science of radiobiology: the effects of radiation, especially of x rays, on cells, tissues, and organs. Since harmful effects are dose-dependent, this fits well with the chapter on health physics, which has been updated.

More questions and problems, with sample solutions, have been added. Besides, the index continues to be user-friendly to facilitate the search for answers in the text.

Many thanks are due Joe Burrage (Diagnostic Imaging, Inc.) for help with modern x-ray generator circuitry in the form of block diagrams, as well as explanations of newer equipment and devices. I am also grateful to Ted Kosnik, Ph.D., for his answers about matters physics, and for plotting the attenuation curves of 100-kV x rays; and to my son Jay E. Selman, M.D., for his review of the chapter on computer science and his very helpful comments.

I am also indebted to the following individuals and companies, thanking them for providing brochures and technical information on equipment and supplies: Larry Spittler, RT (Diagnostic Imaging, Inc.); Nu-Tech (Tyler, Texas); and Paul Oster, RT, CNMT, Chief Technologist of the Nuclear Medicine Laboratory, East Texas Medical Center Hospital, Tyler, Texas.

New illustrations and modifications of previous ones have been meticulously executed by artist Gene Johnson (Tyler, Texas), and for this I am most grateful. At this point I wish to thank again the retired artist, Howard Marlin, for his excellent illustrations in the first eight editions of this book, dating back to 1954, and carried over to the ninth edition.

Finally, and by no means least, I would like to express my sincere gratitude to Michael Thomas, head of Charles C Thomas, Publisher, and his most competent staff, and to Claire Slagle, the superb editor assigned to me, for their cooperation and diligent attention to details in the publication of my book for its Ninth Edition.

Joseph Selman, M.D.

INTRODUCTION

As a student entering an approved School of Radiologic Technology, you have probably heard about the hazards of overexposure to x rays, a special form of light. The key word here is "approved," a designation assuring that the school has in place not only the required teaching program, but also all necessary protective measures to keep your occupational exposure to x rays and related radiation within acceptable limits. Necessary protective maintenance has been established by regulations ("regs") mandated by the relevant State Agency and the U.S.A. Nuclear Regulatory Commission (NRC). I shall briefly summarize here the basic protective features that you will find in a modern radiology department.

The walls of the x-ray rooms have built-in lead or equivalent protection in accordance with State regulations, and have been tested by a certified Health Physicist or other qualified person. A number of protective features have been incorporated in the x-ray equipment as mandated by the Bureau of Radiological Health (BRH). You will note that the console, which controls x-ray tube operation by the radiographer, is located inside a control booth whose walls contain a prescribed thickness of lead (metal) for the radiographer's protection; even the booth's window consists of protective glass of proper thickness.

In fluoroscopy you will see a lead-containing rubber or plastic curtain hanging from the viewing assembly to protect the fluoroscopist; leaded aprons and gloves are also available for your personal protection.

The radiographer must follow all rules governing radiation protection according to the three basic principles: distance, shielding, and time: radiation dose decreases rapidly with increasing distance from the x-ray source; shielding of the hands and body by lead or other protective material; and reducing the time spent in the area during exposure all contribute greatly to dose reduction. Finally, special badges must be worn by all personnel in the x-ray area to detect any breach in safe operating procedures.

In my experience over many years, monitored exposures of radiographers have, with rare exceptions, been well within prescribed limits. Moreover, the risk from occupational exposure to radiation in the radiology department is trivial compared to other occupational, recreational, and highway travel risks. The same applies to the risk in a modern radiology department.

The chapter on Health Physics, which you will study later, provides additional details on the subject of radiation protection of personnel and patients.

CONTENTS

		Page
Preface		vii
Introduction		ix
Chapter		
1.	SIMPLIFIED MATHEMATICS	3
	Arithmetic	3
	Fractions	3
	Percent	5
	Decimal Fractions	5
	Significant Figures	5
	Algebra	6
	Ratio and Proportion	9
	Plane Geometry	10
	Similar Triangles	11
	Graphs and Charts	12
	Large and Small Numbers	13
	Logarithms	14
	Questions and Problems	15
2.	PHYSICS AND THE UNITS OF MEASUREMENT	16
	Standard Units	17
	Fundamental Units	17
	Derived Units	18
	Manipulation of Units	19
	Prefixes Applied to SI Units	21
	Questions and Problems	21
3.	THE PHYSICAL CONCEPT OF ENERGY	22
	Force	22
	Work and Energy	22
	Law of Conservation of Energy	23
	Questions and Problems	25
4.	THE STRUCTURE OF MATTER	26

	Subdivisions of Matter	26
	Atomic Structure–the Electrical Nature of Matter	28
	Atomic Number	30
	Mass Number	30
	Isotopes and Nuclides	30
	The Periodic Table	31
	Chemical Behavior	34
	Ionization	36
	Questions and Problems	38
5.	ELECTROSTATICS	39
	Definition	39
	Electrification	39
	Methods of Electrification	39
	Laws of Electrostatics	41
	Electroscope	43
	Static Discharge	44
	Questions and Problems	45
6.	ELECTRODYNAMICS–ELECTRIC CURRENT	46
	Definition	46
	The Nature of an Electric Current	46
	Sources of Electric Current	47
	Factors in an Electric Circuit	47
	Ohm's Law	50
	Cells and Batteries	50
	Components of Elementary Electric Circuits	51
	Series and Parallel Circuits	52
	Electric Capacitor (Condenser)	56
	The Work and Power of a Direct Current	57
	Questions and Problems	58
7.	MAGNETISM	59
	Definition	59
	Classification of Magnets	59
	Laws of Magnetism	60
	Nature of Magnetism	60
	Magnetic Fields	61
	Characteristics of Lines of Force	62
	Magnetic Induction (Magnetization)	63

Magnetic Permeability and Retentivity 64
Magnetic Classification of Matter 64
Earth's Magnetism 64
Questions and Problems 65
8. ELECTROMAGNETISM 66
Definition 66
Electromagnetic Phenomena 66
The Electromagnet 67
Electromagnetic Induction 68
Direction of Induced Electron Current 69
Self-induction 69
Mutual Induction 71
Questions and Problems 71
9. ELECTRIC GENERATORS AND MOTORS 72
Electric Generator 72
Definition 72
Essential Features 72
Simple Electric Generator 73
Properties of Alternating Currents 75
Direct Current Generator 76
Advantages of Alternating Current 77
Electric Motor 78
Definition 78
Principle 78
Simple Electric Motor 79
Types of Electric Motors 79
Current-measuring Devices 81
Questions and Problems 82
10. PRODUCTION AND CONTROL OF HIGH VOLTAGE–
REGULATION OF CURRENT IN THE X-RAY TUBE 83
Transformer 83
Principle 83
Construction of Transformers 85
Efficiency and Power Losses 86
Control of High Voltage 88
Autotransformer 88
Control of Filament Current and Tube Current 89

Choke Coil   89
Rheostat   90
High-Frequency Control of Current and Voltage   91
Questions and Problems   92
11.   RECTIFICATION AND RECTIFIERS   94
Definition   94
Methods of Rectifying an Alternating Current   95
Rectifiers   100
Rectifier Failure   105
Questions and Problems   106
12.   X RAYS: PRODUCTION AND PROPERTIES   108
How X Rays Were Discovered   108
Nature of X Rays   109
Source of X Rays in Radiology   111
The X-Ray Tube   111
Details of X-Ray Production   112
Electron Interactions with Target Atoms–X-ray Production   113
Target Material   115
Efficiency of X-ray Production   115
Properties of X Rays   115
Specifications of the Physical Characteristics of an X-ray Beam 116
X-ray Exposure (Quantity)   116
X-ray Quality   117
"Hard" and "Soft" X Rays   123
The Interactions of Ionizing Radiation and Matter   124
Relative Importance of Various Types of Interaction   129
Detection of Ionizing Radiation   130
Summary of Radiation Units   131
Exposure–the Roentgen (R)   131
Absorbed Dose–the Gray (Gy)   131
Modification of Kilovoltage X-ray Beams by Filters   132
Questions and Problems   133
13.   X-RAY TUBES   135
Thermionic Diode Tubes   135
Radiographic Tubes   136
Glass Envelope   136
Cathode   136

	Anode	138
	Space Charge Compensation	141
	Factors Governing Tube Life	143
	Questions and Problems	148
14.	X-RAY CIRCUITS	149
	Source of Electric Power	149
	Main Single-Phase X-Ray Circuits	150
	Completed Wiring Diagram	158
	Basic X-Ray Control Panel or Console	158
	Three-Phase Generation of X-Rays	162
	High-Frequency Generation of X Rays	165
	Power Rating of X-Ray Generators and Circuits	166
	Falling-Load Generator	167
	Special Mobile X-Ray Equipment	168
	Battery-Powered Mobile X-Ray Units	168
	Capacitor (Condenser)-Discharge Mobile X-Ray Units	168
	Questions and Problems	171
15.	X-RAY FILM, FILM HOLDERS, AND INTENSIFYING SCREENS	173
	Composition of X-Ray Film	173
	Types of Films	175
	Practical Suggestions in Handling Unexposed Film	176
	Film Exposure Holders	176
	Intensifying Screens	177
	Questions And Problems	186
16.	THE DARKROOM	188
	Introduction	188
	Location of the Darkroom	188
	Building Essentials	189
	Entrance	189
	Size	190
	Ventilation	190
	Lighting	190
	Film Storage Bin	191
	Questions And Problems	191
17.	CHEMISTRY OF RADIOGRAPHY AND FILM PROCESSING	192

Introduction   192
Radiographic Photography   192
Radiographic Chemistry   193
Manual Processing   195
Film Fog, Stains, and Artifacts   196
Automatic Processing   197
Summary of Processor Care   203
Silver Recovery from Fixing Solutions   203
Questions And Problems   203

18.   RADIOGRAPHIC QUALITY   205
Blur   206
Geometric or Focal Spot Blur   206
Focal Spot Evaluation   208
Motion Blur   210
Screen Blur   210
Object Blur   212
Radiographic Density   213
Contrast   217
Radiographic Contrast   217
Subject Contrast   218
Film Contrast   220
Distortion   222
Direct Magnification or Enlargement Radiography
(Macroradiography)   226
Modulation Transfer Function   228
Questions and Problems   230

19.   DEVICES FOR IMPROVING RADIOGRAPHIC QUALITY   232
Scattered Radiation   232
Removal of Scattered Radiation By a Grid   234
Principle of the Radiographic Grid   234
Efficiency of Grids   235
Types of Grids   238
Precautions in the Use of Focused Grids   240
Practical Application of Grids   242
Removal of Scattered Radiation by an Air Gap   246
Reduction of Scattered Radiation by Limitation of the
Primary Beam   247

Other Methods of Enhancing Radiographic Quality   252
   Moving Slit Radiography   252
   The Anode Heel Effect–Anode Cutoff   254
Compensating Filters   254
Summary of Radiographic Exposure   255
Questions and Problems   255

20. FLUOROSCOPY   258
   The Human Eye   258
   Fluoroscopic Image Intensification   259
   Magnification in Image Intensifier   263
   Multiple-Field Intensifiers   263
Viewing the Fluoroscopic Image   264
   Optical Lens System   264
   Television Viewing System   264
   Video Cameras   264
   Television Monitor   266
   Charge-Coupled Device (CCD) TV Camera   267
   Quality of the TV Image   267
   Recording the Fluoroscopic Image   268
   Recording the Video Image   269
   Laser Discs   270
Questions and Problems   271

21. VANISHING EQUIPMENT   273
Stereoscopic Radiography   273
Tomography   276
Xeroradiography (Xerography)   282
Questions and Problems   284

22. MAMMOGRAPHY   285
Quality Standards in Mammography   289
   Qualifications of Radiologic Technologist (Mammographer)   291
   Equipment   291
   Quality Assurance   291
   Quality Assurance Records   291
   Radiologic Technologist Responsibilities   291
Quality Control Tests   292
   Daily   292
   Weekly   292

Monthly   292
Quarterly   293
Semiannually   293
Mobile Mammography Units   294
Questions and Problems   294

23. BASIC COMPUTER SCIENCE   295
Introduction   295
History   295
Data *vs* Information   297
Computer Operations   297
Computer Components   298
Computer Language   303
Binary Number System   303
Summary of Applications in Radiology   304
Questions and Problems   305

24. DIGITAL X-RAY IMAGING   306
Introduction   306
Digital Fluoroscopy   307
Questions and Problems   312

25. COMPUTED TOMOGRAPHY   314
Conventional CT Scanning   314
Spiral (helical) CT Scanning   319
Questions and Problems   322

26. RADIOACTIVITY AND DIAGNOSTIC
     NUCLEAR MEDICINE   323
Introduction   323
Natural Radioactivity   324
Unstable Atoms   324
Radioactive Series   324
Radium   325
Introduction   325
Properties   325
Types of Radiation   325
Radioactive Decay   326
Decay Constant   327
Half-Life   327
Average Life   327

Radon 328
Radioactive Equilibrium 328
Artificial Radioactivity 329
  Isotopes 329
  Artificial Radionuclides 329
  Nuclear Reactor 330
  Nuclear Transformations 332
  Properties of Artificial Radionuclides 332
  Radioactive Decay 333
Applications of Radionuclides in Medicine 335
Radionuclide Instrumentation 336
  Sources of Error in Counting 339
  Efficiency and Sensitivity of Counters 341
  Geometric Factors in Counting 342
  Methods of Counting 343
  Important Medical Radionuclides 344
  Examples of Radionuclides in Medical Diagnosis 346
Imaging With Radionuclides: the Gamma Camera 348
  Collimators 348
  Crystal-Photomultiplier Complex 350
  Single-Photon Emission Computed Tomography (SPECT) 350
  Available Radionuclide Imaging 351
  Calibration of Radiopharmaceuticals and Gamma Camera 353
Questions and Problems 353

27. RADIOBIOLOGY 355
  Definition 355
  History 355
  The Physical Basis of Radiobiology 356
Radiobiologic Effects 357
Structure and Function of Cells 358
  Nucleus 358
  Cytoplasm 359
Cell Reproduction 360
  Mitosis 360
  Meiosis 362
  Structure of DNA 363
  Functions of DNA 364

The Radiobiologic Lesion		366
Modes of Action of Ionizing Radiation		366
Cellular Response To Radiation		368
Nuclear Damage		368
Cytoplasmic Damage		369
Cellular Radiosensitivity		369
Modifying Factors in Radiosensitivity		370
Acute Whole-Body Radiation Syndromes		372
Explanation of Acute Whole-Body Radiation Syndromes		374
Dose-Response Models		375
Sigmoid Dose-Response Curve		375
Linear Dose-Response Curve		376
Linear-Quadratic Dose-Response Curve		377
Injurious Effects of Radiation on Normal Tissues		377
Early Effects: Limited Areas of the Body		377
Late Effects		381
Late Somatic Effects in High-Dose Region		382
Late Somatic Effects in Low-Dose Region		383
Risk Estimates for Genetic Damage		386
Radiation Injury to Embryo and Fetus		387
Questions and Problems		390
28.	PROTECTION IN RADIOLOGY–HEALTH PHYSICS	391
Introduction		391
Background Radiation		392
Dose Equivalent Limit		394
Derivation of Unit For Risk Assessment		394
Dose Equivalent		395
Numerical Dose Equivalent Limits		396
Occupational		396
Fertile or Pregnant Radiation Workers		398
Nonoccupational (General Public) Limit		398
ALARA Concept		399
Personnel Protection From Exposure To X Rays		400
Protective Measures		402
Protective Barriers in Radiography and Fluoroscopy		403
Working Conditions		404
Acceptance (Compliance) Testing		405

Protection Surveys 406
Protection of the Patient in Diagnostic Radiology 407
 Dose Reduction in Radiography 410
 Protection in Mammography 414
 Computed Tomography Scanning 414
 Patient Protection in Fluoroscopy 415
Protection From Electric Shock 416
Protection in Nuclear Medicine 416
Questions and Problems 420
29. NONRADIOLOGIC IMAGING 422
Magnetic Resonance Imaging 422
 Nuclear Magnetic Resonance 422
 From NMR to MRI 429
 Noise in MRI 435
 External Field Magnets for MRI 436
 Surface Coils 436
 A Typical MRI Unit 437
 Hazards of MRI 437
Ultrasound Imaging 438
 Nature of Sound 438
 Properties of Ultrasound 439
 Production of Ultrasound 440
 Ultrasound Beam Characteristics 441
 Echo Reception of Ultrasound 443
 Behavior of Ultrasound in Matter 443
 Ultrasound Image Displays 446
 Question of Biohazard 449
Questions and Problems 449
Appendix—ANSWERS TO PROBLEMS 451
Bibliography 453
Index 457

THE FUNDAMENTALS OF IMAGING PHYSICS AND RADIOBIOLOGY

Chapter 1

SIMPLIFIED MATHEMATICS

Topics Covered in This Chapter

- Arithmetic
    Fractions
    Percent
    Decimal Fractions
    Significant Figures
- Algebra

- Ratio and Proportion
- Plane Geometry
- Graphs and Charts
- Large and Small Numbers
- Logarithms
- Questions and Problems

ALL OF THE PHYSICAL SCIENCES have in common a firm basis in mathematics. This is no less true of radiologic physics, an important branch of the physical sciences. Clearly, then, in approaching a course in radiologic physics you, as a student technologist, should find your path smoothed by an adequate background in the appropriate areas of mathematics.

We shall assume here that you have had at least the required high school exposure to mathematics, although this may vary widely from place to place. However, realizing that much of this material may have become hazy with time, we shall review the simple but necessary aspects of arithmetic, algebra, and plane geometry. Such a review should be beneficial in at least two ways. First, it should make it easier to understand the basic principles and concepts of radiologic physics. Second, it should aid in the solution of such everyday problems as conversion of radiographic techniques, interpretation of tube rating charts, determination of radiographic magnification, and many others that may arise from time to time.

The discussion will be subdivided as follows: (1) arithmetic, (2) algebra, (3) ratio and proportion, (4) geometry, (5) graphs and charts, and (6) large and small numbers. Only fundamental principles will be included.

ARITHMETIC

Arithmetic is calculation or problem solving by means of definite numbers. We shall assume that you are familiar with addition, subtraction, multiplication, and division and shall therefore omit these operations.

Fractions

In arithmetic, *a fraction may be defined as one or more equal parts of a unit*. For example, $1/2$, $1/3$, and $2/5$ are fractions. The quantity below the line is called the **denominator**; it indicates the number of equal parts into which the unit is divided. The quantity above the line is the **numerator**; it indicates the number of equal parts taken. Thus, if a pie were divided into three equal parts, the denominator would be 3; and if two of these parts were taken, the numerator would be 2, so, the two segments would represent $2/3$ of the pie.

Fractions represent **the division of one quantity by another**. This extends the concept of fractions to expressions in which the numerator is larger than the denominator, as in the fraction $5/2$.

If the numerator is smaller than the denominator, as $3/5$, we have a **proper fraction**. If the numerator is larger than the denominator, as $5/3$, we have an **improper fraction**, because $5 \div 3 = 1^2/3$, which is really an integer plus a fraction.

In **adding** fractions, all of which have the **same** denominator, add all the numerators first and then place the sum over the denominator:

$$\frac{2}{7} + \frac{3}{7} + \frac{6}{7} + \frac{5}{7} = \frac{2 + 3 + 6 + 5}{7} = \frac{16}{7}$$

$$\frac{16}{7} = 2\frac{2}{7}$$

Subtraction of fractions having identical denominators follows the same rule:

$$\frac{6}{7} - \frac{4}{7} = \frac{6 - 4}{7} = \frac{2}{7}$$

If fractions are added or subtracted, and the denominators are **different**, then the **least common denominator** must be found. This is the smallest number which is exactly divisible by all the denominators. Thus,

$$\frac{1}{2} + \frac{2}{3} - \frac{3}{4} = ?$$

The smallest number which is divided exactly by each denominator is 12. Place 12 in the denominator of a new fraction:

$$\overline{12}$$

Divide the denominator of each of the fractions in the old equation into 12, and then multiply the answer by the numerator of that fraction; each result is then placed in the numerator of the new fraction:

$$\frac{6 + 8 - 9}{12} = \frac{5}{12}$$

Multiplying fractions means taking their product. To multiply fractions, take the product of the numerators and place it over the product of the denominators,

$$\frac{4}{5} \times \frac{3}{10} = \frac{4 \times 3}{5 \times 10} = \frac{12}{50}$$

The resulting fraction can be reduced by dividing the numerator and the denominator by the same number, in this case, 2:

$$\frac{12}{50} \div \frac{2}{2} = \frac{6}{25}$$

which cannot be further simplified.

Note that when the numerator and the denominator are both multiplied or divided by the same number, the value of the fraction does not change. For instance,

$$\frac{3}{5} \times \frac{2}{2} = \frac{6}{10}$$

is the same as $\frac{3}{5} \times 1 = \frac{3}{5}$

When two fractions are to be divided, as $4/5 \div 3/7$, the fraction that is to be divided is the **dividend**, and the fraction that does the dividing is called the **divisor**. In this case, $4/5$ is the dividend and $3/7$ the divisor. The rule is to invert the divisor (called "taking the reciprocal") and multiply the dividend by it:

$$\frac{4}{5} \div \frac{3}{7}$$

$$\frac{4}{5} \times \frac{7}{3} = \frac{28}{15} = 1\frac{13}{15}$$

Percent

A special type of fraction, ***percent***, is represented by the sign % to indicate that the number standing with it is to be divided by 100. Thus, $95\% = {}^{95}/{}_{100}$. We do not use percentages directly in mathematical computations, but first convert them to fractions or decimals. For instance,

150 × 40% is changed to
$150 \times {}^{40}/{}_{100}$ or $150 \times {}^{2}/{}_{5}$
or 150 × 0.40.

All these expressions equal 60.

Decimal Fractions

Our common method of representing numbers as multiples of ten is embodied in the ***decimal system***. A ***decimal fraction*** has as its denominator 10, or 10 raised to some power such as 100, 1000, 10,000, etc. The denominator is symbolized by a dot in a certain position. For example, the decimal 0.2 = ${}^{2}/{}_{10}$; 0.02 = ${}^{2}/{}_{100}$; 0.002 = ${}^{2}/{}_{1000}$, etc. Decimals can be multiplied or divided, but care must be taken to place the decimal point in the proper position:

$$
\begin{array}{r}
2.24 \\
\times 1.25 \\
\hline
1120 \\
448 \\
224 \\
\hline
2.8000
\end{array}
$$

First, add the total number of digits to the right of the decimal points in the numbers being multiplied, which in this case turns out to be four. Then point off four places from the right in the answer to determine the correct position of the decimal point. The decimal system is used everywhere in science and in the vast majority of countries in daily life.

Figure 1.1. With this calibrated scale we can estimate to the nearest tenth. Thus, the position of the pointer indicates 8.4 units, the 0.4 being the last significant figure.

Significant Figures

The precision (reproducibility of results) of any type of measurement is limited by the design of the measuring instrument. For example, a scale calibrated in grams as shown in Figure 1.1 allows an estimate to the nearest tenth of a gram. Thus, the scale in Figure 1.1 reads 8.4 grams. The last figure, 0.4, is estimated and is the ***last significant figure***–that is, it is the last meaningful digit. Obviously, no greater precision is possible with this particular instrument. To improve precision, the scale would have to show a greater number of subdivisions.

Significant figures are used in various mathematical operations. For example, in addition:

item 1	98.26	grams
item 2	1.350	g
item 3	260.1	g
	359.710	g

Notice that three digits appear after the decimal point in the answer. But in item (3) there is only one digit, 1, after the decimal point; beyond this, the digits are unknown. Therefore, the digits after the 7 in the answer imply more than is known, since the answer can be no more precise than the least precise item being added. In this case, the answer should be properly stated as 359.7. In addition and subtraction the answer can have no more significant figures after the decimal point than the item with the ***least number of significant figures after its decimal point***.

A different situation exists in multiplication and division. Here, the total number of significant figures in the answer equals that in the items having the ***least total number of significant figures***. For example, in

$$25.23 \ \text{cm}$$
$$\underline{\times \ 1.21} \ \text{cm}$$
$$2523$$
$$5046$$
$$\underline{2523 \ \ }$$
$$30.5283 \ \text{cm}^2$$

1.21 has fewer significant figures–three in all–so the answer should have three significant figures and be read as 30.5 (dropping the 0.0283).

In general, to ***round off*** significant figures, observe the following rule: if the digit following the last significant figure is equal to or greater than 5, the last significant figure is increased by 1; if less than 5, it is unchanged. The rule is applied in the following examples:

45.157 is rounded to 45.16
45.155 is rounded to 45.16
45.153 is rounded to 45.15

where the answer is to be expressed in four significant figures.

ALGEBRA

The word ***algebra***, derived from the Arabic language, connotes that branch of mathematics which deals with the relationship of quantities usually represented by letters of the alphabet–Roman, Greek, or Hebrew.

Operations. Mathematical operations with ***letter symbols*** are the same as with ***numerals***, since both are symbolic representations of numbers which, in themselves, are abstract concepts. For example, the concept "four" may be represented by 4, 2^2, 2×2, $2 + 2$, or $3 + 1$; or by the letter x if the value of x is specifically designated to represent "four." In algebra, just as in arithmetic, the fundamental operations include addition, subtraction, multiplication, and division; and there are fractions, proportions, and equations. Algebra provides a method of finding an unknown quantity when the relationship of certain known quantities is specified.

Algebraic operations are indicated by the same symbols as in arithmetic:

+ (plus) add
− (minus) subtract
× (times) multiply
÷ (divided by) divide

= equals

To indicate ***addition*** in algebra, use the general expression

$$x + y \qquad\qquad (1)$$

The symbols x and y, called ***variables***, may represent any number or quantity. Thus, if $x = 4$, and $y = 7$, then, substituting these values in equation (1),

$$4 + 7 = 11$$

Similarly, to indicate ***subtraction*** in algebra, use the general expression

$$x - y$$

If $x = 9$ and $y = 5$ then

$$9 - 5 = 4$$

Notice that algebraic symbols may represent whole numbers, fractions, zero, and negative numbers, among others. Negative numbers are those whose value is less than zero and are designated as $-x$. In algebraic terms, add a positive and a negative number as follows:

$$x + -y$$

If $x = 8$ and $-y = -3$, then

$$8 + -3$$

is the same as

$$8 - 3 = 5$$

The $+$ sign is omitted in the designation of positive numbers, being reserved to indicate the operation of addition.

On the other hand, to subtract a negative quantity from a positive one

$$x - -y$$

If $x = 4$ and $-y = -6$, then

$$4 - -6$$

is the same as

$$4 + 6 = 10$$

Multiplication in algebra follows the same rules as in arithmetic. However, in the multiplication of letter symbols the $\times$ sign is omitted, $x \times y$ being written as xy. If $x = 3$ and $y = 5$, then substituting in the expression xy,

$$3 \times 5 = 15$$

Division in algebra is customarily expressed as a fraction. Thus, $x \div y$ is written as x/y. If $x = 3$ and $y = 5$, then

$$3 \div 5 = {}^3/5$$

When two negative quantities are multiplied or divided, the answer is positive. Thus, $-x \times -y = xy$; and $-x/-y = x/y$. When a positive and a negative quantity are multiplied or divided, the answer is negative; thus, $x \times -y = -xy$, and $x \div -y = -x/y$.

In solving an algebraic expression consisting of a collection of ***terms*** we must perform the indicated ***multiplication and division first***, and then carry out the indicated addition and subtraction. An example will clarify this:

$$ab + c/d - f = ?$$

Suppose $a = 2$, $b = 3$, $c = 4$, $d = 8$, and $f = 5$. Substituting in the preceding expression,

$$2 \times 3 + \frac{4}{8} - 5 = ?$$

Performing ***multiplication and division first***,

$$6 + \frac{4}{8} - 5 = ?$$

Then, performing addition and subtraction,

$$6\frac{4}{8} - 5 = 1\frac{4}{8} = 1\frac{1}{2}$$

A parenthesis inclosing a group of terms indicates that all of the terms inside the parenthesis are to be multiplied by the term outside the parenthesis. This is simplified by performing all the indicated operations in correct sequence–inside the parenthesis first–and then multiplying the result by the quantity outside the parenthesis. For example,

$$6 \, (8 - 4 + 3 \times 2) =$$
$$6 \, (8 - 4 + 6) =$$
$$6 \times 10 = 60$$

Equations. The simpler algebraic equations can be solved without difficulty by application of basic rules. You can easily verify these rules by substituting numerals. In the equation

$$a + b = c + d \qquad (2)$$

$a + b$ is called the ***left side***, and $c + d$ the ***right side***. Each letter is called a term. If any quantity is added to one side of the equation, the same quantity must be added to the other side in order for this to remain an equation. Similarly, if any quantity is subtracted from one side, the same quantity must be subtracted from the other side. To simplify the concept of the equation we may picture it as a see-saw as in Figure 1.2. If persons of equal weight are placed at each end, the board will remain horizontal–the equation is balanced. If a second person is now added to one end of the see-saw, a person of similar weight must be added to the other end in order to keep the board level.

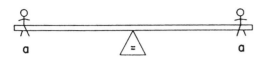

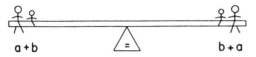

Figure 1.2. Analogy of algebraic equation to a see-saw.

Return again to the simple equation:

$$a + b = c + d$$

in which each term is a variable. If any three of the variables are known, the fourth can be found. Suppose, a is 3, b is 4, c is 1, and d is unknown. Substituting these values in the equation,

$$3 + 4 = 1 + d$$

How can d be found? Simply rearrange the equation so that d is alone, that is, ***the only term in its side***. In this case, subtract 1 from both sides of the equation. However, a mathematical short cut can be used: ***a term may be transposed from one side of an equation to the other side, provided it is given the opposite sign***. Following this rule, 1 becomes a minus 1 when moved to the left side. Thus,

$$3 + 4 - 1 = d$$
$$6 = d$$
$$\text{or,} \qquad d = 6$$

Usually, in solving algebraic equations, rearrange the terms before substituting their numerical values. In the equation

$$a + b = c + d$$

to find d, transpose c to the left side and change its sign.

$$a + b - c = d$$
$$\text{or,} \qquad d = a + b - c$$

(Reversing both sides of an equation does not alter its equality.)

In algebraic equations in which terms are multiplied or divided, analogous rules apply. For example, equation

$$x = y/z$$

may be solved for y by multiplying both sides by z,

$$xz = yz/z$$
$$xz = y$$
$$\text{or,} \quad y = xz$$

The same result may be obtained by moving z from the denominator of the right side to the numerator of the left side. Thus, we have the short-cut rule for cross-multiplication: ***if the denominator of one side of an equation is moved, it multiplies the numerator of the other side and, conversely, if the numerator of one side is moved, it multiplies the denominator of the other side***.

Suppose that in the equation $x = y/z$, x and y are known; then z is solved as follows: move z into the numerator of the opposite side as a multiplier,

$$xz = y$$

and move x into the denominator of the right side as a multiplier, and

$$z = y/x \tag{3}$$

The above rule can be readily tested. Suppose that y is 12, and x is 3. Substituting in equation (3),

$$z = {}^{12}/3$$
$$z = 4$$
$$4 = {}^{12}/3 \tag{4}$$

If we wish to move 3, we must place it in the numerator of the left side of equations (4),

$$4 \times 3 = 12 \qquad (4a)$$

Note that numerical equations (4) and (4a) balance.

Now, referring again to equation (4), sup-pose we wish to move 12. We must place it in the denominator of the left side

$$4/12 = 1/3$$

Again, it is evident that the equation balances.

RATIO AND PROPORTION

A *ratio* is a fixed relationship between two quantities, simply indicating how many times larger or smaller one quantity is relative to another. It has essentially the same meaning as a fraction. One symbol that expresses a ratio is the colon (:). Thus, *a:b* is read "*a* is to *b*." Or, 1:2 is read "1 is to 2." In modern mathematics ratios are usually represented as fractions:

$$a{:}b \text{ is the same as } a/b$$
$$1{:}2 \text{ is the same as } 1/2$$

As noted above, the fraction $1/2$ indicates that the numerator is $1/2$ as large as the denominator. Similarly, $2/3$ indicates that the numerator is $2/3$ as large as the denominator.

The meaning of ratio is important to the technologist, because it underlies the concept of *proportion*, defined as *an expression showing that two given ratios are equal.* Thus, we may have an algebraic proportion,

$$a/b = c/d \qquad (5)$$

which is read "*a* is to *b* as *c* is to *d*." The same idea can also be represented numerically. For example,

$$3/6 = 4/8$$

If any three terms of a proportion are known, the fourth may easily be determined. Suppose in proportion (5) *a* is 2, *b* is 4, *d* is 8, and *c* is to be found. Then,

$$2/4 = c/8$$

Moving 8 to the numerator of the left side (crossmultiplying),

$$\frac{2 \times 8}{4} = c$$

$$c = 16/4 = 4$$

There are three general types of proportions that pertain to radiography.

1. *Direct proportion.* Here, one quantity maintains the same ratio to another quantity as the latter changes. For example, the statement "*a* is proportional to *b*" means that if *b* is doubled, *a* is automatically doubled; if *b* is tripled, *a* is tripled, etc. In order to represent this mathematically, let us assume that the quantity a_1 exists when b_1 exists; and that if b_1 is changed to b_2, then a_1 becomes a_2. Thus,

$$a_1/b_1 = a_2/b_2$$

The numbers below the letters are called "subscripts" and have no significance except to label a_2 as being different from a_1. Such a direct proportion is solved by the method described above.

2. *Inverse proportion.* Here, one quantity varies in the opposite direction from another quantity as the latter changes. Thus, the statement "*a* is inversely proportional to *b*" means that if *b* is doubled *a* is halved, if *b* is tripled, *a* is divided by 3, etc. Such a proportion is set up as follows:

$$\frac{a_1}{1/b_1} = \frac{a_2}{1/b_2}$$

Crossmultiplying,

$$a_1/b_2 = a_2/b_1 \qquad (6)$$

A numerical example should help clarify this. Suppose that when a_1 is 2, b_1 is 4; if a is inversely proportional to b, what will a_1 become if b_1 is changed to 8? In this case, a_2 is the unknown, and b_2 is 8. Substituting in equation (6),

$$^2/8 = a_2/4$$
$$a_2 = {}^8/8 = 1$$

Thus, a is halved when b is doubled.

3. ***Inverse Square Proportion.*** See pages 214-217 (Figure 18.9).

PLANE GEOMETRY

Plane geometry is that branch of mathematics which deals with figures lying entirely in one plane; that is, on a flat surface. Some of the elementary rules of plane geometry will now be listed.

1. A ***straight line*** (in classical geometry) is the shortest distance between two points. It has only one dimension—length.

2. A ***rectangle*** is a plane figure composed of four straight lines meeting at right angles. Its opposite sides are equal. The sum of the lengths of the four sides is called the ***perimeter***. The area of a rectangle is the product of two adjacent sides. Thus, in Figure 1.3 the perimeter is the sum $a + b + a + b = 2a + 2b$. The area is $a \times b$ or ab.

not in the same line and lie in the same plane. The perimeter of a triangle is the sum of the three sides. The area of a triangle is one half the base times the altitude (the altitude is the perpendicular distance from the apex to the base). This is shown in Figure 1.5.

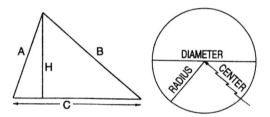

Figure 1.5. Triangle.
Perimeter $= a + b + c$
Area $= {}^1/2ch$

Figure 1.6. Circle.
Circumference $=$
$\pi \times$ diameter $= \pi d$
Area $= \pi \times$ radius$^2 = \pi r^2$

5. A ***circle*** is a closed curved line which is everywhere at an equal distance from one point called the center, lying in the same plane as the center. A straight line from the center to any point on the circle is the ***radius***. A straight line passing through the center and meeting the circle at two points is the ***diameter***, which is obviously equal to twice the radius. The length of the circle is its ***circumference***, obtained by multiplying the diameter by a constant, π (pi). Pi always equals about 3.14. The area enclosed by a

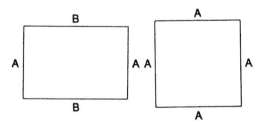

Figure 1.3. Rectangle.
Perimeter $= a + b + a + b$
$= 2a + 2b$

Figure 1.4. Square.
Perimeter $= 4a$
Area $= a \times a = a^2$

3. A ***square*** is a special rectangle in which all four sides are equal, as in Figure 1.4.

4. A ***triangle*** is a figure made up of three straight lines connecting three points that are

circle equals π times the square of the radius. Figure 1.6 shows these relationships.

Similar Triangles

Of great importance in radiology, and closely related to ratio and proportion, is the proportionality of *similar triangles*. Such triangles have an identical shape, although they differ in size. ***In similar triangles the corresponding sides are directly proportional and the corresponding angles are equal***. Thus, similar triangles represent the geometric equivalent of direct proportion.

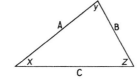

Figure 1.7. Similar triangles. The corresponding angles are *equal*; that is, $x = x$, $y = y$, $z = z$. The corresponding sides are proportional; thus, $a/A = b/B = c/C$.

Figure 1.7 shows two similar triangles. The corresponding sides (those opposite the equal corresponding angles) are proportional, so that

$$a/A = b/B, \text{ or}$$
$$a/A = c/C, \text{ or}$$
$$b/B = c/C$$

A line drawn across any triangle parallel to its base produces two similar triangles, one partly superimposed on the other, as shown in Figure 1.8. Line *BE* has been drawn parallel to the base *CD*. Then triangle *ABE* is similar to triangle *ACD* and their corresponding sides are proportional. This may be simplified by separating the two triangles as in Figure 1.7.

A thorough comprehension of similar triangles is essential to the understanding of photographic and radiographic projection.

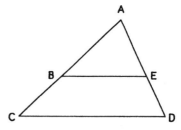

Figure 1.8. Similar triangles. *BE* is parallel to *CD*. Triangle *ABE* is similar to triangle *ACD*. Therefore, $AB/AC = AE/AD = BE/CD$.

Suppose an object is placed in a beam of light originating at a point source and allowed to cast a shadow or image on a surface (see Figure 1.9). The image has to be larger than the object. Knowing the distances of the object and image, respectively, from the light source, as well as the object size, we can determine the image size from the similarity of triangles *ABC* and *ADE*. We do this by setting up the following proportion:

$$\frac{\text{image size}}{\text{object size}} = \frac{\text{image distance}}{\text{object distance}}$$

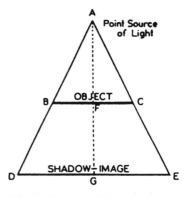

Figure 1.9. Projection of the shadow image of an object by a point source of light.
$DE/BC = AG/AF$.

Substituting the known values in the equation, we can predict the size of the image by solving the equation. Note that if any three of the values are known, the fourth can be readily obtained from the same equation.

This principle also applies in radiography, as will be shown in Chapter 18.

GRAPHS AND CHARTS

For practical purposes, graphs and charts may be regarded as diagrams representing the relationship of two quantities, one of which depends on the other. The dependent factor is called the **dependent variable**. The factor which changes independently is called the **independent variable**.

When data are accumulated, showing how a dependent variable changes with a change in the independent variable, they can either be compiled in a table, or represented by a graph. However, tables do not give intermediate values as do graphs.

The construction and interpretation of a graph may be exemplified as follows: the optimum developing time for x-ray film increases as the temperature of the developing solution decreases (not inversely proportional, however). Table 1.1 reproduces the data for a developer in manual processing. To chart this information on graph paper, first plot the temperature (independent variable) along the horizontal axis of the graph. Then plot the developing time (dependent variable) on the vertical axis, as shown in Figure 1.10. In mathematics, the dependent variable is said to be a **function** of the independent variable. To graph the tabulated data, take the first temperature listed in the table and locate it on the horizontal axis. Trace vertically upward from this point to the horizontal line corresponding to the correct developing time, and mark the intersection with a dot. Repeat this for all the values in the table, and then draw a line that best fits the dots. This constitutes a **line graph** or **curve**.

Having constructed a graph, how do we read it? Suppose that we wish to determine the correct developing time for a temperature of 18.5°C (65°F). Locate 18.5°C (65°F) on the horizontal axis and trace vertically upward to the intersection with the curve; from this point of intersection trace horizon-

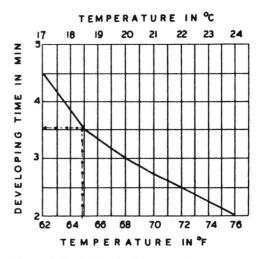

Figure 1.10. Method of using a time-temperature development curve in manual processing. Select the correct temperature on the horizontal axis, in this case 18.5°C (65°F). Follow vertically, as shown by the arrows, to the curve. Then follow horizontally to the time axis where the correct developing time in this case is seen to be 3 1/2 min.

Table 1.1
TIME-TEMPERATURE DEVELOPMENT DATA

Temperature		Developing Time in Minutes
°C	°F	
17	62	4$^1/_2$
18.5	65	3$^1/_2$
20	68	3
21	70	2$^3/_4$
22	72	2$^1/_2$
24	75	2

tally to the vertical axis on the left, where the correct developing time can be read. This is shown in Figure 1.10, where the tracing lines are formed by broken lines with arrows. The horizontal tracing line meets the vertical axis at approximately $3^1/2$ minutes, which is the correct developing time at this temperature. Thus, a properly constructed graph gives the developing times for temperature readings that did not appear in the original table, a process called ***interpolation***.

LARGE AND SMALL NUMBERS

Because ordinary notation is too cumbersome to express the extremely large numbers often encountered in physics, we resort to the ***exponential system*** in ***powers of 10***. For example, the number 10^3 means $10 \times 10 \times 10$. In the quantity 10^3, 10 is the ***base*** and 3 the ***exponent***, and is read as "ten to the third power." In general,

$$10^1 = 10$$
$$10^2 = 100$$
$$10^3 = 1000, \text{ etc}$$

Note that the exponent of 10 indicates the number of zeros required to express the number in ordinary notation.

Now apply this principle: The velocity of light in air or vacuum is 300,000,000 meters per second (m/s). Place a decimal point after 3, following which there are eight zeroes. Therefore, this number can be simply expressed as 3.0×10^8 m/s and read as "three point oh times ten to the eighth." A more complicated number such as 4,310,000,000,000,000 may be simplified in the same way by placing a decimal point after the first digit, 4, and counting the number of ***places*** to the right, in this case fifteen. The corresponding number in powers-of-ten system is 4.31×10^{15}.

Suppose, instead, we are to convert 6.3×10^6 to ***ordinary notation***. Count six places to the right of the decimal point, filling out the places with zeroes. Since the first place is already occupied by a digit, 3, five zeroes must follow. Therefore, the number is 6,800,000. In general, ***any number followed by 10^6 is expressed in millions***.

The same system can be applied to ***extremely small*** numbers, by the use of negative exponents. This is based on

$$10^{-1} = 1/10 = 0.1$$
$$10^{-2} = 1/100 = 0.01$$
$$10^{-3} = 1/1000 = 0.001, \text{ etc.}$$

From this you can see that a negative exponent indicates the number of decimal places to the ***left*** of a digit, ***including the digit***. Thus, 0.004 would be expressed as 4×10^{-3}. To obtain this result, count the number of places to the right of the decimal point, including the first digit,

$$. . .$$
$$0.004$$

The counted number of places (three in this instance) then becomes the negative exponent of 10; thus, 4×10^{-3}. A more complicated number such as 0.00000000000372 would be simply expressed as 3.72×10^{-12}, read as "three point seven two times ten to the minus twelfth."

The simplified system of notation facilitates the multiplication and division of large and small numbers. To multiply such numbers, we multiply the digits and make the exponent of 10 in the answer equal to the algebraic sum of the exponents of 10 in the multipliers. Thus.

$$4 \times 10^3 \times 2 \times 10^2 = 4 \times 2 \times 10^{(3+2)} = 8 \times 10^5$$
$$3 \times 10^6 \times 2 \times 10^{-4} = 3 \times 2 \times 10^{(6-4)} = 6 \times 10^2$$

To divide numbers in this system, we divide the digits and derive the exponent of 10 in the answer by subtracting algebraically the exponent in the divisor from the exponent in the dividend:

$$\frac{6 \times 10^5}{2 \times 10^3} = \frac{6 \times 10^{(5-3)}}{2} = 3 \times 10^2$$

$$\frac{9 \times 10^7}{3 \times 10^{-2}} = \frac{9 \times 10^{(7+2)}}{3} = 3 \times 10^9$$

Note in the above examples that when a power number is moved from the denominator of a fraction to the numerator, or vice versa, we simply change the sign of the exponent. Thus,

$$\frac{8}{3 \times 10^{-5}} = \frac{8 \times 10^5}{3}$$

LOGARITHMS

In several areas of imaging, we shall have to use quantities known as *logarithms*. For our purposes, only the basic aspects of logarithms will be explained. In fact, the preceding section included powers-of-ten; these powers are logarithms "to the base 10," also known as common logs.

As an example of the use of logs, consider the intensity of sound. It so happens that the human ear, in comparing a sound 100 times louder than another, perceives it as being only twice as loud; $100 = 10^2$, so the exponent is the "twice." If the sound is 1000 times the first, the ear perceives it as 3 times louder; $1000 = 10^3$, so now 3 represents the "three times."

It turns out that 2 is the logarithm of 100, stated as $\log 100 = 2$; similarly, $\log 1000 = 3$. In other words, logs are simply the exponent to which 10 must be raised to equal a given number. Note that 10 is the *base* of the system of *common logs*. (There is another system—the *Naperian*, with a different base.) Conversely, the antilogarithm (antilog) of 2 is 100. The following examples will help summarize simple logs:

$10^0 = 1$ $\log 1 = 0$ antilog $0 = 1$
$10^1 = 10$ $\log 10 = 1$ antilog $1 = 10$
$10^2 = 100$ $\log 100 = 2$ antilog $2 = 100$
$10^3 = 1000$ $\log 1000 = 3$ antilog $3 = 1000$

You can see that in each case, for example, $\log 100 = 2$ because 10 has to be raised to the second power to equal 100.

The same applies to negative exponents:

$10^{-1} = 0.1$ $\log 10^{-1} = -1$ antilog $-1 = 0.1$
$10^{-2} = 0.01$ $\log 10^{-2} = -2$ antilog $-2 = 0.01$
$10^{-3} = 0.001$ $\log 10^{-3} = -3$ antilog $-3 = 0.001$

You can see that in each instance, for example, $\log 0.01 = -2$ because 10 has to be raised to the -2 power to equal 0.01.

Why does $10^0 = 1$? Arrange *exponents* in decreasing order:

     . . . . 5 4 3 2 1 **0** -1 -2 -3 -4 -5 . . .

Obviously, each exponent is one less than that preceding, as we go from left to right. What lies between 1 and -1? This must be zero because $1 - 1 = 0$, and $-1 - (1)$ also $= 0$. Thus, $10^0 = 1$. The same applies to all real numbers except the number 0.

Logarithms have proved extremely useful in multiplication and division, especially of large and small numbers, because, to multiply powers of 10, one simply *adds the exponents*; and to divide powers of 10, *one subtracts exponents*. In a chain multiplication with a mixture of positive and negative exponents, all to the base 10, one simply adds the exponents algebraically and uses the resulting exponent as a power of 10, as in the preceding section.

Special tables of logarithms are available to find the logs to the base 10 (ie, common logs) for numbers that are *not* integer powers of 10; for example, let us multiply 43.61 × 25.23 by using logs.

log 43.61 =1.6396
log 25.23 =1.4017
3.0413 total value of exponent

Applying this new exponent to 10,

$10^{3.0413}$ = antilog of 3.0413 = 1100

Actually multiplying the original numbers $43.61 \times 25.23 = 1100.2$. If a 5-place log table had been used, the answer would have shown the .2; this example has been given simply to show how logarithms work. Manual calculations with the aid of log tables have been supplanted by ***pocket calculators*** having a wide range of complexity. Even the multiplication problem just presented can be solved by the use of a simple pocket calculator.

QUESTIONS AND PROBLEMS

1. Reduce the following fractions:
   (a) $4/8$ (b) $9/12$
   (c) $10/150$ (d) $4/5$
2. Solve the following problems:
   (a) $3/5 + 2/5 + 4/5 =$
   (b) $2/3 + 3/4 + 3/7 =$
   (c) $1/2 + 2/3 + 4/5 =$
3. Mr. Jones has 100 bushels of potatoes and sells 75 percent of this crop. How many bushels does he have left?
4. Divide as indicated:
   (a) $4/5 \div 3/5$ (b) $4/9 \div 7/16$
   (c) $8/9 \div 3/5$
5. If $a = b/c$, what is b in terms of a and c? What is c in terms of a and b?
6. Solve the following equations for the unknown term:
   (a) $x + 4 = 7 - 3 + 8$
   (b) $6 + 3 - 1 = 4 + 1 - a$
   (c) $x/3 = 7/9$
   (d) $4/y = \dfrac{6 + 3}{10}$
   (e) $\dfrac{x + 4}{8} = 7 + 2 - 3$
7. Solve the following proportions for the unknown quantity:
   (a) $a/7 = 2/21$ (b) $3/9 = x/15$
   (c) $4/6 = 7/y$ (d) $3/5 = 9/x$
8. The diameter of an x-ray beam is directly proportional to the distance from the tube target. If the diameter of the beam at a 20-in. distance is 10 cm, what will the diameter be at a 40-in. distance?
9. The exposure of a radiograph is directly proportional to the time of exposure. What will happen to the exposure if the time is tripled?
10. What is the area of a circle having a diameter of 4 cm?
11. What is the area of a rectangular plot of ground measuring 20 ft on one side and 30 ft on the adjacent side?
12. Convert 3.6×10^6 to the ordinary number system.
13. Change 424,000 to the exponential system.
14. What is log 10^5 to base 10 (common logs)?
15. State the antilog of 4.
16. (a) Multiply $10^5 \times 10^6$ in the powers-of-ten system. (b) State the log of the answer.
17. (a) Divide 10^8 by 10^5. (b) State the log of the answer.
18. What is antilog $(4 + 5)$?
19. Find the log of 1000×10000.

Chapter 2

PHYSICS AND THE UNITS OF MEASUREMENT

Topics Covered in This Chapter

- Definitions: Physics and Science
- Standard Units
- Fundamental Units
- Derived Units
- SI Units
- Questions and Problems

PHYSICS, AS AN EXACT SCIENCE, requires a precise vocabulary in which each term has a clear and definite meaning. Not only does this simplify the learning process, but it also facilitates the organization of concepts and their accurate communication to others.

In order to appreciate the position of physics within the framework of science, we may use the term *natural science* to include the systematic study of the universe and its contents. Figure 2.1 shows the subdivision of natural science into its major categories. Here you can readily see that while physics is listed as one of the physical sciences, it

actually underlies all of them. In fact, it is ultimately an important contributor even to the biologic sciences.

What, then, is physics? It may be defined as that branch of science which deals with matter and energy, and their relation to each other. It includes mechanics, heat, light, sound, electricity, and magnetism, and the fundamental structure and properties of matter. For our purpose, we shall be interested in those aspects of physics pertaining to the origin, nature, and behavior of x rays and related types of radiation; that is, *radiologic physics*.

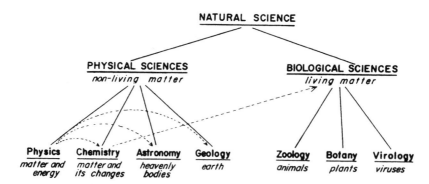

Figure 2.1. Subdivisions of natural science. Notice that physics plays a part in all branches (dashed lines).

16

Thus far, we have used the word science rather freely without definition. ***Science is organized and classified knowledge***. Natural processes go on continually in the universe, with or without our knowledge. But these do not constitute a science until we have applied the scientific method to them. The ***scientific method*** comprises the systematic collection of facts, the study of their interrelationship, and the drawing of valid conclusions from the resulting data. Inherent in the scientific method is the assumption that nature is orderly and that sooner or later we can discover the rules by which she behaves.

Not only do we use the scientific method to describe and classify natural occurrences, but we also seek to correlate them by deriving certain principles known as ***laws***. Scientific laws, based on human experience, are derived by observation of natural phenomena, by laboratory experiments, or by the application of mathematics. Such laws state as clearly and simply as possible, according to our present state of knowledge, that certain events will always follow in the same order, or that certain facts will always be related in the same way. For example, we know by experience that all objects fall to the ground; from this observation was derived the law of gravitation, which applies to the attraction between all bodies of matter.

Scientists are not satisfied with merely discovering the laws of nature. They also attempt to correlate various laws in order to determine their conformity to a ***general pattern***. Thus, wherever possible, natural laws are so tied together that their operation can be expressed as a general concept, often on a mathematical basis Such a broad, unified concept of the laws underlying certain natural phenomena is called a ***theory***. Not only do theories provide a better insight into nature but they also suggest new lines of scientific research, since it is common experience that the solution of one problem in science ultimately gives rise to a host of new problems.

Standard Units

An exact science must include not only the observation and classification of natural occurrences, but also their ***measurement***. In other words, physics as an exact science deals with both the "how" and the "how much" of physical phenomena. In order for these measurements to have the same meaning everywhere, they must be expressed in certain standard units. A ***unit*** is a quantity adopted as a standard of measurement by which other quantities of the ***same*** kind can be measured, such as the meter as an interval of distance, or the second as an interval of time. ***Standard units*** are the basic units in any particular system of measurement; they are used mathematically to describe natural phenomena, derive laws, and predict future events within the realm of nature. For example, by the application of science and mathematics, astronomers can foretell precisely the time of an eclipse of the sun or moon.

Units should have a size compatible with that which is being measured. Thus, we measure the length of a baby in centimeters, but we use a larger unit, the meter, to measure the height of a building; and a still larger unit, the kilometer, to measure the distance between cities. ***And units are multiplied and divided just like numbers***.

The standard units employed in physics consist of two general types, the simpler ***fundamental units*** dealing with ***length***, ***mass*** and ***time***, and the more complicated ***derived units***, obtained by various combinations of the fundamental units.

Fundamental Units

The fundamental units, arbitrarily selected and named, have been so standardized that a given unit has the same meaning everywhere. We now have two widely used systems of measurement, the ***English*** and the ***metric***.

Since the metric system is used internationally in the exact sciences, it will be emphasized here, but the equivalents in the English system will also be indicated because we still use this in everyday life. The metric system is also known as the MKS (meter-kilogram-second) or CGS (centimeter-gram-second) system. These have been replaced in science by the International System of Units, known as the *SI units* which provides for the interconversion of units among all branches of science. The SI units are derived from the *MKS* system, and insofar as the fundamental units are concerned, the two systems are identical.

1. **Length**. The unit of length in the metric system is the *meter*, originally defined as the distance between two scratches on a bar of platinum, kept at Sèvres, France, at the International Bureau of Weights and Measures. (It was redefined in 1960 as 1.65 $\times 10^6$ times the wavelength of the orange-red radiation of krypton 56.) Roughly equal to a yard, the meter (m) can be conveniently subdivided into smaller units such as a centimeter (0.01 or 10^{-2} m) and millimeter (0.001 or 10^{-3} m); or it can be treated as a multiple such as a kilometer (1,000 or 10^3 m). In x-ray physics, we even have occasion to use very small fractions of these units; thus, the angstrom is $1/10,000,000,000$ or 10^{-10} m (or 10^{-8} cm). The following common equivalents should prove useful:

$$1 \text{ meter} = 100 \text{ cm (centimeters)}$$
$$= 1000 \text{ mm (millimeters)}$$
$$1 \text{ Å (angstrom)} = 1/100,000,000 \text{ cm}$$
$$= 10^{-8} \text{ cm} = 0.1 \text{ nm}$$
$$\text{km (kilometer)} = 1,000 \text{ m (meters)}$$
$$= \text{about } 3/5 \text{ mile}$$
$$1 \text{ in.} = 2.54 \text{ cm}$$

2. **Mass**. Representing the quantity of matter in a body (or inertia) its *mass* is determined by weighing, a procedure which measures the force with which the earth attracts the body at some particular geographic loca-

tion. Thus, the more massive the body, the greater the gravitational force and, therefore, the greater its weight. As implied in the above definition, the weight of a body may vary from place to place, but equal masses will have the same weight under identical conditions. The unit of mass is the *kilogram*, which is the weight of a standard piece of platinum-iridium (International Kilogram Prototype) kept at the International Bureau of Standards. All other kilograms are more or less exact copies of this standard unit. A more convenient unit is the gram, which is $1/1,000$ of a kilogram.

$$1 \text{ kg (kilogram)} = 1000 \text{ or } 10^3 \text{ g (grams)}$$
$$1 \text{ kg} = 2.2 \text{ lb}$$
$$28 \text{ g} = 1 \text{ oz}$$

3. **Time**. This is a measure of the duration of events. We are all aware of the occurrence of events and the motion of objects, but to measure duration with relation to our senses alone is inaccurate. The standard unit of time is the *second*, defined as $1/86,400$ of a mean solar day. In other words, the second is a definite fraction of the average time it takes for the earth to make one rotation on its axis.

Derived Units

Of the many derived units, only a few examples of the more common ones will be given here. Those that apply specifically to radiologic physics appear later in the text.

1. **Area** is the measure of a given surface, and depends on length. Thus, a square or rectangle has an area equal to the product of two sides. The area of a circle equals the radius squared times π. Figures 1.3, 1.4, 1.5, and 1.6 in Chapter 1 explain area in detail. In the metric system, area is represented by square meters for larger surfaces and square centimeters for smaller ones. Square centimeters are abbreviated either as sq cm or

cm².

2. **Volume** is a measure of the capacity of a container, and is also derived from length. The volume of a cube equals the product of three sides. In metric units, volume may be expressed in cubic centimeters (cc) or milliliters (ml). One liter exactly equals 1,000 ml, and is very slightly larger than 1,000 cc–approximately one quart.

3. **Density** is the mass per unit volume of a substance, and may be expressed in kg per cubic meter (kg/m³).

4. **Specific gravity** has *no units*. It is the ratio of the density of any material to the density of water. The density of water is 1.

5. **Velocity** is speed in a given direction, and can be expressed in cm per sec, or km per sec, or some other convenient units. Speed is commonly expressed in miles per hr (English system).

6. **Temperature** is a measure of the average energy of motion of the molecules of matter. Two systems are commonly employed today. In the metric system temperature is expressed in degrees Celsius (centigrade), but still the Fahrenheit system is the one generally used in the United States. The following data exemplify the differences between these systems:

CELSIUS
$$0°C = \text{freezing point of water}$$
$$100°C = \text{boiling point of water}$$
FAHRENHEIT
$$32°F = \text{freezing point of water}$$
$$212°F = \text{boiling point of water}$$

Conversion from one system to the other can be achieved by reference to tables, or by use of the following equations:

$$°C = {}^{5}/_{9}\,(°F - 32°F)$$
$$°F = {}^{9}/_{5}\,°C + 32°F$$

Table 2.1 summarizes the units that are relevant to radiology. The abbreviations will be used in this book.

Manipulation of Units

The practical application of units includes:

1. *Conversion of Units in the Same System*–changing quantities between units of different sizes. For example, 1 ft = 12 in.; 1 m = 100 cm. Conversely, 1 in. = $^{1}/_{12}$ ft; 1 cm = 0.01 m. Note that in each case, the larger unit is accompanied by a smaller number than the smaller unit for them to be equivalent; in fact, they are reciprocals. Conversely, 1 ft = 12 in.; 1 m = 100 cm. This principle applies to all units.

2. *Conversion of Units Between Systems*–changing from the metric to the English system, or the reverse. For example, the inch is larger than the centimeter: 1 in. = 2.54 cm. Just as in (1), the larger unit is designated by the smaller number to express equivalence.

3. *Handling Units in Mathematical Operations*–applying them in addition, subtraction, multiplication, and division. Keep in mind that units are treated *exactly like numbers*. For example, we found in Figure 1.4, page 10, that the area of a square is the product of two sides. So if each side is 5 cm in length,

$$5 \text{ cm} \times 5 \text{ cm} = 25 \text{ cm}^2$$

both the numbers and lengths having been multiplied. Another example is the relation between distance, speed, and time:

$$\text{distance} = \text{speed} \times \text{time}$$
$$= \frac{\text{mi}}{\text{hr}} \times \text{hr} = \text{mi}$$

since hr in the numerator and denominator cancel out.

Equations containing a number of identical units can often be simplified by the use of this principle: *units treated like numbers*. Moreover, the solution of such equations can be checked by the appropriateness of the final units.

Table 2.1
SUMMARY OF PHYSICAL UNITS

SI Fundamental Units–MKS System

LENGTH	meter	m
MASS	kilogram	kg
TIME	second	s
ELECTRIC CURRENT	ampere	A

SI Derived Units

			unit name
AREA	m^2		
VOLUME	m^3		
DENSITY (mass/volume)	kg/m^3		
VELOCITY, SPEED	m/s		
ACCELERATION	m/s^2		
FORCE	$m \cdot kg/s^2$	=	newton (N)
WORK, ENERGY	$N \cdot m$	=	joule (J)
POWER	$N \cdot m/s$	=	J/S, watt (W)
ELECTRIC CHARGE	$A \cdot s$	=	coulomb (C)
ELECTRIC CURRENT	C/s	=	ampere (A)
ELECTRIC POTENTIAL	J/A	=	volt (V)
ELECTRIC RESISTANCE	V/A	=	ohm (Ω)
MAGNETIC FLUX	$V \cdot s$	=	weber (Wb)
MAGNETIC FIELD	Wb/m^2	=	tesla (T)
CAPACITANCE	C/V	=	farad (F)
FREQUENCY	cycles/s	=	hertz (Hz)

SI Radiologic Units

EXPOSURE	C/kg	=	roentgen (R)
ABSORBED DOSE	J/kg	=	gray (Gy) = 100 rad
KERMA	J/kg	=	gray = 100 rad
DOSE EQUIVALENT	J/kg	=	sievert (Sv) = 100 rem
ACTIVITY	d/s*	=	becquerel (Bq)

Universal Constants

VELOCITY OF LIGHT	$3(10^8)$ m/s
PLANCK'S CONSTANT	$6.62(10^{-34})$J·s; $4.15(10^{-15})$eV–s**
PI	3.1416

*   disintegrations per second

**   electron volt-second (1 ev = energy of an electron subjected to a potential difference of 1 volt)

Prefixes Applied to SI Units

To simplify the designation of quantities of various sizes, the SI introduced a number of prefixes for standard and derived units. Table 2.2 gives the prefixes for fractions and multiples of units employing the power-of-ten concept. Note that a unit with a small-er prefix always has a larger associated num-ber, than a unit with a larger prefix to express the same value. Thus, 1000 mm = 1 m. (Obviously, 1000 is larger than 1, but mm is 1000 times smaller than m.) These prefixes are of interest to radiologic technol-ogists to a limited degree, but do apply more often in radiotherapy physics.

Table 2.2
POWERS-OF-TEN PREFIXES FOR SUBDIVISIONS AND
MULTIPLES OF DEFINED SI UNITS

Subdivisions				*Multiples*		
			$(10^0 = 1)$			
deci-	(d)	10^{-1}		deka-	(da)	10^1
centi-	(c)	10^{-2}		hecto-	(h)	10^2
milli-	(m)	10^{-3}		kilo-	(k)	10^3
micro-	(μ)	10^{-6}		mega-	(M)	10^6
nano-	(n)	10^{-9}		giga-	(G)	10^9
pico-	(p)	10^{-12}		tera-	(T)	10^{12}
femto-	(f)	10^{-15}		peta-	(P)	10^{15}
atto-	(a)	10^{-18}		exa-	(E)	10^{18}

QUESTIONS AND PROBLEMS

1. Discuss the terms science; law; theory.
2. Why are standard units necessary?
3. Define natural science; physics; radio-logic physics.
4. Which branch of natural science underlies all the others?
5. How were standard units in the metric system established?
6. With what do the fundamental units deal?
7. Discuss the metric system?
8. Explain the relation between mass and weight. Define the unit of mass.
9. What is the scientific unit of time?
10. State the *approximate* equivalent of 1 meter in the English system? One gram? One liter?
11. State millions and billion in powers-of-
10.
12. Change 100 kilovolts to volts.
13. What is 1 milligram (mg) in terms of grams, stated as a decimal? Powers-of-ten?
14. The temperature of a solution is 68°F. Find the equivalent temperature in centigrade units.
15. A patient is found to have a tempera-ture of 40°C. What is his temperature in Fahrenheit units?
16. Find the temperature value which is the same in both the °C and °F sys-tems. (There are two methods, graph-ic and algebraic.)

    Hint: in either conversion equation, let °C = °F, or °F = °C.

Chapter 3

THE PHYSICAL CONCEPT OF ENERGY

Topics Covered in This Chapter

- Force
- Work and Energy
- Law of Conservation of Energy
- Questions and Problems

Force

ALL MATTER is endowed with a property called *inertia*. This may be defined as the tendency of a resting body (of matter) to remain at rest, and the tendency of a body moving at constant speed in a straight line to continue its state of motion. To set a resting object in motion, we must push or pull it; that is, we must apply a *force*. Conversely, to stop a moving object we must apply an opposing force. This concept is embodied in Newton's* First Law.

The general physical concept of force is expressed by Newton's second Law:

$$F = ma \qquad (1)$$

where F is force; m mass; and a acceleration, which is a *change in speed* in a straight line with respect to time. Acceleration can be positive (a speeding up) or negative (a slowing down). Since acceleration has direction as well as intensity, it is a *vector quantity* and so is force. The SI unit of force is the newton (N) = 1 kg·m/s^2 where m is distance in meters, and *not* mass.

Work and Energy

Whenever a force acts upon a body over a distance, *work* is done. For example, to lift an object, we must apply a force to overcome the force of gravity. This applied force multiplied by the distance through which the object is lifted equals the work done. If the object is lifted twice the distance, then twice as much work will be done. Thus,

$$W = Fd \qquad (2)$$

where W is work in joules (jools), F the force in newtons, and d the distance in meters.

In a physical sense, work results from the expenditure of *energy*. What, then, is energy? *Energy may be defined as the actual or potential ability to do work*. It is obvious that work and energy must be measured in the same units, joules, since the work done must equal the amount of available energy.

There are two main types of *mechanical* energy, that is, the energy involved in the operation of various types of machines:

1. **Kinetic Energy**. Every moving body can do work because of its motion. The

---

* Sir Isaac Newton, one of the greatest physicists and mathematicians who, between 1664 and 1666, discovered the law of gravitation and the spectrum of white light, and began to invent the calculus.

energy of such a moving body is called **kinetic energy**. The kinetic energy (K.E.) of a body is expressed by

$$K.E. = 1/2mv^2 \qquad (3)$$

where m is the mass of the body in kg, and v is the velocity in m per s The unit of work is the joule (jool).

For example, a 1000-kg car starts from rest and reaches a velocity of 30 m/s. What is its K.E.?

$$K.E. = 1/2 \times 1000 \times (30)^2$$
$$= 450,000 \text{ or } 4.5 \times 10^5 \text{ J}$$

2. **Potential Energy**. A body may have energy because of its position or its temporarily deformed state. For example, a car parked on a hill will start to roll down once its brakes have been released. It is logical to assume that at the instant before the brakes were released the car possessed stored energy, which becomes energy of motion, or K.E., as the car rolls downhill. A wound-up clock spring and a stretched rubber band (temporarily deformed) are further examples of stored or potential energy (P.E.).

Simply stated, a rock on a cliff has a P.E. defined by the following equation:

$$P.E. = mgh \qquad (4)$$

where *m* is its mass in kg, *g* the gravitational constant 9.8 N(newtons)/kg, and *h* the height in meters (m).

For example, if a rock has a mass of 5 kg and the cliff is 200 m high, its

$$P.E. = 5 \text{ kg} \times 9.8 \text{ N/kg} \times 200 \text{ m}$$
$$= 9,800 \text{ or } 9.8 \times 10^3 \text{ Nm, or joules}$$

Thus, there are two kinds of mechanical energy—kinetic (energy of motion) and potential (stored energy).

Law of Conservation of Energy

What is the relationship between potential energy and kinetic energy? This is readily demonstrated by a simple example. Consider a rock poised on top of a cliff (see Figure 3.1). The rock has a certain amount of stored or potential energy. Where did this energy originate? A definite amount of work would have to be done to lift the rock to its position on top of the cliff against the force of gravity, such work being "stored" in the rock as **potential energy**. Now suppose the rock is shoved over the edge of the cliff; as it falls, its potential energy is gradually converted to kinetic energy. As the potential energy decreases, the kinetic energy increas-

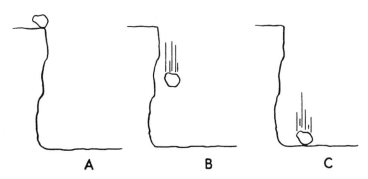

Figure 3.1. The relationship of potential energy to kinetic energy. In *A* the rock is at the top of the cliff, *external work* being required to lift it there against the force of gravity. This work is stored in the rock as *potential energy*. In *B* and *C* the rock is falling and the potential energy is thereby converted to *kinetic energy*. The external work = the potential energy = the kinetic energy.

es correspondingly, so that at the instant the rock reaches the ground its initial potential energy has been completely changed to kinetic energy.

initial energy stored in rock (P.E.)	=	final energy of rock in motion (K.E.)

Notice that the rock has different values of potential energy along its course; in other words, its potential energy gradually decreases as it falls toward the foot of the cliff. If the potential energy at any point above the ground is called an *energy level*, then it follows that the energy levels decrease at successively lower points. This important concept will be applied again later in the discussion of the energy levels of the electron shells in the atom. As exemplified in Figure 3.2, the *difference between any two energy levels equals the amount of potential energy that has been changed to kinetic energy. The rock will not of its own accord move from a lower to a higher energy level; external work is required.* This is another way of saying that *matter will move only from a higher to a lower energy level in the absence of outside work.*

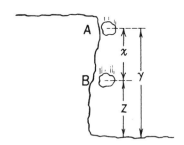

Figure 3.2. Potential energy at *A* − potential energy at *B* = kinetic energy at *B*; or, energy level at *A* − energy level at *B* = kinetic energy at *B*. If potential energy at *A* = *y*, and potential energy at *B* = *z*, then the difference in energy levels is *y* - *z* = *x*. Thus, *x* represents the amount of potential energy converted to kinetic energy when the rock drops from *A* to *B*.

In summary, the work required to lift the rock to the top of the cliff equals the potential energy of the rock at the summit of the cliff; and this, in turn, equals the kinetic energy attained by the rock at the instant it reaches the ground. At the instant the rock strikes the ground its kinetic energy is converted to heat.

The foregoing discussion underlies the *Law of Conservation of Energy*, which states that energy can be neither created nor destroyed, the total amount of energy in the universe being constant. However, energy in one form can be changed into an equivalent amount of energy in other forms.

What are the so-called *"forms of energy?"* They include mechanical, thermal (heat), light, electrical, chemical, atomic, molecular, and nuclear. These can be changed from one form to another. For example, an electric motor converts electrical energy to mechanical energy which is almost equal to the initial electrical energy, any difference being due to the waste of some of the energy due to the production of heat.

electrical energy	=	mechanical energy	+	heat energy

Other examples of the transformation of energy include: the conversion of chemical to electrical energy in a battery; the conversion of heat to mechanical energy in a steam engine; and the conversion of chemical to mechanical energy in a gasoline engine. Each of these transformations obeys the Law of Conservation of Energy; any energy disappearing in one form appears in some other form, and is never destroyed.

The most remarkable transformation of all, that of matter and energy, was formulated by Einstein in his famous equation

$$E = mc^2 \qquad (3)$$

in which *E* is energy, *m* is mass, and *c* is the speed of light in air or vacuum:

$3(10)^8$ m/s [in equation (3) $c^2 = 9(10)^{16}$]

Example. Find the energy equivalent of 1 gram (0.001 kg) of matter.

$$E = 1/2 \times 10^{-3} \text{ kg} \times 9(10^{16}) \text{ (m/s)}^2$$
$$E = 9(10^{13}) \text{ J}$$

Accordingly, mass and energy are mutually convertible, one to the other. Thus, ***matter cannot be destroyed, but can be changed to an equivalent amount of energy, and vice versa.*** In fact, an extremely small amount of matter can give rise to a tremendous amount of energy as in an atom bomb. On this basis, we may expand the conservation law as follows: the sum total of mass and energy in the universe (or in a closed system) remains constant.

QUESTIONS AND PROBLEMS

1. What is the physical concept of work?
2. Define energy; force.
3. Discuss kinetic energy and potential energy. State the equation for kinetic energy and identify the terms and their units.
4. Name the various forms of energy.
5. Define the law of conservation of energy and explain with the aid of an example.
6. What is the source of electrical energy in a battery?
7. Suppose that 1000 energy units of heat are applied to a steam turbine. What is the maximum number of energy units of electricity that can be obtained, neglecting the loss of energy due to friction?
8. Discuss the concept of energy levels.
9. Explain the significance of $E = mc^2$.
10. What is the potential energy of a 20-kg mass at a height or 10 meters?
11. Find the kinetic energy of a 50-kg mass traveling with a velocity of 20 meters per second.

Chapter 4

THE STRUCTURE OF MATTER

Topics Covered in This Chapter

- Subdivisions of Matter
- Atomic Structure–Electrical Nature of Matter
- Atomic Number
- Mass Number
- Isotopes and Nuclides
- Periodic Table
- Chemical Processes
- Ionization
- Questions and Problems

SPECULATION ON THE BASIC STRUCTURE OF MATTER has occupied the mind of man since ancient times. In fact, a Greek philosopher, Democritus, in about 400 B.C., actually suspected that matter is composed of invisible and indivisible particles, which he called *atoms* (from Greek *atomos* = uncut).

One of the methods of studying something is to break it down, a process called *analysis*. For instance, the most satisfactory way to ascertain the mechanism of a clock is to take it apart, study its individual components, and find their relationship to one another. If this has been done properly it should then be possible to reassemble the component parts into a perfectly operating clock. Such a recombination of parts is called *synthesis*.

Subdivisions of Matter

Let us now apply the process of analysis, based on logic and experiment, to the structure of matter. By definition, *matter is anything which occupies space and has inertia*.

Matter is usually found in nature in impure form, as a *mixture* of indefinite composition. An example is *rock salt*, a mixture of certain minerals (sodium chloride, calcium carbonate, magnesium chloride, calcium sulfate, etc) that can be separated in pure form. One of these constituents is ordinary salt–sodium chloride. If a piece of pure salt is now broken into smaller and smaller fragments, there will eventually remain the tiniest conceivable particle that is still salt. This particle, much too small to be visible, is called a *molecule* of salt. *A molecule is the smallest subdivision of a substance having the characteristic properties of that substance*.

The degree of attraction among the molecules of a given body of matter determines its *state*: *solid*, *liquid*, or *gas*. Molecular attraction is strongest in solids, relatively weak in liquids, and weakest in gases.

How can a *molecule* be further subdivided? This cannot be done by ordinary physical means, such as crushing, because the salt molecule is made up of two *atoms*–one of the element sodium and one of the element chlorine–held together by a strong electrochemical force called a *bond* (see Figure 4.1)

26

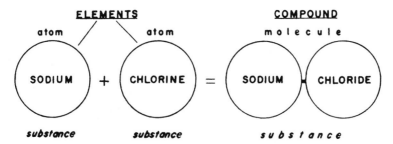

Figure 4.1. Atoms of certain elements are held together by strong electrochemical forces called *bonds*, to form compounds. Elements and compounds are both *substances* because they have a definite, constant structure.

The atom is the smallest particle of an element that has the characteristic properties of that element. Diameters range from about 0.1 to several tenths nm for various atoms (1 nm = 10^{-9} meter or 1 billionth of a meter).

Two new terms have been introduced in the above discussion—*substance* and *element*. These will now be defined. *A substance is any material that has a definite, constant composition*, such as pure salt. (Wood and air are not substances because their composition varies; they are examples of mixtures.) Substances may be *simple* or *complex*. The simple ones, called *elements, cannot be decomposed to simpler substances by ordinary means*. Examples of elements are sodium, iron, lead, oxygen, hydrogen, and chlorine. There are in all ninety-two such naturally occurring elements (see Table 4.1) on pages 32-33.

The complex substances are *compounds*, formed by the *chemical union of two or more elements in definite proportions*. Thus, as we have just seen, salt is a *compound* of equal numbers of atoms of the elements sodium and chlorine in chemical combination (see Figure 5.1); every molecule of salt contains one atom of sodium (Na) and one atom of chlorine (C1). Thus, the compound is called sodium chloride, symbolized by NaCl. Note that *all elements and compounds are also substances* since they have a definite compo-

sition. Compounds may consist of combinations of many atoms.

The commonly occurring gases such as oxygen (O_2), nitrogen (N_2), and hydrogen (H_2) occur in nature in the form of molecules consisting of two atoms. Hence they are known as *diatomic* (two-atom) gases.

Figure 4.2 illustrates the structure of matter down to the atomic level.

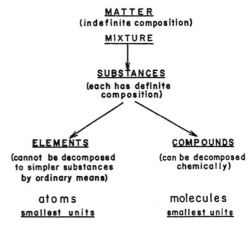

Figure 4.2. Schematic analysis of matter.

In summary, rock salt is an indefinite *mixture* of *substances* each of which has a definite composition. Pure salt is both a *compound* and a *substance* because it has a definite composition, chemically combined atoms of sodium and chlorine in equal numbers. The smallest particle of salt (as well as other compounds) is a ***molecule***, which can be separated electrically into simpler substances–*elements*. Thus, each molecule of salt consists of an atom of sodium and an atom of chlorine (see Figure 4.1). The ***atom*** is the smallest fragment of a particular element that is still recognizable as such. Atoms cannot be further subdivided by ordinary chemical or electrical methods, but can be broken down into smaller particles by special high energy, atom-smashing machines such as the nuclear reactor and the cyclotron.

Atomic Structure–the Electrical Nature of Matter

Let us now examine the makeup of the atom itself. An atom is not a homogeneous particle, but consists of even smaller particles that are present in the atoms of all elements. Although scientists and mathematicians have given us a highly satisfactory concept of atomic structure, it remains tentative and is undergoing almost continuous modification.

Modern theory holds that the atom is fundamentally ***electrical*** in nature. Note that the structure of the atom may be represented by a diagram (not an actual picture) to help correlate certain experimental observations that have been treated mathematically.

In 1913 Niels Bohr, a Danish physicist, first proposed the most widely held theory of atomic structure. Although it has been radically revised since his time, his early model of the atom adequately serves to explain the phenomena that we shall encounter in radiologic physics in this course. We shall therefore represent the atom as a submicroscopic system analogous to our solar system in which planets revolve around a central sun (see Figure 4.3). In the center of the atom lies the positively charged core known as the ***nucleus***, containing almost the entire mass of the atom. Revolving around the nucleus are

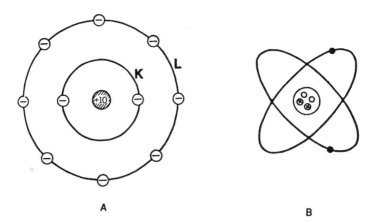

Figure 4.3. *A* shows a greatly simplified atomic model of neon according to the Bohr Theory, with the nucleus at the center surrounded by some of the *shells* of the electrons. The *K* shell can hold no more than two electrons, *L* no more than eight. The number of electrons that additional shells (up to *Q)* can hold has been determined. The electrons revolve very rapidly in their orbits and, at the same time, spin on their axis. Actually, each electron moves in its own path or *orbital* as in *B*, showing the nucleus of a helium atom with its two protons (⊕) and two neutrons (o).

the much lighter *orbital electrons*, each carrying a single *negative charge*. Each electron moves continuously in its own *orbit* or path around the nucleus. However, the electrons whose orbits are at a particular distance from the nucleus are grouped together and designated as belonging to a particular *shell* or *energy level*. These shells are identified by letters of the alphabet, the innermost being called the *K* shell, the next the *L* shell, and so on to *Q* (see Figure 4.3). The atoms of various elements differ in the number of electrons that are normally present in the shells.

There is a specific maximum number of electrons that can occupy a particular shell, listed as follows:

K shell	2 electrons
L shell	8 electrons
M shell	18 electrons
N shell	32 electrons

A shell may contain fewer than the maximum allowable number of electrons; thus, starting with the *M* shell, the outermost shell does not have to contain its maximum complement of electrons before electrons begin appearing in the next higher shell. In fact, an atom tends to be chemically stable when there are eight electrons or an *octet* in the outer shell. For example, the element potassium has two *K* electrons, eight *L* electrons, eight *M* electrons, and one *N* electron, despite the fact that the *M* shell can hold a maximum of eighteen electrons.

In a neutral atom the total number of orbital electrons exactly equals the total number of positive charges in the nucleus. All of the atoms of a given element have the same total number of positive nuclear charges, but the atoms of different elements have different total nuclear charges. *An atom of a given element maintains its identity only if its nuclear charge is unaltered.* If the nuclear charge is altered, as by irradiation with subatomic particles, the element may be transmuted (changed) into an entirely different element.

Nuclear Structure. Of what does the *nucleus* of an atom consist? It is known that the nucleus contains two main kinds of particles or *nucleons*–protons and neutrons. The proton carries a single positive charge (equal and opposite to the negative charge of an electron) and has an extremely small mass, about 1.6×10^{-24} gram, but this is still about 1,828 times the mass of an electron. Ordinarily, protons exist only in the nucleus and can be liberated from the nucleus only under special conditions. Protons of all elements are identical, but *each element has its own characteristic number of nuclear protons*. It must be emphasized that the protons alone give the nucleus its positive charge.

The other important constituent of the nucleus, the *neutron*, is a particle having about the same mass as a proton but carrying no charge; that is, the neutron is electrically *neutral*. (It is incorrect to speak of the neutron as having a neutral charge; rather, it is characterized by absence of charge.) Thus, protons and neutrons compose virtually the entire mass of the nucleus but *only the protons contribute positive charges to the nucleus*. Other nuclear particles have been discovered but will not be considered here.

In summary, we can state that the atoms of various elements consist of the same *building blocks*: protons, neutrons, and electrons. Atoms of various elements differ from one another by virtue of different combinations of these same building blocks. The structure of an atom may be outlined as follows:

1. *Nucleus*–contains one or more:
   a. **Protons**–elementary positive particles with mass about 1.67×10^{-24} g, and diameter about 10^{-13} cm.
   b. **Neutrons**–elementary neutral particles having virtually the same mass as the proton.
2. *Shells*–represent energy levels contain-

ing one or more:

a. **Electrons**–elementary negative particles with mass about 9.11 × 10⁻²⁸ g, and diameter about 4 × 10⁻¹³ cm.

The number of electrons in the shells must equal the number of protons in the nucleus of a *neutral* atom.

Atomic Number

What determines the identity of an element and makes it distinctly different from any other element? The answer lies in the number of nuclear protons. In a neutral atom the number of orbital electrons exactly equals the number of nuclear *protons* (see Figure 4.3). This number is characteristically the same for all atoms of a given element and is different for the atoms of different elements. *The number of protons or positive charges in the nucleus of an atom denotes its atomic number.* Thus, each element has a particular atomic number and, because of the corresponding specific number and arrangement of orbital electrons, its own distinctive chemical properties. Atomic number is designated by symbol Z.

Mass Number

What determines the mass of an atom? We have already indicated that nearly the entire mass resides in the nucleus and, more specifically, in the nuclear protons and neutrons. These particles comprise the units of atomic mass. Therefore, the *total number of protons and neutrons in the nucleus of an atom denotes its mass number* or *atomic mass* (see Figure 4.4). Protons and neutrons are often called *nucleons*. Mass number is designated by symbol A.

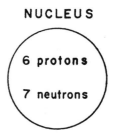

NUCLEUS

6 protons

7 neutrons

Atomic Number = 6

Mass Number = 13

Figure 4.4. Nuclear components.

Isotopes and Nuclides

If we were to examine a sample of a particular element, we would find that it consists of two or more kinds of atoms that differ in mass number, even though they must all have the same atomic number. For example, the familiar element oxygen is a mixture of atoms of mass number 16, 17, and 18, although all these oxygen atoms have the same atomic number–8. Such atoms of the same element having different mass numbers are called *isotopes*. Any particular kind of atom, having a specific number of protons and neutrons, is called a *nuclide*.

Hydrogen, the simplest element, has three isotopes, as shown in Figure 4.5. All three hydrogen nuclides have atomic number 1, but they have different mass numbers–1, 2, and 3. The reason for the difference is immediately evident from the figure; *they differ only in the number of nuclear neutrons*.

The preceding discussion may be generalized as follows: *isotopes may be defined as atoms that have the same number of nuclear protons (equal to the atomic number of the element) but different numbers of nuclear neutrons.* This definition explains the difference in the mass numbers of isotopes.

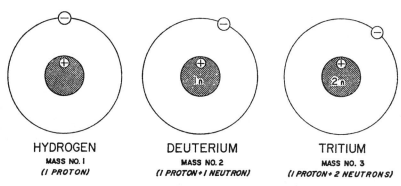

HYDROGEN
MASS NO. I
(I PROTON)

DEUTERIUM
MASS NO. 2
(I PROTON + I NEUTRON)

TRITIUM
MASS NO. 3
(I PROTON + 2 NEUTRONS)

Figure 4.5. Isotopes of hydrogen. These particular isotopes have the same atomic number, 1, but have different mass numbers. They are called *isotopes* of the same element, hydrogen. Their different mass numbers depend on the different numbers of neutrons in the nucleus. Ordinary hydrogen (mass no. 1) is sometimes called *protium*. Tritium (pronounced trit'-é-um) is an artificial isotope and is radioactive.

Note that the atomic number and the mass number **together** specify the nuclear composition of any given atom, thereby designating a particular **nuclide**; the atomic number alone specifies the number of nuclear protons, and hence identifies the element.

Atomic weight refers to the mass of any atom relative to the mass of an atom of carbon 12 isotope taken as 12. The **atomic mass unit (amu)** is therefore defined as one-twelfth the mass of the carbon 12 nucleus. Because elements consist of mixtures of isotopes, the atomic weight is an average and is therefore almost never a whole number. However, since atomic weight, and mass number, are so nearly equal, these terms are often used interchangeably.

The Periodic Table

The preceding discussion will be clarified by actual examples. All of the elements can be arranged in an orderly series, called the **periodic table**, from the lightest to the heaviest (on the basis of atomic weight), or from the lowest atomic number to the highest (see Table 4.1). For instance, ordinary **hydrogen** is the simplest element, since it has only one proton and one orbital electron. The atomic number of ordinary hydrogen is therefore

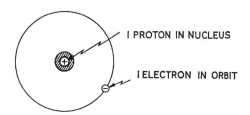

I PROTON IN NUCLEUS

I ELECTRON IN ORBIT

Figure 4.6. Atomic model of ordinary hydrogen.

1, and so is its mass number (see Figure 4.6).

The second element in the periodic series is **helium**, with two protons and two neutrons in the nucleus and two electrons in the *K shell* (see Figure 4.7). The atomic number of this element is 2 and the mass number is 4.

Lithium is the third element. Its atomic number is 3, and one of its isotopes has a mass number of 7 (see Figure 4.8). Note that the first shell has the maximum number of electrons, two, and the third electron is in the next shell. Thus, the *K* shell of lithium has two electrons, and the *L* shell, one.

TABLE 4.1
ATOMIC PERIODIC TABLE

The numbers above each element represent atomic weight.
The numbers below each element represent atomic number.
Elements of particular interest to radiologic technologists are in dark type.
Transuranic elements (that is, beyond uranium) are man made.

Handwritten annotation (top right): Inert Elements

Handwritten annotations (left margin):
(Groups) # of e⁻ in outer shell — similar bonding properties
(Periods) — same # of e⁻ shell, different chemical properties — different # of e⁻ shell properties

Handwritten annotations (bottom left):
K — 2
L — 8
M — 18
N — 32
O — 50
P — 32
Q — 18

Period	Group I	Group II	Group III	Group IV	Group V	Group VI	Group VII			Group VIII
0	1.008 Hydrogen (H) 1									4.003 Helium (He) 2
1	6.940 Lithium (Li) 3	9.02 Beryllium (Be) 4	10.82 Boron (B) 5	12.01 Carbon (C) 6	14.008 Nitrogen (N) 7	16.00 Oxygen (O) 8	19.00 Fluorine (F) 9			20.183 Neon (Ne) 10
2	22.997 Sodium (Na) 11	24.32 Magnesium (Mg) 12	26.97 Aluminum (Al) 13	28.06 Silicon (Si) 14	30.98 Phosphorus (P) 15	32.06 Sulfur (S) 16	35.457 Chlorine (Cl) 17			39.994 Argon (A) 18
3	39.006 Potassium (K) 19	40.08 Calcium (Ca) 20	45.10 Scandium (Sc) 21	47.90 Titanium (Ti) 22	50.95 Vanadium (V) 23	52.01 Chromium (Cr) 24	54.93 Manganese (Mn) 25	55.85 Iron (Fe) 26	58.94 Cobalt (Co) 27	58.69 Nickel (Ni) 28
	63.57 Copper (Cu) 29	65.38 Zinc (Zn) 30	69.72 Gallium (Ga) 31	72.60 Germanium (Ge) 32	74.91 Arsenic (As) 33	79.00 Selenium (Se) 34	79.92 Bromine (Br) 35			83.7 Krypton (Kr) 36
4	85.48 Rubidium (Rb) 37	87.63 Strontium (Sr) 38	88.92 Yttrium (Y) 39	91.22 Zirconium (Zr) 40	92.91 Niobium (Nb) 41	96.0 Molybdenum (Mo) 42	99 Technetium (Tc) 43	101.7 Ruthenium (Ru) 44	102.9 Rhodium (Rh) 45	106.7 Palladium (Pd) 46
	107.88 Silver (Ag) 47	112.41 Cadmium (Cd) 48	118.70 Indium (In) 49	121.77 Tin (Sn) 50	127.6 Antimony (Sb) 51	126.93 Tellurium (Te) 52	126.92 Iodine (I) 53			131.3 Xenon (Xe) 54
	132.9 Cesium (Cs) 55	137.4 Barium (Ba) 56	Rare Earths 57–71	178.6 Hafnium (Hf) 72	180.9 Tantalum (Ta) 73	183.9 Tungsten (W) 74	186.3 Rhenium (Re) 75	190.2 Osmium (Os) 76	193.1 Iridium (Ir) 77	195.2 Platinum (Pt) 78

5	197.2 *Gold* *(Au)* 79	200.6 Mercury (Hg) 80	204.4 Thallium (Tl) 81	207.2 *Lead* *(Pb)* 82	209.0 *Bismuth* *(Bi)* 83	210 Polonium (Po) 84	211 Astatine (At) 85	222 *Radon* *(Rn)* 86
6	224 Virginium (Vi) 87	226.05 *Radium* *(Ra)* 88	Actinide° Series 89–103					

°*Actinide Series*

227 Actinium (Ac) 89	232 Thorium (Th) 90	231 Protoactinium (Pa) 91	238.1 Uranium (U) 92
(237) Neptunium (Np) 93	(242) Plutonium (Pu) 94	(243) Americium (Am) 95	(247) Curium (Cm) 96
(249) Berkelium (Bk) 97	(251) Californium (Cf) 98	(254) Einsteinium (Es) 99	(257) Fermium (Fm) 100
(256) Mendelevium (Md) 101	(254) Nobelium (No) 102	(257) Lawrencium (Lw) 103	

(Number in parenthesis is mass number of most stable nuclide.)

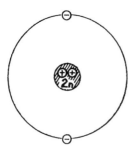

Figure 4.7. Atomic model of helium; two protons in the nucleus and two electrons in *K* shell.

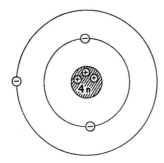

Figure 4.8. Atomic model of lithium.

Chemical Behavior

When the elements are tabulated in order of increasing atomic number (periodic table), they fall into *eight vertical groups* and *seven horizontal periods* (see Table 4.1). It turns out that the *vertical groups* represent families of elements with surprisingly similar chemical properties. In other words, they participate similarly in chemical reactions. This relationship of the various elements was first discovered in 1870 by Mendeleev, the eminent Russian chemist. The *periods* (horizontal rows) consist of elements with the same number of electron shells, but with different chemical properties.

Why do the elements in any vertical group have similar chemical properties? These depend on the number of electrons in their *outermost shells*. For instance, lithium (Li), sodium (Na), and potassium (K), shown in Table 4.1, belong to the same family—Group I. Their atomic structure is shown in Figure 4.9. The neutrons are omitted here because they do not influence chemical behavior. The diagram shows that although these elements have different numbers of shells, they have one thing in common: *the outermost shell of each has one electron* which it can readily give up, leaving the atom with eight electrons in its outermost shell—the stable octet (eight) configuration for sodium and potassium, and duet for lithium. This determines the similarity in their chemical behavior.

Consider now vertical group VII: fluorine (F), chlorine (Cl), and bromine (Br). As shown in Figure 4.10, these elements have different numbers of shells, but are similar in that they all have *one less electron* in the outermost shell than the number of electrons needed to saturate it to a chemically stable octet configuration. Therefore, these elements resemble each other in their chemical properties.

The number of electrons in the outermost shell determines their combining ability or *valence*. The family of elements represented by lithium, sodium, and potassium has one electron in the outermost shell, and the valence is + 1, because these elements can *give* that electron to an element which needs an electron to saturate its outermost shell and form an octet. On the other hand, the elements in the fluorine, chlorine, bromine family, as described above, have one less electron in the outermost shell than the number needed to form an octet. Their valence is –1. These elements can *accept* an electron from the lithium family. The atom that gives up the electron now has one excess positive charge, whereas the atom that accepts the electron has one excess negative charge. Because of their opposite charges, such charged atoms—*ions*—strongly attract each other to form *compounds*. The attraction

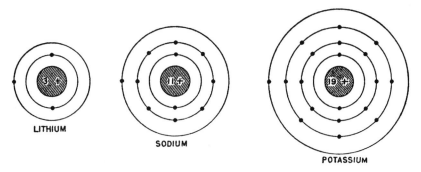

Figure 4.9. Elements with valence of + 1.

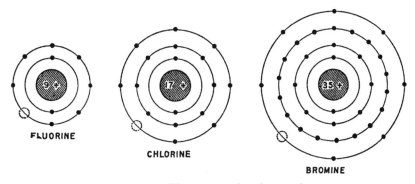

Figure 4.10. Elements with valence of -1.

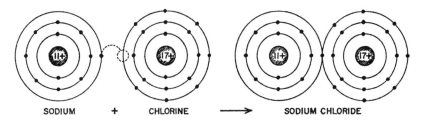

Figure 4.11. Combination of sodium and chlorine chemically to form sodium chloride. The chemical bond is *ionic*, common to the formation of almost all inorganic compounds.

between the two ions is a ***chemical bond***, this type being called an ***ionic bond*** (see Figure 4.11).

Another type, the ***covalent bond***, involves the ***sharing*** of outer orbital electrons. This occurs in the water molecule, which consists of two atoms of hydrogen bonded to one atom of oxygen by a sharing of the hydrogen valence electron as shown in Figure 4.12. With this type of bond a shared electron divides its time orbiting around the hydrogen nucleus and the oxygen nucleus, slightly more around the latter. Therefore, the oxygen side of the water molecule is slightly negative with respect to the hydrogen side, so we call this ***polar covalent bonding***.

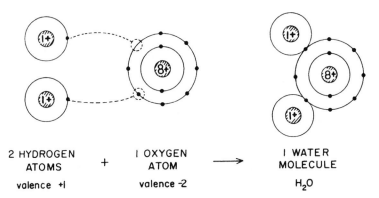

Figure 4.12. Chemical combination of hydrogen and oxygen to form water, by sharing of two electrons—that is, a *covalent bond.* Note polarization of molecule—positive on left, negative on right.

Suppose a particular element normally were to have its outer shell saturated with electrons, or in octet configuration. It could not enter into a chemical reaction, since it would be unable to give up, accept, or share an electron. Referring again to Table 4.1, notice that there is such a family of elements—Group O—known as the *inert elements.* They include helium, neon, argon, etc., as shown in Figure 4.13. (It has been possible, recently, to bring about chemical union of inert elements in very small quantities.)

Valence explains why elements always combine in *definite proportions.* Thus, sodium chloride is always formed by the union of one atom of sodium and one atom of chlo-

rine. Water is always formed by the combination of one atom of oxygen with two atoms of hydrogen.

We may therefore conclude that *the chemical properties of an element depend on its valence,* which in turn depends on the number of electrons in the outermost shell. But protons and neutrons do not participate in chemical reactions, although they go along for the ride, as it were.

Ionization

If an electron is removed from one of the shells of a neutral atom, or if an electron is added to a neutral atom, the atom does not lose its identity because it retains the same nuclear charge so its atomic number persists. However, the atom is now electrically charged. Thus, if an electron is *removed,* the atom becomes *positively charged* because one of its nuclear charges is now unbalanced. If an electron is *added* to a neutral atom, the atom becomes *negatively charged.* Such charged atoms will drift when placed in an electric field, as will be discussed later, and are called *ions* (from Greek *ienai* = to go). The process of converting atoms to ions is termed *ionization, which is produced exclusively through the addition or removal*

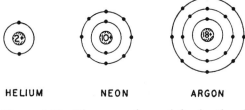

HELIUM   NEON   ARGON

Figure 4.13. Three members of the family of inert elements, with valence of 0. The outer shell in each case is completely filled with electrons. Therefore, these elements do not ordinarily enter into chemical reactions.

of orbital electrons.

1. **Exposure of Matter to X Rays or Gamma Rays**. These may remove electrons from the atoms of matter lying in their path. Such atoms become positively charged, that is, positive ions. The removed electrons may ionize other atoms, combine with neutral atoms to produce negative ions, or recombine with positive ions to form neutral atoms.

2. **Exposure of Matter to a Stream of Electrons**. The fastmoving electrons in an electron beam or the electrons released from atoms by x rays as in (1) may interact with and remove orbital electrons from other atoms. An atom, on losing an electron, now has an excess positive charge, but it is soon neutralized by picking up a free electron from the surrounding air or water. Ionization by x rays or fast electrons is of extremely short duration–the ion becomes neutralized almost instantaneously.

3. **Spontaneous Decay of Radionuclides**. The radiation emitted by radioactive nuclides during decay includes charged particles and gamma rays, which can produce ionization.

4. **Exposure of Certain Elements to Light**. When light strikes the surface of certain metals such as cesium or potassium, electrons are liberated from their atoms and emitted from the surface of the metal. This principle is utilized in the photoelectric cell, which is the main component of the x-ray phototimer.

5. **Chemical Ionization**. If two neutral atoms such as sodium and chlorine are brought together, the sodium atom with valence +1 gives up its outermost or valence electron to the chlorine atom to fill its outermost shell. The sodium atom becomes a positive sodium ion (deficiency of one electron), while the chlorine becomes a negative chlorine ion (excess of one electron). When the compound sodium chloride is dissolved in water its ions separate so that the water now contains sodium ions (Na^+) and chlorine ions (Cl^-). If the poles of a battery are connected to electrodes and these are immersed in the solution, the Na^+ ions move toward the negative electrode (cathode) as *cations*, while the Cl^- ions move toward the positive electrode (anode) as *anions*, as shown in Figure 4.14. On arriving at the *cathode*, each Na^+ ion picks up an electron and becomes a neutral sodium atom. Each Cl^- ion gives up its extra electron to the *anode* and becomes a neutral chlorine atom.

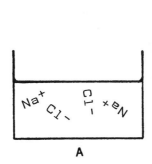 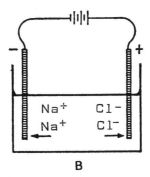

Figure 4.14. Chemical ionization and electrolysis. In *A* is shown a solution of ordinary salt (sodium chloride = NaCl) whose ions have separated to form equal numbers of Na^+ and Cl^- ions. In *B* a pair of electrodes has been immersed in the solution, the Na^+ ions move toward the negative electrode (cathode) and the Cl^- ions drift toward the positive electrode (anode), a process called *electrolysis*.

These effects of an electric current are called *electrolysis*.

6. **Thermionic Emission**. Heating a metal to incandescence (glowing hot) entails the release of electrons from its surface. These electrons are the loosely bound, outermost orbital electrons of the atoms of the metal; heating the metal imparts sufficient kinetic energy to the electrons to facilitate their escape from the surface, analogous to the evaporation of water molecules from a body of water. Such ejection of electrons is called *thermionic emission*, or *thermionic effect*; it is essential to the operation of an x-ray tube.

Ionization is an extremely important process. Many chemical reactions take place between ions in solution. Ionization of air by x rays underlies the measurement of the exposure rate of an x-ray beam. Ionization of body tissues indirectly through preliminary release of electrons is the fundamental mode of action of x and gamma rays in therapy.

QUESTIONS AND PROBLEMS

1. Define element; compound; substance.
2. State briefly the Bohr concept of atomic structure.
3. Draw a model of the hydrogen atom and label the parts.
4. What constitutes the nucleus of an atom?
5. Define orbit; shell; octet.
6. How does the mass of a proton compare with that of an electron?
7. A neutral atom has twelve electrons revolving in the orbits around the nucleus. What is its atomic number?
8. An atom has a nucleus containing six protons and two neutrons. What is its atomic number (Z)? Mass number (A)?
9. Define isotope, and give an example. Define nuclide.
10. How do elements combine chemically in fixed or definite proportions?
11. What is an ion? By what methods can ionization be produced?
12. Explain the similarity of chemical behavior of certain elements.
13. Discuss two types of chemical bonds. Give an example of a compound formed by each type.
14. Where are the following located in the atom: proton, neutron, electron? Why are they called the building blocks of matter?

Chapter 5

ELECTROSTATICS

Topics Covered in This Chapter

- Definition
- Methods of Electrification
- Laws of Electrostatics
- Electroscope
- Questions and Problems

Definition

THE TERM *electrostatics* is defined as *the branch of physics that deals with stationary or resting electric charges.* Another name for resting charges is *static electricity*.

Electrification

As explained in Chapter 4, all atoms contain electrons in motion about a nucleus. If one or more of these electrons are removed, the atom is left with an excess positive charge. Should the removed electron become attached to a neutral atom, then the latter will become negatively charged. Thus, *there are only two kinds of electric charges, positive and negative*.

The same principle holds for a body of matter from whose atoms electrons may be removed or to which electrons may be added by a process called *electrification*. So an electrified or charged body has either an excess or a deficiency of electrons, being negatively charged in the first instance and positively charged in the second.

Methods of Electrification

How can we electrify a body? We can do so in three ways:

1. **Electrification by Friction**. You have probably observed after walking over a woolen carpet that touching a metal door knob will cause a spark to jump between your hand and the metal. Again, you may have noted that in combing your hair a crackling sound is heard as the comb seems to draw the hair to it. These are examples of *electrification by friction*, that is, the removal of electrons from one object by rubbing it with another of a different kind. This phenomenon was known to the ancient Greeks who observed that amber, when rubbed with fur, attracts small bits of straw or dried leaves. Since the Greek word for amber is *elektron*, the origin of the root word for *electricity* is obvious.

On the basis of the two kinds of charges developed by friction, we designate one *negative* and the other *positive*. When a glass rod is rubbed with silk, some electrons are removed from the rod, leaving it positively charged; this is *positive electrification*. At the same time, the silk acquires an excess of electrons by *negative electrification*.

If amber is rubbed with fur, the amber picks up electrons and becomes negatively charged, thereby undergoing negative electrification, while the fur undergoes positive electrification.

2. **Electrification by Contact**. When a body charged by friction touches an uncharged object, the latter acquires a like charge–*electrification by contact*. If the first object is negatively charged, it will give up some of its electrons to the second object, imparting to it a negative charge. If, on the other hand, the first object has a positive charge, it will remove electrons from the second object, leaving it positively charged. *We can conclude that a charged object confers the same kind of charge on any uncharged body with which it comes in contact*.

3. **Electrification by Induction**. Matter can be classified on the basis of its electrical conductivity, that is, how readily it allows the flow of electrons. The materials in the first group are called *nonconductors* or *insulators* because electrons do not flow freely in them; examples are plastics, hard rubber, and glass. The materials in the second class are called *conductors* because they allow a free flow of electrons; they are exemplified by *metals*, such as silver, copper, and aluminum. Electrification by induction utilizes metallic conductors. Another class, intermediate in conducting ability, comprises the *semiconductors* (see pages 100-105).

Surrounding every charged body there is a region in which a force is exerted on another charged body. This region or zone is called an *electric field*. An uncharged metallic object experiences a shift of electrons when brought into the electric field of a charged object, a process called *electrification by induction*. Note that *only the electrons move*. In Figure 5.1, if the negatively charged rod, B, is brought *near* the originally uncharged metal body, A, but not touching it, the negative field near B will repel

electrons on A distally, leaving the end of A nearest B positively charged. The reverse is also true; if B were a positively charged rod, the end of A nearest B would become negatively charged. As B, in either case, is withdrawn from the electric field of A, the electrons on A are redistributed to their original position and A is no longer charged.

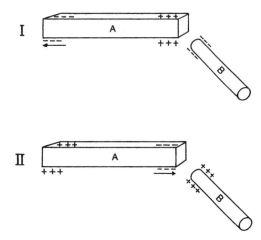

Figure 5.1. Temporary electrification by induction. *A* is a metallic rod. *B* is a charged body. In I electrons on *A* have been repelled to the left by the negative charge on *B*. In II the electrons on *A* have been attracted to the right by the positive charge on *B* (see Figure 5.3).

It is possible to charge a piece of metal by induction so that it will *retain its charge* even after the charging body has been removed. In Figure 5.2, *A* is shown first as it is affected by the electric field of *B*. Connecting the negative end of *A* temporarily to a water pipe (ground) causes the excess electrons to pass down to ground. If the ground connection is broken and *B* is *then* removed, *A* remains positively charged. It must be emphasized that *in electrification by induction, the charged body confers the opposite kind of charge on the metallic body which is placed in its field*.

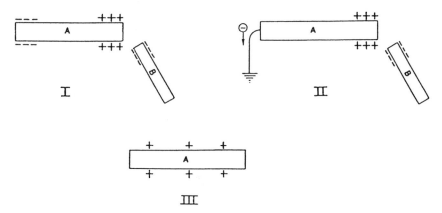

Figure 5.2. Semipermanent electrification by induction. *A* is a metallic rod. *B* is a charged body.

What is meant by ground? Since the earth is essentially an infinite reservoir of electrons, any charged body can be neutralized (same number of positive and negative charges) if it is *grounded*, that is, connected to wet earth by a conductor. Thus, grounding a positively charged body causes electrons to move up from the ground to neutralize the body. On the other hand, grounding a negatively charged body drives its excess electrons to ground, again neutralizing the body. In both instances, the charged body is neutralized by being brought to the same electrical potential energy level as ground. *Ground potential equals zero*. The symbol for ground is ⏚.

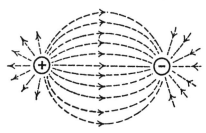

Figure 5.3. Electric fields. The electric lines of force are assumed to be directed from a positive charge to a negative charge. A negative charge, such as an electron, placed in the field would drift toward the positive charge.

Laws of Electrostatics

There are five fundamental laws of electrostatics:

1. **Like Charges Repel Each Other; Unlike Charges Attract Each Other**. As noted previously, every charge has around it an *electric field of force*. Such a field exists between two electric charges, and is assumed to be directed from a positive to a negative charge (see Figure 5.3). There is a resulting force of *attraction* between two *oppositely* charged bodies. But with two *similarly* charged bodies, their electric lines of force repel each other, as in Figure 5.4, so there is a *repelling* force between them. Figure 5.5 shows this law demonstrated experimentally. (Pith is the light, spongy material in the stems of certain plants.)

2. **The Electrostatic Force Between Two Charges is Directly Proportional to the Product of Their Quantities, and Inversely Proportional to the Square of the Distance Between Them**. The force may be one of attraction or repulsion, depending on whether the charges are different or alike. For example, if we have two

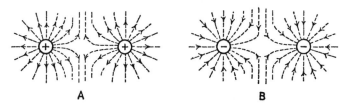

Figure 5.4. When like charges are brought near each other, their electric fields exert a force of repulsion or separation between the like charges.

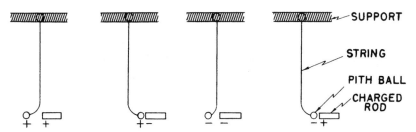

Figure 5.5. Like charges repel. Unlike charges attract.

unlike charges and the strength of one is doubled, the force of attraction between them is doubled. If both are doubled, the force of attraction between them is $2 \times 2 = 4$ times as great. If the distance between the original charges is doubled, the force between them is $1/2 \times 1/2 = 1/4$ as great. If the distance is halved, the force is $2 \times 2 = 4$ times as great.

The force F between them, if they are point charges, is expressed by Coulomb's Law

$$F = k \frac{Q_1 Q_2}{D^2}$$

where F is in newtons, Q_1 and Q_2 in coulombs, and D in meters. k is a constant that depends on the medium. Being that force is a vector quantity, this law applies both when the charges are alike (repulsion), or when they are opposite (attraction). A vector quantity has both size and direction.

3. **Electric Charges Reside Only on External Surfaces of Conductors**. When a hollow metal ball is charged, all of the

charges are on the outside surface, while the inside remains uncharged, as is evident from Figure 5.6. This is explained by the mutual repulsion of like charges which tend to move as far away from each other as possible. The greatest distance obviously exists between them if they move to the outer surface of the object.

4. **The Concentration of Charges on a Curved Surface of a Conductor Is Greatest Where the Curvature Is Greatest**. This is shown in Figure 5.7. The greatest possible concentration of charges

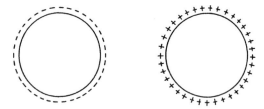

Figure 5.6. Electric charges tend to distribute themselves on the outer surface of a hollow metallic sphere.

occurs on a point which represents an extremely high degree of curvature; in fact, charges become so crowded on a ***point*** that they easily leak away.

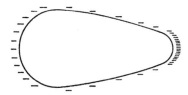

Figure 5.7. Electric charges tend to concentrate in region of sharpest curvature.

5. Only Negative Charges (Electrons) Can Move in Solid Conductors.

The positive charges are actually charged atoms which do not drift in solid conductors.

Electroscope

An electroscope detects the presence and sign of an electric charge. Under suitable conditions, it can be calibrated to determine the size of a charge. A simple electroscope is shown as a diagram in Figure 5.8.

We can charge an electroscope either by contact or by induction.

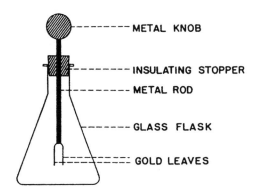

Figure 5.8. Gold leaf electroscope. When an electric charge is imparted to the electroscope, the leaves repel each other and spread apart.

1. Charging an Electroscope by Contact.

Figure 5.9 shows how this is done. If a ***negatively*** charged rod is allowed to touch the metal knob, as in A, electrons will move from the rod to the knob, stem, and gold leaves of the electroscope. Since the leaves now have the same negative charge, they repel each other and spread apart. The electroscope is now charged ***negatively***. But if a ***positively*** charged rod is brought into contact with the metal knob of the electroscope, some electrons will move from the electroscope to the rod, leaving the electroscope with a ***positive*** charge; the leaves will diverge just as when they were negatively charged.

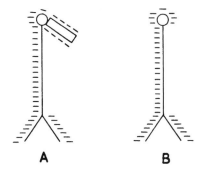

Figure 5.9. Charging an electroscope by contact.

2. Charging an Electroscope by Induction.

In Figure 5.10, a negatively charged rod is brought ***near***, but not touching, the metal knob of the electroscope. The rod's negative electrostatic field repels electrons from the knob down to the leaves, causing them to separate. If the knob is then grounded while the rod remains near it, the leaves will collapse because they have lost their charge. If the ground connection is next broken and the rod is ***then removed***, there will be a deficiency of electrons in the electroscope (the excess electrons having passed to ground), leaving the electroscope with a positive charge. This causes the

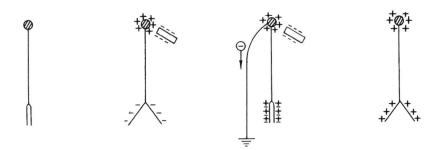

Figure 5.10. Charging an electroscope by induction. The successive steps are indicated in the diagrams from left to right.

leaves to spread. In general, the larger the charge, the greater will be the separation of the gold leaves.

An electroscope can be used not only to detect electric charges, but also to determine their sign, that is, whether they are positive or negative. When a positively charged body is brought near the knob of a negatively charged electroscope, some of the electrons will be attracted from the leaves up to the knob, and the leaves will come together. But when a negatively charged body is brought near the knob of a negatively charged electroscope, some of the electrons from the knob will be repelled down to the already negative leaves and they will diverge even further. We can therefore deduce that a charged body brought near a similarly charged electroscope causes the leaves to diverge farther. But a charged body brought near an electroscope that has an opposite charge causes the leaves to come together.

A charged electroscope can be used to detect the presence of x rays because of the ionization they produce in air. The resulting ions neutralize the charge on the electroscope, causing the leaves to collapse.

Static Discharge

Suppose we have two oppositely charged bodies separated by an air space or insulator.

The surplus electrons on the negatively charged body naturally tend to move over to the positively charged body, but this may be prevented by the air space or insulator between the bodies. However, if the electrons continue to be piled up on the negatively charged body, the electric field becomes correspondingly stronger. Eventually, continued buildup of electric charge causes breakdown of the insulator as the electrons jump from the more negative to the less negative body, producing a *spark*.

A difference of electrical potential energy exists between two oppositely charged bodies, or between two bodies having like charges of different sizes. The point where electrons are in excess is at a higher negative potential than the point at which they are deficient. Therefore, the electrons move "downhill," from the point of higher to the point of lower negative potential, analogous to the difference in mechanical potential energy in the example cited on page 24.

The above discussion explains the familiar *lightning discharge*. A thunder cloud may have a positive charge on its upper side and a negative charge on its lower side, or the reverse, built up by its motion through the air. The earth beneath develops an opposite charge by electrostatic induction. When the difference of electrical potential between the bottom of a negatively charged cloud and the earth exceeds a certain critical value, electrons rush between the two in the form of a bolt of lightning. Since this usually involves a sudden transfer of excess electrons to the earth, they may rush back toward the cloud, then back to the earth, repeating the process of declining oscillation many times within a few millionths of a second.

QUESTIONS AND PROBLEMS

1. Discuss electrification. By what three methods can it be accomplished?
2. What happens if two similarly charged bodies are brought close together? Oppositely charged bodies?
3. How many kinds of electricity are there? Which one is capable of drifting in a solid conductor?
4. If a negatively charged object is touched to a neutral object, what kind of charge does the latter acquire?
5. State the five laws of electrostatics.
6. Of what value is the electroscope? How does it work?
7. What is meant by an electric field?
8. Describe electrification by contact; by induction.
9. Explain static discharge and give an example.
10. Define conductor; insulator; semiconductor. Give examples.

Chapter 6

ELECTRODYNAMICS–ELECTRIC CURRENT

Topics Covered in This Chapter

- Definition
- Nature of Electric Current
- Sources of Electric Current
- Factors in Electric Circuit
- Ohm's Law
- Cells and Batteries

- Components of Elementary Electric Circuits
- Series and Parallel Circuits
- Electric Capacitor (condenser)
- Work and Power of Electric Current
- Questions and Problems

Definition

THE PRECEDING chapter dealt with the characteristics of stationary electric charges. Under certain conditions, electric charges can be made to drift through a suitable material termed a *conductor*. *The science of electric charges in motion is known as electrodynamics, or current electricity*.

The Nature of an Electric Current

Fundamentally, *an electric current consists of a flow of charged particles*. This may occur under the following conditions:

1. **In a Vacuum**–electrons may jump a gap between two oppositely charged electrodes in a vacuum tube. This will be discussed in detail later in connection with x-ray tubes.

2. **In a Gas**. You have all seen neon signs. These are glass tubes containing neon gas at low pressure. Two metal electrodes are sealed in the ends of the tubes. Application of voltage across the electrodes

causes free electrons present in the neon gas to speed toward the anode, thereby ionizing and exciting the neon gas. This results in the emission of pink light.

3. **In an Ionic Solution**–as already described on page 37, upon being dissolved in water a salt separates into its component ions. If electrodes are then immersed in the solution and connected to a source of direct current, the positive ions move toward the cathode while the negative ions move toward the anode. Such migration of ions in opposite directions characterizes the flow of an electric current in an ionic solution (see Figure 4.14 on page 37).

4. **In a Metallic Conductor**–this will be our main concern in the present chapter, and in later ones as well. *An electric current in a solid conductor consists of a flow of electrons only*. In the atoms of metallic conductors, the valence (outermost) electrons are "free" and, under certain conditions, these free electrons can be made to drift through the conductor in its *conduction band* (see pages 100-101). An electron does not move instantaneously from one end of a wire to the

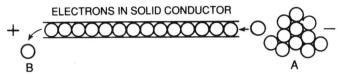

Figure 6.1. Diagrammatic representation of electron flow in a solid conductor. As an electron enters the conductor at *A* where there is an excess of electrons another electron leaves at *B* where there is a deficiency of electrons.

other, the individual electron drift being relatively slow—less than 1 mm per sec. However, the electrical impulse or **current** moves with tremendous speed, approaching that of light in a vacuum (3×10^8 meters per sec). To make electrons flow, the conductor must have an excess of electrons at one end, and a relative deficiency at the other. The net result is that electrons will then flow from the point of excess (higher negative potential) to the point of deficiency (lower negative potential), just as a body will fall from a point of higher potential energy to a point of lower potential energy (see pages 22-23 and Figure 6.1).

An *electric circuit is defined as the path over which the current flows*. Electrons will not flow in a nonconductor such as glass or plastic because of the gap between the valence band and the conduction band and, in semiconductors, they flow only under certain conditions. Figure 6.2 shows diagrammatically the components of a simple electric circuit.

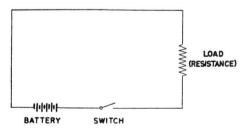

Figure 6.2. Simple resistance circuit. The load represents the total resistance of electrical devices in the circuit.

Sources of Electric Current

Various methods are available for piling up electrons at one point and simultaneously removing them at another point, thereby setting up a difference of electrical potential and causing an electric current to flow in a conductor. They all require the expenditure of energy as exemplified by the following two main types of devices:

1. *Cells* or *batteries* which convert *chemical energy* to electrical energy.
2. *Dynamo* or *generator* which converts *mechanical energy* to electrical energy (described on pages 73-75).

Other sources of energy include solar (sunlight), atomic (nuclear), wind, and geothermal (underground natural steam). These are in various stages of development and use.

Factors in an Electric Circuit

Three main factors characterize a simple electric circuit, which is carrying a *steady direct current*, that is, a current of constant strength flowing always in the same direction. They include *potential difference, current*, and *resistance*. These factors can best be understood by comparing them with those governing the flow of water through a pipe. They need modification for alternating current (see pages 66, 77-78).

1. **Potential Difference**. We may define potential difference as a *difference in electri-*

cal potential energy between two points in an electric circuit, due to an excess of electrons at one point relative to the other. A piling up of electrons at one point and simultaneous removal at another point are brought about through the use of a battery or generator. The potential difference between two points actually represents the amount of *work* expended in moving a unit charge from one point to the other. This resembles the situation shown in Figure 6.3 in which there are two containers of water at different levels connected by a pipe closed by a valve. A difference of pressure exists between the water in both containers, proportional to the difference in height *d* of the water levels. As soon as the valve is opened, the difference in pressure causes the water to flow from container *A* to container *B* until their levels become equalized and the pressure difference disappears.

Similarly, an excess of electrons at one end of a conductor causes a difference of electrical pressure or a *potential difference* to be set up, *even if the conductor is interrupted by an open switch* as in Figure 6.2. Closing the switch causes electrons to rush from the point of excess (higher negative potential) to the point of deficiency (lower negative potential). Thus, just as with the water analogy, it is *the difference of potential that drives the electrons*. If the difference of potential can be maintained by a battery or generator, electron current will persist as long as the switch is closed.

As the current flows, electrical potential gradually falls along the circuit. Similarly, with water flowing through a pipe, the water pressure decreases gradually along the pipe.

It would be well to review the section in Chapter 3 on the difference of potential energy between two points at different levels above ground, and the movement of a rock from a higher to a lower energy level. The analogy with the difference in water pressure and the difference of electrical potential is obvious.

The term *electromotive force* (emf) applies to the maximum difference of potential between the terminals of a battery or generator. As noted above, this is not really a force but rather a quantity of work or energy needed to move a unit electric charge through the circuit. In fact, 1 volt = 1 joule/coulomb (SI units).

The unit of potential difference (or emf) is the *volt, defined as that potential difference which will maintain a current of one ampere*

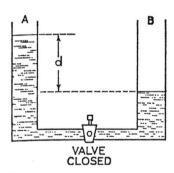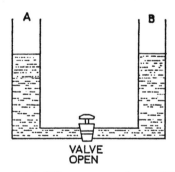

Figure 6.3. Water pressure analogy of electrical potential difference or voltage. With the valve closed, there is a difference of pressure between the water in arms *A* and *B*, proportional to the difference in water levels, *d*. (This is analogous to the potential difference between the poles of a battery on open circuit with no current flowing.) When the valve is opened, water flows from arm *A* into arm *B* until the difference in pressure disappears. (In a closed electric circuit, current continues to flow as long as potential difference is applied.)

in a circuit whose resistance is one ohm.

2. **Current**. The second factor in an electric circuit is *current*, *defined as the amount of electric charge flowing per second*. (In a water pipe, there is a definite amount of water flowing per second past a given point). Since an electric current consists of a flow of electrons, the more electrons flowing per second the "stronger" is the current. Similarly, the more water flowing per second in a pipe, the stronger is the current of water.

The unit of current is the *ampere*, *which may be defined as one coulomb quantity of electric charge flowing per second* (ie, 6.25×10^{18} free electrons per sec). A current of one ampere will deposit a standard weight of silver per second from a solution of silver nitrate by electrolysis (to be exact, 0.001118 g per sec).

3. **Resistance**. The third factor in an electric circuit is *resistance*, a property of the materials making up the circuit itself. *Electrical resistance is that property of the circuit which opposes or hinders the flow of an electric current*; it does *not* slow current flow, which remains about equal to the speed of light.

Resistance in an electric circuit is analogous to that in the flow of water in a pipe because of the friction of the molecules of water against the molecules of the pipe material. Note that the very large resistance of a nonconductor prevents the flow of current.

The unit of electrical resistance is the *ohm* (symbol Ω, Greek *omega)*, defined by the International Electrical Congress of 1893 as follows: *one ohm is the resistance of a standard volume of mercury under standard conditions* (to be exact, it is the resistance to a steady current by a column of mercury weighing 14.45 g, having a length of 106.3 cm and a uniform cross-sectional area, at 0°C). The term *conductance* denotes the reciprocal of resistance; that is, the ability of a circuit to conduct a current. At one time its unit was the mho (inverse of ohm), but the SI unit is the siemens (symbol, G).

On what does the resistance of a conductor depend? It depends on four things: material, length, cross-sectional area, and temperature.

a. *Material*. Certain materials, known as conductors, readily allow the flow of electrons. They are best exemplified by metals, such as copper and silver. Although copper is a poorer conductor than silver, it is much cheaper and therefore widely used in electric wires. Other materials, such as glass, wood, and plastic, being nonconductors, have virtually no free electrons and hence offer tremendous resistance to the flow of electricity. Such nonconducting materials are called *insulators* or *dielectrics*. Intermediate in resistance are the *semiconductors* (see pages 101-106).

b. *Length*. A long conductor, such as a long wire, has more resistance than a short one. In fact, the *resistance is directly proportional to the length of the conductor*. Similarly, the resistance of a pipe to the flow of water increases with its length.

c. *Cross-sectional Area*. A conductor with a large cross-sectional area has a lower resistance than one with a small cross-sectional area. More specifically, *the resistance of a conductor is inversely proportional to the cross-sectional area*. Similarly, a water pipe having a large cross-sectional area offers less resistance than a small pipe.

d. *Temperature*. With metallic conductors, the *resistance becomes greater as the temperature rises*.

The following equation summarizes the determining factors in resistance:

$$R = \rho \frac{L}{A} \qquad (1)$$

where R is the resistance in ohms; ρ is the resistivity, a proportionality factor depending on the material of the conductor and the temperature; L is the length in meters; and A is the cross-sectional area in m^2.

Ohm's Law

Georg Simon Ohm (1787-1854), a German physicist, discovered that when a steady direct current is flowing in a resistance circuit (ie, one in which resistance is the only obstacle to current flow), there is a definite relation between the potential difference (volts), current (amperes), and resistance (ohms). This is expressed in **Ohm's Law**, which states that *the value of the current in a metallic circuit equals the potential difference divided by the resistance*:

$$I = \frac{V}{R} \qquad (2)$$

where I^* = current in amperes
V = potential difference in volts
R = resistance in ohms

If any two of these values are known, the third may be found by substitution in the equation and solving for the unknown.

Example 1. A circuit carrying a current of 30 A has a resistance of 4 Ω. Find the voltage across the circuit.

Solution. Rearrange Ohm's Law:

$$V = IR$$
$$= 30A \times 4 = 120 \text{ V}$$

Example 2. Find the resistance of a circuit when the current is 12 A and the potential difference is 240 V.

Solution. Rearranging Ohm's Law:

$$R = V/I$$
$$= 240 \text{ V}/12 \text{ A} = 20 \text{ Ω}$$

Two rules must be observed in applying this law:

1. When Ohm's Law is applied to a **portion of a circuit**, the current in that portion of the circuit equals the voltage across that portion of the circuit divided by the resistance of that portion of the circuit.

2. When Ohm's Law is applied to the **whole circuit**, the current in the circuit equals the total voltage across the circuit divided by the total resistance of the circuit.

Cells and Batteries

Let us turn now to some of the **chemical devices** that act as sources of electric current. There are two main types, the **dry cell** and the **wet cell**.

1. **Dry Cell.** An ordinary dry cell, such as that used in a flashlight, consists of a carbon rod surrounded by manganese dioxide and immersed in a paste composed of ammonium chloride, zinc chloride, cellulose, and a small amount of water. These ingredients are placed in a zinc can, the top of which is closed by a layer of asphalt varnish (see Figure 6.4).

The **alkaline-type** dry cell consists of a manganese dioxide rod immersed in an alkaline potassium hydroxide paste contained in a zinc cylinder. The manganese dioxide rod forms the anode, and the zinc container the cathode. An alkaline cell has a useful life about ten times that of an ordinary dry cell.

Dry cells are actually moist. Note that the central rod nowhere comes in contact with

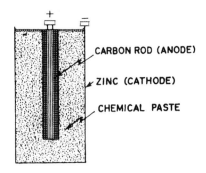

Figure 6.4. Diagram of an ordinary dry cell.

---

* I is intensity of electron flow in amperes = coulombs/s.

the zinc can. A chemical reaction sets up a potential difference of 1.5 volts between the battery terminals or electrodes; that is, between the zinc and the central rod. The zinc becomes the negative electrode (cathode), while the central rod becomes the positive electrode (anode).

2. **Wet Cell**. An example of the wet cell is the lead storage cell used in automobiles. It consists of a hard rubber or plastic case containing sulfuric acid in which are immersed two electrodes, one lead and the other lead oxide. As the result of a chemical reaction with the sulfuric acid, the lead becomes negative while the lead oxide becomes positive. Each cell in a storage battery produces a potential difference of about 2 volts. Automobile batteries usually have three or six wet cells.

Of special interest is the sealed *rechargeable nickel-cadmium* or *nicad* cell. It has a nickel oxide anode and a cadmium cathode. The plates are each wrapped in a separator and immersed in potassium hydroxide.

Components of Elementary Electric Circuits

Certain fundamental principles having been presented, we can now examine the simplest type of electric circuit; that is, one supplied by a *battery* as the source of potential difference. The current is a *steady direct current* (DC) flowing in *one direction only*. Since the only hindrance to the flow of current in such a circuit is resistance, it is often termed a *resistance circuit*.

Essential Parts. The simplest electric circuit has three components: (1) battery, (2) conductor, and (3) load, or resistance. The circuit may be represented by a diagram which shows, incidentally, some of the more common electrical symbols (see Figure 6.2).

Current flows only when the switch is *closed*. If the circuit is broken at any point,

it is then called an *open circuit*. When completed it becomes a *closed circuit*. However, keep in mind that there is a potential difference even on open circuit (don't touch!).

Polarity of Circuits. The current in a circuit supplied by a battery has a definite direction, or *polarity*. Physicists assume the direction of current flow to be from positive to negative. However, to facilitate understanding x-ray tube operation, we shall assume the *direction of current flow to be the same as the flow of electrons; that is, from the cathode (–), through the external circuit, to the anode (+)* (see Figure 6.5).

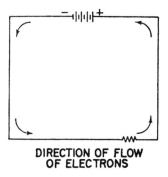

DIRECTION OF FLOW OF ELECTRONS

Figure 6.5. The direction of the *electron current* (that is, *electron flow*) will be used in this book. This is the reverse of the conventional direction.

Connection of Meters. The polarity of DC circuits requires the proper connection of DC measuring devices or meters. Since these devices also have *polarity*, the two terminals on a meter are labeled (+) and (–) and, to avoid damage, must be so connected that the (+) terminal is on the (+) side of the circuit and the (–) terminal on the (–) side.

A *voltmeter* measures, in volts, the potential difference between any two points in a circuit. (Potential difference is often referred to as *voltage drop*.) Therefore, the voltmeter is always connected in *parallel*; that is, it is

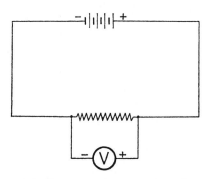

Figure 6.6. Correct connection of a voltmeter–in parallel.

placed in a small circuit which is a branch of the main circuit, as in Figure 6.6. Obviously, a voltmeter connected successively between various pairs of points in a circuit where the resistances may differ, will give different voltage readings–they will be larger across the higher resistance and smaller across the lower resistance. Similarly, the drop in pressure between any pair of points in a water pipe will be different if the bore of the pipe is not uniform, there being a greater fall in pressure where the pipe is narrower, that is, where there is greater resistance to flow of water.

An ***ammeter*** (ampere + meter) measures, in amperes, the quantity of electric charge flowing per second. It is connected directly into the circuit, that is, in series (see Figure 6.7). When so connected, ***an ammeter will read the same no matter where it is placed in***

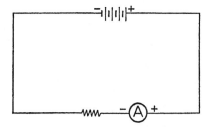

Figure 6.7. Correct connection of an ammeter–in series.

a series circuit.

Ammeters and voltmeters look alike externally and consist of a box with a glass window enclosing the mechanism, in which a pointer moves over a calibrated scale as shown in Figure 6.8. The instrument is always labeled "voltmeter" or "ammeter." The construction of these devices will be described later.

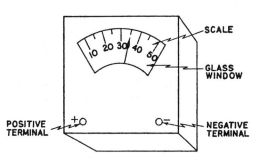

Figure 6.8. Voltmeter or ammeter, depending on internal circuitry.

Series and Parallel Circuits

Electric appliances (toasters, flatirons, fans, etc) and electric sources (batteries, generators) may be connected in a circuit in two principal ways. One is called ***a series circuit, which may be defined as an electric circuit whose component parts are arranged end-to-end so that the current passes consecutively through each part*** (see Figure 6.9). The other arrangement is called a ***parallel circuit, wherein the component parts are connected as branches of the main circuit so that the current is divided among them*** (see Figure 6.11).

1. **Series Circuit**. When ***current-consuming devices*** such as electric appliances are connected in series as shown in Figure 6.9, the total resistance is equal to the sum of the separate resistances. Thus, if R = total resistance of the whole circuit, and r_1, r_2, and r_3 represent the resistances of various

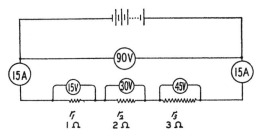

Figure 6.9. Resistors in series. Ohm's Law applies to the entire circuit, and to any portion of it. The total voltage drop, 90 V is equal to the sum of the voltage drops across all the resistors, 15 V + 30 V + 45 V. The current, 15 amp, is the same everywhere in the circuit. The voltage drop across each resistor is proportional to the resistance, because $V = IR$ and I is constant in this circuit.

devices connected in series, and disregarding the small internal resistance of the battery, then,

$$R = r_1 + r_2 + r_3 \qquad (3)$$

Suppose in this case, $r_1 = 1$ ohm, $r_2 = 2$ ohms, and $r_3 = 3$ ohms, and the ammeter reads 15 amp. (Remember that the current is the same everywhere in a series circuit.) What is the voltage across the **entire** circuit? First substitute the given values of resistance in equation (3),

$$R = 1 + 2 + 3 = 6 \text{ ohms}$$

Then apply Ohm's Law

$$I = V/R$$
$$\text{or}$$
$$V = IR \qquad (4)$$

and substitute the values of I and R (obtained above)

$$V = 15 \text{ amp} \times 6 \text{ ohms} = 90 \text{ volts}$$

There is another approach to this problem—the application of Ohm's Law to each part of the circuit. Again, recall that the current I remains constant throughout the circuit. Since $V = IR$ the voltage across each

resistor in Figure 6.9 is equal to the current times the resistance of each resistor:

$$v_1 = Ir_1 = 15 \times 1 = 15 \text{ volts}$$
$$v_2 = Ir_2 = 15 \times 2 = 30$$
$$v_3 = Ir_3 = 15 \times 3 = 45$$

The total voltage is then

$$V = v_1 + v_2 + v_3$$
$$V = 15 + 30 + 45 = 90 \text{ volts}$$

which is the same result obtained above by applying Ohm's Law to the entire circuit.

What is the real significance of the equation $V = IR$? It actually states that the **voltage across a resistor equals the current times the resistance**. But, besides this, it indicates that there is a fall or drop in potential across a resistor, and as the resistance increases so does the potential drop. Sometimes this is called a voltage drop or, what is the same thing, an **IR drop**. Keep in mind that the terms potential difference (in volts), voltage, voltage drop, and *IR* drop are used interchangeably.

In summary, then, voltage is a measure of the potential difference across a resistor. The voltage drop across the resistor increases as the resistance increases. It is obvious from the above examples that the voltage drop across an entire series circuit must equal the sum of the voltage drops across each resistor.

If **current-producing devices** such as electric cells are connected in series, **the total voltage equals the sum of the individual voltages**. In such a circuit, the positive pole of one cell is connected to the negative pole of the next, as in Figure 6.10. If V is the total voltage, and v_1, v_2, v_3, and v_4 are the voltages of the individual cells, then

$$V = v_1 + v_2 + v_3 + v_4 \qquad (5)$$

The total current in such a circuit can be computed from the total voltage and the total resistance simply by applying Ohm's Law.

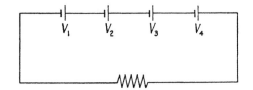

Figure 6.10. Electric cells in series.

2. **Parallel Circuit**. Figure 6.11 shows a parallel circuit; careful comparison of *A* and *B* reveals that the circuits are essentially identical, although they have been drawn differently. The conductors themselves have so little resistance that it may be disregarded. In a parallel circuit the **voltage drop is the same across all branches, regardless of their resistance**. Thus, the voltage drop across r_1 is the same as that across r_2 and that across r_3, and the voltage drop in each case is therefore equal to *V*, the voltage supplied by the battery. However, application of Ohm's Law to each portion of the circuit will show that the current in each branch is **inversely** proportional to the resistance of that branch. In other words, the largest current will flow in the branch having the least resistance.

What is the resistance of a circuit with several parallel branches? It is obvious that in such a circuit, the current divides among the several branches or paths, so that the total resistance must be less than that in an individual path. Another way of looking at this is as follows: as more resistors are added in **parallel**, there is an increase in their total cross-sectional area so that the total resistance decreases.

The total resistance of a **parallel circuit** is expressed by the following equation:

$$1/R = 1/r_1 + 1/r_2 + 1/r_3 + \ldots 1/r_n \quad (6)$$

where *R* is the total resistance and r_1, r_2, etc are the resistances of the separate branches.

Thus, the total resistance of a parallel circuit can be found by the rule based on equation (6): **the reciprocal of the whole resistance equals the sum of the reciprocals of the separate resistances in a parallel circuit**.

A typical **example** is presented in Figure 6.12 to clarify this rule. If the resistances in the branches of a parallel circuit are 1, 2, and 3 ohms respectively, what is the total resistance of the circuit? Substituting the values of *r* in equation (6), we find that the total resistance is less than any of the resistances in the individual branches.

$$1/R = 1/1 + 1/2 + 1/3$$

$$1/R = \frac{6 + 3 + 2}{6}$$

$$1/R = 11/6$$

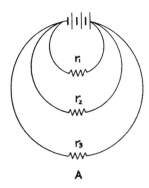

A

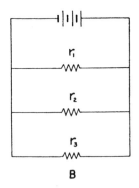

B

Figure 6.11. *A* and *B* are different methods of representing a parallel circuit. Note that they are fundamentally identical.

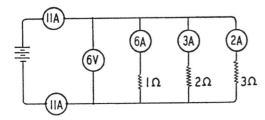

Figure 6.12. Resistors in parallel. The voltage drop, 6 V, is the same across the whole circuit and *across each branch.* The current differs in the branches, the smallest current flowing in the branch having the largest resistance. The total current, 11 amp, equals the sum of the branch currents: 6 amp + 3 amp + 2 amp. The total resistance is found by $1/R = 1/r_1, + 1/r_2, + 1/r_3$.

Inverting both sides of the equation,

$$R = 6/11 = 0.55 \text{ ohm}$$

Suppose, in the above example, that the potential difference across the circuit is 6 volts. How do we find the current in each branch and in the main lines? Since this is a parallel circuit, the voltage drop across each branch must be the same–6 volts. Then, applying Ohm's Law to each branch,

$$i = v/r$$

$$i_1 = 6/1 = 6 \text{ amp}$$
$$i_2 = 6/2 = 3 \text{ amp}$$
$$i_3 = 6/3 = 2 \text{ amp}$$

The *current in the main lines of a parallel circuit equals the sum of the branch currents,*

$$I = i_1 + i_2 + i_3 \qquad (7)$$

Substituting the values of the branch currents just obtained,

$$I = 6 + 3 + 2 = 11 \text{ amp}$$

To verify the answer, we apply Ohm's Law to the entire circuit. Since the total current is 11 amp and the total resistance is 0.55 ohm,

$$V = IR$$

$$V = 11 \times 0.55 = 6 \text{ volts}$$

As expected, the current divides among the branches so that the **smallest current flows in the branch having the greatest resistance**. Since *V* is equal in all branches, if *r* in a given branch is greater, *i* must be smaller, and *vice versa*, in accordance with $V = IR$.

With **current-producing devices** connected in parallel, the total voltage is the same as that of a single source, if all the sources are alike. Figure 6.13 shows this type of connection. The total current, that is, in the main line, equals the sum of the currents provided by all the sources.

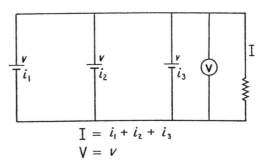

Figure 6.13. Cells in parallel. Total current, $I = i_1 + i_2 + i_3$. Voltage across entire circuit, $V = v$.

In a series circuit, electrical failure of one appliance, such as a light bulb, causes all other lights in the circuit to go out. Besides, as more appliances are added in a series circuit, each appliance receives less voltage, since the total voltage drops must be divided among more appliances; thus, as more electric lights are added, all become dimmer. This situation would be intolerable in ordinary home and commercial lighting.

The parallel circuit is used in electrical wiring of homes, factories, and other buildings. With this, failure of one appliance does not prevent operation of the others, since they are on branches of the main circuit. As more branches are added, the total resistance decreases and the total amperage increases while the voltage remains unchanged; thus, the more appliances added, the greater the amperage in the ***main circuit***. Adding too many appliances may cause the amperage in the main line to become excessive and ***overload*** the circuit. Under these conditions the wiring system could become hot enough to ruin one or more appliances, or start a fire. This danger can be avoided by connecting protective devices such as circuit breakers or fuses in the circuit. A ***fuse*** contains a wire which melts more easily than the wires of the circuit being protected (see Figure 14.3). When the amperage exceeds the safe maximum, the fuse melts, thereby opening the circuit and stopping the flow of current before the circuit can become overheated. Such a fuse, in which the protective wire has melted, is called a ***blown fuse***. Contact between two non-insulated wires carrying a current results in a ***short circuit***; there is a marked fall in resistance and a corresponding rise in amperage which is usually sufficient to blow a fuse. The ***circuit breaker***, another type of protective device, will be discussed later.

Electric Capacitor (Condenser)

As described above, electrical energy is produced during a chemical reaction in a dry or wet cell in the form of ***stored energy***. Another method of storing electrical energy depends on the distribution of electric charge entirely on the surface of a conductive body. Such a charged body stores electrical energy if it is insulated to prevent the charge from "leaking" away. This principle is applied in the ***electric capacitor***.

One of the simplest types is the ***parallel-plate capacitor***, consisting of a pair of flat metallic plates arranged parallel to each other and ***separated*** by a small space containing air or special insulating material. Capacitor operation is shown in Figure 6.14. Connecting a capacitor to a source of direct current, such as a battery causes electrons to move from the negative terminal of the battery to the plate to which it is connected. At the same time, an equal number of electrons pass from the opposite plate to the positive terminal of the battery, as indicated in Figure 6.14A. ***No electrons pass from one plate to the other across the space between them***, unless there is a breakdown of the insulator due to application of excessively high voltage. When the capacitor acquires a full charge, the difference of potential (voltage) across its plates reaches a maximum that is equal and opposite in direction to that of the charging current. The current now ceases to flow in the circuit.

If the capacitor is then disconnected from the battery, it retains its electrical charge until the plates are connected by a conductor as in Figure 6.14B, whereupon the capacitor

CAPACITOR

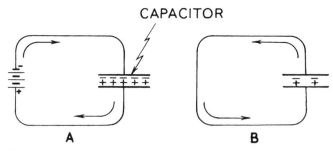

A   B

Figure 6.14. Parallel plate capacitor. In *A* the capacitor is connected to a battery and is being charged. In *B* the capacitor has been disconnected and its plates connected by a wire, whereupon it loses its charge. The arrows indicate the direction of electron flow in the wire.

discharges and becomes neutral. The direction of the discharging current is **opposite** to that of the original charging current, which stops flowing as soon as the capacitor discharges completely. (With an alternating current the capacitor plates repeatedly change polarity so the current continues to flow.)

How much charge can a capacitor store? This is given by

$$Q = CV \qquad (8)$$

where Q is the charge in coulombs, C the capacitance in farads, and V the applied voltage.

Rearranging equation (8) yields

$$C = \frac{Q}{V} \qquad (9)$$

which states that capacitance is the accumulated charge per volt in farads. The farad is a very large unit, the microfarad (millionth of a farad) being more practical.

The larger the area of the plates, and the smaller the space between them, the greater is the capacitance or electrical storing ability. The insulator, or **dielectric**, between the plates also determines the capacitance for a given plate size and spacing; if the **dielectric constant** (insulating ability) of air is taken as 1, then wax paper has a dielectric constant of 2, and glass a constant of about 7. These materials will increase the capacitance of a given capacitor two times, and seven times, respectively, as compared with an air dielectric. In practice, a capacitor is usually made up of multiple parallel plates to increase the capacitance.

The Work and Power of a Direct Current

Electrical energy, just as any other form of energy, is capable of performing work. According to the Law of Conservation of Energy, a given amount of electrical energy is convertible to a definite amount of work and/or heat.

The **power** of a **steady direct current**, or the **electric energy expended per second**, is simply the product of the current and voltage; that is, the power equation:

$$P = IV \qquad (10)$$

power = amperage × voltage
(in **watts**)

The unit of power is the **watt**, 746 watts being equal to 1 horsepower.

We know that an electric current produces heat in a circuit. To find the power loss in terms of the heat produced per second, we convert equation (10) to a form which contains the term resistance, since this remains constant regardless of the voltage or current. Applying Ohm's Law,

$$I = V/R$$

$$V = IR$$

Substituting IR for V in equation (10),

$$P = I \times IR$$
$$\textbf{Power Loss} = \textbf{I}^2\textbf{R watts} \qquad (11)$$
(heat production/sec)

where I is the current in amperes and R the resistance in ohms.

Two conclusions may be drawn from equation (11). First, the power loss due to the heating effect of the current is proportional to the resistance; thus, that portion of the circuit having the greatest resistance will sustain the greatest amount of heat production.

Second, the power loss in the form of heat increases very rapidly with an increase in the current. Since **the power loss is proportional to the square of the current**, doubling the current increases the power loss by a factor of 2^2, or four times. Tripling the current increases the power loss by a factor of 3^2, or nine times. The importance of this principle will be discussed later.

QUESTIONS AND PROBLEMS

1. How does an electric current flow along a wire? In a salt solution? Under what other conditions do electrons flow?
2. Describe the two main sources of electric current.
3. Define potential difference. What is the unit of measurement? State the difference between potential difference, electromotive force, and voltage?
4. What is meant by electrical resistance, and on what does it depend?
5. Define electric current strength and state its unit.
6. In which direction does an electric current flow? What is meant by the "conventional" direction?
7. State and explain Ohm's Law.
8. An electric circuit has, in series, appliances with these resistances: 4 ohms, 10 ohms, 6 ohms, 40 ohms. Find the value of the current when the applied potential difference is 120 volts?
9. What voltage is required to produce a current of 30 amperes in a circuit having a resistance of 5 ohms?
10. What type of electric circuit is used in house wiring? Why?
11. A circuit has the following resistors connected in *parallel*: 3 ohms, 6 ohms, 18 ohms. What is the total resistance of the circuit? With 12 volts applied across the circuit, what is the amperage in each branch and in the main lines?
12. State the power equation. Name the unit of power?
13. A circuit has a direct current of 2 amperes and a electromotive force of 12 volts. Find the power delivered to the circuit.
14. What is the function of an electric capacitor? Describe its construction and explain its operation with the aid of a diagram. Contrast its behavior in a direct current with that in an alternating current.

Chapter 7

MAGNETISM

Topics Covered in This Chapter

- Definition
- Classification of Magnets
- Laws of Magnetism
- Nature of Magnetism
- Magnetic Fields

- Characteristics of Lines of Force
- Magnetic Induction (magnetization)
- Magnetic Force
- Magnetic Classification of Matter
- Questions and Problems

Definition

Magnetism may be defined as the ability *of certain materials to attract iron, cobalt, or nickel.* In general, any material that attracts these materials is called a *magnet.* The various types of magnets will be described in the next section.

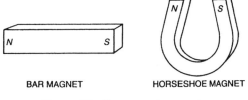

BAR MAGNET   HORSESHOE MAGNET

Figure 7.1. Permanent magnets.

Classification of Magnets

There are three main types of magnets: *natural magnets, artificial permanent magnets,* and *electromagnets.*

1. **Natural magnets** include, first, the *earth* itself, a gigantic magnet which, in common with all others, can deflect the needle of a compass. Another example is the ore *lodestone,* consisting of iron oxide (magnetite) that has become magnetized by lying in the earth's magnetic field for an extraordinarily long period of time. In some parts of the world there are actually magnetic mountains consisting of such ores. The ancient Greeks recognized the magnetism of lodestone more than 2500 years ago.

2. **Artificial permanent magnets** usually consist of a piece of *hard steel* in the shape of a horseshoe or bar which has been *artificially* magnetized so that it is capable of attracting iron (see Figure 7.1). The *magnetic compass* (see Figure 7.2) is simply a small

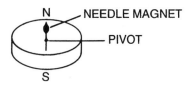

Figure 7.2. Diagram of a magnetic compass. The needle is actually a permanent magnet, the same end of which always points north (approximately the geographic north pole).

permanent magnetic needle swinging freely on a pivot at its center, used to detect the presence of magnetic materials and to locate the earth's magnetic poles. A widely used alloy of aluminum, nickel, and cobalt—***alnico***—can be made into an extremely powerful permanent magnet that has many times the lifting ability of a steel magnet, although aluminum itself cannot be magnetized.

3. **Electromagnets** are temporary magnets produced by means of an electric current, as will be described later.

Laws of Magnetism

The *three fundamental laws of magnetism* include:

1. *Every magnet has two poles,* one at each end. This concept is based on the observation that when a magnet is inserted into a container of iron filings and then removed, iron particles will cling to its ends or poles, as illustrated in Figure 7.3. When

Figure 7.3.   Polarity of a magnet. This is demonstrated by immersing first one end, then the other, of a magnet into a box of iron filings which cling mainly at the ends or *poles.*

allowed to swing freely (as in a magnetic compass), one pole of the magnet points north and the other end south, the poles being called, respectively, **north** and **south** poles.

2. *Like magnetic poles repel* each other; ***unlike poles attract*** each other. Thus, if two bar magnets are brought together with their south poles facing each other as in Figure

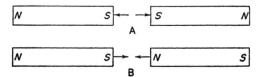

Figure 7.4.   *(A)* Like poles repel.   *(B)* Unlike poles attract.

7.4A, there is a force which tends to push them apart. If, on the other hand, the magnets are brought together with opposite poles facing each other, a force tends to pull them together, as in Figure 7.4B.

3. The ***force of attraction or repulsion*** between two magnets was at one time thought to follow the the inverse square law of electric charges and gravitation, but this is fallacious and no longer accepted.

Nature of Magnetism

Why are some materials magnetic (iron-attracting) whereas others are not? The most widely accepted explanation, an outgrowth of Weber's theory of magnetism first proposed in the early nineteenth century, is based on observations such as the following:

1. Breaking a magnet results in each fragment becoming a ***whole magnet*** with its own north and south poles, as in Figure 7.5.

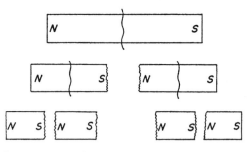

Figure 7.5.   When a magnet is broken, each part becomes a whole magnet with a north and a south pole.

2. Heating or hammering a piece of steel while it lies near a magnet causes the steel to become a magnet (ie, magnetized).

3. Stroking an iron bar repeatedly with one end of a magnet causes the bar to become magnetized.

4. Gently jarring a test tube of iron filings near a magnet causes them to become magnetized. If the test tube is now shaken, vigorously, it loses its magnetism; that is, becomes demagnetized.

These observations suggested that elements such as iron, nickel, and cobalt consist of extremely small, discrete magnets. But only with the advent of modern atomic science has the true nature of such elements been fully explained. According to atomic theory, the atoms of magnetic materials normally have two more electrons spinning in one direction than in the opposite, creating *magnet dipoles*–magnetic moments. Billions of such dipoles occur in microscopically visible *domains,* in which all the dipoles are aligned in the same direction.

A *magnetic element* can exist either in the magnetized or nonmagnetized state. In a *nonmagnetized* piece of iron, the domains are in a state of random disorder as in Figure 7.6A, so the iron does not have magnetic poles. But in *magnetized* iron the domains are aligned in orderly fashion, all their north poles pointing in one direction, and their south poles in the opposite direction, as shown in Figure 7.6B.

Tiny single-domain needle-shaped iron or alnico particles are used for high-resolution recording on magnetic tapes and discs. Of interest is the presence of single-domain iron particles in the bodies (eg, brains) of migratory animals such as birds, whales, and dolphins, wherein they serve as natural directional compasses.

In *nonmagnetic elements* such as copper, just as many electrons spin in one direction as in the other, so their opposing magnetic effects cancel out. Therefore, in such elements the atoms are not magnetic, no domains are formed, and they *cannot* be magnetized.

Magnetic Fields

What is there about a magnet that causes it to attract iron? Surrounding a magnet there is a zone of influence called a *magnetic field*, representing a summation of the fields of all the domains. The magnetic field can be demonstrated by placing a piece of cardboard over a magnet and sprinkling iron filings on the cardboard. If the latter is gently tapped, the filings arrange themselves into a pattern shown in Figure 7.7A, consisting of *magnetic field* lines; these constitute *magnetic flux*–the more closely spaced these lines, the stronger will be the field. In fact, the *strength of a magnetic field is propor-*

A. NON-MAGNETIZED IRON

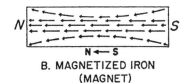

B. MAGNETIZED IRON
(MAGNET)

Figure 7.6. Magnetic domains in a magnetic substance. In *A*, when a ferromagnetic substance such as iron is not magnetized, the domains are arranged haphazardly and there is no net polarity. In *B* the iron has been magnetized and the domains are arranged in orderly fashion to produce opposite poles.

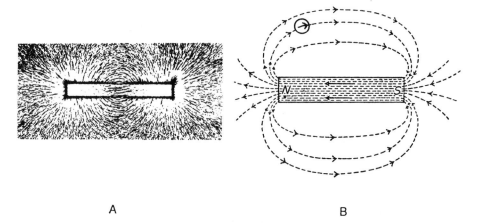

A   B

Figure 7.7. *A.* Iron filings sprinkled on a piece of cardboard assume a pattern that corresponds to the lines of force around a bar magnet. *B.* The lines of force have been mapped by a small magnetic compass as it is moved around the magnet. (*A* is from Heath Physics by David G. Martindale, Robert W. Heath, William W. Conrad, Robert R. Macnaughton, and Mark A. Carle. Copyright 1994 by D. C. Heath and Company. All rights reserved. Reprinted by permission of McDougal, Littell, Inc.)

*tional to the number of lines per square centimeter.**

The field lines can be mapped out by means of a small magnetic compass. Starting at the north pole of a bar magnet, one simply moves the compass around it toward its south pole and observes the behavior of the needle which acts as though it were following a line through space from the north to the south pole of the magnet, as shown in Figure 7.7B.

Characteristics of Lines of Force

Magnetic lines may be pictured as follows:

1. They are directed from the north pole to the south pole of a magnet as curved lines in the surrounding space. Within the magnet they are directed from the south pole to the north pole (see Figure 7.7B).

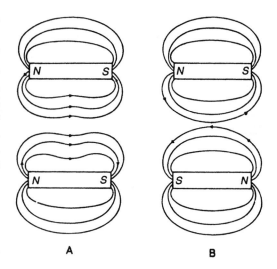

A   B

Figure 7.8. Interaction of magnetic lines of force, depending on their relative directions. In *A* the lines are in the same direction and the resultant force tends to separate the magnets. In *B* the lines of force are in opposite directions and the magnets attract each other.

* According to modern physics theory, the magnetic "lines" are distortions in space by the magnetic field.

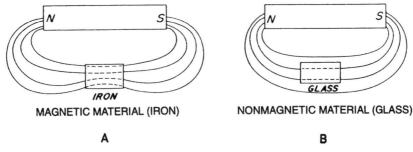

Figure 7.9. The effect of magnetic and nonmagnetic materials on a magnetic field. In *A*, the iron "concentrates" the lines of force because of its magnetic permeability: this results from the summation of the lines of force associated with induced magnetism in the iron, and the lines of force of the magnetic field already present. In *B*, glass, a nonmagnetic material cannot become magnetized and so it has no effect on the magnetic field.

2. Lines of force seem to repel each other when they are in the same direction, and attract each other when they are in opposite directions (see Figure 7.8).

3. The field is distorted by magnetic materials, but is not affected by nonmagnetic materials (see Figure 7.9).

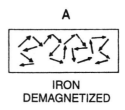

Magnetic Induction (Magnetization)

In an ordinary piece of iron the magnetic domains are jumbled helter-skelter by molecular motion and so the iron does not behave as a magnet (see Figure 7.10A). When such a piece of nonmagnetized iron is brought near one pole of a magnet, the end of the iron nearest the pole assumes the *opposite polarity* as a result of *magnetic induction;* the magnetic field of the pole exerts a force on the magnetic domains of the iron, making them line up in an orderly arrangement as in Figure 7.10B. The iron thus becomes temporarily magnetized and behaves as a magnet while lying in the field. When the iron is removed from the field, its magnetic domains return to their disorderly arrangement and the iron no longer is magnetized (ie, it no longer behaves as a magnet).

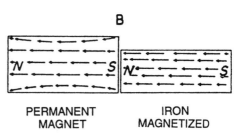

Figure 7.10. Magnetic induction. *A.* The magnetic domains in the iron are normally jumbled, so the iron does not behave as a magnet. *B.* The iron has been brought into the magnetic field of a magnet; the magnetic domains in the iron line up in an orderly arrangement and the iron is now magnetized. Note that the end of the iron near one pole of the magnet assumes the opposite polarity. When the iron is removed from the magnetic field of the permanent magnet, its magnetic domains return to their disorderly arrangement and the iron no longer behaves as a magnet.

In summary, then, whenever a magnetic material (iron, cobalt, or nickel) is brought near a magnet, it first becomes magnetized (magnetic induction), its end nearer the magnetic pole acquiring the opposite polarity. This results in a force of attraction between the two.

Magnetic Permeability and Retentivity

Some materials are more susceptible than others to magnetic induction; that is, their magnetic domains can be readily lined up when placed in a magnetic field. The ease with which a given material can be magnetized in this way is designated as its *magnetic permeability.*

Some materials such as steel do not readily become demagnetized after they have been removed from the magnetizing field; that is, once the magnetic domains have been aligned they tend to remain so. The ability of a magnet to resist demagnetization is called *magnetic retentivity*.

Examples of the above include:

1. "Soft" iron–high permeability, low retentivity.
2. "Hard" steel–low permeability, high retentivity.

Thus, a metal which is easily magnetized is also easily demagnetized (ordinary iron), whereas a metal which is difficult to magnetize is also difficult to demagnetize (hard steel).

Magnetic Classification of Matter

Matter can be classified on the basis of its behavior when brought into a magnetic field.

1. **Ferromagnetic materials** (ordinarily called magnetic materials) include *iron, cobalt,* and *nickel.* They are strongly attract-ed by a magnet because of their extremely high permeability, that is, their great susceptibility to magnetic induction. Ferromagnetic substances become *permanently* magnetized under certain conditions. For example, steel is a form of iron that has been hardened by special treatment, as a result of which its magnetic permeability decreases while its retentivity increases; it is therefore susceptible to permanent magnetization. Another example is the alloy *alnico*, which can be strongly magnetized (see page 59); it has been used to remove magnetic foreign bodies such as nails and bobby pins from the stomach, under fluoroscopic control.

2. **Paramagnetic materials** are feebly attracted by a magnet. Platinum, an example of such a material, has extremely low magnetic permeability.

3. **Nonmagnetic materials** are not attracted by a magnet because they are not susceptible to magnetic induction. In such materials, just as many electrons spin in one direction as in the other so that the atoms are nonmagnetic and therefore do not form magnetic domains. Examples of nonmagnetic materials include wood, glass, and plastic.

4. **Diamagnetic materials** are actually repelled by a magnet, although this action is feeble. Only a few elements exhibit this property, among them beryllium and bismuth.

Earth's Magnetism

One can detect earth's magnetism by using a *magnetic compass* (see Figure 7.2), which consists of a horizontal *magnetized steel needle* that swings freely about a vertical axis at its center. When brought near a magnet the needle, since it is a magnet, is deflected according to the laws of magnetism. Thus, the south pole of the compass needle turns toward a magnetic north pole, while the north pole of the compass needle turns toward a magnetic south pole. Notice that the north pole of a compass (or any other magnet) points approximately northward and is therefore a "north-seeking" pole.

The earth behaves as a spherical magnet, but may be regarded as having a virtual magnet at its center (see Figure 7.11), with its south pole beneath the north magnetic pole N_M and its north pole beneath the south magnetic pole S_M. N_G and S_G are the geographic poles. On the earth's surface, the angle between the direction of geographic north and magnetic north (compass reading) is called *magnetic declination;* this varies with the observer's location. Magnetic field lines surround the earth like those around a bar magnet (refer back to Figure 7.7), but the earth's magnetic field is very weak–about 6×10^{-5} T (tesla, SI unit of magnetic field strength). It is of interest that earth's magnetic field was elucidated by the physician-scientist William Gilbert, M.D., in England in the year 1600!

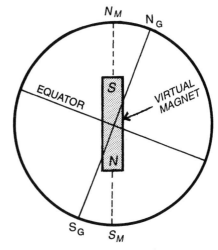

Figure 7.11. Earth's magnetic field behaves as though a virtual bar magnet exists at its center, aligned with the N_M – S_M magnetic axis. The geographic axis N_G – S_G lies at a varying angle to the magnetic axis. Note that a magnetic south pole is under the north magnetic pole N_M and a magnetic north pole under the south magnetic pole S_M.

QUESTIONS AND PROBLEMS

1. Define magnetism in your own terms.
2. Name and discuss briefly the three main types of magnets.
3. What is the present concept of the intimate structure of a magnet? Show by diagram. State three facts which support this theory.
4. State the laws of magnetism.
5. Explain magnetic induction. How does a magnet attract a piece of iron?
6. Compare ordinary iron and hard steel with regard to their magnetization and demagnetization.
7. Define magnetic field; magnetic flux.
8. What happens to a magnetic field when a piece of ordinary iron is placed in it?
9. Indicate the direction of magnetic lines of force in space? Within the magnet?
10. How is magnetism detected?
11. State the magnetic classification of matter.
12. Define the following terms pertaining to magnetism:
    a. Domain
    b. Permeability
    c. Retentivity
    d. Flux
    e. Field
    f. Poles
    g. Compass
    h. Paramagnetism
    i. Ferromagnetism
    j. Diamagnetism
13. Explain the earth's magnetism, including the relationship between the magnetic and geographic axes, as well as the earth's polarity.

Chapter 8

ELECTROMAGNETISM

Topics Covered in This Chapter

- Definition
- Electromagnetic Phenomena
- Electromagnet
- Electromagnetic Induction

- Direction of Induced Current
- Self-Induction
- Questions and Problems

Definition

EVERY ELECTRIC CURRENT is accompanied by magnetic effects. *Electromagnetism is defined as the branch of physics that deals with the relationship between electricity and magnetism.*

Electromagnetic Phenomena

In 1820, Oersted, a Danish scientist, discovered that a magnetic compass needle turns when placed near copper wire carrying a *direct current* but returns to its normal position when the current is discontinued. Since this indicated the presence of a magnetic field, Oersted concluded that *a magnetic field always surrounds a conductor in which an electric current is flowing*. Actually, a magnetic field surrounds any stream of charged particles, whether in a metallic conductor or in space.

What is the direction of this magnetic field? It is specified by the *left thumb rule: if the wire is grasped in the left hand with the thumb pointing in the direction of the electron current [(–) to (+)], then the fingers*

encircling the wire will indicate the direction of the magnetic lines around the current (see Figure 8.1).

Figure 8.1. Left hand thumb rule. The thumb points in the direction of the electron current. The fingers designate the direction of the magnetic field around the conductor.

These magnetic lines can be demonstrated by Davy's Experiment. A wire is pushed through a hole in a piece of cardboard and connected to a source of direct current. If iron filings are sprinkled on the cardboard, and the cardboard is then gently tapped, the iron particles arrange themselves in concentric circles around the wire. This is illustrat-

66

ed in Figure 8.2, in which the concentric circles represent the magnetic lines of force, or *magnetic flux*, surrounding the electric current in the wire.

Note carefully this very important fact: *a magnetic field exists around a wire only while an electric current is flowing*. The magnetic field disappears on open circuit because this terminates the current.

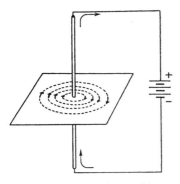

Figure 8.2. The magnetic lines of force around a conductor carrying a direct current. The dotted lines represent the magnetic flux when the electron current is in the indicated direction.

A magnetic field is present also around an alternating current, but because of the rapid reversal of field direction, it cannot be demonstrated by a compass or by Davy's Experiment.

The Electromagnet

While a current flows in a wire *helix* (coil), one end of the helix behaves as though it were the north pole and the opposite end the south pole of a magnet. By applying the thumb rule to each turn of the wire in the coil (or the whole coil) we can predict the direction of the magnetic flux inside the coil, and therefore which end of the coil becomes the north pole. Such a helix carrying an electric current is known as a *solenoid* (see Figure 8.3). In modern x-ray

equipment, solenoids are used to lock various movable parts, such as tube stand or Bucky tray, in almost any desired position.

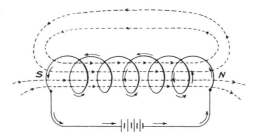

Figure 8.3. A helix (coil) carrying an electric current is called a *solenoid*. This shows the magnetic field associated with a solenoid.

As noted earlier (page 64), certain metals, such as iron, possess a high degree of magnetic permeability. Placing an iron rod inside a solenoid increases the strength of the associated magnetic flux because the iron becomes magnetized by the magnetic field of the solenoid; *a solenoid with an iron core is an electromagnet* (see Figure 8.4).

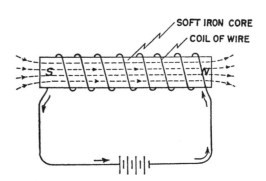

Figure 8.4. Electromagnet, a solenoid into which has been inserted an iron core to increase the strength of the magnetic field. If the current were reversed in the illustration, the magnetic poles would be reversed. If the current were interrupted, the electromagnet would lose its magnetism because of the low retentivity of the soft iron core.

Electromagnets have many practical applications, the best known being the electric bell and the telegraph. In x-ray equipment, electromagnets are used in remote control switches and relays of various kinds. These will be discussed later under the appropriate headings.

Electromagnetic Induction

You have just learned that electrons moving in a conductor are surrounded by a magnetic field. Under suitable conditions the reverse is also true; *moving magnetic fields can induce electron currents.* In 1831, Michael Faraday discovered how to convert mechanical energy to electrical energy. On rapidly passing a closed wire between the poles of a horseshoe magnet, he found that an electric current was generated in the wire. This is the principle of *electromagnetic induction,* stated more precisely as follows: *whenever a conductor cuts across magnetic flux, or is cut by magnetic flux, an electromotive force (emf) or potential difference is induced in the conductor.* The practical unit of emf is the volt, which is also the unit of potential difference. Interaction between a line of force and a single loop conductor is called a *flux linkage.*

The magnitude or size of the induced emf depends on the number of lines of force cut per second, or the number of flux linkages per second. This is determined by four factors:

1. The *speed* with which a conductor cuts the magnetic lines of force or magnetic flux; *the more flux lines cut per second, the higher the induced emf.*

2. The *strength* of the magnetic field, or the degree of crowding of the magnetic flux lines–*the stronger the magnetic field the higher the induced emf.* A moment's thought will reveal that this is essentially the same as statement (1); the closer the spacing of the magnetic lines, the more flux cut per second.

3. The *angle* between the conductor and the direction of the magnetic field. As this angle approaches 90° the spaces between the lines of force become smaller, progressively more flux lines are cut per second, and therefore a correspondingly larger emf is induced, as shown in Figure 8.5.

4. The number of *"turns"* in the conductor if it is wound into a coil, because the lines cut by each turn of the coil are additive; that

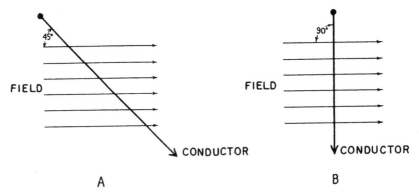

Figure 8.5. *A* and *B* are conductors seen end-on. The solid lines show the direction of motion of the conductors across the magnetic field. *A* cuts the magnetic flux at 45°, while *B* cuts it at 90°. The space between the flux lines is smaller in *B* so more lines of force are cut per second and the induced voltage is greater than in *A*.

is, there are more flux linkages per second. The induced emf is directly proportional to the number of turns in the coil.

Bear in mind that *an emf is induced only while the wire is moving across the field.* If the motion of the wire ceases, the induced emf drops to zero. But the motion of the wire need only be relative; thus, *if the wire is stationary and the magnetic flux passes across it, an emf will be induced in it* just as well, and the above four principles apply. Finally, *if the wire is stationary and the magnetic flux changes, varies in strength—more flux linkages per s—an emf will be induced in the wire.*

An emf is induced in a conductor moving relative to a magnetic field *even if the circuit is open* because emf represents the potential difference between the ends of the conductor. However, no current flows unless the circuit is closed.

In summary, we may state that there are three ways to induce an emf or voltage in a wire by electromagnetic induction:

1. The wire may move across a stationary magnetic flux, or

2. The magnetic flux may move across a stationary wire, or

3. The magnetic flux may vary in strength while a stationary wire lies in it. This principle is extremely important in the electric transformer, to be described later.

We may conclude that *whenever a conductor and a magnetic flux move with relation to each other, an emf is induced in the conductor, the magnitude of the induced emf being directly proportional to the number of magnetic lines per second crossing the conductor.*

Direction of Induced Electron Current

An induced electron current resulting from an induced emf has a definite direction

depending on the relationship between the motion of the conductor and the magnetic field. This is stated in the *left hand rule* or *dynamo rule: first, hold the thumb and first two fingers of the left hand so that each is at right angles to the others, as in Figure 8.6. Next, let the index finger point in the direction of the magnetic flux, and the thumb point in the direction the conductor is moving. The middle finger will then point in the direction of the induced electron current.*

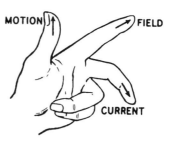

Figure 8.6. Left hand dynamo rule. This shows the direction of *electron* current flow.

The relationships among the direction of the magnetic flux, the motion of the conductor (wire or coil), and the induced electron current are summarized in Figure 8.7. This principle underlies the electric generator, which will be discussed in detail in the next chapter.

Self-induction

Suppose we have a coil in a *direct current* circuit whose switch is open. Now suppose we close the switch; at this instant a magnetic field is set up about the coil, building up rapidly to a maximum, where it remains *while the primary current continues to flow.* During this *brief interval* in which the magnetic field "grows up" around the coil, immediately after closure of the switch, the magnetic flux is really expanding, thereby *cutting the coil itself.* By the rules of electro-

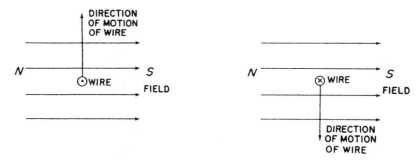

Figure 8.7. The direction of the electron current induced in a wire. The wire is shown in cross section. ⊗ indicates that the current is moving into this page. O indicates that it is moving out of this page, toward the reader.

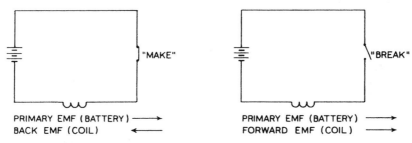

Figure 8.8. Self-induction. On "make" (instant of switch closure) there is a *momentary* self-induced emf bucking or opposing the primary battery emf. On the "break" (instant of switch opening), the self-induced emf is reversed momentarily while the battery emf is unchanged.

magnetic induction, an emf is induced in the coil in a direction ***opposite*** to the ***applied emf,*** a phenomenon known as ***self-induction.*** Thus, at the instant of switch closure, the self-induced current tends to ***oppose*** the flow of primary current in the coil (see Figure 8.8), an example of Lenz's Law (see page 90).

While the switch is closed and the primary current is flowing steadily in one direction, the surrounding magnetic field is also constant and therefore does not move relative to the coil. Consequently, ***the self-induced current falls to zero*** (Figure 8.9).

If the switch is then opened, the magnetic field collapses, sweeping across the coil in a reverse direction from that occurring when the switch was closed. As a result, the self-induced emf reverses and tends to main-

tain the current in the coil momentarily after the switch has been opened.

In summary, then, closing the switch results in momentary induction of an emf opposing the applied emf, while opening the switch causes momentary reversal of the self-induced emf which is now in the same direction as the applied emf. Figure 8.8 describes self-induction, and Figure 8.9 shows graphically the relation of self-induced emf to switch opening and closing.

An iron core inserted into the coil produces an increase in the self-induced emf which opposes or ***bucks*** the applied emf. The size of the bucking effect depends on the depth of insertion of the core.

Although we have explained the principle of self-inductance by means of a direct current, in actual practice self-induction devices

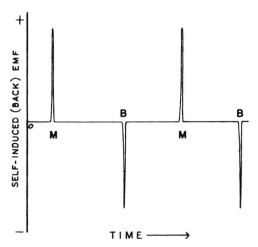

Figure 8.9. Self-induction of emf occurs only at the instant of "make" or "break" of a direct current. While a constant current flows in one direction, no emf is induced. *M* = make. *B* = break.

such as the choke coil operate only with **alternating current** (AC) as described on pages 89-90. The rapid reversal in direction and magnitude of AC produces an effect similar to the manual opening and closing of a switch with DC.

Mutual Induction

Placing a pair of insulated wire coils side by side and applying an alternating current to one will induce an emf in the other, despite the insulation (see Figure 10.1, page 84). This would be expected because the varying magnetic field associated with the alternating current links with the second coil. Application of this principle of electromagnetic mutual induction will be presented later in the description of the electric transformer, an essential component of an x-ray machine.

QUESTIONS AND PROBLEMS

1. Define electromagnetism in your own words. How can you easily determine its presence?
2. What is the left thumb rule? How is it applied?
3. Describe an electromagnet. In what two ways can you determine its polarity?
4. Define and discuss electromagnetic induction?
5. Discuss the four factors that determine the magnitude (size) of an induced emf? Name two terms used interchangeably with emf.
6. In Figure 8.10, what is the direction of the electromagnetically induced current?
7. Two coils of wire are lying parallel to

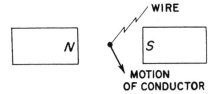

Figure 8.10.

each other. If one is connected to the terminals of a battery, what will happen in the second coil? What happens when the switch is opened? Closed?
8. Describe the difference between a helix, a solenoid, and an electromagnet.
9. Compare self-inductance of AC and DC circuits.

Chapter 9

ELECTRIC GENERATORS AND MOTORS

Topics Covered in This Chapter

- Electric Generator
    Definition
    Essential Features
    Simple Electric Generator
    Properties of Alternating Current (AC)
      Circuits
    Direct Current Generator
- Electric Motor
    Definition
    Motor Principle
    Simple Electric Motor
    Types of Electric Motors
    Current-Measuring Devices
- Questions and Problems

ELECTRIC GENERATOR

Definition

$\mathbf{A}$*n electric generator or dynamo is a device that converts mechanical energy to electrical energy by electromagnetic induction.*

Essential Features

As described in the last chapter, electromagnetic induction is the setting up of an emf in a conductor as it cuts, or is cut by, magnetic flux. In actual practice, an electric generator has two components to bring about the continual cutting of magnetic flux by a conductor:

1. *A powerful electromagnet*–the field magnet–to set up the magnetic field.
2. *An armature,* consisting of a coil of wire that is rotated mechanically in the magnetic field. The mechanical energy needed

to turn the armature is usually supplied by a steam turbine which is heated by oil, gas, coal, or a nuclear reactor; or by a waterfall, as in Figure 9.1.

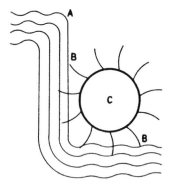

Figure 9.1. The waterfall at *A,* dropping on the blades *B* of a turbine, rotates it counterclockwise about the axis *C.* An armature connected to axis *C* will rotate with the turbine.

72

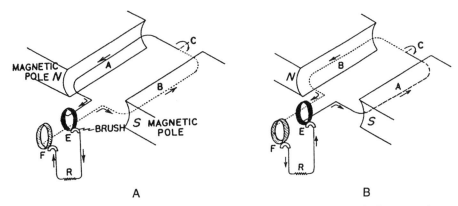

Figure 9.2. Simple alternating current generator. Armature *AB* rotates clockwise in the magnetic flux (which passes from *N* to *S*). In *A*, side *A* moves up through the field while side *B* moves down. Arrows show the direction of the induced electron current which leaves the generator through slip ring *E*, passes through external circuit *R*, and completes the circuit through slip ring *F*. As the armature continues to rotate, sides *A* and *B* reverse positions as in *B* and the current is now reversed in the armature and in the external circuit *R*.

Simple Electric Generator

An ***elementary form*** of electric generator may be represented by a single loop of wire (armature) rotated mechanically between the poles of a magnet (see Figure 9.2). What are the size and direction of the current produced by such a generator? Examination of the relationship between the armature and the magnetic field in which it is rotated provides the answer. In Figure 9.2 the armature is represented by the wire loop *AB* whose ends are ***connected separately*** to the metal ***slip rings*** *E* and *F* which rotate with the armature in the direction indicated by *C*. Carbon ***"brushes"*** rest firmly against each slip ring, at the same time letting the rings rotate. In this way the circuit is completed through the external circuit containing *R* (resistance, representing load or current-consuming devices). Magnetic flux passes from the *N* pole to the *S* pole.

When the armature is rotating clockwise, as in Figure 9.2A, with *A* passing up through the magnetic field, an emf is induced in it in the direction indicated by the straight arrow

(according to the left hand rule). At the same time, *B* is passing down through the field, so that its emf is in the direction of the straight arrow near *B*. The net effect is that the current leaves *A* through slip ring *E*, passes through the external circuit *R*, and enters slip ring *F*.

As the armature coil rotates, *A* and *B* exchange positions, as shown in Figure 9.2B. Now *A* is moving down through the field while *B* is moving up, and the current ***leaves slip ring F*** and ***enters slip ring E***. Note that the current is now ***reversed in both the coil and the external circuit***. This reversal takes place repeatedly as the armature continues to rotate. A current which periodically reverses its direction is called an ***alternating current***, usually abbreviated AC.

Let us now examine the magnitude of the induced voltage ***from instant to instant*** as the coil turns in the field. In Figure 9.3, the generator is simply shown end-on, *A* and *B* representing the cross section of the wire armature, rotating in the direction of the arrows around axis *C*. At the instant the armature is in position (1) it is moving paral-

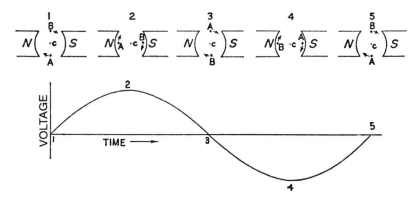

Figure 9.3. The relation of emf or current to the various positions of the rotating coil (armature) of an *AC* generator. *A* and *B* represent the single-loop coil in cross-section (see Figure 9.2). The induced emf depends on the angle at which the coil cuts the magnetic flux at any particular instant. *C* is the axis of rotation of the armature coil.

lel to the magnetic field and therefore not cutting magnetic flux. Because no emf is induced at this instant, the graph shows armature voltage to be zero. As the armature continues to move toward position (2) it cuts progressively more magnetic flux lines per second until, at position (2), it moves perpendicular (90°) to the flux lines. Now the induced emf in the armature reaches a maximum, as represented by the peak of the curve. At (3) the armature is again moving parallel to the flux and the induced emf reaches zero. At (4), with the armature again cutting the flux at an angle of 90° but in the opposite direction, the induced emf is reversed, and the peak of the curve falls below the horizontal axis of the graph. Finally, the cycle is completed when the armature returns to its original position.

The alternating current curve embodies certain useful information. Such a curve, presented in Figure 9.3, shows the relationship of the AC to the successive positions of the generator armature as it rotates between the magnetic poles. This is a *sine curve* because it depends on the sine of the angle, from instant to instant, between the plane of

the armature and a plane perpendicular to the magnetic field.

In Figure 9.4 the AC curve is shown again in its usual form. The *YY'* axis (vertical) may represent either the induced emf or the current while the *X* axis (horizontal) represents time. The distance between two successive ***corresponding points*** on the curve, for example *AB*, represents ***one cycle*** of the AC and indicates that the armature has made one complete revolution. In common use in the United States is 60-cycle (60-Hz*) AC, consisting of sixty complete cycles such as *AB* in each second, and corresponding to sixty revolutions of the armature per second. ***The number of cycles per second is called the frequency of an alternating current.*** Note that ***each cycle consists of two alternations,*** that is, the voltage starts at zero, rises gradually to a maximum in one direction (shown above the *X* axis), gradually returns to zero, increases to a maximum in the opposite direction (shown below the *X* axis), and finally returns to zero. Thus, there are 120 alternations per second with a 60-cycle current, which means that the ***current flows back and forth in the conductor 120 times in***

_____

* Hz (hertz) is frequency in cycles per sec.

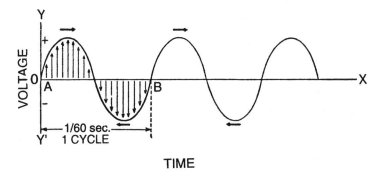

TIME

Figure 9.4. Sine curve representing a 60-cycle (60-Hz) alternating current (AC). The thick horizontal arrows show the direction of the *current* during different parts of the AC cycle. Note that the current reverses its direction every 1/120 sec. The thin vertical arrows indicate the variation in voltage or amperage from instant to instant.

each second. The single sine curve represents a *single phase AC.*

Since the AC voltage and amperage vary from instant to instant, they cannot be expressed by a single value in the same way as a direct current which flows steadily in one direction (see Figure 9.5). Instead, we must use the concept of *effective* or *root mean square (rms) value of an alternating current,* defined as *that direct current which has the same heating effect in a resistor as the alternating current in question.* For example, if a steady direct current of 1 ampere produces the same heating effect in a 1-ohm resistor as does a given AC, then the AC may be said to have an effective value of 1 amp. If we have a pure sine wave as in Figure 9.4, then the relationship between the peak and the *effective* or *root mean square value* of the current becomes:

rms current $= 0.707 \times$ maximum current

or

maximum current $= 1.41 \times$ rms current

The *effective* or *rms voltage* has the same relationship to the *maximum voltage:*

rms voltage $= 0.707 \times$ maximum voltage

or

maximum voltage $= 1.41 \times$ rms voltage

Properties of Alternating Currents

In Chapter 7 we found that a circuit carrying a *steady direct current* (DC) has a property called resistance, which tends to hinder the flow of current. A *coil* of wire connected in series in such a circuit holds back the flow of current at the instant of switch closure, over and above its electrical resistance. This results from the *inductance* of the coil, which sets up a *back emf* (see pages 70 and 90). Therefore, the direct current (DC) rapidly rises to a plateau and remains constant (see

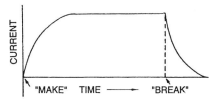

Figure 9.5. Curve representing a steady direct current with a coil in the circuit. The circuit is closed at "make." The current rises rapidly to a maximum and maintains that value at a constant level until the circuit is opened at "break," when it rapidly drops to zero.

Figure 9.5), since the associated magnetic field becomes constant and there is no longer a back emf. But with an *alternating current* the rapid change in magnitude and direction of the current causes a corresponding fluctuation in the strength and direction of the magnetic field set up around a coil in the circuit. Because of the coil's inductance, the persistently changing magnetic field around the coil supplied by an AC sets up a *back emf* in a direction opposite the applied alternating voltage, first in one direction and then in the other. This "bucking" tendency of inductance is called *inductive reactance,* and is measured in ohms because it opposes the flow of current.

A *capacitor* in a DC circuit allows current to flow only until it has reached full charge, whereupon the current ceases to flow. However, if a capacitor is added to an AC circuit, it periodically reverses polarity as the current reverses. This offers a certain amount of hindrance to the flow of current, known as *capacitive reactance*.

The apparent total resistance of an AC circuit is called its *impedance,* which depends on the inductive reactance, capacitive reactance, and true electrical resistance. The calculation of the impedance from these three factors is *not* that of simple addition,

and will not be pursued further.

Ohm's Law applies to AC if the above discussed properties are taken into consideration,

$$\text{effective current} = \frac{\overline{\text{effective voltage}}}{\text{impedance}}$$

$$I = \frac{V}{Z} \qquad (1)$$

where I is the effective current measured in amperes, V is the effective voltage measured in volts, and Z is the impedance measured in ohms.

Direct Current Generator

The principle of the AC generator also applies to the DC generator, but in the latter a *commutator* instead of slip rings feeds the generated current into the external circuit. Whenever the A and B segments of the armature exchange positions as shown in Figure 9.6, the corresponding half of the commutator always supplies current to the external circuit in the same direction even though there is an AC in the armature itself. The commutator consists of a split metal ring with the halves separated by an insulator, each half being permanently fixed to its cor-

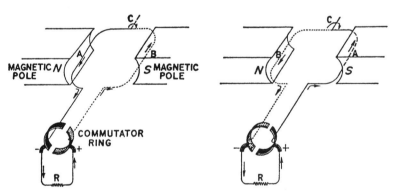

Figure 9.6. Simple direct current (DC) generator. Note that a commutator ring ensures that current is always supplied to the external circuit in the same direction. (Compare with Figure 9.2, which illustrates the AC generator.)

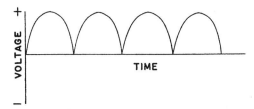

Figure 9.7. Voltage waveform of a direct current curve obtained from a simple DC generator such as that shown in Figure 9.6. Note that the curve differs from a sine curve in that the parts of the curve that lie below the horizontal axis are turned up to lie above the axis. This is a *pulsating direct current*, in contrast to the steady DC supplied by a battery.

responding end of the wire.

The wave form of the current obtained from such a generator is represented in Figure 9.7. Note that the part of the sine wave which with AC would be below the horizontal line lies above the line. This is a *pulsating direct current.* In actual practice, a DC generator consists of numerous coils and a commutator which is divided into a corresponding number of segments. Since the coils are so arranged that some are always cutting magnetic flux during rotation of the armature, emf is always being induced and never falls to zero, as in the simple one-coil DC generator illustrated above. Consequently, the voltage is steadier than that generated by a single-coil armature. Figure 9.8 shows the DC curve obtained with a

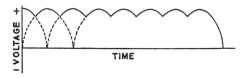

Figure 9.8. Direct current waveform produced by a multiple-coil generator. The solid line curve represents the voltage fluctuation which is much less than that produced by any one of the coils alone (shown by the broken line).

multiple coil generator.

Advantages of Alternating Current

For at least two reasons, AC is employed much more frequently than DC.

1. AC is needed to operate transformers which are indispensable in radiographic equipment as well as in many areas of industry. A transformer will not operate adequately on voltage with the waveform shown in Figure 9.8.

2. The electric current generated at the power plant is often transmitted over a great distance. It has been shown earlier that with **direct current** the power loss (in the form of heat) is related to the current and resistance as follows:

$$P = I^2R \qquad (2)$$

where $P =$ loss of power in the form of heat, measured in watts
$I =$ current in amperes
$R =$ resistance in ohms

According to this equation, the power loss is proportional to the *square of the amperage,* so that if a current is transmitted at a very *low* amperage there will be a much smaller loss of electric power than if transmitted at high amperage. An example will illustrate this point.

Suppose we wish to transmit 50,000 watts of electric power over a distance of ten miles (a total of twenty miles of wire), the wire having a resistance of 0.15 ohm per mile. The total **R** is $0.15 \times 20 = 3$ ohms. If we use **DC** at 500 volts and 100 amps (500 V × 100 A = 50,000 watts), the power loss is obtained by the power equation:

$$\textit{Power loss} = I^2R$$
$$= 100 \times 100 \times 3$$
$$= 30,000 \text{ watts}$$

Thus, three-fifths of the power is wasted as a result of the heating effect of the current in the transmission wires!

If, instead, we used **AC** at 5000 volts and 10 amperes (again 50,000 watts), the power loss would be much less. The above equation for power consumption cannot be applied accurately to AC without a correction known as the **power factor,** but it does give a rough approximation. Neglecting the power factor,

$$Power\ loss = I^2R$$
$$= 10 \times 10 \times 3$$
$$= 300\ watts$$

as contrasted with a power loss of 30,000 watts with DC transmitted at lower voltage and higher amperage.

Upon reaching its destination the high voltage (often called high tension) must be reduced to useful values, ordinarily 120 volts for small appliances and lighting, and 120 or 240 volts for x-ray equipment. This is accomplished by means of a **transformer,** an electromagnetic device which will be described in detail in the next chapter.

ELECTRIC MOTOR

Definition

An electric motor is a device that converts electrical energy to mechanical energy. In other words, an electric current can be made to do work by means of a motor.

Principle

Whenever a conductor carrying an electric current is placed in a magnetic field, there is a force or side-thrust on the conductor. It should be emphasized that two conditions must prevail: *the conductor must carry a current, and it must be located in a magnetic field.* The conductor experiences a force that is directly proportional to its length or number of turns, the strength of the magnetic field, and the size of the current. The force is increased when the conductor is in the form of a coil.

The motor effect results from the interaction between the magnetic field around the electron current in the conductor, and the other magnetic field. Therefore, a magnetic field acts on a current-carrying wire as shown in Figure 9.9, where it is represented in cross section and the following conditions prevail:

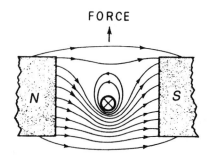

FORCE

Figure 9.9. The central dark circle represents a wire in cross section carrying a current "into the page." As explained in the text, the wire is forced upward by the weaker magnetic field above (flux lines in opposite directions) and by the stronger magnetic field below (flux lines in the same direction). (Adapted from *The New Physics and Chemistry Dictionary and Handbook* by Robert W. Marks. Copyright © 1967 by Bantam Books, Inc.)

1. The electric current in the wire is directed away from the reader in this particular case.

2. Magnetic flux surrounds the wire as shown by the small arrows (associated with the current in the wire).

3. Magnetic flux provided by the **field** magnets is directed to the right between the poles of the magnets.

Now look at the upper side of the wire; there the magnetic flux surrounding the wire is directed opposite to that of the flux between the field magnets. Such oppositely directed flux lines "attract" each other. At the same time, note that on the lower side of the wire the surrounding magnetic flux is in the same direction as the flux between the field magnets. Flux lines in the same direction "repel" each other. The net effect is for the wire to be thrust upward by the interaction of the two magnetic fields, as shown by the vertical arrow.

This principle is embodied in the *right hand or motor rule* which mirrors the left hand rule. *The right hand or motor rule states that if the thumb and first two fingers of the right hand are held at right angles to each other; and if the index finger indicates the direction of the magnetic field, and the middle finger the direction of the electron current in the conductor, then the thumb will point in the direction that the conductor will move.* Applying this rule to Figure 9.9, we find that the same result is obtained as before: the wire is thrust upward. Note that according to this rule a conductor experiences a force that is perpendicular to both the magnetic field and the electron current.

Simple Electric Motor

The diagrams shown in the section on the electric generator apply also to the electric motor, except that current is supplied at the brushes instead of being withdrawn. An AC motor has *slip rings,* whereas a DC motor has a *commutator.*

In its elementary form, an electric motor may be represented, as in Figure 9.10, by a cross-sectional diagram of a single coil armature carrying an electric current and lying in a magnetic field. Applying the *right hand rule,* we find that as *A* is thrust upward *B* is thrust downward. The net result is that the

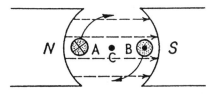

Figure 9.10. Cross section of an elementary motor. *A* and *B* are the coil wire seen end-on. *C* is the axis of rotation. The curved arrows show the direction in which the coil armature rotates when the proper current is fed to the armature.

armature turns in a clockwise direction about axis *C.*

At this point we should mention the *back emf* of a motor. When the coil of a motor rotates in a magnetic field, an emf is induced in the coil by electromagnetic induction. This self-induced emf is in a direction *opposite* to the direction of the emf already applied to the coil and is therefore called *back emf.* Thus, an electric motor acts at the same time as a generator, sending some current back to the main lines. This is why a motor consumes less current than a heating device.

Types of Electric Motors

There are two main types of electric motors:

1. **Direct current motor.** This is the reverse of a direct current generator.
2. **Alternating current motors.** The two kinds of alternating current motors include *synchronous* and *induction*.

The *synchronous motor,* supplied by AC, is built like a single-phase AC generator. However, current is fed to the armature, causing it to rotate in the magnetic field. Certain conditions are necessary for the operation of a synchronous motor. Its armature must rotate at the same speed (revolutions per minute) as the armature of the AC generator supplying the current, or at some fixed multiple of that speed. In other words, the speed of this type of motor must be *synchronous with the speed of the generator.*

Special devices are required to bring the speed of the motor armature up to the required number of revolutions per minute. The synchronous motor is used in electric clocks and synchronous timers. The main *disadvantage* of the synchronous motor lies in the limited speeds with which it will operate, that is, it must be in step with the generator. At the same time, a fixed speed may be of *advantage* in certain types of equipment, such as electric clocks.

The *induction motor* has a *stator* with an even number of stationary electromagnets distributed around the periphery. In the center is a rotating part or *rotor,* consisting of bars of copper arranged around a cylindrical iron core, and resembling a squirrel cage. There are no slip rings, commutators, or brushes because this type of motor works on a different principle from other types of motors and generators. The stator is supplied by a *multiphase* current, which means that the AC in one opposite pair of coils is out of phase (out of step) with that in the next pair of coils. Thus, successive pairs of coils set up magnetic fields which successively reach their peak strength, acting as though they were rotating, and dragging along the rotor. The drag on the rotor is caused by eddy currents induced in its copper bars by the "moving" magnetic field. Each copper bar, now carrying an induced current while lying in a magnetic field, experiences a push or force according to the motor principle. This force causes rotation of the rotor.

Although an induction motor is usually designed for modified single-phase or

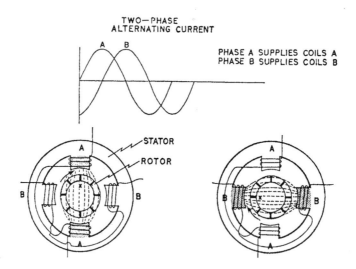

Figure 9.11. Principle of the induction motor. In phase *A* (left diagram) the magnetic flux is directed downward as shown. In the next $^1/_4$ cycle, phase *B* activates its stators and the magnetic flux is now directed from right to left (the flux in the preceding $^1/_4$ cycle having disappeared). In the next phase, the magnetic flux is directed upward, and finally from left to right. Eddy currents are set up in the rotor by electromagnetic induction. This process continues as long as current is supplied. In effect, then, we have a magnetic field that rotates clockwise, dragging the rotor with it and making it rotate. A split-phase modification is used with rotating anodes.

three-phase operation, its principle is more simply explained in Figure 9.11 by a two-phase generator with an armature consisting of two coils perpendicular to each other. The current produced by such a generator may be represented by two curves, one of which is one-fourth cycle ahead of the other, as shown in the upper part of the figure. This two-phase current is applied to the stator to induce a rotating magnetic field. The rotor is shown in cross section at the center.

A *split* single-phase induction motor rotates the anode in a x-ray tube; such a motor uses a capacitor to split the single phase into two (see Panichello, p. 30).

Current-measuring Devices

An important application of the *motor principle* pertains to the construction of instruments for the measurement of current and voltage. The basic device employing this principle is the *moving coil galvanometer,* which consists essentially of a coil of fine copper wire suspended between the poles of a horseshoe field magnet. One end of the coil is fixed and the other is attached to a hairspring helix. When a *direct current* is passed through the coil as it lies in the magnetic field of the permanent magnet, the coil moves according to the motor principle. In this case, the construction of the instrument (see Figure 9.12) is such that the coil must rotate through an arc that is usually less than one-half circle (180°). Rotation takes place against the spring, so that when the circuit is opened the coil returns to the zero position. Since the degree of rotation of the coil is proportional to the current, the attached pointer moving over a *calibrated scale* measures the current.

A galvanometer with a low resistance in parallel is capable of measuring *current* when connected in *series* in a DC circuit; if it has been previously calibrated to read amperes, it is an *ammeter* (see Figure 6.8, p. 52).

On the other hand, a galvanometer protected by a high resistance in series measures *potential difference* when connected in *parallel* with the circuit being measured; if it has been previously calibrated to indicate volts, it is a voltmeter (see Figure 6.8, p. 52).

Thus, modifications of the same basic instrument–the galvanometer–allow measurement of amperage or voltage in a DC circuit, provided the instrument has been calibrated against known values of amperage

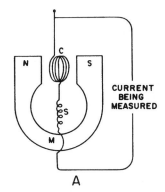
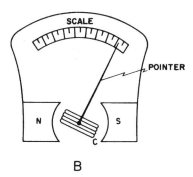

Figure 9.12. Moving coil galvanometer. In *A,* side view. In *B,* adapted as direct reading ammeter or voltmeter.

or voltage, respectively.

Further modification is necessary to measure the voltage and amperage of an ***alternating current,*** because the coil would "freeze" due to the rapid alternation of the current in the coil. (Recall that a 60-cycle AC sustains 120 reversals of direction per second.) Instead of using a horseshoe magnet, one must provide a magnetic field which reverses its direction in step with the current entering the coil. To do this, the AC being measured is, at the same time, used to supply the coils of a pair of field electromagnets. Since the current alternates simultaneously in the rotating coil and in the field coils, the rotation of the coil and hence the deflection of the pointer are unaffected by the alternations of the current, but are still affected by the size of the current. This device, called an ***electrodynamometer,*** can be designed to read AC voltage or amperage. The same type of instrument can be modified to measure electric power in a circuit, reading the values directly in watts.

QUESTIONS AND PROBLEMS

1. What is an electric generator? The generator rule?
2. List some of the commonly used sources of energy in a generator.
3. Describe, with the aid of a diagram, a simple electric generator. Show how it produces an alternating current and relate it to the AC sine curve. Label the curve in detail.
4. How does an AC generator differ from a DC generator?
5. Compare the current curves of single phase AC, single phase DC, and an electric battery.
6. Why is AC generally preferred to DC? Prove by means of an example.
7. Define the effective (RMS) value of alternating current or voltage. How is it derived from peak values?
8. Find the RMS value of 80 kVp.
9. The effective value of an alternating current is 30 A. What is the peak amperage?
10. Find the power loss due to heat in DC circuit with a resistance of 20 ohms and a current of 30 amperes.
11. Define an electric motor. What is the motor principle?
12. Discuss the right hand or motor rule.
13. Under what conditions will a conductor, placed in a magnetic field, tend to move up through the field?
14. What are the two main types of motors?
15. Describe the principle and construction of an induction motor.
16. What is a synchronous motor? How does it differ from an induction motor?
17. Discuss the importance of the induction motor in radiographic equipment.
18. State the underlying principle of the galvanometer. Show with the aid of a simple diagram the construction of a DC voltmeter.

Chapter 10

PRODUCTION AND CONTROL OF HIGH VOLTAGE–REGULATION OF CURRENT IN THE X-RAY TUBE

Topics Covered in This Chapter

- Transformer
    Principle
    Construction of Transformers
    Efficiency and Power Losses
- Regulation of High Voltage
    Autotransformer
- Control of Filament Current and Tube
    Current
    Choke Coil
    Rheostat
    High-Frequency Control
- Questions and Problems

TRANSFORMER

ELECTRIC POWER to the consumer is usually supplied at 120 to 240 volts, but much higher voltage is needed to give the electrons in an x-ray tube enough energy to generate x rays. Such high voltages are produced by the *transformer,* also referred to as an *x-ray generator.* It serves as a major component of an x-ray machine. In this chapter we shall discuss only the single-phase circuit carrying 60-cycle (60-Hz) alternating current (AC).

Principle

A transformer is an electromagnetic device which changes an *alternating current* from low voltage to high voltage, or from high voltage to low voltage, without loss of an appreciable amount of electrical energy (usually less than 5 percent). The transformer transfers electrical energy from one circuit to another *without the use of moving parts or any electrical contact between the two circuits,* employing the principle of *electromagnetic mutual induction.*

The simplest type of transformer–the *air core* transformer–consists of two highly insulated coils of wire lying side by side. One of these, the *primary coil,* is supplied with an alternating current (AC). The other or *secondary coil* develops AC by mutual induction (see Figure 10.1). The primary coil is the *input side* of the transformer, whereas the secondary coil is the *output side.*

How does a transformer function? An AC in the *primary* coil sets up, in and around it, a magnetic field that varies rapidly in direction and strength, just as does the AC itself. This changing magnetic flux *cuts*

83

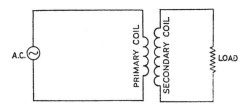

Figure 10.1. Air core transformer. An electromotive force is set up in the secondary coil by electromagnetic *mutual induction.*

or **links with** the secondary coil, inducing in it an alternating emf by the process of electromagnetic induction. But the resulting AC in the secondary coil sets up an induced emf in the primary coil (opposite in direction to the primary current). Since emf is induced in each coil by the changing magnetic flux in the other, we call this process **mutual induction,** which is obviously a special case of electromagnetic induction.

Note here, again, the important principle of electromagnetic induction: the emf induced in any coil is directly proportional to the number of turns in the coil that cuts, is cut by, or links with, a given magnetic flux. This also applies to the transformer. Suppose the primary and secondary coils each have one turn. The AC in the primary coil sets up a magnetic field that cuts the one turn in the secondary coil; therefore, the emf is the same in both coils (neglecting any loss of energy during the process). But suppose the primary coil has one turn and the secondary has two turns. Now the **same** magnetic flux set up by the AC in the single-turn primary coil cuts the double-turn secondary coil, **doubling the emf in the latter.** But if the primary coil has two turns and the secondary one turn, the secondary emf will be **half** the primary emf.

The preceding discussion is embodied in the **transformer law: the emf induced in the secondary coil is to the emf in the primary coil, as the number of turns in the secondary coil is to the number of turns in the primary**

coil, expressed simply in practical units in the following equation:

$$\frac{V_S}{V_P} = \frac{N_S}{N_P} \qquad (1)$$

where V_S = voltage in secondary coil
V_P = voltage in primary coil
N_S = no. of turns in the secondary coil
N_P = no. of turns in the primary coil

Stated in words, this means that if the number of turns in the secondary coil is twice the number in the primary coil, then the voltage in the secondary will be twice the voltage in the primary. If the number of turns in the secondary coil is three times that in the primary coil, then the voltage induced in the secondary will be three times as great as in the primary, etc. Such a transformer, having more turns in the secondary coil than in the primary, puts out a **higher** voltage than is supplied to it and is therefore a **step-up transformer.** N_S/N_P is the step-up ratio.

Example 1. A certain transformer has a step-up ratio of 500. If the emf in the primary is 200 V, find the secondary kV.

Solution. According to the transformer law,

$$\frac{V_S}{V_P} = \frac{N_S}{N_P}$$

In this problem, step-up ratio $\frac{N_S}{N_P} = 500$

Therefore, $\frac{V_S}{V_P} = 500$

and $V_S = V_P \times 500 = 200 \text{ V} \times 500$
$V_S = 100{,}000 \ V_p$ or $100 \ kV$

Example 2. What primary voltage is needed to obtain a secondary of 100 kV, if the step-up ratio is 500? Invert equation (1), giving

$$\frac{V_P}{V_S} = \frac{N_P}{N_S}$$

Then $\dfrac{V_P}{100 \text{ kV}} = \dfrac{1}{500}$

$$V_P = \dfrac{100,000 \text{ V}}{500} = 200 \text{ V}$$

If the secondary coil has fewer turns than the primary coil, the output voltage will be *less* than the input voltage, and the transformer is a *step-down transformer*.

What happens to the value of the *current* in a transformer? According to the Law of Conservation of Energy, there can be no more energy coming out of the transformer than is put in, and similarly, the power output (energy per unit time) can be no greater than the power input. Since the power in an electric circuit equals voltage multiplied by amperage, then (neglecting the power factor),

$$I_S V_S = I_P V_P \qquad (2)$$

where I_S = current in amperes in the secondary coil

I_P = current in the primary coil

V_S = voltage in the secondary coil

V_P = voltage in the primary coil

Rearranging this equation, we get the proportion:

$$\dfrac{I_S}{I_P} = \dfrac{V_P}{V_S} \qquad (3)$$

In other words, if the voltage is increased, as in a step-up transformer, the amperage is decreased; and if the voltage is decreased, as in a step-down transformer, the amperage is decreased. Thus, a *step-up transformer increases the voltage, but decreases the amperage in an inverse ratio.*

Example 3. The current in the primary circuit of a transformer is 25 A. If the voltage step-up ratio is 500, find the current in the secondary circuit.

Solution. The ratio of the secondary to the primary currents is the *inverse* of the voltage step-up ratio, or 1/500. Therefore,

$$\dfrac{I_S}{I_P} = \dfrac{1}{500}$$

$$I_S = \dfrac{I_P}{500}$$

$$I_S = \dfrac{25}{500} = 0.05 \text{ A or 50 mA}$$

The power output of an x-ray transformer is rated in kVA (kilovolt-amperes or kilowatts). Thus, an x-ray generator that delivers a tube current of 200 mA at 140 kVp is rated at 28 kVA or kW (obtained from 140 kVp × 0.2 A).

In x-ray diagnostic equipment, the transformer with a step-up ratio of 500 takes an input of, for example, 80 to 240 V and multiplies it to 40,000 to 120,000 V (40 to 120 kVp). This provides the high potential difference needed to drive the electrons through the x-ray tube with sufficient speed to generate x rays. At the same time, it reduces the current to thousandths of an ampere (mA). (With three-phase equipment, 150 kV becomes possible.)

Construction of Transformers

Four main types of transformers will now be described.

1. **Air Core Transformer.** This consists simply of two insulated coils lying side by side as shown in Figure 10.1 on page 84.

2. **Open Core Transformer.** An iron core inserted into a coil of wire carrying an electric current markedly intensifies the magnetic flux within the coil because of the magnetization of the core. Therefore, a transformer becomes more efficient if each *insulated* coil has an iron core, as shown in Figure 10.2. This is known as an *open core transformer.* Although more efficient than an air core transformer, the open core type is still subject to an appreciable waste of power due to loss of magnetic flux at the ends of the

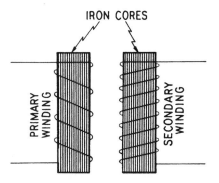

Figure 10.2. Open core transformer. There is some loss of power due to leakage of magnetic flux in the air at the ends of the iron cores.

cores, a condition termed **leakage flux.**

3. **Closed Core Transformer.** Here the heavily insulated coils, often called *windings,* are wound around a square or circular iron ring or core as in Figure 10.3. With this type of transformer, the closed core provides a continuous path for the magnetic flux, so that only a small fraction of the magnetic energy is lost by leakage. An AC in the primary winding sets up an alternating magnetic flux in the iron core. Because this same flux links with (or cuts) both the primary and the secondary windings, it induces the *same voltage per turn in both windings* (coils). Therefore, the induced emf in *each* winding is proportional to the number of turns in that winding. Thus, the transformer law applies, namely, $V_S/V_P = N_S/N_P$ (see also pages 84-85).

The core is **laminated,** that is, it is made of layers of metal plates. Eddy currents (see page 87) are set up in the core by electromagnetic induction during transformer operation. Such eddy currents take energy from the current in the transformer coils, causing a waste of power known as **transformer loss.** Lamination of the core hinders the formation of these eddy currents and thereby increases the efficiency of the transformer; in other words, there is less power wasted. Silicon steel is often used in laminated cores because its high electrical resistance further reduces eddy current power loss. Since the closed core transformer is much more efficient than the open core, it is the *one most commonly used in x-ray-generating equipment.* The highly insulated coils are submerged in a metal box containing a special type of oil for additional insulation, which also helps keep the transformer cool by improving the dissipation of heat produced during operation.

4. *Shell-type Transformer.* This, the most advanced type, is used as a commercial or power transformer. Here, too, a laminated core is used, consisting of a pile of sheets of silicon steel, each having two rectangular holes, as illustrated in Figure 10.4. The primary and secondary coils are both wound around the central section of the core for maximum efficiency. These windings must be highly insulated from each other by a special coating. In addition, the entire transformer is submerged in a tank filled with a special type of oil for maximum insulation and cooling.

Efficiency and Power Losses

The efficiency of a transformer is the ratio of the power output to the power input in a load circuit, ie, a closed circuit containing

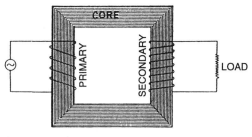

Figure 10.3. Closed core transformer. This type of iron core provides a continuous path for the magnetic flux, thereby minimizing flux leakage.

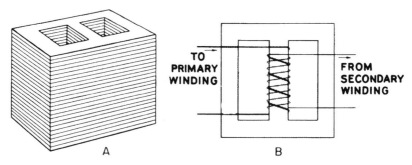

Figure 10.4. *A,* core of shell-type transformer. *B,* shell-type transformer, top view. Note the windings of the primary and secondary coils around the *same* central section of the core to provide high efficiency. Very heavy insulation is required.

resistance:

$$\% \text{ efficiency} = \frac{\text{power output}}{\text{power input}} \times 100$$

Ideally, the power input and power output would be equal, with an efficiency of 100 percent. However, in practice the efficiency is more likely to be about 95 percent or more; that is, there is up to about 5 percent less power output than power input, the lost energy appearing as heat. In contrast to other types of electrical and mechanical equipment, this represents a small waste of energy, and the transformer is therefore a highly efficient device.

We shall now summarize the various kinds of *losses of electrical power* in a transformer and the methods of minimizing them to improve efficiency.

1. *Copper Losses.* These pertain to the loss of electrical power due to the *resistance* of the coils, a power loss that can be reduced by using copper wire of adequate diameter. In a step-up transformer the primary winding carries a large current, so its wire must be thicker than the wire in the secondary coil which carries a small current (milliamperes). Recall that a thick wire, having less resistance than a thin one, can carry more current with less waste of energy in the form of heat *(Power Loss = I^2R).*

2. **Eddy Current Losses.** The alternating magnetic flux set up in the transformer *core* by the AC in the windings induces electrical *eddy currents* (swirling currents) in the core itself by electromagnetic induction. Eddy currents, in turn, produce heat in the core because electrical power loss due to heat = I^2R. The wasted electric power must come from that supplied to the transformer, thereby contributing to the loss of efficiency. Eddy currents can be minimized by the use of *laminated (layered) silicon steel plates,* highly insulated from each other by a special coating. Lamination and high-resistance silicon steel increase the electrical resistance of

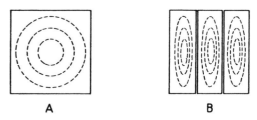

Figure 10.5. *A.* Dashed lines suggest the paths taken by eddy currents in a solid iron core. *B.* With the same volume of iron having three laminations (layers), the total paths for the eddy currents have been significantly increased; this means increased resistance and therefore smaller eddy current effect.

the core, thereby decreasing the size of the eddy currents according to Ohm's Law (see Figure 10.5).

3. *Hysteresis Losses.* Since the transformer operates on, and puts out AC, the magnetic domains in the *core* are repeatedly rearranging themselves as the core is magnetized first in one direction and then the other by the AC in the windings. This produces heat in the core, thereby wasting electrical power. Such a loss of power, called *hysteresis loss,* is reduced by the laminated silicon steel core.

CONTROL OF HIGH VOLTAGE

Properly designed x-ray equipment must provide some means of regulating the *input voltage* to permit the choice of a variety of kilovoltages to be applied to the x-ray tube. This allows the technologist to obtain x rays of appropriate penetrating power for a particular technic. The *autotransformer* is a device that varies input voltage to a transformer to control its output voltage.

Autotransformer

According to the description of the step-up transformer, you should recall that there is a *fixed ratio* of voltage output to voltage input. Furthermore, the voltage input is fixed at 120 or 240 volts, depending on the power supply and type of equipment.

Without any further modification, we would have available only a single kilovoltage, a situation that would be intolerable.

How can we modify the basic equipment so as to obtain the required range of kilovoltages? This is accomplished simply by *varying the voltage input to the transformer primary.* Thus, if the ratio of the step-up transformer is 500 to 1 and the input to the primary coil is 100 volts, then the transformer will put out $500 \times 100 = 50,000$ volts (50 kVp). If the input is 180 volts, the output will be $500 \times 180 = 90,000$ volts (90 kVp). To obtain the required variety of input voltages we connect a device known as an *autotransformer* between the source of AC and the primary side of the transformer. As you will see, the autotransformer is a *variable transformer* (see Figure 10.6).

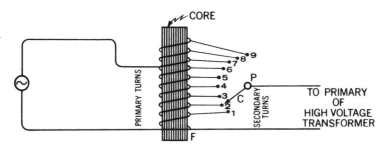

Figure 10.6. Diagram of an autotransformer (kV selector), which has only one winding or coil. The *autotransformer* secondary voltage stands in the same ratio to the primary voltage, as the ratio of the number of tapped turns (in this case, 2) is to the number of primary turns (7); that is, $2/7$, so the voltage is stepped down. If C is turned above 7, the ratio exceeds 1 and the voltage is stepped up.

Construction. The autotransformer consists of a single coil of insulated wire wound around a *large iron core.* At regular intervals along the coil insulation has been omitted and the bare points connected or tapped off to metal buttons. A movable contactor, *C,* varies the number of turns included in the secondary circuit of the autotransformer, thereby varying its output voltage. Thus, the autotransformer serves indirectly as a *kV selector.*

Principle. The autotransformer is an electromagnetic device which operates on the principle of *self-induction. A single coil or winding serves as both the primary and secondary, the number of turns being adjustable.* (This is in sharp contrast to the transformer, wherein the number of turns cannot be changed.) By moving the contactor to tap more or fewer turns on the secondary side of the autotransformer, we can vary the ratio of the number of secondary turns to the number of primary turns, thereby varying the ratio of voltage output to voltage input. This relationship, similar to that in the transformer, is embodied in the *autotransformer law:*

$$\frac{\text{secondary voltage}}{\text{primary voltage}} = \frac{\text{number of tapped turns}}{\text{number of primary turns}}$$

Thus, in Figure 10.6, if the primary voltage is 240,

$$\frac{\text{autotransformer secondary voltage}}{240} = \frac{2}{7}$$

autotransformer secondary voltage = 68.6 volts

The secondary voltage of the autotransformer is applied to the primary side of the main step-up transformer (see Figures 11.6 and 15.12).

An autotransformer can be used only where there is a relatively small difference between its input and output voltage. Furthermore, *the position of the contactor must not be changed while the exposure switch is closed because sparking may occur between the metal buttons, thereby damaging them.*

CONTROL OF FILAMENT CURRENT AND TUBE CURRENT

We have just learned that control of high voltage by means of an autotransformer allows convenient kV selection. Of equal importance is the capability of x-ray equipment to vary tube current or milliamperage (mA).

The construction and operation of an x-ray tube will be explained in detail later (see pages 111-112). At present it is enough to say that an x-ray tube has two circuits: (1) a filament circuit carrying the current needed to heat the filament and (2) the tube circuit itself carrying the tube current which passes between the electrodes of the x-ray tube and produces x rays. Of utmost significance is the fact that *a small change in filament current produces a large change in tube current* (see Figure 14.2). Therefore, by regulating the filament current we can control the x-ray output in terms of R/min.

Three options allow control of filament current: choke coil, rheostat, and high-frequency circuit.

Choke Coil

Principle. The choke coil, an electromagnetic device operating on the principle of *self-inductance,* requires *alternating current.*

In Chapter 9 we learned that an electric current in a coil sets up a magnetic field so that one end of the coil becomes a north pole and the opposite end a south pole. If the magnetic field varies in direction and strength, as when it is generated by an *alternating current,* the magnetic flux links with or cuts the coil itself and induces a voltage which *opposes* the voltage already applied across the coil. This, the *back emf of self-induction,* exemplifies Lenz's Law which holds that the electron current induced by a changing magnetic field, always flows in such a direction that *its* induced magnetic field opposes the action of the original inducing field. It follows from the Law of Con-ser-vation of Energy.

The introduction of an *iron core* into the coil intensifies the magnetic field and increases the inductive reactance of the coil (analogous to, but not identical with, resistance) producing the following effects:

1. A *larger voltage drop* across the choke coil.

2. A *smaller remaining voltage* for the rest of the circuit.

3. A *decrease in current* due to the larger inductance of the circuit.

These effects become more pronounced with deeper insertion of the core into the coil. The principle of the choke coil is explained further in Figure 10.7.

The choke coil was previously used to regulate filament current in the x-ray tube, to permit control of tube current (mA). However, it has been largely replaced by a rheostat or high-frequency control.

Rheostat

A rheostat is a *variable resistor,* designed to regulate *filament current* in radiographic equipment, as shown in Figure 10.8. A variable resistor controls not only the voltage but also the current according to Ohm's Law, $I = V/R$. An increase in the resistance causes the current in the filament circuit to decrease, resulting in less heating of the filament, a smaller electron emission, and a much smaller tube current. But at the same time, the increased rheostat resistance entails an increased voltage drop across the rheostat, thereby leaving a smaller voltage for the

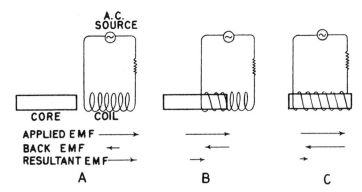

Figure 10.7. Operation of a choke coil. Note that AC is required. Arrow length indicates the emf in that direction. In *A,* with the core completely out, back emf is minimal and voltage drop across the coil is small, so the resultant emf is only slightly less than the applied emf. In *B,* partial insertion of the core increases back emf and voltage drop across the coil. In *C,* with the core completely inserted, back emf of self-induction is maximal, producing a large voltage drop across the coil with a small resultant emf available for the rest of the circuit, a*t the same time, the current is decreased.* The adjustable position of the core permits selection of the desired emf and current.

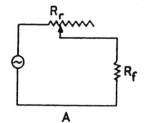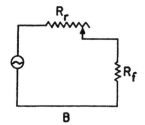

Figure 10.8. Diagram of a filament circuit where R_r is the resistance setting of the rheostat, and R_f is the resistance of the tube filament. In *A*, with a smaller value of rheostat resistance, a large current flows in the circuit including the filament, ie, filament circuit. In *B*, with a larger value of rheostat resistance, there is a small current in the circuit. In each instance there is a corresponding change in the filament voltage.

filament, provided the total voltage drop across the entire filament circuit is kept constant. The rheostat is no longer used in modern state-of-the-art x-ray equipment for control of filament current.

HIGH-FREQUENCY CONTROL OF CURRENT AND VOLTAGE

For many years attempts were made to design x-ray equipment that would generate x rays at constant potential to eliminate the kV fluctuations inherent in 60-cycle (hertz or Hz) single-phase machines. You should recall that with half-wave rectification there are 60 voltage peaks/s, and with full-wave rectification 120 peaks/s; between these peaks, x-ray energies range downward to zero and are increasingly absorbed in the tissues without contributing to the x-ray image. It would therefore be advantageous to reduce the tissue-to-film dose ratio.

In the 1970s an x-ray *therapy* unit called the *Vanguard* (Picker X-Ray Corp.) was introduced, generating 280-kV x rays at more nearly constant potential than other x-ray machines. The *Vanguard* converted a three-phase current to a high-frequency (HF) 1.2 kHz (1200 cycles/sec) current; but this system lost out to the already growing popularity of the cobalt-60 therapy units. At about the same time, 6 to 70 kHz generators were being devised for general radiography,

but fell victim to the imported three-phase units which, although expensive, were widely accepted for angiography and later for general radiography.

With the advent of high-frequency (H-F) generators, it became possible to obtain nearly constant voltage output with ripple of less than two percent. Such H-F units operate on a storage battery, 115-120 VAC (AC voltage) or 220-240 VAC* single phase, depending on their design, but most modern x-ray machines use *3-phase power supply* for

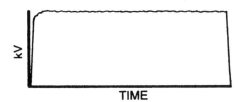

Figure 10.9. Waveform of kilovoltage in a high-frequency circuit. Note the slight ripple, which represents almost constant potential operation.

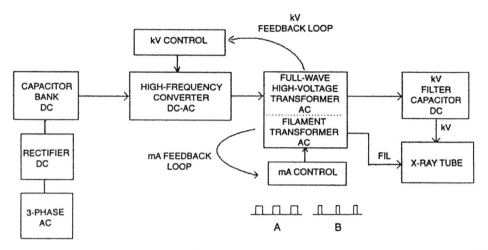

Figure 10.10. Block diagram of a high-frequency circuit to show control of tube current and kilovoltage. Microprocessor-controlled feedback maintains instantaneous regulation of mA and kV.

their H-F generators, which provides nearly constant kV (see Figure 10.9) and increased x-ray output. Input voltage as high as 480 VAC is available for H-F generators with 3-phase power supply. Moreover, feedback circuits under microprocessor control provide instantaneous regulation of kV and mA (see Figure 10.10), allowing shorter exposure times and greater consistency in image quality (see also pages 165-166). Most mammography units employ H-F circuits because of their outstanding reliability.

QUESTIONS AND PROBLEMS

1. Why is a high voltage needed in x-ray equipment?
2. What device is used to change low voltage to high voltage?
3. Discuss the transformer law. A transformer has 100 turns in the primary coil and 100,000 turns in the secondary. If the input is 110 volts, what is the output voltage?
4. Show by diagram a closed core transformer, first as a step-up transformer, second as a step-down transformer.
5. How are eddy currents reduced in a transformer? Why should they be kept to a minimum?
6. Why is a core used in a transformer?
7. What two types of insulation are used in a transformer and why?
8. Discuss the purpose and principle of an autotransformer. How does it operate? What is the autotransformer law?
9. Why is the autotransformer preferred in the control of high voltage in an x-ray machine?
10. The main electrical supply to an x-ray machine is 240 volts. If the x-ray transformer has a step-up ratio of 1000 and we wish to obtain 100 kVp what will be the step-down ratio of the autotransformer?
11. Describe a rheostat. In what type of x-ray equipment is it most widely used

today? How does it operate?

12. Explain the principle of the choke coil. By what devices has it been replaced in radiographic equipment?

13. Which type of current (AC or DC) is needed for the operation of choke coil; autotransformer; rheostat?

14. Why is it necessary to regulate the filament current of an x-ray tube? How is

tube current related to filament current?

15. Describe the operation of a high-frequency filament circuit. What are its advantages over the rheostat?

16. Suppose a technique calls for 85 kVp. If the transformer has a step-up ratio of 1000, what primary voltage will be needed?

Chapter 11

RECTIFICATION AND RECTIFIERS

Topics Covered in This Chapter

- Definition
- Voltage and Current Curves–Single-phase AC
- Self-rectification
- Half-wave Rectification
- Full-wave Rectification
- Solid State Rectifiers
- Questions and Problems

Definition

AS POINTED OUT BEFORE, an alternating current (AC) periodically reverses direction and varies in strength. While an x-ray tube can operate on *any* high voltage AC, it *operates most efficiently when supplied by high voltage direct current* (DC)–a current that is *unidirectional,* flowing always in the same direction. *Rectification may be defined as the process of changing alternat-*

ing current to direct current.

There are two possible ways of rectifying the high voltage AC supplied by the secondary side of the step-up transformer: (1) by suppressing that half of the AC cycle represented by the portion of the curve that lies below the line (ie, the negative half cycle) or (2) by changing the negative half cycle to a positive one. The resulting curves are shown in Figure 11.1.

As an aid to understanding the purpose

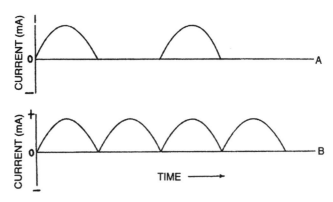

Figure 11.1. Current in the two basic types of rectified circuits. *A*. With half-wave rectification, the negative half lying below the line is *suppressed*. *B*. With full-wave rectification, the negative half lies above the line and becomes positive. In either case, a *pulsating* direct current is produced.

and process of rectification, we shall summarize the structure and operation of x-ray tubes, although Chapter 14 contains a detailed account of this important subject.

An *x-ray tube* consists of a glass bulb from which the air has been evacuated as completely as possible. Sealed into the ends of the bulb–or *tube,* as it is better known–are two terminals or electrodes: the *cathode* and the *anode.* The cathode consists of a thin tungsten wire (filament) which, when heated to incandescence (white hot) by a *low voltage current,* gives off electrons. These form a cloud or *space charge* near the filament. If, now, a *high voltage* from the secondary of the transformer is applied between the cathode and anode so as to produce a large negative charge on the cathode and a large positive charge on the anode, the resulting strong electric field drives the space charge electrons toward the anode at tremendous speeds. When the electrons strike the anode, their *kinetic energy* is converted in small measure to x rays, but mainly to heat production by agitating the atoms in the target. During ordinary operation, this electron stream can pass in *only one direction, from cathode to anode.* If the applied AC has first been rectified so that the resulting current is always directed from cathode to anode, the tube can withstand greater energy loading (ie, exposure factors) than if the current has not been rectified.

Methods of Rectifying an Alternating Current

There are two main systems of rectification: (1) self-rectification and (2) vacuum tube or solid state diode rectification. Vacuum tube diodes, also called *valve tubes* were used for many years in radiographic equipment. However, solid state diodes (see pages 100-105) have supplanted valve tubes in modern diagnostic units, both single-phase and three-phase. *All rectifying systems are connected between the secondary side of the x-ray transformer and the x-ray tube.*

1. **Self-rectification.** In this, the simplest type (see Figure 11.1A), the high voltage is applied *directly* to the terminals of the x-ray tube. Under ordinary conditions, an x-ray tube allows passage of electrons only from the cathode to the anode during the positive half cycle of the AC curve, when the anode is positively charged. This half of the voltage cycle is the *useful voltage* or *forward bias.*

During the negative half cycle the "anode" is negative and the "cathode" is positive, but despite the presence of a high voltage across the tube (however, in the "wrong" direction) no current will *normally* flow because there is no space charge near the anode. At the same time, the reverse voltage–also known as the *inverse voltage* or *reverse bias*–is actually higher than the useful voltage. This results from *transformer regulation,* a condition due to power loss in the transformer core from a slight fall in kV during forward bias. Such a fall in kV does not occur during negative or reverse bias because there is no current in the tube.

The principal disadvantage of self-rectification is the low heat-loading capacity (ie, limitation to low exposure factors) of an x-ray tube in such a circuit. In a self-rectified circuit the anode must never be heated to the point of electron emission because then, during the negative half cycle, the inverse voltage would drive the electrons in the wrong direction–toward the filament–causing the filament to melt and ruining the tube. Therefore, with self-rectification, exposure factors (kV and mAs) must be limited to lower values (lower tube rating) than with full-wave rectification.

An important point to remember is that the milliammeter indicates *average mA.* In a self-rectified circuit, because only one-half the cycle is used, the *peak* mA and consequent anode heat must be greater, for the

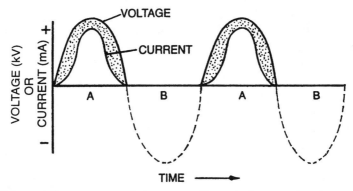

Figure 11.2. Voltage *and* current waveforms in a self-rectified circuit (single-pulse rectification). During the first half-cycle, *A,* current passes through the tube, the applied voltage being *effective* or *useful.* During the second half-cycle, *B,* no current flows from cathode to anode because the voltage is *inverse* or *ineffective,* but is slightly higher than the effective voltage. Note the difference in the shapes of the current and voltage waveforms (not shown in Figure 11.1).

same milliammeter reading than in a full-wave rectified circuit in which both halves of the cycle are used.

Figure 11.2 describes the voltage or current through the tube in a self-rectified circuit. This type of rectification is also known as *self-half-wave* or single-pulse rectification (one voltage peak per cycle).

Self-rectification was formerly used in mobile x-ray units, but it has been replaced by more sophisticated methods of rectification, to be described next.

2. **Diode Rectification.** A diode nor-mally passes current in one direction only, that is, ***electrons flow from cathode to anode.*** Although it is customary in solid state circuitry to show the current passing from anode to cathode, we shall consistently trace the ***flow of electrons*** in x-ray circuits in this course since it is conceptually simpler.

Either one or two diodes can provide half-wave rectification, also known as ***one-pulse*** rectification (one voltage peak per cycle). Figure 11.3 shows a circuit containing ***one rectifier diode.*** During the ***positive half*** of the AC cycle, the polarity of the sec-

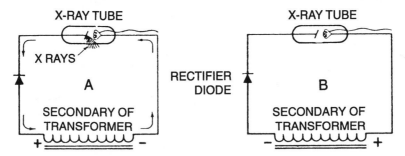

Figure 11.3. Half-wave rectification with a single rectifier diode. In *A,* electrons flow readily in both the x-ray tube and rectifier diode; this is *forward bias.* In *B,* during the next half cycle, the polarity of the transformer has reversed and electrons cannot flow from anode to cathode in the rectifier diode, thereby protecting the x-ray tube from the inverse voltage, also called *reverse bias.*

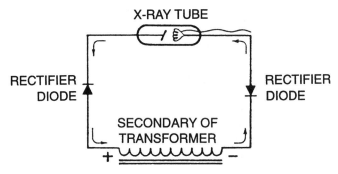

Figure 11.4. Half-wave rectification with two rectifier diodes. When the polarity of the transformer reverses, no current flows in the circuit.

ondary coil of the transformer is that shown in the figure, the direction of the electron current being from cathode to anode. The current flows through the rectifier and the x-ray tube as indicated by the arrows. During the negative half cycle, when the polarity of the transformer secondary is reversed, the rectifier diode is nonconductive; in fact, it behaves as an open switch, thereby preventing application of high voltage to the x-ray tube. Thus, the rectifier *suppresses the inverse voltage*, diminishing the possibility of reverse flow of electrons in the x-ray tube, should the anode become hot enough during operation to be a source of electron emission. Such a reverse flow of electrons could easily destroy the filament. Therefore, the diode allows greater loading of the x-ray tube than is possible with self-rectification, although we obtain half-wave or one-pulse rectification in either case (see Figure 11.2).

With *two rectifier diodes*, as shown in Figure 11.4, the high voltage during the negative half cycle is divided among the diodes and the x-ray tube, thereby increasing the efficiency of the system and improving the heat-loading capacity of the x-ray tube. The wave form of two-diode rectified current resembles that of self-rectified current (see Figure 11.2); in other words, it, too, is single-

or one-pulse rectification.

Four rectifier diodes can be arranged to provide *full-wave rectification* as shown in Figure 11.5; this is *two-pulse rectification* since there are two voltage peaks per cycle. In *A*, the current flows from the negative pole of the transformer through the successively numbered portions of the circuit to the cathode of the x-ray tube, eventually returning to the positive end of the transformer. (Remember that each rectifier diode permits the current to flow only in one direction, from cathode to anode.) In *B*, the polarity of the transformer reverses due to alternation of the current. The current again follows the successively numbered portions of the circuit to the cathode of the x-ray tube, finally reaching the positive end of the transformer. Note that at some time during its course through the circuit, the current arrives at the junction of two rectifier cathodes, as at point *S* in *A*; the current passes through one of the diodes because passage through the other diode would lead it back to a point of higher potential and this is physically impossible (analogy: a ball will not, of itself, roll uphill to a point of higher potential energy).

You should encounter no difficulty in learning the proper connection of the rectifier diodes once you have mastered the following simple steps:

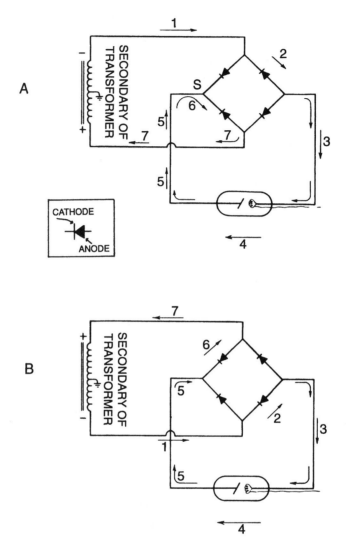

Figure 11.5. Full-wave rectification with four rectifier diodes. By following the numbered segments of the circuits in *A* and *B,* you can see that regardless of the polarity of the transformer engendered by the AC, the electron current always reaches the x-ray tube in the same direction; that is, both half cycles are used. This is called two-pulse rectification because of the two voltage peaks per cycle.

(1) First arrange a circuit without indicating the diodes, as in Figure 11.6.

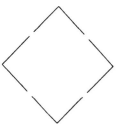

Figure 11.6.

(2) Then insert two diodes joined as in Figure 11.7.

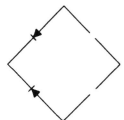

Figure 11.7.

(3) Now connect the **cathodes** to the **anode** of the x-ray tube as in Figure 11.8.

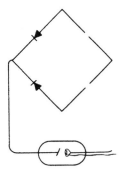

Figure 11.8.

(4) Next, insert two diodes on the opposite side of the circuit, with their anodes joined, and then connect them with the **cathode** of the x-ray tube, as in Figure 11.9.

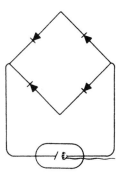

Figure 11.9.

(5) Finally, connect the remaining terminals to the terminals of the transformer, as in Figure 11.10.

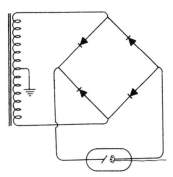

Figure 11.10.

The above steps are necessary because there is only one correct way to connect the rectifier diodes to each other and to the x-ray tube and transformer. If this relationship is altered in any way, the system will fail to rectify. *Notice that in each half cycle, only one parallel pair of rectifier diodes is conductive*

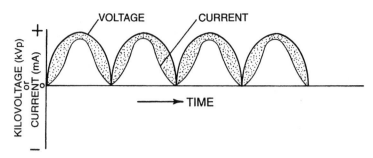

Figure 11.11. Voltage and current waveforms with four-diode rectification. This is full-wave or two-pulse rectification since there are two voltage (and current) peaks per full cycle. Note again the difference between the voltage and current waveforms.

(see Figure 11.5). The waveform of the full-wave rectified current appears in Figure 11.11.

There is one disadvantage in four-diode rectification. Using the entire AC wave results in the production of a high percentage of low energy x rays (low kVp). But these can be minimized either by (1) adequate aluminum filtration, or (2) three-phase or high-frequency x-ray generators.

Full-wave two-pulse rectification permits a large increase in the tube rating or heat-loading capacity (relative to half-wave or single-pulse rectification), which means that larger exposures can be used without damaging the tube. But, in any case, a tube *must* be used in accordance with the rating furnished by the manufacturer for that particular tube.

Rectifiers

At one time, specially designed vacuum diode tubes served as rectifiers. These have been replaced by *solid state diode rectifiers,* which employ a special class of materials known as *semiconductors.*

To understand its basic principles we must look into the conduction of an electric current in solids from a slightly more advanced point of view than in Chapter 6.

One of the ways in which we can classify matter depends on its *conductivity,* defined as the ease with which electrons move about within it. Accordingly, there are three classes of solids: (1) *conductors* (excellent conductivity—electrons can be made to drift readily from point to point), (2) *nonconductors* or *insulators* (very poor conductivity—high resistance to movement of electrons), and (3) *semiconductors* (intermediate conductivity).

What accounts for the difference in conductivity of these three classes of matter? Energy-state diagrams come to our aid. An *energy state* is the energy of a particular electron, corresponding to its energy level in the atom. In Figure 11.12 we see the energy states that an electron may have in a single atom. Since a piece of matter is an aggrega-

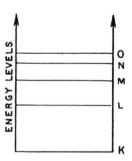

Figure 11.12. Possible energy states of an electron in a single atom. The higher energy states are in the direction of the arrows.

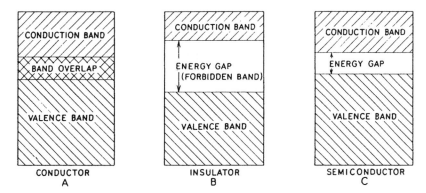

Figure 11.13. Energy level diagrams of three types of matter. In a conductor *(A)* electrons move freely from the valence band to the conduction band. In an insulator *(B)* there is a large energy gap between the valence and conduction bands preventing free movement of electrons. In a semiconductor *(C)* a small energy gap between the valence and conduction bands makes it possible to raise electrons to the conducting band, thereby controlling resistance.

tion of a fantastic number of atoms usually arranged in a crystal lattice, that is, a regular pattern, their energy levels merge into a series of *energy bands.*

There are three types of energy bands—*valence, forbidden,* and **conduction.** No electron can remain in the forbidden band; this band represents the energy needed to raise an electron from the valence band to the conduction band where the electron can move freely. The various bands are shown in Figure 11.13.

In a *metallic conductor,* shown in Figure 11.13A, the outermost or *valence* electrons are very loosely bound; in fact, the valence band actually overlaps the conduction band. Thus, an extremely weak electric field (applied emf) causes the electrons in the valence band to drift in a particular direction as an electric current. Because electrons can be made to flow in a metal with the application of so little energy, electric current in a conductor is not so easily controllable as in a semiconductor .

In an *insulator,* shown in Figure 11.13B, all electron-containing bands are completely filled. Since a very large energy gap exists between the valence band and the conduc-

tion band in insulators, an extremely large electric field would be needed to raise an electron to the conduction band, so that, for all practical purposes, an insulator does not conduct an electric current (we may say that it has virtually infinite resistance).

Now we come to the third class of solids—*semiconductors,* which are intermediate between conductors and insulators in their ability to conduct an electric current. In Figure 11.13C a small energy gap exists between the valence band and the conduction band, so only a small amount of energy is needed in semiconductors to raise the valence electrons to the conduction band. Note that this energy gap is much smaller in semiconductors than in insulators (compare Figure 11.13B with 11.13C). Because the valence and conduction bands do not overlap in semiconductors, current flow can be regulated; for example, the addition of certain impurities to a semiconductor modifies its conductivity. In fact, its conductivity is controllable by the kind and amount of impurity, which creates imperfections in the crystal lattice.

Of the available semiconductors, we shall limit our discussion to *silicon* (an element in

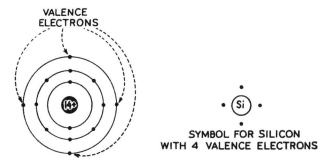

Figure 11.14. The silicon atom has four valence (outer shell) electrons as indicated by the broken lines.

ordinary sand–silicon dioxide) because it is now being used in the manufacture of solid state rectifiers for x-ray equipment.

Basic to an understanding of semiconductors is the rule that, in combining chemically, most atoms share their valence electrons in such a way that the outer shell of each contains eight electrons. This is known as the ***octet rule.***

The silicon atom has four valence electrons in its outermost shell, as shown in Figure 11.14. Silicon atoms aggregate into crystals by sharing these valence electrons with each other–socalled ***covalent bonding*** – as shown in Figure 11.15. Such bonds, in effect, saturate the outer or valence shells, thereby satisfying the octet rule and enhancing the stability of the crystal structure. Pure

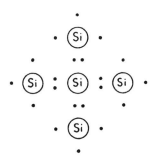

Figure 11.15. Diagram of the crystal lattice of silicon atoms which share valence electrons in pairs by means of *covalent bonds*. Each silicon atom is surrounded by four others (actually, in three dimensions). Note that the central Si atom, by electron sharing, now has an *octet*– eight valence electrons. In a real crystal there are tremendous numbers of Si atoms forming covalent bonds.

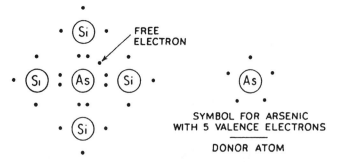

Figure 11.16. *N*-type silicon. Addition of a minute amount of arsenic with its five valence electrons introduces a free electron into the crystal. Under the influence of an applied emf, this free electron can be raised to the conduction band and made to drift through the crystal as part of an electric current.

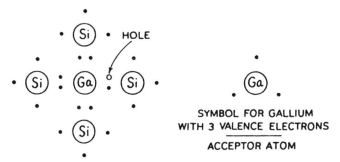

Figure 11.17. *P*-type silicon. Addition of a minute amount of gallium with its three valence electrons results in a deficit of one shared electron in the crystal lattice. An absent electron is equivalent to a positive charge, here called a "hole"; this can accept an electron.

silicon is an example of an *intrinsic* semiconductor.

If a small amount of *arsenic* is added to silicon, it contributes to the crystal lattice an *extra* electron, not needed for bonding silicon and arsenic atoms (see Figure 11.16). This is a loosely bound electron which can easily be raised to the conduction band by a small applied electric field. Arsenic is known as a *donor* atom because it furnishes electrons. Silicon "doped" with arsenic is called *n-type silicon* (*n* stands for negative). Its conductivity can be controlled by the amount of arsenic doping–the more arsenic, the more electrons available and the greater the conductivity.

On the other hand, if *gallium* with its three valence electrons is added to the silicon as shown in Figure 11.17, there is a shortage of one electron in the crystal structure called a *"hole."* Such a hole *behaves like a positive charge,* gallium-doped silicon being known as *p-type silicon* (*p* stands for positive). Since the hole can accept an electron, gallium is an *acceptor* atom.

How do these two types of semiconductors work? In Figure 11.18A, representing an n-type semiconductor, electrons drift in an applied electric field just as they do in a conductor. But in a p-type semiconductor, shown in Figure 11.18B, an applied electric field causes electrons to move toward holes, at the same time leaving new holes which are then, in turn, filled by other electrons.

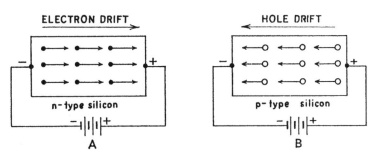

Figure 11.18. *A,* behavior on *n*-type silicon when an electric field (emf) is applied. Free electrons are raised to the conduction band and drift through the crystal just as in a conductor. *B,* in *p*-type silicon, electrons enter from the negative terminal of the battery and combine with holes which may be regarded as drifting toward the negative terminal.

Thus, the holes, in effect, drift in the opposite direction from that of the electrons as though the holes were positive charges.

Suppose, now, we were to combine n-type and p-type silicon at a *junction* as in Figure 11.19. Some electrons from the n-type silicon combine with holes of the p-type near the junction, leaving positive and negative ions, respectively, as shown in the figure. A potential difference exists

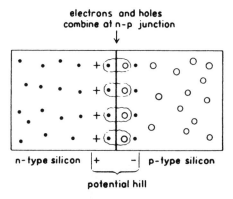

Figure 11.19. Behavior of an *n-p* junction. When *n*-type and *p*-type silicon are brought into contact, holes and electrons combine near the junction leaving, respectively, negatively and positively charged ions. A potential "hill" (potential difference) is built up at the junction.

between these oppositely charged ions, called the ***barrier voltage*** or ***potential hill.***

Let us see how an n-p junction rectifies– that is, ***conducts current in only one direction.*** Suppose an emf is applied across the junction in the direction shown in Figure 11.20A, with the negative terminal of the source at the n-end, and the positive terminal at the p-end. Electrons in the n-type silicon move toward the junction and, at the same time, holes in the p-type silicon also move toward the junction where the electrons and holes combine. As a result, the n-type silicon accepts electrons from the negative terminal of the battery, while the p-type silicon passes electrons to the positive terminal. Thus, the diode (ie, the n-p junction) continues to conduct current as long as the battery terminals are connected as, shown–a ***forward biased diode.***

Now if the battery terminals are reversed, as in Figure 11.20B, electrons leave the n-type silicon, and holes leave the p-type silicon. Since they are instantaneously depleted, no current flows across the diode, which is now ***reverse-biased.*** So if an alternating current is applied to an n-p junction it undergoes rectification, being conducted across the junction only when the negative half of the AC cycle is directed from *n* to *p*. The

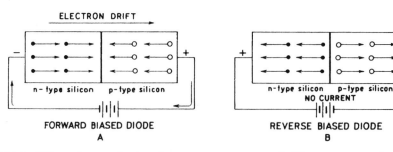

Figure 11.20. Effect of an applied emf on an *n-p* junction. *A.* Electrons enter the *n*-type silicon from the source, while electrons leave the *p*-type silicon and pass to the source, completing the circuit. *B.* With the battery terminals reversed, electrons leave the *n*-type silicon, and holes leave the *p*-type silicon; both are immediately depleted and no current flows. When *AC* is applied, current flows only during the forward-biased half of the cycle, so the *n-p* junction behaves as a rectifier.

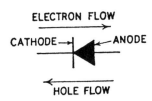

Figure 11.21. Symbol for a solid state diode rectifier.

symbol for a solid state diode rectifier is shown in Figure 11.21.

A practical silicon rectifier for x-ray equipment consists of a stack of individual diodes, each containing an n-p junction. These diodes, called **modules,** are connected in series, the complete assembly being enclosed in a sealed ceramic tube, which contains an insulating liquid. Each module can withstand a maximum of 1000 volts, so a 150-kVp generator requires 150 stacked modules. These rectifiers are of the **controlled avalanche type,** to minimize individual module failure. A silicon rectifier, usual-

Figure 11.22. Silicon rectifier for x-ray equipment, often called a "stick" rectifier. From about 8 to 10 inches in length.

ly called a "stick" rectifier in x-ray equipment, appears in Figure 11.22.

The advantages of the silicon rectifier over the valve tube may be listed as follows:

1. Compact size.
2. No filaments, hence no filament transformers needed for solid state rectifiers. This is especially important in three-phase equipment where twelve rectifiers are ordinarily used.
3. Low forward voltage drop–less than 300 volts at maximum current rating of 1000 mA.
4. Low reverse current–0.002 mA at

maximum voltage.
5. Long life due to rugged construction.

Because of these advantages, solid state rectifiers have replaced valve tubes in modern radiographic units.

Rectifier Failure

A failed or "open" rectifier diode causes the milliammeter, if present, to register one-half the selected value and radiographs to be consistently underexposed. With three-phase high-frequency units, the microprocessor will sense the excess- or *overcurrent* and, depending on the equipment, indicate *error* or *appropriate code number* on the console. Some x-ray machines automatically shut down under these conditions.

When a rectifier shorts out, a loud "bang" occurs in the transformer, requiring immediate attention of the service engineer before the machine is turned on again.

The simplest device for testing the competence of rectifiers with **single-phase equipment** is the **spinning top,** a flat metal top

Figure 11.23. Diagram of a lead spinning top used to determine the integrity of the rectifiers or exposure timer in a single-phase half- or full-wave rectified circuit.

which has a small hole punched near one edge (see Figure 11.23). It is placed on a film exposure holder containing an x-ray film, and made to spin during an x-ray exposure of one-tenth second. If all four diodes are operating, the circuit is fully rectified and there should be 120 pulses/s with 120 corre-

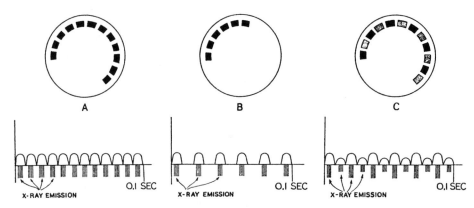

Figure 11.24. Spinning top test. In *A,* the radiograph of the spinning top shows twelve spots in 1/10 s, representing 120 kV peaks/s; this is full-wave single-phase rectification. In *B,* there are six spots in 1/10 s representing 60 kV peaks/s; this is half-wave single-phase rectification, which may occur when one diode rectifier is completely inoperative. In *C,* one rectifier diode is malfunctioning (ie, partly operative), so there is less voltage left for the x-ray tube; therefore, every other spot appears lighter. (Data kindly furnished by Willis R. Preston.)

sponding peaks of x-ray output on 60-cycle current. Therefore, in one-tenth second there will be twelve peaks and the image of the spinning top on the radiograph will show twelve spots, as in Figure 11.24A. If only six dark spots appear in the radiograph of the top, as in Figure 11.24B, there were only six pulses in one-tenth second or sixty pulses per second, indicating that the circuit is half-wave rectified, provided the exposure timer is known to be operating correctly.

The spinning top may be used to test the accuracy of the x-ray exposure timer if all four rectifier diodes are known to be func-

tioning properly, in single-phase equipment. However, with *12-pulse three-phase equipment,* the image produced by a spinning top is a dark band with intermittent indistinct darker pulses that do not permit accurate evaluation of the timer. An oscilloscope may be used with such equipment, as it provides an image of the actual pulses; in a properly operating 12-pulse circuit, there will be 12 pulses in 1/60 sec. For utmost accuracy (down to about 1 millisecond) special automated solid-state radiation detectors are available.

QUESTIONS AND PROBLEMS

1. Define rectification.
2. Why is it desirable to rectify the current for x-ray generation?
3. Name two main types of rectification.
4. How are x rays produced?
5. Show with the aid of a diagram:
   (a) single-diode rectification and
   (b) four-diode rectification.

6. Draw and label the current curves in a self-rectified circuit; a single-diode rectified circuit; a four-diode rectified circuit. Which are half-wave rectified? Full-wave rectified?
7. With the aid of a diagram, compare the rectified voltage and current curves.
8. Define inverse voltage; reverse bias;

useful voltage. What is their importance?

9. What is meant by single-pulse rectification? Two-pulse rectification? During which part of the voltage curve are x rays produced?

10. What property of diodes makes them suitable as rectifiers?

11. In what type of equipment might you find self-rectification used?

12. Why are diodes not conductive during the inverse half of the AC cycle?

13. Show by aid of a diagram a rectifier using four diodes. Draw the wave form of the rectified current. Why is it called full-wave or two-pulse?

14. What are semiconductors? How is their conductivity controlled?

15. Describe the principle of the n-p junction.

16. What intrinsic semiconductor is most often used in solid state rectifiers in medical radiography?

17. Discuss solid state rectifiers on the basis of principle and construction.

18. Describe the spinning top method and state its purpose.

19. What method is now preferred for the purpose mentioned in question 18?

Chapter 12

X RAYS: PRODUCTION AND PROPERTIES

Topics Covered in This Chapter

- Discovery of X rays
- Nature–Electromagnetic Waves and Photons (particles)
- Sources
  Natural
  Artificial–X-Ray-Tube
- Conditions Necessary for X-Ray Production
- Electron Interactions in Target Atoms
- Target Material
- Efficiency of X-Ray Production
  Dependence on Target's Atomic Number
- Physical Specifications of an X-Ray Beam
  Quantity Based on Ionization or Energy Release in Air
  Quality Based on Penetrating Ability
- or Photon Energy
- Polyenergetic Nature
- X-Ray Spectra
- X-Ray Attenuation in Matter–Half-Value Layer (HVL)
- Interactions of X Rays With Matter
  Classical
  Photoelectric
  Compton
  Pair Production
- Relative Importance of Interactions With Matter
- Beam Filtration
- Detection of X Rays
- Relation of Exposure and Absorbed Dose
- Questions and Problems

How X Rays Were Discovered

FOR A NUMBER of years before the discovery of x rays physicists had been observing high voltage discharges in vacuum tubes. In 1895 Wilhelm Conrad Roentgen, a German physicist, was studying these phenomena in a Crookes tube operated at high voltage in a darkened room. Suddenly, he noticed the *fluorescence* (glowing) of a barium platinocyanide screen lying several feet from the end of the tube. He soon realized that the fluorescence was caused by a hitherto unknown type of invisible radiation. Moreover, he found that this radiation could pass completely through solid materials such as paper, cardboard, and wood, since they did not prevent fluorescence when placed between the tube and the barium platinocyanide screen. However, the rays could be stopped by denser materials such as lead.

Thus, by sheer accident, Roentgen had discovered a new type of radiation, which exited the Crookes tube and passed through certain solids opaque to ordinary light. He quickly realized the potentiality of the new radiation in medicine when he placed his wife's hand between the tube and a photographic plate. Imagine his excitement on seeing the bones with rings on this first *pub-lished* radiograph (see Frontispiece).*

Roentgen gave the name *x rays* to his newly found, invisible, penetrating radiation because the letter *x* represents the unknown in mathematics, although within the next few months he had ascertained most of the properties of x rays.

The discovery of x rays changed the course of medical history, to say nothing of its impact on science and industry. Roentgen's name is often directly linked with x rays in that they are also known as *roentgen rays. Roentgenology* is that branch of medicine dealing with the use of x or roentgen rays in diagnosis and treatment. The term *radiology* includes not only x rays but also natural and artificial radionuclides, computerized tomography, magnetic resonance imaging, and ultrasonography. *Radiography* deals with the art and science of recording x- ray images on film, paper, magnetic tape, and magnetic disc. *Fluoroscopy* includes the observation of x-ray images on a screen coated with fluorescent material, either directly or with the aid of an intensifying system.

Nature of X Rays

An x-ray beam consists of a group of rays that are related to white light, ultraviolet, infrared, and other similar types of radiant energy. They are all classified as *electromagnetic radiation*–wavelike fluctuations of electric and magnetic fields set up in space by oscillating (vibrating) electrons. A mechanical analogy would be, for example, the waves produced on the surface of water by a dropped stone.

The electric and magnetic fields fluctuate perpendicular to their direction as well as to

Figure 12.1. The electrical and magnetic components of an electromagnetic wave oscillate (vibrate) in mutually perpendicular planes. From J. Selman: *The Basic Physics of Radiation Therapy*, 3rd Ed. Springfield, Thomas, 1990.

each other, as shown diagrammatically in Figure 12.1. As they are identical in form, we usually depict only one of them as in Figure 12.2.

All electromagnetic waves such as radio, heat, light, x rays and gamma rays have the same general form and travel with the same

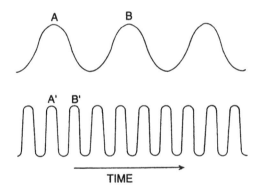

Figure 12.2. Electromagnetic waves. The upper and lower wave trains differ in wavelength, that is, the distance between two crests such as *A* to *B* and *A'* to *B'*. The lower wave thus has a shorter wavelength and therefore more crests or cycles in a given period of time, than does the upper wave, because their speeds are identical. Thus, the wave with the shorter wavelength has greater frequency or number of cycles/s (hertz, Hz).

---

* Arthur W. Goodspeed in Philadelphia, while demonstrating a Crookes tube on 2/22/1890, made the first radiograph but could not explain it until Roentgen discovered x rays on 11/8/1895. Roentgen deservedly won the Nobel prize in physics in 1901.

constant speed as light–3×10^8 meters, 3×10^{10} cm, or 186,000 miles per s in a vacuum or air. However, they differ in **wavelength**–the distance between two successive crests in the wave, such as A to B in Figure 12.2. The number of crests or cycles per second is the **frequency** of the wave, the unit of frequency being the **hertz** defined as one cycle/s. You can see, by comparing the upper and lower waves in Figure 12.2, that if the wavelength is decreased, the frequency must increase correspondingly. This can also be shown in another way. The speed with which the wave travels is equal to the frequency multiplied by the wavelength:

$$c = f\lambda \qquad (1)$$

where $c =$ speed of x rays or light in vacuum or air $(3 \times 10^8 \text{ m/s})$
$f =$ frequency in hertz or Hz (cycles/s)
λ (Greek *lambda*) = wavelength in m (meter)

Since c, the speed of all electromagnetic waves, is constant in a given material, an increase in frequency must always be accompanied by a corresponding decrease in wavelength and, conversely, a decrease in frequency by an increase in wavelength. **Frequency is inversely proportional to wavelength.**

The wavelengths of x rays are extremely short; for instance, in **ordinary radiography,** their useful range extends from about 0.1 to 0.5Å. Recall that $1Å = 10^{-10}$ or one ten-billionth m = 0.1 nm (nanometer). The ultrashort wavelengths are associated with enormous frequencies–from 3×10^{19} to 6×10^{18} hertz (Hz).

The various kinds of electromagnetic radiation, ranging from radio waves with wavelengths measured in several thousands of meters, down to high energy x and gamma rays with wavelengths measured in nanometers $(10^{-9}$ meter), are shown in Figure 12.3.

As will be explained later in more detail (see pp. 118-119), each electromagnetic wave

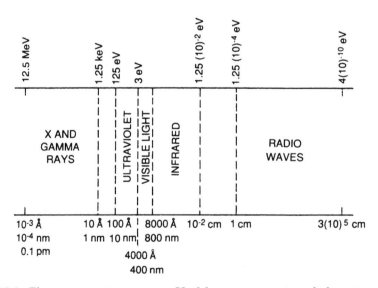

Figure 12.3. Electromagnetic spectrum. Useful energy range in radiology is as follows:

Radiography	40 to 120 kVp
Cobalt 60 gamma rays	1.25 MeV
Linear accelerator	4 to 25 MV

also behaves as a bit of energy that depends on the frequency of the wave. Such an energy bit is called a *photon* or *quantum.* It is preferable to designate x and gamma rays according to their photon energy rather than their wavelength.

Source of X Rays in Radiology

The X-ray Tube

Whenever a stream of fast moving electrons suddenly undergoes deceleration (ie, reduction of speed), x rays are produced. While these conditions may prevail in nature, they can also be brought about artificially by means of an *x-ray tube.* Gamma rays, which are identical to x rays, are emitted by the *nuclei* of certain radionuclides.

At this point we shall outline the main features of a typical x-ray tube, reserving a more detailed discussion of the subject for a later section.

Historically, the early x-ray tubes were cold cathode gas tubes. Such a tube consisted of a glass bulb in which a partial vacuum had been produced, leaving a small amount of gas. In it were sealed two electrodes, one negative (cathode) and the other positive (anode). The cathode terminal was *not* heated. Application of a high voltage across the terminals caused ionization of the gas in the tube with release of a stream of electrons which produced x rays upon striking the positive terminal (anode). The gas tube is now obsolete not only because of its inefficiency but also because tube mA could not be changed independently of kV, making it difficult to control x-ray quantity and quality.

In 1913 W. D. Coolidge, at the General Electric Company Laboratories, invented a new type of x-ray tube based on a radically different principle (Edison effect); this was the *hot cathode diode tube,* which revolutionized radiographic technic because it made possible the independent control of mA and kV. The principles of construction and operation are surprisingly simple and must be clearly understood by all those concerned with the use of x rays.

An x-ray tube has the following components (see Figure 12.4):

1. A *glass envelope* or *tube* from which the air has been evacuated as completely as possible. Air must be removed not only from he interior of the tube, but from the glass and metal parts as well, by prolonged baking before the tube is sealed. This process is called *degassing.* A high vacuum is necessary for two reasons:

> a. To prevent collision of the high speed electrons with gas molecules, which would cause significant slowing of the electrons (see below).
> b. To prevent oxidation and burning-out of the filament.

2. A *hot filament* supplied with a separate low-voltage heating current. The filament serves as the *cathode,* or negative electrode, when the high voltage is correctly applied.

3. A *target* which serves as the *anode,* or positive electrode, when high voltage is correctly applied.

4. A *high voltage* applied across the electrodes, charging the filament negatively and the target positively.

Notice that there are *two circuits in the x-ray tube:* one, a low voltage heating circuit through the cathode itself; and the other, a high-voltage circuit between the cathode and anode to drive the electrons (see Figure 12.4). At the cathode end, the high voltage conductor is connected to one side of the filament circuit.

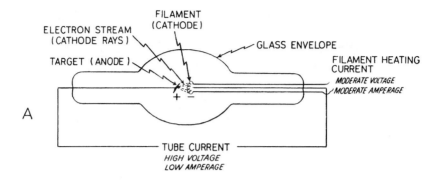

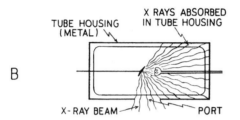

Figure 12.4. Coolidge x-ray tube. *A* shows the two circuits and the general features. The inset at *B* indicates that x rays are emitted from the target in all directions, but only those approaching the port (window) pass out as the *useful beam,* the remaining x rays being absorbed in the protective housing.

DETAILS OF X-RAY PRODUCTION

Whenever (1) *fast-moving electrons undergo rapid deceleration* or (2) *electrons drop from an outer shell to a hole in an inner atomic shell, x rays are produced.* The x-ray tube is a device for obtaining free electrons, then speeding them up, and finally stopping them. In addition to these basic processes, the electrons must also be concentrated on a small area of the anode known as the *focal spot.* These four conditions required for the production of x rays in a hot cathode tube will now be described in some detail because of their extreme importance.

1. **Separation of Electrons.** In common with all other atoms, the tungsten atoms in the filament have orbital electrons circulating around a central nucleus. How can these electrons be liberated from their atoms? The *filament current* supplying the filament causes it to become glowing hot or *incandescent,* with resulting separation of some of its outer orbital electrons. Such electrons, escaping from the filament, form a cloud or *space charge* nearby. The electrons liberated in this manner are called *thermions,* and the process of their liberation through the heating of a conductor by an electric current is called *thermionic emission.*

2. **Production of High Speed Electrons.** Now a high potential difference (kV) applied across the tube gives the filament a very high *negative charge* (cathode) and the target an equally high *positive charge* (anode). The resulting strong electric field causes the space charge electrons to rush at an *extremely high speed through the tube*

from cathode to anode. This electron stream is expressed in milliamperes (mA), while the total charge transferred in the process is in milliampere-seconds (mAs). The speed of the electrons at 80 kVp is about 80 percent of the speed of light!

3. **Focusing of Electrons.** The electron stream in the tube is confined to a very narrow beam and is concentrated on a small spot on the anode face–*the focal spot*–by a negatively charged metal focusing cup in which is suspended the filament; the narrower the electron beam the smaller the focal spot and the sharper the x-ray images.

4. **Stopping of High Speed Electrons in Target.** When the fast electrons enter the target of the x-ray tube, their *kinetic energy* changes to heat, light, and x rays. X-ray tube efficiency is such that *at 80 kVp only about 0.6 percent of this energy is converted to x rays while the remaining 99.4 percent appears as heat.* (See page 115 for a discussion of efficiency of x-ray production.) In fact, only about one part in a thousand of the kinetic energy of the electrons eventually results in radiologically useful x rays! As shown in Figure 12.4B, x rays are emitted in *all* directions, but only those leaving the window of the x-ray tube comprise the *useful beam.*

You must avoid confusing the electrons flowing in the tube, with the x rays emerging from the tube. Imagine a boy throwing stones at the side of a barn. The stones are analogous to the electron beam in the x-ray tube, whereas the emerging sound waves are analogous to the x rays that come from the target when it is struck by electrons.

Electron Interactions with Target Atoms–X-ray Production

When the fast electron stream enters the tube target, the electrons interact with target atoms producing x rays by the following two processes:

1. **Brems Radiation.** Upon approaching the strongly positive *nuclear field* of a target atom, the negatively charged high-speed electron is deviated from its initial path because of the attraction between these opposite charges (see Figure 12.5). As a result, the electron slows down or *decelerates,* thereby losing some of its kinetic energy; the *lost kinetic energy is radiated as an x ray of equivalent energy.* The term *bremsstrahlung* or *braking radiation* has been applied to this process. Brems radiation is *polyenergetic,* that is, nonuniform in energy and wavelength because the amount of braking or deceleration varies among electrons according to their speed and how closely they approach the nucleus. With each different deceleration, a corresponding

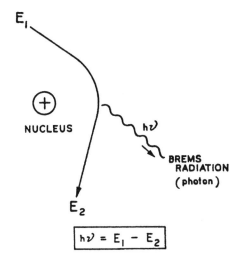

Figure 12.5. Production of bremsstrahlung (brems radiation). An electron with energy E_1 on approaching an atomic nucleus, experiences an attractive electrostatic force which causes it to change direction along a curved path. In changing direction the electron radiates energy h which must come from the kinetic energy of the electron. Therefore the electron moves away from the nucleus with energy E_2. The radiated brems photon energy $hv = E_1 - E_2$.

amount of kinetic energy is converted to x rays of equivalent energy; just as the deceleration varies, so does the energy of the x rays. An electron that happens to approach the nucleus head-on is completely stopped by the nuclear electrostatic field. In this special case of the brems effect, all of the kinetic energy of the electron is converted to an x ray of equivalent energy. The deceleration of electrons depends also on the ***atomic number of the target.*** Thus, targets of higher atomic number are more efficient producers of brems radiation (eg, rhenium with atomic number 75 as opposed to tungsten with atomic number 74). It should be pointed out that electrons also interact with, and are repelled by, the electrons swarming around the target nuclei within the tungsten atoms. In this process heat is transferred to the target atoms; such heat production far exceeds x-ray production at ordinary values of tube potential (kVp).

2. **Characteristic Radiation.** This is the other process of x-ray production in an x-ray tube. An electron with a sufficient minimum kinetic energy may interact with an ***inner orbital electron*** (for example, in *K* or *L* shell) of a target atom, ejecting it from its orbit (see Figure 12.6). To free this electron, work has to be done to overcome the attractive force of the nucleus. The atom is now unstable,

being ***ionized*** (electron missing from atom) and in an ***excited state*** (electron vacancy in a shell). Immediately, the space or "hole" vacated by the electron is filled by electron transition from one of the outer shells. Since, in the first place, energy was put into the atom to free the electron, a like amount of energy must be given off when an electron from a higher energy level enters to fill the hole in the shell (satisfying the Law of Conservation of Energy). This energy is emitted as a ***characteristic x ray*** because its energy is unique to the target element and the involved shells. Moreover there is a cascade of characteristic rays because, for example, the replacing electron leaves a hole which must, in turn, be filled by an electron from a still higher energy level or shell, and so on.

We may now summarize the nature of the ***primary radiation,*** that is, the x rays emerging from the target:

a. ***General Radiation***—"white," polyenergetic radiation with a continuous range of energies (and wavelengths) due to deceleration of electrons by strongly positive electric fields of nuclei in target atoms. This is ***bremsstrahlung,****constituting about 90 percent of emitted x rays when 80 to 100 kV is applied

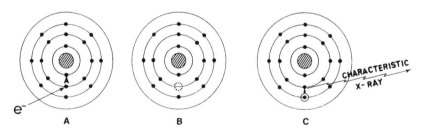

Figure 12.6. Production of characteristic radiation. *A.* The incoming *electron* "collides" with an inner shell electron. *B.* The atom is in an "excited state," an electron now being displaced from its shell. *C.* An electron jumps from an outer shell to replace the ejected electron; this is accompanied by characteristic radiation, as the atom returns to its normal state.

* Bremsstrahlung (German) means "braking radiation." It is sometimes called "brems radiation."

to the tube.

b. ***Characteristic Radiation***–consists of limited, discrete energies (and wavelengths), constituting about 10 percent of emitted x rays when the tube potential is in the range of 80 to 100 kV. No characteristic radiation is produced at tube potentials less than 69 kV with a tungsten target.

Target Material

The target metal must be selected with two aims in mind. First, it must have a very ***high melting point*** to withstand the extremely high temperature to which it is subjected. Second, it must have a ***high atomic number*** because (1) the resulting ***characteristic*** radiation is of high energy and therefore more penetrating and (2) there is increased production of brems radiation. Tungsten, a metal with a melting point of 3370°C and an atomic number of 74, generally satisfies the above requirements. Rhenium (atomic number 75) and tungsten, coated on a molybdenum-tungsten disc, form the target in rotating anode radiographic tubes for more efficient x-ray production. To facilitate the dissipation of heat from the anode in ***stationary*** anode tubes (eg, dental x-ray tubes), a tungsten button is imbedded in a block of copper, which is a better conductor of heat than tungsten. The electron beam in the tube is focused on a small spot on this tungsten target. A molybdenum target is used in mammographic units to produce low-energy x rays (see pages 286-287).

Efficiency of X-ray Production

As already indicated, only a small fraction of the energy of the cathode-ray electrons undergoes conversion to x rays in the target. Most of the electrons' energy is degraded to heat as the result of collisions with outer shell electrons of the tungsten atoms. X rays are produced by the previously described process of brems radiation and, if the applied potential (kV) is high enough, by the additional process of characteristic radiation.

The efficiency of x-ray production may be defined as the percentage of the electrons' kinetic energy that is converted to x rays. The efficiency is directly proportional to the atomic number of the target and the applied potential, as expressed approximately by the following equation:

$$\% \text{ } \textbf{efficiency} = \textbf{K} \times \textbf{Z} \times \textbf{kVp} \qquad (2)$$

where K is a constant $= 1 \times 10^{-4}$, Z is the atomic number of the target, and kVp is the peak kilovoltage. For example, with a tungsten target ($Z = 74$) and 80 kVp.

$$\text{efficiency} = 1 \times 10^{-4} \times 74 \times 80$$
$$= 0.6\%$$

This means that of the total kinetic energy of the electron stream, only about 0.6 percent appears as x rays and the remaining 99.4 percent as heat in the anode when the applied potential is 80 kVp.

Properties of X Rays

We shall now list some of the more important properties of x rays:

1. ***Highly penetrating, invisible rays,*** which belong to the general category of electromagnetic radiation and behave both as ***waves and particles.***
2. ***Electrically neutral,*** so they cannot be deflected by electric or magnetic fields.
3. ***Polyenergetic,*** having a wide span of energies and wavelengths; the useful energy range in conventional radiography is about 25 to 120 kVp (maximum photon energy 25 to 120 keV).
4. Liberate ***minute*** amounts of heat on passing through matter.
5. Travel ordinarily in ***straight lines.***

6. Travel only at the ***same speed as light***, 3×10^8 m/s or 186,000 m/s in a vacuum.

7. Ionize gases ***indirectly*** because of their ability to remove orbital electrons from atoms.

8. Cause ***fluorescence*** of certain crystals, making possible their use in fluoroscopy, and in radiographic intensifying screens.

9. ***Cannot be focused*** by a lens.

10. ***Affect photographic film,*** producing a latent image which can be developed chemically.

11. Produce ***chemical and biologic changes*** by ionization and excitation.

12. Produce ***secondary and scattered radiation.***

SPECIFICATIONS OF THE PHYSICAL CHARACTERISTICS OF AN X-RAY BEAM

One may describe a given x-ray beam on the basis of two properties–***intensity*** and ***quality.*** Strictly speaking, the intensity of the radiation at a given point in the beam refers to the quantity of radiant energy flowing per second through a unit area of the surface perpendicular to the direction of the beam at the designated point. However, because it is simple and adequate for diagnostic radiology, the term ***output*** has been adopted in this book to indicate x-ray ***intensity,*** simply stated in R/min, or mR/min.

The quality of an x-ray beam refers to its penetrating ability. The concepts of exposure and quality will be taken up next.

X-ray Exposure (Quantity)

Under certain controllable conditions, the degree of ionization (see page 131) produced by radiation in matter is proportional to the quantity of radiation of a particular type. Therefore, ionization would be expected to serve as an appropriate basis for measuring x- and gamma-ray quantity in terms of exposure. Indeed, in 1928 the Second International Congress of Radiology established the ***roentgen,*** now symbolized by R, as the international unit of exposure based on ***ionization in air.***

After development of a special free-air ionization chamber by Lauristen S. Taylor in 1932, the roentgen was redefined in 1938, based on a specified mass in air (equivalent to 1 cc under standard conditions).

A revised definition of the roentgen as the unit of exposure was adopted by the International Commission on Radiation Units and Measurements (ICRU) in 1962, as expressed in the following equation:

$$X = \Delta Q/\Delta m \qquad (3)$$

where X is the exposure in R, and ΔQ is the sum of all the electric charges on all the ions of ***one sign*** produced in air, when all the electrons released by photons (x and gamma rays) in a mass of air, Δm, are completely stopped in air.

Note that the most recent definition of exposure (equation [3]) implies two steps in the process: (1) photons first release electrons by interaction with atoms in the air, and (2) these primary electrons go on to produce ions whose total charge of one sign (positive or negative) is a measure of exposure in air. As specified in the definition, all the electrons released in (2) must give up all their kinetic energy in the formation of ions in air. Thus, the conditions laid down in the definition of 1938 were stated more precisely in 1962, so the value of the R is really unchanged in the newer definition. On this basis, the roentgen may be expressed as an amount of charge released per unit mass of air, specifically,

$$1 \text{ R} = 2.58 \times 10^{-4} \text{ coulomb/kg air}$$

In the modern International System of Units–SI–the **exposure unit** replaces the roentgen, and is an exposure of one coulomb per kilogram; this unit is still not widely used.

The radiation exposure per unit time, for example R/min, at a given location is the **output.**

$$\underset{(R/min)}{\text{output}} = \frac{\text{exposure in R}}{\text{time in min}} \quad (4)$$

The **total exposure** in R is obtained by multiplying the output by the exposure time:

$$\underset{(R)}{\text{exposure}} = \underset{(R/min)}{\text{output}} \times \underset{(min)}{\text{time}}$$

(Sprawls defines x-ray *intensity* as *exposure rate per unit area* receiving the radiation, in $R/min/cm^2$. This is closer to the strict definition of radiation intensity–energy flow per unit time per unit area, which is complicated by the polyenergetic nature of an x-ray beam. Since this information is superfluous in diagnostic radiology, the simpler but adequate concept of *output* will suffice in this book as a measure of x-ray intensity.)

Exposure is not measured routinely in the practice of radiology, except for protection purposes. These appear in Chapter 28.

The output of an x-ray beam may be altered by varying four factors: (1) tube current, (2) tube potential, (3) distance, (4) filtration, and (5) target metal (discussed on pages 111, 115).

1. **Tube Current.** An increase in the filament current liberates more electrons **per second** at the filament. As a result, the tube current (mA) also increases and more electrons strike the target per second, increasing the amount of x radiation emitted per second. Thus, **the output is directly proportional to the mA,** if the other factors remain constant. Keep in mind that a change in mA does **not** affect the quality of the radiation.

2. **Tube Potential.** A rise in **kilovoltage**

(kV) increases output (R/min) because the electrons in the tube are speeded up and produce more x-ray photons/s at the target. The resulting x-ray beam also contains photons of higher energy than those produced at lower kV. An increase in tube potential increases both the output and penetrating ability of an x-ray beam, but that this relationship is not directly proportional. In fact, output is approximately proportional to kVp^2. As will be shown later (see page 163, the exposure rate is higher with three-phase x-ray generators than with single-phase for the same applied kV, unless mAs is adjusted downward.

3. **Distance.** X-ray exposure depends strongly on the distance from the target to the ion chamber because the x rays proceed in a spreading, cone-shaped beam. As the distance from the target increases, the number of x-ray photons in each square centimeter of the beam cross-section decreases, so the exposure decreases. Since an ion chamber has a definite size, it measures the radiation output in a constant small volume of the beam. **Output is governed by the inverse square law of radiation,** to be explained in greater detail in Chapter 18.

4. **Filtration.** The thicker the filter and the higher its atomic number, the greater will be the reduction in the output beyond the filter. At the same time, **the beam is hardened due to the greater removal of soft (low energy) than of hard (high energy) rays by the filter,** except at the absorption edge (see page 132). Preferential removal of lower energy, poorly penetrating photons by a suitable filter serves to reduce patient exposure (see pages 411-412).

X-ray Quality

Definition. The quality of an x-ray beam refers to its ability to penetrate matter. In radiology, we may define penetrating ability as the fraction of a particular x-ray beam that

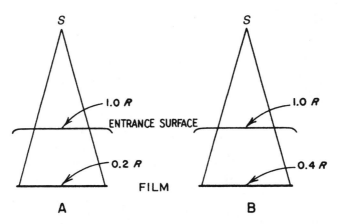

Figure 12.7. Concept of penetrating ability of an x-ray beam in radiography. *A.* For an assumed exposure of 1.0 R at the entrance surface of the patient, the exposure at the film might be 0.2 R because of beam attenuation (absorption, scattering, distance) in the patient; this represents a transmission or *penetrating ability* of 0.2/1 × 100 = 20%. *B.* For the same entrance exposure of 1.0 R, the exposure at the film might be 0.4 R, representing a transmission or penetrating ability of 40%. Thus, beam *B* is more penetrating than beam *A*.

can pass through a given part of the body and onto an x-ray film; or that can enter the body and deposit energy in a tumor-containing volume at a particular depth.

However, we need a more precise concept of ***penetrating ability*** than that just stated. As shown in Figure 12.7A, for a diagnostic x-ray beam, suppose the measured exposure at the point where the beam enters the body were 1 R and the point where the beam leaves the body (at the film) were 0.2 R, then the transmission or penetrating ability would be 0.2/1 × 100 = 20 percent. With a more penetrating beam giving an exposure of 1 R at the entrance point as in 12.7B, the exposure at the film might be 0.4 R, giving a transmission or penetrating ability of 0.4/1 × 100 = 40 percent. Thus, the beam in 12.7B is more penetrating than that in 12.7A.

X-ray Energy and Quality. To understand the relationship between x-ray energy and quality (ie, penetrating ability) we must first look into the meaning of x-ray energy. An individual x ray is a bit of energy called a ***photon*** or ***quantum,*** represented by the equation formulated by the German physi-

cist Max Planck:

$$E = hf \qquad (5)$$

where E is photon energy in electron volts (eV), h Planck's constant (4.15×10^{-15} eV-s), and f the photon frequency in hertz (Hz, cycles/s). This equation tells us that the energy of a given photon is directly proportional to the frequency of the associated wave. Thus, x rays of high frequency have higher energy than x rays of low frequency. But since the frequency times the wavelength of an x ray equals a constant (speed of light), as shown in equation (1) on page 110, a wave of high frequency must have a short wavelength; and, conversely, a wave of low frequency must have a long wavelength.

Example. An x-ray photon has a frequency of 10^{19} Hz. Find its energy in eV.

Solution. Substituting in equation (5),

$$E = 4.15(10^{-15}) \text{ eV-s} \times 10^{19}/\text{s}$$
$$= 4.15(10^{4}) \text{ eV, or } 41.5 \text{ keV}$$

[since second (s) in the numerator cancels s in the denominator]

Table 12.1
INTERRELATIONSHIP OF FACTORS IN RADIATION QUALITY
(SPEED OF RADIATION IS CONSTANT)
$c = frequency \times wavelength$

Frequency	Wavelength	Energy	Quality
high	short	high	good penetrating ability
low	long	low	poor penetrating ability

In summary, the quality of x rays may be defined as their penetrating ability, which is determined by their energy. Hence, we may summarize the statements in the preceding paragraph as follows: a highly penetrating x-ray beam consists of photons which, on the average, have high energy, high frequency, and short wavelength. Conversely, a poorly penetrating x-ray beam consists of photons which, on the average, have low energy, low frequency, and long wavelength. This relationship is shown in Table 12.1.

What influences the *energy* of x rays? We learned earlier that x rays consist of brems and characteristic radiation resulting from the interactions of high speed electrons with target atoms. The larger the kinetic energy of the electrons in the x-ray tube, the higher will be the energy of the resulting x radiation. Since the kinetic energy of the electrons depends on their velocity which in turn depends on the potential applied to the tube, it follows that an increase in kV will ultimately produce x rays of higher energy. This discussion may be summarized as follows: the higher the kV at any instant, the greater the velocity of the electrons, the greater their kinetic energy, the greater the energy of the x rays produced, and the greater the penetrating ability of the x-ray beam. We may therefore conclude that the *penetrating ability of x rays is increased by increasing the kilovoltage, and decreased by decreasing the kilovoltage.* However, this relationship is *not* a simple direct proportion.

How is the energy of an x-ray beam stated? Since the energy of x rays depends on the speed of the electrons in the tube, which depends on the applied potential, we customarily use the *peak* (maximum) voltage (kVp) for the energy of the beam. But in this book we shall simply use the term kV, except where needed for emphasis. However, we must bear in mind that the beam is actually *polyenergetic*, consisting of photons whose energies range from a minimum to the peak value, in this case 100 kV (often stated as kVp).

On the other hand, since the energy of *monoenergetic* radiation such as characteristic x rays and gamma rays is uniform for any given beam, it may be expressed in units of electron volts. One *electron volt* is the energy acquired by an electron when it is accelerated through a potential difference of 1 volt. Thus, the energy of monoenergetic radiation or of a particular photon is stated in electron volts (eV) or in some multiple thereof, such as kiloelectron volts (keV), million electron volts (MeV), or billion electron volts (GeV)*; for example, the gamma rays emitted by technetium 99m all have the same energy–140 thousand electron volts, usually stated as 140 keV.

Insofar as equivalence is concerned, an x-ray beam having a particular kVp has a quality resembling a monoenergetic x-ray beam of about one-third to one-half the peak energy. Thus, a polyenergetic 120-kV x-ray beam is roughly equivalent to 50-keV monoenergetic x rays.

---

* GeV = gigaelectron volts, where giga means billion, or 10^9.

Causes of the Polyenergetic Nature of an X-ray Beam. As we have already noted, an ordinary x-ray beam is ***polyenergetic,*** consisting of a tremendous number of photons that vary in energy and penetrating ability. Since photon energy is directly proportional to frequency and inversely proportional to wavelength, we may say that an ordinary x-ray beam is polyenergetic, consisting of ***innumerable rays of different energies and wavelengths.*** Analogy with white light may help clarify this. White light comprises a mixture of various colors. When passed through a glass prism, white light separates into its components which are identical to the colors of the rainbow, or the ***spectrum.*** These colors differ in frequency and wavelength. For instance, the wavelength of "red" is about 7000 Å (700 nm*), whereas the wavelength of "violet" is about 4000 Å (400 nm). A light beam made up of a single pure color is called ***monochromatic.*** Similarly, an x-ray beam consists of different x-ray "colors"–energies or wavelengths–although they are invisible. X rays can be diffracted by certain crystals and the photons separated according to their energy or wavelength.

We shall now explain the four major reasons for the polyenergetic nature of an x-ray beam:

1. ***Fluctuating Kilovoltage.*** The applied kVp varies according to the wave form of the pulsating current (see Figure 0.0). Since changes in kV are manifested by changes in the energy of the resulting x rays, there must necessarily be a range of photon energies corresponding to the fluctuation of kV.

2. ***Processes in X-ray Production.*** The x-ray beam leaving the target consists of ***general radiation*** (brems radiation resulting from deceleration of electrons by target nuclear fields), and ***characteristic radiation*** (arising from electron transfers within the atoms of the target). The general and characteristic x rays constituting the primary beam, therefore, have a variety of energies.

3. ***Multiple Electron Interactions with Target Atoms.*** Electrons traveling from cathode to anode in the x-ray tube undergo varying numbers of encounters with target atoms before being completely stopped. In these encounters x rays are produced.

4. ***Off-focus Radiation.*** Some electrons may strike anode metal other than the focus (target), causing the emission of x rays. This ***off-focus radiation*** has been reduced by special design of modern tubes. Further decrease in off-focus radiation that happens to enter the x-ray beam can be achieved by special beam-limiting devices (collimators), which fit close to the tube aperture, thereby improving recorded detail (image sharpness).

X-ray Spectra. A particular x-ray beam can be precisely characterized by sorting out its photons according to their energy. This concept may be clarified by an analogy. Suppose we wish to classify a population of high school senior boys (or girls) based on their height. We would first measure all the boys (or girls) in the designated population and group them according to height. These data would then be set down as in Table 12.2 and plotted graphically as in Figure 12.8, the number of boys or girls in each group being the dependent variable, and the height the independent variable.

Similarly, we can sort out the rays or photons in a given x-ray beam by means of a special instrument known as an x-ray spectrometer, determining the ***spectrum,*** or the range of intensities or relative numbers, of x rays of various energies. Plotting such data generates a ***spectral distribution curve.*** Figure 12.9 gives these curves for x-ray beams produced in a tungsten target by three different applied peak tube potentials–30, 40, and 80 kV, respectively.

You can see that in each case the curve

---

* 1 nm = 10^{-9} m, or 1 billionth m

Table 12.2
HEIGHT DISTRIBUTION OF A CLASS OF
SENIOR HIGH SCHOOL BOYS

Height	Number of Boys
in.	
63	1
64	2
65	4
66	7
67	13
68	14
69	15
70	15
71	14
72	10
73	7
74	1

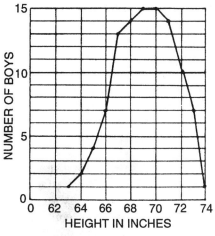

Figure 12.8. Distribution curve of a class of high school senior boys according to height. Note the peak incidence at 68 to 70 inches. A similar curve would be obtained with girls but the peak in the curve would be shifted to the left, at an earlier age.

crosses the horizontal axis at a point corresponding to the peak or maximum applied kV. Thus, ***the energy of the most energetic***

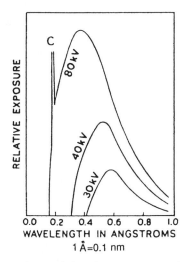

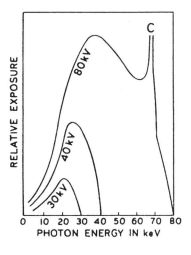

Figure 12.9. Spectral distribution curves of x rays generated by peak tube potentials of 30, 40, and 80 kV, respectively. Those on the left show relative exposure rates as a function of wavelength, and those on the right as a function of photon energy. The latter is preferred. *Characteristic radiation* appears as a peak at *C* (actually, a group of closely spaced peaks), requiring an applied potential of at least 69 kV. The remaining curves (indicated by 30 kV, 40 kV, and 50 kV) represent *general radiation*. Note that in each case the maximum photon energy in the beam is the same as the applied peak kV (figure on the right).

and most penetrating photons in a particular x-ray beam depends solely on the applied peak kV.

A comparison between these curves shows that an increase in kVp causes an increase not only in the output of higher energy photons, but also in output of photons of all other energies in the beam, as indicated by the areas under the respective curves.

If at least 69 kVp is applied across the tube, the curve is modified by the appearance of a sharp **peak** at 69 kV representing the **characteristic radiation** emitted by the tungsten target. This is shown in the 80-kVp curve.

Photon energy is related to wavelength by the following equation.

$$\textbf{\textit{photon energy}} = \frac{\textbf{\textit{1.24}}}{\lambda} \text{ keV} \qquad (5)$$

where λ is the wavelength in nanometers (nm). Thus, the energy of a photon hf whose associated wavelength is 0.02 nm, is obtained from equation (5):

$$hf = \frac{1.24}{0.02} = 62 \text{ keV}$$

The minimum wavelength λ_{min} of the radiation in an x-ray beam can be derived from equation (5):

$$\lambda_{min} = \frac{1.24}{hf_{max}} \text{ nm} \qquad (6)$$

So minimum wavelength is inversely related to maximum photon energy.

In Figure 12.10 we see that if the kV is kept constant, an increase in mA simply increases x-ray output. But there is no shift in the curves—the maximum photon energy remains the same because it depends only on the applied peak kV. This is also evident from equation (5).

Specification of X-ray Quality– Half-value Layer and Kilovoltage. Because of the polyenergetic nature of the

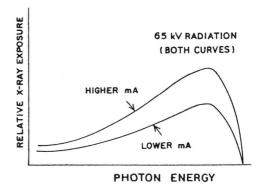

Figure 12.10.   Effect of tube mA on relative exposure rate.   With the kV held constant, a higher mA produces a larger exposure rate for all photon energies in the beam.   However, the mA has no effect on the maximum photon energy, which is the same for both curves.

photons in an x-ray beam, the spectral distribution curves described in the preceding section depict most completely the distribution of photon energies in any given beam. However, this degree of detail is not required for our purpose in medical radiology. Instead, we use two methods of specifying beam quality, which include (1) the **half-value layer** and (2) the **accelerating potential** (applied voltage).

1. **Half-value Layer (HVL).** The HVL of an x-ray beam is defined as that thickness of a specified metal that reduces the radiation output to one-half its initial value. The metal in diagnostic radiology is a thin aluminum plate called a **filter.** Using a calibrated radiation detector, at a fixed distance from the tube port, make a two-minute exposure and record it. This reading includes the **inherent filtration** of the tube—the filtration due to the glass window and the oil in the housing—and the built-in filter.

Then make a series of two-minute measurements with additional thicknesses of filter, and tabulate the results as in Table 12.3. The results are plotted on a graph as in Figure 12.11 after being normalized to 1.000

Table 12.3
ATTENUATION VALUES OF A 100–kVp BEAM IN ALUMINUM (Al)
AND DERIVATION OF THE BEAM'S HALF–VALUE LAYER (HVL) *

Filtration (mm Al)	Electrometer Reading **	Normalized Value ***
0	0.873	1.000
0.5	0.697	0.798
2	0.45	0.515
3	0.354	0.405
5	0.24	0.275
5.75	0.208	0.238

*.Data measured by Ted Kosnik, PhD, Chief Physicist, East Texas Medical Center Cancer Institute, Tyler, Texas.

** Readings in units of electric charge, obtained with an electrometer (exposure proportional to electric charge).

*** Normalization $= \dfrac{\text{reading with filter}}{\text{reading without filter}}$

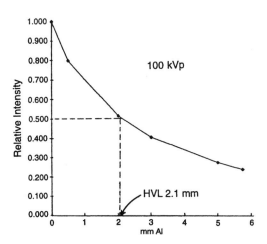

Figure 12.11. Attenuation curve of a 100-kVp beam in aluminum. The relative intensity is normalized from electrometer readings of x-ray output at a fixed distance from the focal spot, with various thicknesses of aluminum filter in the beam as shown on the horizontal axis. A horizontal line is drawn from 0.500 intensity, to the curve: and a line from this point vertically down to the horizontal axis shows the HVL–2.1 mm Al. Scattering material must be removed from the vicinity of experimental setup. (Courtesy of Ted Kosnik, PhD, Chief Physicist, East Texas Medical Center Cancer Institute.)

with no added filter. To normalize, multiply each reading by 1.000/unfiltered-beam reading, in this case, 1.000/0.873. Finally, locate the value on the graph corresponding to a relative output of 0.5, which is 2.1 mm Al, the HVL of this particular beam.

2. ***Applied Kilovoltage.*** In radiography, the kVp is ordinarily used to designate beam quality since the HVL does not serve a useful purpose. However, it is extremely important for ***protection*** of the patient that radiographic beams have the required HVL (see pages 411-412).

"Hard" and "Soft" X Rays

X rays may be roughly classified as ***"hard"*** or ***"soft"*** on the basis of their penetrating ability. This simply expresses the quality of a beam in nonscientific terms.

1. ***Hard x rays*** are higher energy photons with relatively high penetrating ability. Hard x-ray beams may be obtained with

   a. Higher generating voltage.

   b. Filters such as copper, which absorb relatively more of the less penetrating rays. The higher the atomic number of the filter and the greater its thickness, the greater is its hardening effect on the x-ray beam. This applies to orthovoltage beams.

2. **Soft x rays** are lower energy photons with relatively low penetrating ability. Soft x-ray beams may be obtained with

    a. Lower generating voltage.

    b. Filters of lower atomic number, such as aluminum.

THE INTERACTIONS OF IONIZING RADIATION AND MATTER

On passing through a body of matter, an x-ray beam undergoes **attenuation,** that is, a progressive decrease in the initial number of photons. Attenuation consists of two related processes: (1) **absorption** of some of the photons by various kinds of interactions with the atoms of the body in the path of the beam (see pages 125-130) and (2) emission of **scattered** and **secondary** radiation resulting from interactions in (1). A third factor is the loss of beam intensity due to the **inverse square law.**

Scattered radiation refers to those x-ray photons that have undergone a **change in direction** after interacting with atoms.

Secondary radiation comprises the characteristic radiation **emitted** by atoms after having absorbed x-ray photons.

Most of the x-ray photons do not interact with atoms at all, but pass through the body unchanged. This results from two factors: (1) x rays are electrically neutral so there is no electric force between them and orbital electrons, and (2) atoms contain mainly empty space.

Dual Nature of X Rays. The various types of interactions between x rays and atoms make up one of the most fascinating chapters in x-ray physics. As stated earlier, x rays are electromagnetic waves, but they also behave as particles–tiny bits of energy called **photons** or **quanta** (plural of quantum). While seeming to contradict the electromagnetic nature of x rays, the phenomena of radiation absorption and emission can be explained only by the apparently contradictory theories that a given x ray exists at the same time as both a **particle** of **energy** and a **wave.**

This **dual nature** of electromagnetic radiation is now an accepted fact in physics. It tells us how characteristic radiation is produced in an x-ray tube when an electron drops from a higher to a lower energy level in the atom of the target metal (see Figure 12.6). Such a characteristic x ray is a quantum or photon of energy, and its energy and frequency are peculiar to the target element. Thus, an **atom of a given element can emit definite characteristic photons only,** and these are different from those of any other element. The energy of a characteristic photon represents the difference in the energy levels of the shells between which the electron has shifted.

On the basis of this concept, known as the **quantum theory,** an x-ray beam consists of showers of photons, or bits of energy (see page 118), traveling with the speed of light and having no electric charge. Also known as the **corpuscular theory of radiation,** it applies equally to other forms of electromagnetic radiation such as light and gamma rays.

X-ray Interactions with Matter. As a basis for understanding the interactions of ionizing radiation and matter we must know something about the electronic **energy levels** or **shells** within the atom. These are analogous to the mechanical model of the rock on the cliff (see pages 23-24). Since the nucleus carries a positive charge, it exerts an electric force of attraction on the orbital electrons, the force being greater the nearer the electron shell is to the nucleus. Hence, more work would be required to remove an electron from the *K* shell and out of range of the nuclear electric field than would be required to remove an electron from one of the outer shells. Inasmuch as work must be done in moving an electron away from the nucleus in the direction of the outer shells, *the shells are at progressively higher energy levels the farther they are located from the nucleus.* Thus, the *K* shell represents the lowest energy level, while the *Q* shell represents the highest.

The energy required to remove an electron from a particular shell and beyond the range of the nuclear positive electrostatic field is called the *binding energy* of that shell. Therefore, the binding energy is largest for K- shell electrons because of their proximity to the positively charged nucleus, and decreases progressively for succeeding shells, consistent with Coulomb's Law. The binding energy is characteristic of a given element and its particular shell; thus, it is 70,000 eV (70 keV) for the *K* shell of tungsten, but only 500 eV (0.5 keV) for the *K* shell of the average atom in the soft tissues of the body (tungsten has a much larger positive nuclear charge).

Owing to the high energy needed to remove electrons from an inner shell, they are said to **bound.** But almost no energy is required to liberate electrons from an outer shell, so they are called *free* or *valence electrons.*

Let us now detail the possible events following the penetration of matter by x-ray photons. Any of four kinds of interactions may occur between the photons and atoms in their path, depending on the atomic number of the atoms and the energy of the photons: (1) photoelectric interaction, (2) coherent or classical scattering, (3) Compton interaction with modified scattering, and (4) pair production. Only the first three are relevant to diagnostic radiology. Remember that atoms consist mainly of empty space, so most photons pass through without interacting.

1. **Photoelectric Interaction (Photoelectric Effect).** This type of interaction is most likely to occur when the energy–hv–of the incident (incoming) x-ray photon is *slightly* greater than the binding energy (see above) of the electrons in one of the inner shells such as the *K* or *L*. The incident photon gives up all of its energy to the atom; in other words, the photon is *truly absorbed* and disappears during the interaction.

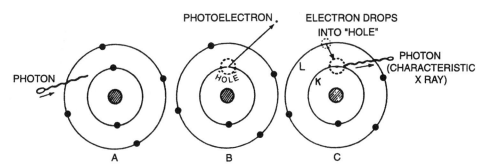

Figure 12.12. Photoelectric interaction with true absorption. *A.* Incident photon loses all its energy on entering an atom, being *absorbed* in the process. The atom responds by ejecting an *inner shell* electron which becomes a photoelectron. *B.* Atom is in an *excited state* (also ionized–electron missing). *C.* Electron from a higher energy level fills vacancy or *hole* in the *K*-shell, a *characteristic* x-ray photon being emitted.

Immediately, the atom responds by ejecting an electron, usually from the *K* or *L* shell, leaving a hole in that shell (see Figure 12.12). Now the atom is ***ionized positively*** (why?) and in an ***excited state.*** Note that the energy of the incident photon ultimately went to (1) free the electron from its shell and (2) set it in motion as a ***photoelectron,*** which ionizes atoms nearby and releases secondary electrons.

The energy exchange during a photoelectric interaction is beautifully expressed in Einstein's equation:

$$hf_{photon} = W_K + \tfrac{1}{2}mv_e^2 \quad (7)$$

energy of   binding energy   kinetic energy
incident   of   of
photon   *K* shell   *K* photoelectron

where *hf* is the energy of the incident (entering) photon, W_K is the binding energy or ***work function*** of the *K* shell of the element concerned, and $\tfrac{1}{2}mv_e^2$ is the kinetic energy of the photoelectron.

The same equation applies to photoelectric interactions in other shells. However, when photoelectric interaction occurs with electrons in ***outer*** shells, the incident photons are more likely to be in the ultraviolet or visible light region.

What happens next? Suppose that a *K* electron has been ejected. The hole in the *K* shell is ***immediately*** filled by transition of an electron from one of the outer shells, usually the *L* shell, because they are at higher energy levels than the *K* shell (see pages 124-125). During the transition of an *L* electron to the hole in the *K* shell the energy difference between these shells is radiated as a photon known as a ***fluorescent*** or ***characteristic x ray*** because its energy (also its frequency and wavelength) is typical of the shells and the element concerned. Thus, the characteristic x rays of copper differ from those of lead.

In pursuing the sequence of events, you may ask, "What happens to the hole in the *L* shell?" It is promptly filled by an electron

from an outer shell, that is, from a still higher energy level. This continues as each successive hole is filled by electron transition from some higher energy level, until the atom loses its excited state, returning to ***normal*** or ***ground state.*** The total energy carried by the characteristic photons in a ***single*** photoelectric interaction equals the binding energy of the shell from which the electron was initially expelled–in this case, the *K* shell.

In summary, then, the energy of the incoming photon in the ***photoelectric interaction*** involving the *K* shell has the following fate:

a. Photon enters atom and completely disappears.

b. A *K*-shell electron is ejected, leaving a hole.

c. Atom has excess energy–excited state.

d. A part of photon's energy was used to liberate electron and the rest to give it kinetic energy; ejected electron is a photoelectron, which releases ***secondary electrons*** by ionizing other atoms.

e. Hole in *K* shell is filled by electron transition from a shell farther out, accompanied by emission of a characteristic x-ray photon.

f. Holes in successive shells are filled by electron transitions from shells still farther out, each transition being accompanied by a corresponding characteristic photon; this sequence is a ***cascade.***

g. Sum of energies of all the characteristic photons equals binding energy of shell from which the photoelectron originated, in this case, the *K* shell.

Thus, the photoelectric effect gives rise to two kinds of secondary radiation–photoelectrons and characteristic x rays. Because of their negative charge, photoelectrons ionize atoms in their path. However, they usually have very low energy. For example, a

100-keV photon releases, in soft tissue, a 99.5-keV photoelectron which is absorbed in only about 1 mm of tissue. Still, this is one of the important ways in which x rays transfer energy to tissues and produce biologic changes in them. The characteristic x rays also have very low energy–about 0.5 keV– and are locally absorbed.

The photoelectric effect is very important in radiography because the chance of its occurrence varies **directly** with Z^3 (atomic number[3]) of the irradiated tissue, and **indirectly** with photon energy[3]. This means that if the average atomic number of organ A is twice that of organ B, the likelihood of a photoelectric interaction in organ A would be 2^3 or 8 times greater in organ B. But doubling the photon energy of photons irradiating an organ would decrease the chance of a photoelectric interaction to $(1/2)^3$ or $1/8$.

Example 1. Tissue A has an average atomic number *(Z)* of 7, and tissue B has a Z of 20. Find the relative probability of a photoelectric interaction of B relative to A.

Solution. The ratio of B/A is $20/7 = 2.86$. Now take the cube of 2.86, which is 23.4, representing the number of times greater the probability of a photoelectric interaction in tissue B.

Example 2. A given tissue is irradiated by 40-keV photons. Compare the probability of a photoelectric interaction of 80-keV vs 40-keV photon.

Solution. In this case, the relationship is the ratio $(40/80)^3 = (1/2)^3 = 1/8$. So at 80kV the chance of a photoelectric interaction is $1/8$ that at 40 kV.

Bone has a high atomic number relative to soft tissue and photon energy is low; hence, the large difference in photoelectric absorption by these structures, a major factor in **radiographic contrast**–the differential darkening of various areas of a radiograph. But differences in tissue thickness and density also contribute to contrast (see pages 218-

219).

2. **Coherent or Unmodified (Classical or Thomson) Scattering.** If a *very low energy* x-ray photon interacts with a firmly bound orbital electron, it may set the electron into vibration. This produces an electromagnetic wave identical in energy to that of the incident photon, but *differing in direction* (see Figure 12.13). Thus, in effect, the entering photon has been **scattered without undergoing any change in wavelength, frequency, or energy.** Unmodified scattering occurs mainly with x rays whose energy is well below the range that is useful in clinical radiology.

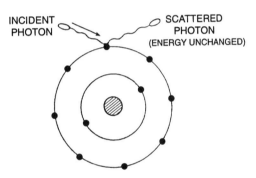

Figure 12.13. Unmodified classical, or Thomson scattering. The entering photon has undergone a change of direction only.

3. **Compton Interaction with Modified Scattering.** If an incident (entering) photon of sufficient energy encounters a *loosely bound,* outer shell electron it may dislodge the electron and proceed in a different direction, as shown in Figure 12.14. The dislodged electron is called a **Compton** or **recoil electron.** It acquires a certain amount of kinetic energy which must be subtracted from the energy of the entering photon, in accordance with the Law of Conservation of Energy. Consequently, the energy of the incident photon is decreased, so it comes out of the atom with less energy (also, longer

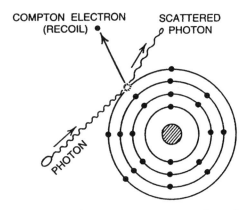

Figure 12.14. Compton interaction with modified scattering. *Part* of the photon's energy has been used up in removing a loosely bound orbital electron. Therefore, the emerging photon has *less energy,* and it has also undergone a *change in direction.*

wavelength and lower frequency). This phenomenon, discovered in 1922 by A. H. Compton, is called the **Compton Effect.** The emerging photon, having undergone a change of direction, is called a **scattered photon.** The characteristic radiation, arising from electron transitions between outer shells, may be disregarded in radiology because of its extremely low energy.

The Compton Effect may be expressed as follows:

$$hv_i \quad = \quad KE_r \quad + \quad hv_s \quad (8)$$

energy of   energy given   energy of
incident photon   to recoil electron   scattered photon

The energy of the scattered photon depends on the angle of scatter relative to its initial direction—the greater the angle, the lower the energy. The scattered photon interacts like other photons with atoms.

As the energy of the incident photon **increases,** the probability of occurrence of the Compton interaction **decreases.** But, at the same time, the **scattered photons** (1) have more energy and (2) tend to be scattered more and more in a forward direction, increasing the likelihood of their passing

completely through the body and reaching the film. This constitutes the **scattered radiation** which, in radiography, impairs image contrast by an overall fogging effect–a form of noise that degrades recorded detail.

Another problem with Compton scattered photons is the exposure hazard they create to personnel, especially during fluoroscopy, mobile radiography, and cardiac catheterization. These require special protective measures, which will be described on pages 404-405.

Despite the decreased chance of occurrence of the Compton Effect at higher photon energies, the probability of the photoelectric effect decreases even more sharply, so as the energy of the incident photon increases above 70 keV (about 180-kV x rays) the Compton Effect becomes the predominant type of interaction.

4. **Pair Production.** In this interaction, beyond the realm of diagnostic radiology, a **megavoltage** photon with energy of at least 1.02 million electronvolts (MeV), upon approaching a nucleus, may disappear and give birth to a **pair**–a negative electron or **negatron,** and a positive electron or **positron** (see Figure 12.15). These charged particles share unequally, as kinetic energy, the photon's energy in excess of 1.02 MeV. The positron, as it comes to rest, combines with any negative electron and disappears giving rise to two photons with an energy of 0.511 MeV, moving in opposite directions. This is the **annihilation reaction.** The two processes exemplify, first, the conversion of energy to matter, and then matter to energy according to Einstein's equation $E = mc^2$. Note that the "magic" number 0.51 MeV is simply the energy equivalent of the mass of an electron at rest. Pair production becomes significant at about 10 MeV, but not until about 24 MeV does it predominate over the Compton Effect insofar as energy absorption is concerned. Pair production is important in **megavoltage x-ray therapy.**

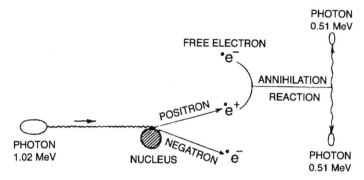

Figure 12.15. Pair production and annihilation reaction.

Secondary and Scattered Radiation Consists of:

1. Primary electrons
   a. Photoelectrons
   b. Compton or recoil electrons
   c. Positron-negatron pairs
2. Secondary and scattered x rays
   a. Characteristic
   b. Coherent or unmodified scattered
   c. Modified scattered (Compton Effect)
   d. Annihilation radiation

The secondary and scattered photons are less energetic (softer) than the incident photons, except for the relatively insignificant factor of unmodified scattering. The production of primary electrons accounts for the absorption of radiation by matter, since this removes energy from the primary beam. Hardening of an x-ray beam by a *filter* results mainly from photoelectric absorption of *relatively more low energy than high energy photons,* so that the emerging beam retains a relatively greater *percentage* of higher energy, more penetrating photons.

Relative Importance of Various Types of Interaction

The various types of interactions between x or gamma radiation and matter will now be discussed as they apply to diagnostic and therapeutic radiology, especially with regard to their relative importance.

In the diagnostic range of 25 to 150 kVp the photoelectric effect predominates insofar as energy absorption from the beam is concerned. Because photoelectric absorption is about four to six times greater in bone than in an equal mass of soft tissue in the radiographic kV range, this type of interaction is responsible for much of the *radiologic contrast* (manifested by differences in film darkness) between these tissues. Differences in density (grams/cc) of various tissues also contribute to radiologic contrast.

On the other hand, the Compton Effect makes its presence known mainly by the *scattered radiation* resulting from interaction with the atoms of the tissues, table top, etc. In fact, in the radiographic range, scattered radiation may contribute as much as 80 to 90 percent of the beam in soft tissues.

As will be shown later, the scattered radiation approaches the film from many directions and degrades radiographic quality, making it necessary to use grids and collimators. As kV is increased the scattered radiation becomes progressively more damaging to radiographic quality because (1) its energy increases and it is therefore more likely to pass completely through the body and reach the film and (2) it is scattered more and more forward toward the film. Note that the characteristic radiation arising in the photoelectric process is also scattered in many directions, but is so soft that it is absorbed locally in the tissues.

In radiation therapy, the electrons released by high-dose ionizing radiation expend sufficient energy to cause tissue damage. The therapeutic effect depends on poorer recovery of cancer than normal cells.

The photoelectric effect predominates below about 140 kVp. Primary photons in the incident beam are completely absorbed upon interaction with the tissue atoms, setting in motion photoelectrons which ionize atoms of the tissue along their paths.

As the applied voltage is increased from about 150 kV to 3 MV, the probability of the Compton interaction decreases, but the probability of the photoelectric interaction decreases even more, so that the Compton Effect becomes predominant.

Although pair production begins at about 1.02 MeV, it does not contribute significantly to energy absorption until about 10 MeV. Only at 24 MeV and above does ionization by negatrons and positrons become the main process of energy absorption.

Table 12.4 gives a classification of x-ray beams.

Detection of Ionizing Radiation

There are several ways in which the effects of ionizing radiation make its presence known:

1. **Photographic Effect**—causes changes in a photographic emulsion so that it can be developed chemically.

2. **Luminescent Effect**—causes certain materials to glow in the dark. These include zinc cadmium sulfide, calcium tungstate, lead barium sulfate, and the newer rare earth phosphors such as lanthanum and gadolinium.

3. **Ionizing Effect**—ionizes gases and discharges certain electrical instruments such as the electroscope.

4. **Thermoluminescent Effect**—undergoes conversion to light in certain crystals after they have been heated.

5. **Chemical Effects**—changes the color of certain dyes.

6. **Physiologic Effect**—reddens the skin, destroys tissues, and causes genetic damage.

Table 12.4
CLASSIFICATION OF BEAMS

X rays	Energy	Filter	Distance
			cm
Surface Therapy (Grenz rays)*	10 - 20 kVp	0	20
Radiography	25-120 kVp	2.5 mm Al	100
Superficial Therapy	100 kVp	0 - 2 mm Al	20 - 50
Orthovoltage Therapy*	200 - 400 kVp	0.5 - 2 mm Cu	50
Megavoltage Therapy	4 - 25 MeV	0	100
Cobalt 60 Therapy (gamma rays)	1.25 MeV eq.	0	80

*Obsolete

SUMMARY OF RADIATION UNITS

The changes produced in tissues by ionizing radiation result from the transfer of energy to the atoms and molecules of the tissues by the processes of *ionization* and *excitation.*

We must emphasize here that x and gamma rays represent *indirectly ionizing* radiation. While it is true that they liberate electrons by one or more of the interaction processes, it is these primary electrons which play, by far, the major role in causing subsequent ionization and excitation of atoms. In other words, the ionizing effect of x and gamma photons occurs indirectly by means of the *primary electrons* they first release in the tissues–photons transfer energy to orbital electrons, which, in turn, transfer energy to atoms to produce pairs of oppositely charged ions, or *ion pairs.* In pair production, electrons arise by materialization of very high energy photons.

The amount of energy transferred by ionizing radiation per unit length of path is called *linear energy transfer–LET.* We saw in the sections on the interactions of radiation with matter that they cause ionization. With a given type of radiation, such as x rays, the degree of ionization and the resulting tissue effects are related to the amount of energy absorbed at a particular site. This *absorbed dose* of radiation depends on the quantity (exposure in R) and quality (energy) of the radiation, and on the nature of the tissue (muscle, bone, etc). We know that more hard than soft radiation is needed for a given tissue effect as, for example, skin reddening (erythema). Such dependence on beam hardness or quality–socalled *wavelength* or *energy dependence*–is due to differences in the number and distribution of ions liberated by radiation of various energies, that is, differences in LET.

Exposure–the Roentgen

As described earlier, we use the *roentgen* (R) as the unit of *exposure,* based on the ionizing ability of the radiation. Thus, exposure is a measure of the amount of radiation delivered to a particular site. Sprawls recommends the surface integral exposure (SIE) as a measure of radiation received by the skin surface during an exposure, expressed by R × cm² or *R-cm².* Also, we must state beam *quality,* namely, the peak kV applied to the tube as well as the half-value layer in the orthovoltage region, or the energy of the radiation in MV or MeV in the megavoltage region. The mA, time, and distance should also be included where applicable, although they have no bearing on the quality of the x-ray beam. The total duration of a therapy course in terms of days or weeks has a distinct influence on the effect of treatment, since a given radiation exposure at one sitting has a much more pronounced effect than the same total exposure given in fractions over a period of days or weeks.

Absorbed Dose–the Gray (Gy)

The preceding paragraph dealt with exposure in roentgens, no mention being made of the dose *absorbed* in the tissues. Since only the energy actually absorbed in the irradiated volume is responsible for the ensuing biologic changes, *we may define the absorbed dose of radiation as the amount of energy absorbed per gram of irradiated matter.* The unit of absorbed dose, the *rad (radiation absorbed dose),* defined as an *energy absorption of 100 ergs per gram.* This applies to any type of radiation (x rays, gamma rays, beta rays, etc) and to any kind of matter, including tissues. In the International System of units the rad has been replaced by the **gray** (Gy) = 1 joule/kg; thus, 1 Gy = 100 rad. On this basis, 1 centigray (cGy) = 1 rad. Absorbed dose is now expressed in **Gy** or **cGy.**

The **roentgen,** it should be recalled, is limited to x rays and gamma rays up to 3 million volts (MV).

What is the relationship of the unit of absorbed dose–cGy–to the unit of exposure–R? The answer is not a simple one, because it depends on a number of factors. Although exposure and absorbed dose are numerically *about* the same in radiography, conversion factors must be used in therapy.

Modification of Kilovoltage X-ray Beams by Filters

The use of filters with x-ray beams has been mentioned several times, and a more complete discussion will now be presented. A filter is simply a sheet of metal placed in the path of a polyenergetic x-ray beam with an HVL up to about 4 mm Cu, to increase its average penetrating ability; that is, **a filter increases the average hardness of a beam.** This is brought about through the interaction of the x-ray photons with electrons in the atoms of the filter material (see section on interaction of radiation with matter). Recall that these interactions consist of absorption and scattering of photons. As a result, a **greater fraction of low energy** (long wavelength) **than of high energy** (short wavelength) x rays is **removed by a filter,** although some higher energy photons are also removed because of the "edge effect"–photons with energy close to the binding energy of the *K* or *L* shells are more likely to be absorbed by photoelectric interaction with *K* or *L* electrons than are photons of lower energy. In general, though, a kilovoltage

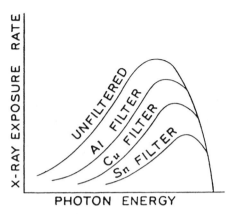

Figure 12.16. Family of curves for kilovoltage x-ray beams filtered by various added filters, as compared with an unfiltered beam. Note that the maximum photon energy is not altered since it depends on the peak kV only. The peak of the curve is shifted toward a higher photon energy as the atomic number of the filter increases; at the same time, the total exposure rate (area under the curve) decreases.

x-ray beam undergoes an increase in half-value layer on passing through a filter.

Notice that filtration does *not* change the maximum energy of a polyenergetic beam (see Figure 12.16); this depends solely on the **peak kilovoltage** applied to the tube. However, the smallest energy a photon can have in an x-ray beam leaving tube depends on the **inherent filtration** of the tube; that is, the filtration afforded by the glass envelope and the cooling oil layer through which the beam must pass after leaving the target.

The inherent filtration of a tube is usually expressed in the equivalent thickness of aluminum or copper. Any additional filtration one chooses to place in the beam is called **added filtration.** The **total filtration** of an x-ray beam obviously represents the sum of the inherent filtration plus the added filtration.

$$\frac{\text{total}}{\text{filtration}} = \frac{\text{inherent}}{\text{filtration}} + \frac{\text{added}}{\text{filtration}} \quad (15)$$

Filtration affects both the ***exposure rate*** and the ***quality*** (half-value layer) of a kilovoltage x-ray beam:

1. ***Exposure rate*** is always decreased by filtration because some photons are removed by absorption and scattering. In fact, the thicker the filter and the higher its atomic number, the smaller will be the exposure rate.

2. ***Quality*** is changed by filtration. An increase in either the atomic number or the thickness of a filter increases the average penetrating ability of a beam; more specifically, the half-value layer increases. The type of filter should match the energy of the radiation. With radiation produced at 50 to 100 kV, aluminum is most suitable as a filter. Copper is used as the ***primary filter*** in the 120 to 400 kV range, but an aluminum filter must be placed between the copper and the patient as a ***secondary filter*** to absorb the characteristic radiation (8 keV) emitted by the copper. The resulting aluminum characteristic radiation (only 1.5 keV) is absorbed in a few centimeters of air.

As will be further explained in Chapter 28, a radiographic beam must be filtered by a sufficient thickness of aluminum (2.5 mm) to remove selectively an adequate amount of low-energy x rays that would otherwise be absorbed in the body without contributing to the radiographic image.

QUESTIONS AND PROBLEMS

1. Which properties of x rays led to their discovery?
2. What are roentgen rays? What are the two apparently contradictory theories as to their nature?
3. How are the frequency and wavelength of radiation related? What is the speed of x rays? Gamma rays?
4. How do soft rays differ from hard rays?
5. Describe the four essential components of a modern x-ray tube. Discuss the principles of x-ray production.
6. Discuss in detail the production of x rays when the electron stream hits the target.
7. What is characteristic radiation and how does it arise?
8. Discuss the differences between characteristic and general radiation. What is their relative importance in radiography?
9. Name ten properties of x rays.
10. What is meant by the quality of roentgen radiation? What factors influence

it and in what manner? How can it be measured and specified (ICRU recommendations)?
11. Explain fully why a beam of x rays emerging from the target is polyenergetic. What are monoenergetic x rays?
12. What determines the maximum photon energy in an x-ray beam? State the equation that derives photon wavelength from energy.
13. Define and explain half-value layer and filters. What purpose does HVL serve?
14. Why is tungsten so widely used as the target material?
15. How is the quantity of radiation measured?
16. What factors influence x ray output (R/min)?
17. Discuss the four types of interactions of ionizing photon radiation and matter, showing how the corpuscular theory radiation applies.
18. What is the relative importance of the

various interactions of photons with matter in radiography and radiotherapy?

19. Of what do scattered and secondary radiations consist?

20. How does a filter influence the maximum photon energy in a radiographic beam? The output in R/min?

21. Find the efficiency of x-ray production of a tungsten target at 90 kVp.

22. The output of an x-ray beam at 100 cm is 50 R/min. What is the total exposure in 2 min?

23. An x-ray beam at a given distance produces an exposure of 25 R in 2 min. Find the output.

24. What is the energy of an incident photon in a photoelectric interaction if the binding energy of the K-shell is 70 keV and the photoelectron's energy is 30 keV?

25. A tissue is irradiated with 50-keV photons. Find the probability of a photoelectric interaction as compared with 60 keV photons.

26. Calculate the SIE (surface integral dose) when a 20×15 cm^2 area receives an average exposure of 100 mR.

Chapter 13

X-RAY TUBES

Topics Covered in This Chapter

- Thermionic Diode Tubes
- Radiographic Tubes
- X-Ray Production
- X-Ray Tube Construction
    - Glass Envelope
    - Cathode (filament)
    - Anode (target)
- Apparent and Effective Focal Spot Sizes
    - Line-focus Principle
- Space Charge Compensation
- Saturation Current
- Tube Current *vs* Filament Current
- Factors Governing Tube Life
    - Tube Rating Chart
    - Anode Thermal Capacity
    - Anode Cooling Curve
- Practical Steps to Prolong Tube Life
- Questions and Problems

THERMIONIC DIODE TUBES

IN CHAPTER 12 we learned that x rays are produced whenever a stream of fast-moving electrons undergoes rapid deceleration (loss of speed). These conditions prevail during the operation of a special type of *thermionic vacuum tube*–the hot filament or Coolidge x-ray tube (see Figure 12.4). Before describing the detailed construction of such a tube, let us review its underlying principles:

1. *A hot metal filament gives off electrons* by a process called *thermionic emission.* The filament is heated by a separate *filament current.* The rate of electron emission increases with an increase in the temperature of the filament which, in turn, is governed by the filament current (specified in amperes).

2. *If no kilovoltage is applied,* the emitted electrons remain near the filament as an electron cloud or *space charge.*

3. *When kilovoltage is applied* between the filament and target, putting a negative charge on the filament (cathode) and a positive charge on the target (anode), *space charge electrons are driven over to the anode at high speed by the large potential difference.* The maximum and average speeds of the electrons increase as the peak kilovoltage is increased. The electron stream crossing the gap between the cathode and anode constitutes the *tube current,* measured in milliamperes (mA).

4. *If the applied kilovoltage and resulting electron speed are high enough,* x rays are produced when the electrons enter the target, their *kinetic energy* being converted to heat (ordinarily more than 99 percent) and x rays (less than 1 percent).

A typical x-ray tube is a thermionic diode consisting of a tungsten filament cathode, a tungsten target anode, sealed in an evacuated glass tube enclosure. There are two circuits: one to *heat the filament,* and the other to *drive the space charge electrons* to the anode and the high-voltage anode cable.

The structural and other details of radiographic and therapy tubes will now be described.

RADIOGRAPHIC TUBES

In this section we shall examine in detail the essential features of radiographic tubes used in medical x-ray diagnosis.

Glass Envelope

A cathode and anode are enclosed in a thick Pyrex glass *tube* or *envelope,* with a thinner *window* where the x rays emerge. The tube contains as perfect a vacuum as possible. In fact, the tube is baked during manufacture to expel air and other gases that may have been trapped in the glass or metal parts, a process called *degassing.* As already noted, the vacuum offers an unobstructed path for the electron stream (tube current) and also retards burnout of the filament. Radiographic tubes are usually cylindrical in shape. Immersion in *insulating oil* in a suitable metal housing prevents high voltage sparkover between the terminals, thereby making possible the design of smaller tubes.

Cathode

The negative terminal or *cathode assembly* consists of (1) the *filament,* (2) its *supporting wires,* and (3) the *focusing cup* (see Figure 13.1). Loosely speaking, the terms "cathode" and "filament" are used interchangeably. The filament itself is a small coil of tungsten wire, the same metal that is used in electric light bulb filaments. In most radiographic tubes the filament measures about 2 mm in diameter and 10 mm or less in length, and is mounted on two stout wires which support it and also carry electric cur-

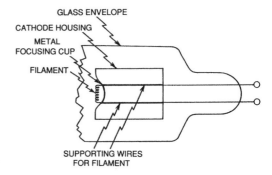

Figure 13.1. Details of the cathode of an x-ray tube (side view).

rent. These wires lead through one end of the glass tube to be connected to the proper electrical source. A *low voltage filament current* is sent through the wires to heat the filament, and one of these wires is also connected to the *high voltage source,* which provides the high negative potential needed to drive the electrons toward the anode at great speed. A negatively charged concave metal cup behind the filament confines the electrons to a narrow beam and focuses them as the *focal spot* on the tungsten target.

The filament current, serving to heat the filament and provide a source of electrons within the tube, usually operates at about 10 volts and 3 to 5 amperes. An increase in the filament current raises the temperature of the filament and the rate of electron emission causing a sharp rise in tube current (mA) as shown in Figure 13.2.

Some tubes, such as those used in older mobile equipment, have a single filament. However, modern radiographic tubes, in

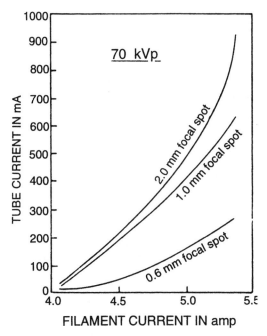

Figure 13.2. Relation of tube current (mA) to filament current. Note that a small increase in filament current produces a large increase in mA. Furthermore, mA is larger with a larger tube focus for a given filament current.

both stationary and mobile equipment, are provided with two filaments mounted side by side. The filaments usually differ in size,

producing focal spots of two different sizes on the target. Such tubes are called d*ouble focus* tubes, although they are really double filament tubes. Note that only one filament is activated for a given exposure, being selected at the control panel. Supporting the two filaments are three stout wires, one of which is connected in common with both filaments. In Figure 13.3, if terminals 1 and 3 are connected to the low voltage source, the large filament lights up. If, instead, terminals 2 and 3 are connected to the low voltage source, the small filament lights up. The high voltage, applied through wire 3, which is common to both filaments, gives either one a high negative potential.

Filament evaporation. As the tube ages, the filament gradually evaporates as the result of heating during use. Consequently, a number of changes take place in the tube.

1. ***Thinning of the Filament.*** Loss of metal (tungsten) by evaporation causes the filament to become progressively thinner. Its electrical resistance therefore increases (R inversely proportional to cross sectional area). However, filament circuitry is designed to maintain constant current (I). Thus, I^2R increases, thereby raising the filament temperature with resulting increase in

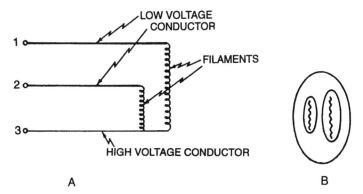

Figure 13.3. *A.* Simplified diagram of the connection of two filaments of an x-ray tube (double focal spot tube) to three wires, one of which is common to both filaments. *B.* The filaments are mounted side by side, facing the target.

tube current (mA). This may be corrected by a service person or by internal circuitry, which adjusts the filament current ***downward*** just enough to obtain the proper electron emission for the required tube current. It will be found that *as the tube ages, a progressively lower filament current setting is required for a desired tube milliamperage.*

2. ***Deposition of Tungsten on the Glass Envelope.*** As the filament evaporates, some of the vaporized tungsten deposits on the internal wall of the glass tube. When this occurs on the window of the tube, it acts as an additional filter, reducing x-ray output. Furthermore, the tungsten deposit, being metallic, tends to attract electrons from the electron stream, eventually puncturing the tube. In fact, this is the most common cause of tube failure.

Anode

There are two main types of anodes: (1) stationary and (2) rotating.

1. **Stationary Anode.** Such an anode is used in dentistry. A gap of a few centimeters separates the cathode from the anode which consists of a block of copper in which is imbedded the target, a small tungsten button. These metals are selected for definite reasons:

a. *Tungsten*–so placed that it forms the actual target of the x-ray tube, the high speed electrons striking it directly. The target is the area on the anode in which useful x rays are actually produced. Since more than 99 percent of the electrons' kinetic energy changes to heat at the target, the ***high melting point*** of tungsten–3370° C–makes it especially suitable as a target metal, as does its ability to conduct heat. Furthermore, ***tube efficiency,***

that is, the percentage of the electrons' energy appearing in the form of x rays, is directly proportional to the atomic number of the target and the applied kV (see page 115) (mammographic x-ray tubes have molybdenum, rhodium, tungsten, or vanadium targets with proper filters).

b. ***Copper***–a much better conductor of heat than tungsten, carrying the heat away from the target more rapidly and thereby protecting it from overheating, within certain limits. Even greater heat-loading capacity can be achieved by circulating air, water, or oil around the anode, permitting larger exposures than would otherwise be possible.

Note again the electron stream bombards a limited area on the target called the ***focus*** or ***focal spot.*** X rays are emitted from the entire surface of the actual focus; in fact, there are innumerable points in the focal area from which x rays are projected in all directions. Using the smallest practicable focal spot enhances the sharpness of the radiographic image. Therefore, radiographic tubes are provided with a small filament and a metal focusing cup. However, as the size of the focal spot decreases it experiences a greater concentration of heat with the attendant danger of overloading and melting the target. Tubes are so designed that the focal spot is small enough to give satisfactory radiographic detail, but not so small as to restrict their practical usefulness.

Anodes of radiographic tubes exhibit the ***line-focus effect**** because their plane is angled toward the film or image receptor (see Figure 13.4) so that the projected image is smaller than the actual image. This also applies to rotating anodes. The actual focus is a rectangle as in *B;* assume this rectangle, *T,* to measure about 2 mm × 6 mm. In *A,* the side view of the tube, note that the anode

---

* Commonly called "line-focus principle," I prefer the term "line-focus effect" because it is incidental to the fact that the anode surface must be angled to direct the x-ray beam toward the image receptor. However, anodes are deliberately designed with steeper angles to produce smaller effective focal spot sizes, based on the line-focus effect.

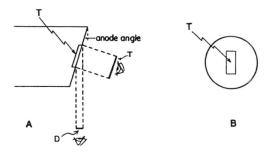

Figure 13.4. Line-focus effect. In *A*, the target is shown in side view, and *D* is the *projected, apparent,* or *effective focus.* In *B*, the target is shown in face view with the actual rectangular focus, *T*, at its center. The same effect applies to rotating anodes.

target, the **effective** or **apparent focal area** as projected toward the film is only *D*, which is about 2 mm × 2 mm. Perhaps it would become clearer if you were to imagine yourself lying supine in the position of the film and looking upward at the target–the focal spot would appear as a small square rather than a rectangle, which is foreshortened, just as a pencil appears shorter when it is held obliquely vertical in front of the eyes than it is when held parallel. Furthermore, the effective size of the focal spot varies with the **direction** in which it is projected. As shown in Figure 13.5, it becomes smaller toward the anode end of the tube.

surface is inclined, making an angle of 17° with the vertical. Since the film will be placed at some distance directly below the

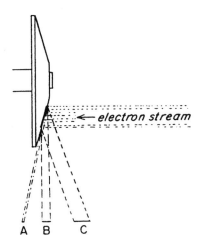

Figure 13.5. Effect of angle of emission of x rays on apparent or effective focal spot size. An anatomic detail at *A* (toward the anode end of the tube) will have a sharper image because in this direction the effective focal spot size is smaller than at *C* where the effective focal spot size is larger. *A, B,* and *C* represent the effective sizes of the same focal spot projected at different angles, that is, in different directions with respect to the anode surface. Rotating anode shown.

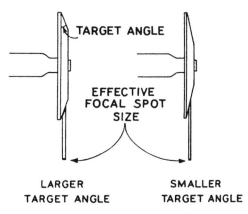

Figure 13.6. Effect of target (anode) angle on apparent or effective focal spot size. Note that a decrease in the target angle gives a smaller effective size when the actual focal spot size remains unchanged. Rotating anodes shown.

Another important relation is shown in Figure 13.6, the effective size of the focal spot diminishes as the anode angle becomes smaller (anode steeper). However, this also results in a more pronounced heel effect (see page 254.

2. **Rotating Anode.** In 1936 there became available a radically different kind of anode of ingenious design–the **rotating anode.** As its name indicates, this anode,

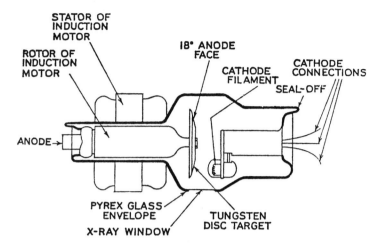

Figure 13.7. Rotating anode tube.

attached to the shaft of a small induction motor, ***rotates during x-ray production.*** Rotating anodes were previously constructed of solid tungsten in the form of a heavy disc. Later, an improved type of rotating anode was introduced, consisting of a molybdenum ("moly") disc coated with a ***tungsten-rhenium alloy*** to serve as the ***focal track target.*** Rhenium diminishes roughening of the focal track with use, thereby ensuring high x-ray output, or ***emissivity.*** For heavy-duty tubes, a layer of graphite under the moly disc reduces the weight of the large anode and hastens the dissipation of heat.

Rotating anodes usually measure 7.6 cm (3 in.) in diameter, although 12.7 cm (5 in.) models are available for heavy duty; rotational speeds are 3600 rpm and 10,000 rpm, respectively (rpm = rotations per min). The latter are required for special procedures such as angiography. A diagram of a rotating anode tube appears in Figure 13.7.

As already mentioned, rotating anodes display the ***line-focus effect*** because they must have a beveled edge as shown earlier in Figure 13.6. The cathode is so positioned that the electron beam strikes the focal track on the beveled edge, the degree of beveling

being called the ***target*** or ***anode angle.*** Anodes have various angles–10°, 12°, or 17°. These affect the size of the focal spot as shown above in Figure 13.6, but they also permit heavier exposures at very short exposure times without tube damage; short-exposure rating *increases* as anode angle *decreases* (see Table 13.1).

Table 13.1
EFFECT OF TARGET ANGLE OF A
ROTATING ANODE ON SHORT-EXPOSURE
RATING RELATIVE TO THAT OF A
STANDARD TARGET ANGLE OF 17°

Target Angle	Approximate Increase in Short Exposure Rating
12°	30%
10°	60%

Outside the tube is the stator of the induction motor, the rotor being attached to the anode stem ***inside.*** Rotation of the rotor depends on electromagnetic induction (review pages 68-69). Self-lubricating bearings coated with ***metallic barium*** or ***silver*** minimize friction, thereby prolonging tube life.

How does such an anode operate? During exposure the anode rotates rapidly while being bombarded by the electron stream. Although ordinarily rated at 3600 rpm, this is more likely to be 3000 rpm due to slippage of the rotor. During rotation the anode constantly "turns a new face" to the electron beam, so that the heating effect of the beam does not concentrate at one point, as in the stationary anode, but spreads over a large area called the *focal track* on the beveled face of the anode as shown in Figure 13.8. However, the effective focus remains

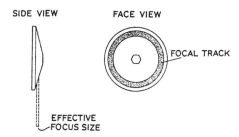

Figure 13.8. Line-focus effect with a rotating anode. The *focal track* is the actual area of impact of the electrons, or the actual focal area. Note how much smaller the apparent focal area is than the focal track. For example, a typical focal track is about 250 mm in length (circumference) and its width about 3 mm for a 1 mm^2 apparent focal spot. Thus, the area bombarded by electrons at any instant is $1 \times 3 = 3$ mm^2, while the heat is spread over an area of $3 \times 250 = 750$ mm^2.

constant in position relative to the film because of the extremely smooth motion of the anode. As a result, it is possible to have an effective focus of 0.6 mm or even smaller–0.1 and 0.3 mm fractional focus tube. Furthermore, it becomes feasible to subject a tube to high milliamperage. For instance, with the old stationary anode tubes, the upper limit for a 4 mm focus was 200 mA, whereas with a rotating anode, the upper limit for a 2 mm focus is ordinarily 500 mA,

the maximum exposure time being approximately the same in both cases. Special tubes have been designed to tolerate a current as large as 1000 mA for very short exposures, operating on three-phase current.

Rotating anode tubes usually have two filaments to provide 0.6 and 1 mm focal spots; the smaller one is used to obtain fine radiographic definition, whereas the larger one can tolerate heavier loading in shorter exposure time to minimize motion.

Special rotating anode tubes with 0.3 mm focal spots, and even smaller, have been designed for *magnification radiography.* Called *fractional focus tubes,* they permit placing the film at a longer distance from the part to permit direct magnification of the radiographic image, as described on pages 226-228.

You can see from the above discussion that rotating anode tubes with a relatively high thermal capacity can successfully energize extremely small focal spots. This has made possible the superb recorded detail in radiography.

Space Charge Compensation

The low-voltage current supplied to the x-ray tube filament heats it to incandescence (white hot), causing the emission of electrons (thermionic emission). The rate of electron emission, or the number emitted per second, depends on the temperature of the filament which, in turn, is governed by the current supplied to it. In fact, *a small increase in filament current produces a large increase in the rate of electron emission and resulting space charge.*

If *no* high voltage is applied across the tube, the electrons remain in the vicinity of the hot filament as a *space charge.* This gives rise to the *space charge effect* which may be simply explained as follows: since the space charge consists of electrons, it is negative and therefore tends to hold back the further emission of electrons (like charges repel). In fact, some electrons are actually repelled from the space charge back into the filament. At any particular value of the filament current, an equilibrium is reached between the rate of electron emission by the filament and the rate of return to the filament from the space charge. Thus, *the space charge has a definite size for a given value of the filament current in a given tube,* and as just indicated, a small change in filament current produces a large change in the size of the space charge.

If a sufficiently high voltage (kVp) is now applied across the tube, imparting a large negative charge to the filament and an equally large positive charge to the anode, the space charge electrons are driven toward the anode at a speed that depends on the instantaneous kV (but *not* directly proportional). When tube current (mA) is plotted as a function of kVp, a family of *characteristic curves* is obtained as shown in Figure 13.9. As expected, there is a different curve for each value of filament current (amp). Further-

more, the larger the filament current, the greater will be the resulting mA.

Let us now examine the characteristic curves. Note that the lowest curve (representing a filament current of 4.2 amp) has an initial curved portion and then flattens out at about 100 kVp, after which a further increase in kVp causes no additional increase in mA. This is explained as follows: a progressive increase in kVp drives more and more of the space charge electrons per second to the anode; that is, there is an increase in mA. This is called the *space-charge limited region* of tube operation. But above 100 kVp all of the space charge electrons are driven to the anode immediately after emission from the filament, so that further increase in kV cannot increase mA; hence, this portion of the curve is flat, representing *saturation current* or the *temperature-limited region.* Here the mA can be increased only by increasing the filament current, which increases filament temperature.

The remaining three curves in Figure 13.9 do not have a flat portion below 150 kVp, so at the corresponding filament currents the tube operates only in the *space-charge limited region.* Operation of an x-ray tube in the space-charge limited region is undesirable in radiography because kV cannot be changed independently of mA—when the technologist changes the kV setting there is an unavoidable change in mA. To correct this situation a *space charge compensator* introduced into the filament circuit automatically lowers the filament current by just the right amount, as the kVp is raised, to keep the mA constant over the useful kV range. Now the technologist has full control of tube operation, being able to change kVp without producing any appreciable effect on mA.

In modern automatic equipment with monitor-type mA selection, the space charge compensator has an additional function; it also selects the correct filament current to provide the desired space charge and result-

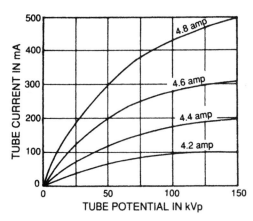

Figure 13.9. Family of characteristic curves of a typical rotating anode tube *(Machlett Dynamax "60"®, Dunlee).* Note the progressive rise in mA as kV is increased. A space charge compensator is used to flatten the curves so that kV can be changed without a change in mA as explained in the text.

ing mA. As will be explained later, filament and tube currents are often controlled by high-frequency generators, governed by microprocessors.

Factors Governing Tube Life

Overheating the filament or anode disc inevitably shortens tube life. While there are specific differences between the filament and anode in this regard, they have in common the tendency to vaporize and deposit tungsten on the glass wall of the tube. This leads to internal electrical stresses resulting in instability of operation and possible puncture of the tube wall. Furthermore, the tungsten deposit increases inherent filtration and so decreases x-ray output and slightly hardens the beam.

We shall evaluate the factors that influence tube life as they pertain to the filament, anode, and housing. The x-ray tube itself, as distinct from the housing, is called the *insert.*

Filament Factors. The large tube currents (mA) often demanded in radiography require a large filament current and temperature. Failure of a radiographic tube may result from "burning out" of the filament due to application of excessive current, which raises filament temperature beyond the rated limit. But tube failure can also occur from prolonged heating of the filament at normal current loads, just as an electric light filament burns out after a period of normal use. To extend tube life, a special *booster circuit* limits filament current to a low or standby level until the x-ray exposure switch is closed. At this instant, the booster circuit automatically raises the filament current sufficiently to provide the desired tube current. At the same time, the technologist must avoid using the "boost-and-hold" method of making exposure (as in radiography of infants) unless it is absolutely necessary; excessive boost time unquestionably shortens tube life.

Undue heating of the filament by excessively high mAs causes evaporation of the filament metal with resulting decrease in its diameter; a reduction of only 10 percent will cause the filament to break, ending tube life. As mentioned above, coating of the glass wall of the tube by vaporized filament tungsten may lead to puncture by electric *sparkover,* the most frequent cause of tube failure.

Anode Factors. The ability of the anode disc to accumulate, store, and discharge heat, limits the maximum allowable techniques. These three anode factors will now be taken up individually.

1. *Radiographic Tube Rating Chart.* If the rate of temperature rise in the anode disc exceeds a limiting value, localized melting or cracking occurs as a result of thermal stresses set up in the metal by the difference in temperature between the focal spot or track at the surface, and the depths of the disc. With a slower rate of temperature rise, enough time may elapse to permit heat conduction to equalize the temperature throughout the disc. Melting of the metal, short of actually destroying the disc, causes it to vaporize and coat the glass wall of the insert, eventually resulting in its puncture. Furthermore, roughening of the focal track area and cracking of the anode disc reduces x-ray output because, as they leave the anode, the photons have to penetrate the uneven surface of the anode. Figure 13.10 shows the appearance of heat-damaged focal tracks and anode discs. Finally, warping of the anode disc causes decreased output and noisy operation.

The safe limits of exposure factors for a "cold" anode can easily be found by referring to the radiographic tube rating chart provided by the manufacturer for each type of tube and for a given kind of operation. Such a chart indicates the maximum "safe" exposure time for any selected combination of kVp and mA for a single exposure and a

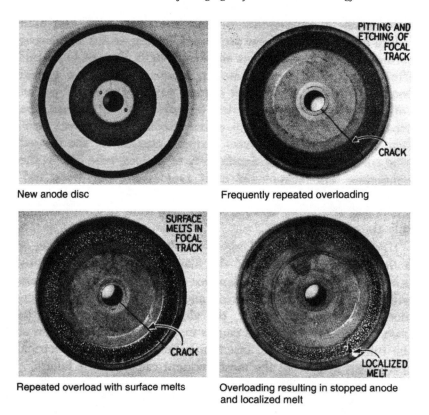

New anode disc

Frequently repeated overloading

Repeated overload with surface melts

Overloading resulting in stopped anode and localized melt

Figure 13.10. Surface appearance of anode disc under various conditions. *Adapted from Cathode Press, courtesy of the Machlett Laboratories, Inc.*

relatively cool tube. The chart in Figure 13.11 is fictitious and is shown only by way of an example. It includes a series of curves representing various mA values of tube current. The vertical axis of the graph represents kVp, and the horizontal axis maximum allowable exposure time in sec. Such a chart is simple to use. Assume that it applies to a particular tube of *given focal spot size,* and with a *given type of rectification.* Suppose we are to make an exposure of 300 mA and 65 kVp, and we wish to determine the maximum allowable exposure time. Starting at 65 on the vertical axis, follow the horizontal line to its intersection with the 300 mA curve. Then from this point of intersection, follow vertically downward to the horizontal line, crossing it at three seconds, the maxi-

mum rating or tolerable exposure time with these factors. You must be extremely careful to use the correct tube rating chart for the particular tube and for the following conditions, to avoid overheating the anode. Fortunately, modern state-of-the-art equipment has exposure *limiters* to protect the x-ray tube.

a. *Type of Rectification.* Maximum tube rating is less with half-wave (single-pulse) than full-wave rectification. This was important with early x-ray machines.

b. *Size of Tube Focus.* Maximum tube rating is less with a small focal spot or track than with a large one.

c. *Tube Design.* This affects the maximum tube rating; therefore, use the appropriate

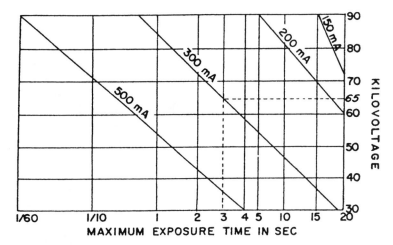

Figure 13.11. Tube-rating chart, arbitrary values. The chart for any particular tube must be obtained from the manufacturer.

tube rating chart supplied by the manufacturer for the particular tube. For example, a tube with high-speed disc rotation has a higher rating than a conventional tube. Furthermore, smaller anode angles permit higher tube ratings for very short exposures.

d. *Cold versus Hot Tube.* The chart applies only to a relatively cold tube, one that has not been subjected to heavy loading just before use.

e. *Type of Power Supply.* Tube rating will be different for single phase and three phase, and for short or long exposures.

f. *Application.* A special chart must be used for rapid sequence radiography, as in angiography and tomography.

The tube rating chart for the particular tube in use should be mounted in the control booth, especially if the equipment lacks **automatic exposure limiters;** these prevent activation of the x-ray tube if it were to be overloaded by the selected technical factors.

2. **Anode Thermal Capacity.** With multiple exposures the anode assembly gradually accumulates heat. (This is not the same as the maximum exposure factors for a "cold" anode, as determined from a tube rating chart.) We must therefore deal with the capacity of an anode to store heat without cracking, melting, or warping, any of which may reduce x-ray output (see above). Overheating and vaporization of disc metal result in coating of tungsten on the glass wall of the insert with the attendant harmful effects described earlier. Furthermore, excessive accumulation of heat may reach the anode stem and rotor bearings. The latter may become sufficiently deformed to slow down or stop the rotor, a condition that is obviously fatal to the insert. The thermal capacity of an anode is measured in **heat units** (H.U.), defined by the following equation for **single-phase** operation:

$$\textbf{H.U.} = \textbf{kVp} \times \textbf{mA} \times \textbf{time in s} \quad (1)$$

Because kV is nearly constant in a 3-phase circuit, 6-pulse operation requires equation (1) to be multiplied by the factor **1.35;** and for 3-phase 12-pulse **1.41.** However, these factors should not be used to convert single-phase charts to three-phase operation; the applicable tube rating chart must be used for each type of tube and equipment as noted on page 143.

Note that heat units are not fictitious, but really represent energy:

$$1 \text{ kVp} \times 1 \text{ ma} \times 1 \text{ s} = 1 \text{ watt-s } \textit{(nominal)}$$

which is a unit of energy. (Actually, 1 H.U. is equivalent to an energy of 0.71 watt-sec, where 0.71 is the RMS conversion factor.) The **thermal capacity** of the anodes of various diagnostic tubes ranges from about 70,000 H.U. to about 400,000 H.U., depending on the **size of the anode** (ie, anode mass) and the cooling method.

The equation defining heat unit storage also demonstrates that an x-ray tube operates more efficiently at higher kilovoltages. For example, a given radiographic technic calling for an exposure of 60 kVp, 100 mA, and 1 s would develop the following number of heat units in the anode:

$$\text{H.U.} = 60 \times 100 \times 1$$
$$\text{H.U.} = 6000$$

If the kVp is increased by 10 (ie, about 15 percent) and the time reduced one-half, the exposure in terms of visible radiographic effect would remain essentially the same, but now let us see how many heat units would be developed in the anode:

$$\text{H.U.} = 70 \times 100 \times {}^1/_2$$
$$\text{H.U.} = 3500$$

Thus, with an increase in kVp the heat units in the anode have been decreased almost one-half, even though the exposure has remained virtually unchanged.

The **short-exposure rating,** or the ability of a focal track to withstand a large heat input in a small fraction of a second, does not necessarily follow the heat storage capacity of the anode. For example, as described on page 164, the short-exposure rating is higher with three-phase than with single-phase current for a tube with the same thermal capacity. Furthermore, as shown earlier in Table 13.1, the short-exposure rating of a rotating anode tube increases significantly as the tar-

get angle decreases. Still a third factor in this problem is the speed of rotation of the anode; thus, with high speed anodes–10,000 rpm–the short-exposure rating is increased about 70 percent over that with 3000 rpm.

In general, *an increase in the actual area of the focal spot or track increases the instantaneous rating of the anode, whereas an increase in the anode mass increases its thermal capacity.*

After prolonged use, or with minor overloading, the anode disc becomes roughened by **pitting** or **etching.** When severe, this reduces the x-ray output of the tube. The use of **rhenium-tungsten-molybdenum** instead of pure tungsten as the anode material aids in minimizing this problem.

In the ordinary oil-immersed tube, adequate **cooling** results from the transfer of heat to the oil, tube, housing, and surrounding air by simple convection and conduction. For heavier loads, as in special procedures and high voltage radiography, cooling is improved by mounting a small fan outside the tube housing. Some specially designed tubes are cooled by a forced oil circulating system, the oil flowing into the hollow anode and then between the tube and housing; this provides a very high rate of heat dissipation. Thermal capacity may be enhanced by the use of rotating anodes with surface areas 100 percent greater than usual. The oil incidentally provides efficient insulation, while the metal housing helps make the tube rayproof by shielding out unwanted radiation.

3. *Anode Cooling Curve.* The rate at which the anode cools, or the **heat dissipation rate,** starting at any particular value of heat storage (heat units), is given by the manufacturer's anode cooling curve appropriate to the tube under consideration. Figure 13.12 shows a typical anode cooling curve. At no time should another exposure be made if the additional heat produced in the anode would increase the stored heat beyond the instantaneous rating. Thus, start-

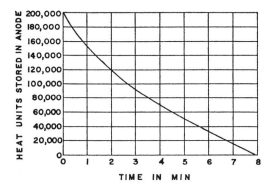

Figure 13.12. Anode cooling curve. With this tube, if the anode has the maximum 200,000 heat units already stored, then at the end of two minutes the stored heat will have decreased to 120,000 H.U. It would then be safe to load the anode with additional 80,000 H.U. for a total maximum of 200,000 H U. For complete evaluation, a housing cooling curve should also be consulted. *(Based on data furnished by courtesy of Machlett Laboratories, Inc.)*

ing with a known amount of heat already present in the anode, we must allow enough time to elapse for it to cool down sufficiently to permit the additional heat load. An example is given in the legend to Figure 13.12.

Housing Factors. Often overlooked, the ability of the metal housing to store and dissipate heat contributes to tube life. Thus, the temperature of the housing must be kept below 90°C (200°F), both to prevent damage to the housing itself and to permit adequate cooling of the insert. In fact, excessive heat may bring the oil to the boiling point and even cause the housing to *explode.* This is a real hazard with rapid serial exposures in angiography and cineradiography. An air blower may help cool the housing, but is effective only if the housing cools more slowly than the anode disc. Regular servicing of the blower is mandatory to assure proper function, especially for an under-table tube. Special cooling curves, similar to those for anode cooling rate, are available for the housing.

Summary of Practical Steps in Extending Useful Tube Life. A number of precautions in the use of radiographic tubes contribute significantly to prolongation of their useful life. These have been described in detail but will now be summarized.

1. *Filament life* is shortened by application of large filament currents (and resulting high temperatures). Therefore the "boost" or preparation time during which the filament temperature is held at full value should be kept as short as possible.

2. *Instantaneous ratings,* as derived from tube rating charts, should never be exceeded. Be certain to use the correct chart for the particular tube, focal spot size, type of rectification, and type of current (single or three-phase).

3. The *anode should be warmed up* before being subjected to a large load; use warmup times suggested by the manufacturer. Heat produced by a large power load on a cold anode causes uneven expansion that may crack the anode disc.

4. The *life of the bearings* in a rotating anode tube is shortened as the total rotation time is increased. Therefore the anode should not be run unnecessarily. This becomes a special problem in radiography of infants when the rotor is kept activated while the technologist awaits the opportune instant to make the exposure. In addition, the bearings may be damaged by heat overload of the anode.

5. *Excessive temperature of oil in the tube housing may shorten tube life.* Therefore adequate cooling of the tube housing must be provided; for example, by use of an auxiliary fan which may be provided at the time of installation.

6. *Rapid sequence exposures* as in angiography and cineradiography may easily exceed the tube rating. Therefore the proper rapid sequence rating charts and anode cooling curves should be consulted. In fact,

it is good practice to work up a chart showing the maximum number of exposures in the time intervals for various procedures which may pose a hazard for the x-ray tube; this chart should be posted in the control booth for ready reference.

7. ***Avoid overheating the filament and anode*** because even minor overheating, if repeated a sufficient number of times, causes evaporation of tungsten which is deposited on the glass wall of the insert. Eventually the insert may be punctured by sparkover, especially with application of high kV. In fact, tube puncture ranks high as a cause of tube failure in ordinary routine radiography, whereas anode warping or cracking occur more often in rapid sequence radiography.

QUESTIONS AND PROBLEMS

1. Describe, with aid of a diagram, a double focus, stationary anode tube.
2. Explain the principle of the rotating anode tube and show its important structural features by means of a simple diagram.
3. Enumerate its advantages over a stationary anode tube.
4. As a radiographic tube ages, what adjustment must be made in the filament current, and why?
5. What effect does evaporation of the filament have on tube operation? tube life? x-ray output?
6. Describe the line focus effect and discuss its importance in radiography. To what types of tubes does it apply?
7. Discuss saturation current.
8. How does the space charge effect influence the operation of a radiographic tube? How does a change in filament current affect space charge?
9. To what part of the tube does heat storage refer? In what units is it expressed?
10. Using the chart in Figure 13.11, determine the tube rating ("safe" exposure time) at 70 kV and 500 mA.
11. Discuss the importance of anode heat storage; anode cooling rate; housing cooling rate; anode pitting; anode evaporation.
12. Why does a radiographic tube anode have a greater heat loading capacity with full-wave rectification than with half-wave rectification?
13. Summarize the steps that should be taken to prolong tube life.
14. Compare the values of peak milliamperage and milliammeter readings in a full-wave and half-wave rectified circuit.
15. How are radiographic tubes cooled?
16. Describe the spinning top method and state its purpose. What is used instead of a spinning top to check timer accuracy in twelve-pulse three-phase equipment, and why?
17. What are the main causes of x-ray tube failure? How can they be minimized?
18. In Figure 13.12, at 3 1/2 min, how many heat units remain stored in the anode? How much additional heat could then be stored in the anode?

Chapter 14

X-RAY CIRCUITS

Topics Covered in This Chapter

- Source of Electric Power
- Main X-Ray Circuits
  Primary
  Secondary
- Complete Wiring Diagram of Single-Phase Circuit
- Primitive Control Panel
- Modern Control Panel
- Three-phase Generation of X Rays
- High-Frequency Generator
- Falling-load Generator
- Mobile X-Ray Machines
- Questions and Problems

THE MAJOR ITEMS of an x-ray unit have been covered in some detail, but they have been discussed individually without relation to each other in an x-ray machine. There now remains the task of combining them to create a functioning x-ray machine. A number of auxiliary devices are necessary for the proper operation of x-ray equipment, and these will also be described.

Source of Electric Power

The electric power company supplies a high voltage alternating current to the *pole transformer* outside the building. This transformer, which is mounted atop a pole, is a step-down transformer. It reduces the high voltage to 120 to 240 volts, depending on the type of equipment to be served, and is called the *line voltage.*

A *three-wire system* conducts the current into the building, the two outer wires being "hot" and the middle one "grounded" (neu-

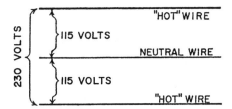

Figure 14.1. Three-wire system which brings electric power into the building. By connecting the electrical appliance or equipment to the appropriate wires as indicated, one may obtain 120 volts or 240 volts.

tral), as shown in Figure 14.1. At the instant that one hot wire is 120 volts above ground, the other is 120 volts below ground, thereby producing a potential difference of 240 volts. Thus, connection to the two hot wires provides a 240-volt source, whereas connection to either hot wire and the neutral wire provides 120 volts.

Most x-ray units now operate on 240 volts, although some of the less advanced equipment, and especially the older mobile units, require only 120 volts.

Note that the current delivered to the Radiology Department is **alternating current** (AC) because it is necessary for the operation of the x-ray transformer. Under unusual circumstances when only direct current (DC) is available, it must first be converted to AC so that it can be stepped up by the transformer.

The Main Single-phase X-ray Circuits

To simplify the discussion of the electrical connections of an x-ray unit, we may regard the circuit as being divided at the x-ray transformer. The part of the circuit connected to the primary coil of the transformer is the **primary** or **low voltage** circuit, whereas the part connected to the secondary coil is the **secondary** or **high voltage** circuit. This scheme is perfectly natural, since the two coils of the transformer are electrically insulated from each other, and since the profound difference in voltage in the two circuits requires separate consideration. However, some parts of the equipment are connected to both the low and high voltage sides.

1. **Primary Circuit.** Included in this portion of the circuit is all the equipment that is connected between the electrical source and the primary coil of the x-ray transformer. Each essential component will be described in proper sequence.

a. **Main Switch.** This is usually a double-blade, single-throw switch, illustrated in Figure 14.2.

b. **Fuses.** Each conductor leaving the main switch is provided with a **fuse**–an insulated cylinder with a metal cap at each end, in the center of which a thin strip of metal of low melting point connects the caps. Overloading the circuit with excessive amperage causes the metal strip within the fuse to melt because of the heating effect of the current, thereby breaking the circuit.

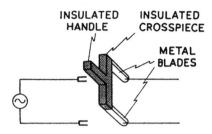

Figure 14.2. Double-blade single-throw switch. When the switch is closed toward the left, the metal blades contact the corresponding wire and the circuit is completed to the x-ray machine.

Figure 14.3 shows the components and electrical connection of such a fuse, which serves to protect equipment from overload and reduce fire hazard. A blown fuse is easily replaced, but the underlying cause should promptly be ascertained and corrected.

c. **Autotransformer.** Discussed in detail in Chapter 10, the autotransformer changes the voltage supplied to the primary of the transformer, thereby providing various kilovoltages for the x-ray tube.

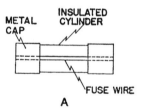

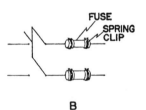

Figure 14.3. *A.* Diagram of a fuse. *B.* Its connection in a circuit. When the circuit is overloaded (excessive amperage) the wire in the center of the fuse melts, breaking the circuit.

d. *Prereading Kilovoltmeter.* Since kVp selection must be made (by means of an autotransformer) before a radiographic exposure, we must have a metering device that can indicate the desired kVp. This function is served by the *prereading kilovoltmeter* which is an AC voltmeter connected in parallel with the autotransformer, that is, across the primary circuit between the autotransformer and the transformer (see Figure 14.12). How, then, is it possible for a voltmeter in the primary circuit to indicate kVp across the secondary circuit? The answer lies in the calibration (standardization) of the prereading kilovoltmeter in the primary circuit by the manufacturer, against actual peak kV values in the secondary circuit by means of spark gap measurements. These depend on the fact that a given kVp will cause a spark to jump between two metal spheres separated by a particular distance or *air gap* (at standard temperature, atmospheric pressure, and relative humidity). The kilovoltages obtained by such spark gap measurements are recorded directly on the scale of the prereading kilovoltmeter for each corresponding autotransformer setting. In other words, the prereading kilovoltmeter, since it is in the primary circuit, does not measure kVp directly, but rather indicates what the kVp will be in the secondary circuit because it has been initially calibrated to do so at the factory. The main advantage of the prereading kilovoltmeter is that *slow* fluctuations in line voltage can easily be corrected. If, for instance, the line voltage should drop, the autotransformer controls are adjusted manually or automatically until the correct kVp shows on the prereading kilovoltmeter.

Some types of x-ray equipment do not employ a prereading kilovoltmeter. Instead, the desired kVp is obtained by means of kV-labeled *pushbuttons* which select the appropriate transformer settings. The calibration of these settings is similar to that of a prereading kilovoltmeter. But these fixed pushbutton settings indicate kVp accurately only if the line voltage is the same as it was during the actual calibration. Since the line voltage may fluctuate slowly, depending on the other electrical equipment in use, a means of detecting such fluctuation is provided by a *compensator voltmeter* connected in parallel across a portion of the primary side of the autotransformer. A mark on the compensator voltmeter scale indicates the correct line voltage. If the pointer is above or below this mark on the meter, a *line voltage compensator* permits adjustment of the line voltage to the correct value. The compensator varies the number of turns on the primary side of the autotransformer until the compensator meter needle indicates that the correct line voltage has been restored; only then will the kVp be correct. High-frequency equipment provides automatic kVp control (see page 165).

e. *Timer and X-ray Exposure Switches.* These switches control the current to the primary coil of the transformer, serving to complete the x-ray exposure. The switches themselves are modifications of an ordinary pushbutton, shown in Figure 14.4, and can be used for either hand or foot operation. But these alone cannot withstand the high amperage in the primary circuit (eg, as high as 90 A with a 200 mA unit), or prevent electric shock. Therefore, these switches operate a *remote control switch,* which in turn closes the primary circuit. Modern x-ray machines use a microprocessor (computer) to operate remote control switching.

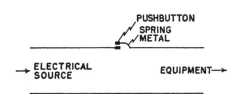

Figure 14.4. Pushbutton switch. Depressing the button brings the metal contactors together, thereby closing the circuit.

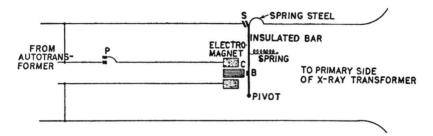

Figure 14.5. Remote control switch. The pushbutton, *P,* when depressed, completes the circuit through the coil of the remote control switch. This magnetizes the core, *C,* which attracts the metal button, *B,* on the insulated bar. This closes the primary circuit by bringing together the contacts at *S.* When the pushbutton is released, the reverse occurs, breaking the circuit.

f. **Remote Control Switch.** Operated by either hand or foot switch, its basic design can best be appreciated by studying Figure 14.5. Its design calls for opening and closing near or zero kVp to avoid power surge.

g. **Timer.** Included in the exposure switch circuit is a timer which can be set to start and stop an exposure of preselected duration. There are four types of exposure timers. They include (1) mechanical, (2) electronic, (3) milliampere-second meter, and (4) phototimer.

(1) **Mechanical timer.** As the name implies, such a hand-held timer has a clock mechanism with a pointer that can be rotated, by means of a knob, around a scale calibrated in seconds. Turning the knob to the selected time winds up a spring; and activating the exposure button releases the spring and terminates the exposure at the preselected time interval. Mechanical timers are only accurate to about 0.25s and are limited to dental x-ray machines where a high degree of accuracy is not required.

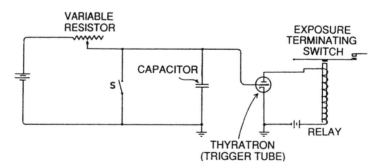

Figure 14.6. Simplified circuitry of an electronic timer. When switch *S* is closed, charge cannot build up on the capacitor. Activation of the exposure switch (in another part of the circuit) starts the exposure and opens switch *S,* allowing a charge to build up on the capacitor. When this reaches a preset value, the capacitor suddenly discharges through the thyratron tube, activating the exposure termination switch. This may be either an electromagnetic switch, as shown, or a *thyristor,* a solid state device functioning like a thyratron but allowing exposures as short as 1 ms with a 1-ms delay. The time for build-up of capacitor charge depends on *RC,* which is the product of resistance by capacitance. Since *C* is constant, we select the exposure time by adjusting *R* by means of a rheostat (variable resistor).

(2) ***Electronic Timer***–provides extremely accurate exposures down to 1 millisecond ($1/1000$ s). Its circuitry, shown in Figure 14.6, consists essentially of a rheostat (variable resistor), a capacitor, and a ***thyratron*** which is a special gas-filled triode that becomes conductive at a particular, critical voltage. When the x-ray exposure switch is closed, switch *S* is simultaneously opened, allowing current to build up a charge on the capacitor. When the capacitor charge reaches a critical value, discharge suddenly occurs through the thyratron, activating the exposure-terminating switch and thereby ending the exposure. The time required to build up the necessary charge on the capacitor governs the time of the x-ray exposure; it depends on the product *RC* (resistance × capacitance) and can be varied by adjusting the resistance of the circuit by means of the rheostat. The capacitance C remains constant in this particular circuit. In six-pulse and twelve-pulse three-phase equipment the thyratron, and the electromagnetic relay with its mechanical contactor, are replaced by a ***thyristor,*** which is a solid state device that functions like a thyratron but can provide much shorter and more accurate exposures. These may be as short as 1 millisecond with a 1-millisecond (ms) = $1/1000$ s delay between activation of the exposure switch and the actual start of the exposure.

(3) ***Milliampere-second (mAs) Meter.*** This is, strictly speaking, not really a timer since it reads out mAs–the product of mA × exposure time. It is described on page 157.

(4) ***Automatic Exposure Control***–includes a phototimer circuit. With this system, developed by Morgan and Hodges in the early 1940s, a film is exposed to the x-ray beam and as soon as it has received the correct amount of radiation for the desired degree of film darkening (density), the exposure terminates automatically. This requires a special type of diode known as a ***phototube;*** its cathode is coated with an alkali metal such as potassium or cesium, which has the peculiar property of ***giving off electrons when struck by light.*** If the anode of the phototube has, in the meanwhile, been given a positive charge from an outside source, the electrons emitted by the cathode are attracted to the anode, constituting a current in the phototube. A fluorescent screen between the phototube and the x-ray source, we can make the phototube indirectly sensitive to x-rays, the brightness of the screen depending on the intensity of the x radiation. As shown in Figure 14.7, the phototube-fluorescent screen combination is placed behind the cassette whose back is radiotransparent (x-ray transmitting). When a predetermined quantity of radiation has reached the fluorescent screen, depending on the part being radiographed, the resulting current in the phototube operates a capacitor, thyratron, and relay circuit which activates a contactor to terminate the exposure.

Automatic exposure control in modern x-ray units has abandoned the phototube, replacing it with a thin, radiolucent, ***parallel-plate ionization chamber*** positioned between the patient and film cassette. This also eliminates the need of a fluorescent screen and phototube. Because the charge liberated in the ion chamber varies directly with the desired radiographic density, the chamber is calibrated initially (upon installation) to terminate the exposure via the remainder of the circuit as in Figure 14.7, when the proper amount of radiation has been delivered.

Automatic exposure control entails extremely accurate reproduction of radiographic density, provided the anatomic part is carefully centered. The most useful application of automatic exposure control is in spotfilm and chest radiography, and in photofluorography, although it is also available now for general radiography.

h. ***Backup Timer and Minimum Exposure Time.*** Modern equipment designed for auto-

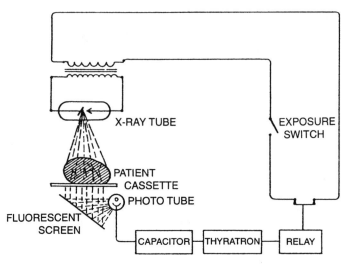

Figure 14.7. Automatic exposure control. When the exposure switch is closed, x rays passing through the patient and reaching the fluorescent screen of the phototimer assembly cause the screen to emit light which activates the phototube. Current then flows in the phototube and charges the capacitor while the exposure is in progress. When the predetermined voltage is built up on the capacitor, the thyratron becomes conductive, activating the relay which opens the exposure circuit and terminates the exposure even while the manual exposure switch is still closed. In most automatic exposure systems available now, the fluorescent screen and phototube have been replaced by a *parallel plate ionization chamber.*

matic exposure control has a special manual backup timer that is set to terminate the exposure in the event of failure in the automatic circuit, to avoid the hazard of overloading the tube or overexposing the patient. According to the Code of Federal Regulations (21CFR), the maximum allowable exposure time shall be equal to or less than $1/60$ sec, or the time interval to deliver 5 mAs, whichever is greater. The maximum allowable mA × time shall be limited to 600 mAs per exposure, except when the tube potential is less than 50 kVp, in which case the limit shall be 2000 mAs per exposure. In equipment with pushbutton selection of the anatomic part, the backup timer is set automatically for each selection by internal circuitry, but HEW regulations still apply as above.

A manual *backup timer* is set for a specific time, usually about 0.5 s. As an example,

if the exposure called for 140 mAs at a 200 mA setting, the time would be 140 mAs/200 mA = 0.7 s, and the exposure would be terminated at 0.5 s by the backup timer. Thus, the film would be underexposed. This follows from the relation

$$\frac{mAs}{mA} = time\ (s)$$

because units can be handled in mathematics in the same way as numbers.

Radiographic equipment timer circuits have a *minimum exposure time,* which varies with the type of equipment, ranging from about 0.05 s (50 ms) in older equipment, to 0.005 s (5 ms) in later models. (With the introduction of thyristors, minimum exposure times will be in fractions of milliseconds.) For example, with a equipment having a 0.05 s minimum exposure time, if the exposure required 2 mAs at 200 mA, the time would be 2 mAs/200 mA = 0.01 s. Since this is less than the minimum exposure time (0.05 s) the exposure would terminate at 0.05 s and the film would be overexposed. But if the exposure required 12 mAs, the time would be 12 mAs/200 mA = 0.06 s as this is longer than the minimum exposure time (0.05 s the exposure time would terminate correctly after a total of 0.06 s).

i. *Circuit Breaker.* Additional protection against overloading the circuit is provided by a circuit breaker which can easily be reset, while a blown fuse has to be replaced. It is usually connected in series with the exposure switch, timer, and remote control switch. Figure 14.8 illustrates schematically the operation of a magnetic circuit breaker, shown in the exposure circuit, 2. When the exposure switch is closed, circuit 2 is completed through the timer, the electromagnet of the remote control switch, and the circuit breaker contacts which are in the closed position. This activates the electromagnet of the remote control switch closing the primary circuit, 1, that leads to the primary side of the x-ray transformer. If there should be a momentary surge in the primary current,

this will, through circuit 3, increase the strength of the electromagnet in the circuit breaker to the point where it will open circuit 2, thereby interrupting the current to the electromagnet of the remote control switch, opening the remote control switch, and breaking the primary circuit, 1. The circuit breaker must then be reset manually before another exposure can be made.

j. *Filament Circuit of X-ray Tube.* The primary circuit supplies the *heating* current for the filament of the x-ray tube, but this current must first be reduced to 3 to 5 amp and 6 to 12 volts by a *rheostat* (variable resistor); in accordance with Ohm's law, the greater the resistance, the smaller the current (amp). Most radiographic units are equipped with *pushbutton type* mA selectors which also choose the desired focal spot size. The pushbuttons operate through individual preset resistors which provide the correct filament current for each tube current, such as 100, 200, and 500 mA. A more advanced type of filament current regulator is the *high-frequency control,* described on page 165. In series with this is an *oil immersed step-down transformer,* which further

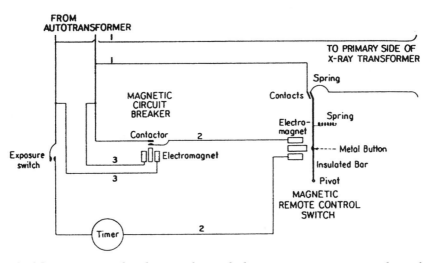

Figure 14.8. Magnetic circuit breaker together with the magnetic remote control switch, as they are connected in the primary circuit.

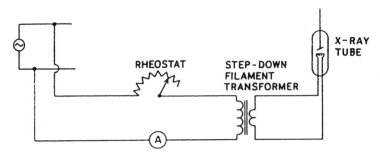

Figure 14.9. Filament circuit. The rheostat varies the filament current (and voltage), thereby controlling filament temperature, electron emission, and tube current (mA). Thus, it is an *mA selector*. The step-down transformer further helps reduce the voltage and also insulates the primary circuit from the high voltage in the secondary circuit. Instead of a rheostat, modern equipment uses a high-frequency circuit.

reduces the voltage to the required value, as shown in Figure 14.9. Because this *filament transformer* has its secondary coil in the high voltage circuit and its primary coil in the low voltage circuit of the x-ray unit, it must be oil immersed to provide sufficient insulation between these two circuits.

There is often included in the filament circuit a *filament stabilizer* for the x-ray tube because a relatively small change in the filament voltage or current causes a large change in electron emission and consequent tube current (mA). The filament stabilizer corrects for instantaneous fluctuations in line voltage that may be caused by momentary demand elsewhere on the line, such as in the starting of an elevator or air conditioner. It may be so effective that a variation in line voltage of 10 percent will cause no greater change than 1/2 percent in the filament voltage. The stabilizer consists of a capacitor and a small, modified split transformer which are so arranged that they compensate for a rise or fall in line voltage and provide a more uniform filament voltage, maintaining the filament current more nearly constant.

k. *Filament Ammeter.* In order to measure the filament current, and hence the amount of heat developed in the filament, an ammeter is connected in series in the *filament circuit* of old style equipment. By pre-

vious calibration, we can establish the readings of this meter that correspond to desired mA in the x-ray tube at a given kV. In modern radiographic units a *space charge compensator* automatically adjusts the filament current to maintain constant mA over a wide range of kV; furthermore, pushbutton selection of mA automatically provides the correct filament current for the desired tube current.

l. *Primary Coil of X-ray Transformer.* Although the high voltage transformer operates as a unit, we may regard its primary winding (coil) as a component of the primary circuit. Both the primary and the secondary coils are immersed in special transformer oil for added insulation.

In general, the wires in the primary circuit must be relatively large, because of the high amperage. Circuit breakers and fuses should be conveniently located for easy resetting or replacement.

2. **Secondary Circuit.** This includes the secondary coil of the transformer and all devices to which it is connected electrically. The wire conductors in the secondary circuit have a smaller diameter (than those in the primary circuit) because they carry a smaller current (mA).

a. *Secondary Coil of X-ray Transformer.* As we have indicated earlier, this consists of many turns of electrically insulated wire that is thinner than the wire in the primary coil because of the very small current in the secondary circuit. The transformer **steps up** the primary voltage to get the high voltage needed to operate the x-ray tube, the step-up ratio being about 500:1 or 1000:1.

b. **Milliammeter.** In order to measure the mA in the x-ray tube, a **milliammeter** is connected in series in the high voltage circuit. Since the milliammeter is grounded together with the midpoint of the secondary coil of the x-ray transformer, it is at zero potential and can therefore be safely mounted in the control panel without hazard to personnel.

Attention must be called to the fact that the **tube current is measured in milliamperes by the milliammeter placed in the high voltage circuit, while the filament current is measured by an ammeter placed in the low voltage filament circuit.** The milliammeter measures average values and gives no indication of the peak values of tube current.

c. **Milliampere-second or mAs Meter.** It is desirable to have in series with a regular ammeter, a mAs meter. Due to the large mass of the rotating mechanism, it turns so slowly that the attached indicator needle registers the product of mA and time, that is, it indicates **milliampere-seconds** (mAs). This device is needed because the ordinary milliammeter does not have time to register the true mA at exposures of less than $1/10$ sec. The ballistic meter therefore serves to measure mAs at very short exposure times, and because of its great sensitivity, it must be used in conjunction with an electronic timer. With exposures longer than $1/10$ s an ordinary milliammeter registers average mA.

d. **Rectifier.** In all but self-rectified units, a system of rectification is included to change AC supplied by the transformer to DC. As already noted, rectification enhances the heating capacity of the x-ray tube, permitting larger exposures.

e. **Cables.** Since they conduct high-voltage current from the rectifier to the x-ray tube and would entail a shock hazard, three expedients eliminate this problem:

(1). *Insulation.* The cables are covered by an inside layer of highly insulating material, enclosed by a flexible plastic cover (see Figure 14.10).

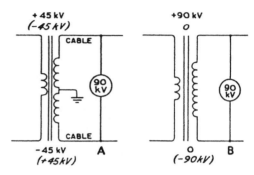

Figure 14.10. *A.* The transformer secondary has been grounded at its midpoint. With the upper end of the secondary coil positive, and the lower end negative *final kV = 45 kV – (–45 kV) = 90 kV.* Each cable needs insulation for only 45 kV above or below ground. A similar situation prevails when the polarity of the secondary reverses in the next half cycle (values in parenthesis). *B.* The secondary has not been grounded at its midpoint. Total secondary voltage is again 90 kV, but it now represents the difference between 90 kV at one end of the secondary, and 0 at the other end. Each cable must therefore be insulated for the full 90 kV above or below ground.

(2). *Grounding.* Just beneath the protective external plastic cover, there is a surrounding woven wire sheath that has been securely *grounded.*

(3). *Secondary Winding of Step-up Transformer Grounded at its Midpoint.* This reduces the amount of insulation and weight of the cables that would otherwise be required, explained as follows: grounding the transformer secondary at its midpoint does not change total kVp because one-half is below,

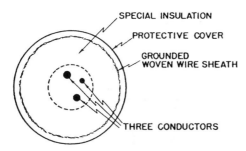

Figure 14.11. Cross section of a shockproof cable.

and the other half above ground potential (see Figure 14.11). For example, if there is 90 kVp across the transformer and the center is not grounded, the kVp, will fluctuate between 90 kVp above ground and 90 kVp below ground according to the AC sine wave. If the center of the secondary coil is grounded, the peak kVp will still be 90, but in this case one-half or 45 kVp will be above ground and one-half will be 45 kVp below. The difference between 45 above zero and 45 below zero is still 90. Under these conditions each cable has to be insulated for only 45 kVp (rather than 90 kVp), representing a significant saving in weight and cost.

The same type of cable is used for both terminals of the x-ray tube. When the **cathode** cable is connected to the cathode end of a double-focus x-ray tube, all three conductors make contact with the two filaments (see Figure 13.3). The other end of this cable makes contact with both the filament circuit and one end of the transformer secondary. When the **anode** cable is connected to the anode, connection is made with only one of the conductors within the cable. At the other end of the cable, the three conductors join a single conductor which connects with the opposite end of the transformer secondary. Thus, the three conductors in the anode cable serve as a single conductor, which is all that is needed for the anode.

f. **X-ray Tube.** The ultimate goal of the

x-ray equipment is the operation of an x-ray tube, the last piece of equipment connected in the high voltage circuit.

Completed Wiring Diagram

In order to visualize clearly the relationship of the various parts of a single-phase full-wave rectified x-ray unit, we must connect them correctly in the circuit. Figure 14.12 presents schematically a full-wave rectified x-ray unit employing four rectifier diodes. Most of the auxiliary items, though desirable or even necessary for proper operation, have been purposely omitted to avoid complicating the diagram. The hook-up of these accessory devices has been indicated in the preceding sections. We can readily modify this basic diagram; thus, for self-rectification we simply omit the rectifier and connect the transformer to the tube terminals.

Basic X-ray Control Panel or Console

An x-ray machine requires not only suitable control devices for easy selection of kVp, mA, and time, but also readily accessible meters to check the operation of the equipment. The controls and meters are mounted compactly in a *control panel*–a separate unit which is connected electrically to the x-ray equipment and is comparable in function to the dashboard of an automobile.

X-ray control panels vary greatly in complexity, depending on the design of the individual x-ray machine. If we realize that all control panels, regardless of their intricacy, are similar in their basic design, the problem of operating unfamiliar equipment becomes a relatively simple matter. The more complicated units are variations of the basic pattern.

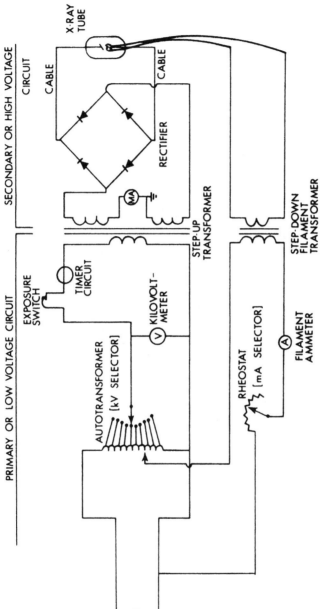

Figure 14.12. Simplified wiring of a single phase x-ray unit, with full-wave rectification.

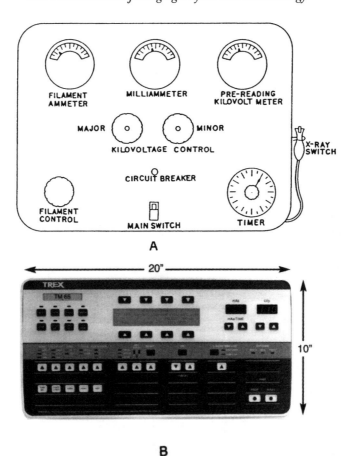

Figure 14.13. *A.* Primitive x-ray control panel in the 1920s to 1940s. The filament control knob operates the rheostat to select tube current, which is measured by the grounded filament amme- ter. kVp is selected by major and minor control knobs. The milliammeter indicates average mA. *B.* Example of a state-of-the-art high-frequency, three-phase control panel, which is micro- processor controlled; it is much smaller than the primitive panel, and yet includes automatic expo- sure control with icons indicating the anatomic part, kVp and mAs, radiographed part, SID, and other information. (Courtesy of Continental-Trex Medical Corp.)

Figure 14.13A shows a "bare bones," nonautomatic control panel comprising all the essential items; familiarity with this dia- gram will make it much easier to understand the operation of the more complex types of equipment. The three meters shown in the figure include the milliammeter, filament ammeter, and kilovoltmeter. The mil- liammeter registers, in mA, the current in the secondary circuit, including the x-ray tube.

A filament ammeter indicates filament (heating) current, which ultimately deter- mines mA. Turning the control knob in the lower left hand corner of the diagram, labeled "filament control," varies the fila- ment current. This control actually operates a rheostat in the filament circuit to vary resis- tance.

A kilovoltmeter does not actually measure kVp directly, but is connected across the primary circuit and therefore measures volts. However, these readings have been calibrated beforehand against known tube kilovoltages and the meter scale has been marked accordingly in kVp (see page 151). A kilovoltmeter serves also as a compensating voltmeter because any drop in the line voltage produces a lower kVp reading on the meter. The technologist can adjust this manually by turning the "major" and "minor" kVp controls which operate the autotransformer, a device that varies the voltage input to the primary of the x-ray transformer. The major control varies the kVp in steps of ten, while the minor varies it in steps of one.

An important modification of this basic plan is found in the type of control panel in which the kilovoltmeter is replaced either by *kilovoltage selector* control knobs or *pushbuttons* operating through the autotransformer to provide a range of kVp. The appropriate kVp settings have been arrived at by previous calibration at the factory. In other words, when the settings read "56," it means that if the kVp were actually measured it would be 56 kVp, provided the line voltage corresponds to that at the time of original calibration. With this system a line voltage meter and a line voltage compensator must be included to correct for variations in line voltage.

Modern radiographic equipment includes a *milliamperage selector.* Instead of a filament control knob, there is an array of pushbuttons which activate the proper resistor to obtain the correct filament current for the desired mA. The pushbuttons also select the focal spot. For example, the pushbutton setting for large focal spot and 200 mA should automatically activate the large focal spot and deliver a tube current of 200 mA.

The timer, indicated in the lower right-hand corner of the diagram, controls the duration of the x-ray exposure which is initi-ated by the pushbutton x-ray switch. In addition, there is a main switch; when this is in the "on" position, the primary voltage is applied to the autotransformer, and also to the kilovoltmeter which is connected in parallel across the primary circuit. At the same time, the current also flows through the *filament only* of the x-ray tube and registers on the filament ammeter. Thus, *when the main switch is closed, only the filament ammeter and the kilovoltmeter should normally be activated.* Then, when the x-ray exposure switch is closed, the primary voltage applied to the primary side of the x-ray transformer is stepped up by the transformer, and current finally passes through the x-ray tube as indicated on the milliammeter. Thus, *the milliammeter does not register until the x-ray switch is closed.*

The construction and operation of the circuit breaker have already been discussed.

Most diagnostic equipment is supplied with a special radiographic-fluoroscopic *changeover switch,* enabling the radiologist to do spotfilm work automatically. Shifting the spotfilm device into the radiographic position activates the *changeover switch* and automatically selects the exposure factors for radiography. When the spotfilm is withdrawn from the radiographic position, the exposure factors automatically return to fluoroscopy. Phototiming is essential for timing spotfilm exposures.

You will find it advantageous to correlate the sections on the various items of x-ray equipment with the operation of the basic control panel, thereby developing a much clearer concept of the function of the various parts.

Modern equipment has become much more complex in that microprocessor control has greatly automated and simplified control operation and selection of techniques. The technologist is faced with a wide range of available consoles (eg, see Figure 14.13B).

THREE-PHASE GENERATION OF X-RAYS

The circuitry of three-phase equipment is necessarily more complex than that of single-phase. At present, there are three main types of three-phase circuits: six-pulse-six-rectifier, six-pulse-twelve-rectifier, and twelve-pulse-twelve-rectifier. All use *solid state diode rectifiers.*

A three-phase generator operates on *three-phase current,* which consists of three single-phase currents out of step with each other by one-third cycle or 120° (see Figure 14.14). Therefore, the transformer primary and secondary coils must each have three windings. These are arranged in either a *delta* or a *star* ("wye") configuration (see Figure 14.15). Three-phase circuits all have *delta-wound primary coils,* but differ in the form of the secondary windings. In Figure 14.16 is shown a wiring diagram of a twelve-pulse-twelve-rectifier generator having a balanced circuit.

With three-phase equipment, three autotransformers are needed for kV selection, one for each phase. Furthermore, because of its nearly constant value, the high voltage circuit cannot be closed and opened at or near zero potential as in the conventional single-phase unit. This introduces the danger of *power surge* which can damage the equipment. Therefore, special thyratron timer circuits have been designed to make the contactors open and close sequentially instead of simultaneously. Power surge is further suppressed by the solid state rectifiers, but in modern equipment thyratrons and mechanical contactors have been replaced by silicon-controlled rectifiers (SCRs) to facilitate switching and provide very short exposure times (see page 105).

Three-phase generators are available with ratings up to 1000 mA at 80 kVp (ie, 80 kW) and with exposure times as short as 1 ms ($1/1000$ s). Furthermore, as shown in Figure 14.17, the voltage is the resultant of three out-of-phase single-phase voltages and never reaches zero. In fact, the voltage shows a small fluctuation or *ripple* and may be regarded as nearly *constant potential.* Al-

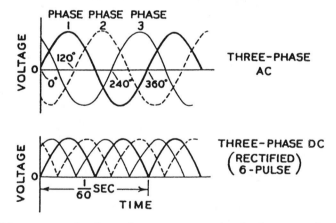

Figure 14.14. Voltage curves for three-phase power supply. In the upper diagram phase *1* lags phase *2* by 120°, and phase *2* lags phase *3* by 120°. In the lower diagram the three-phase current has been rectified to give six pulses in $1/60$ sec, with a 13 percent variation in voltage, or *ripple.* Special circuitry (not shown) provides twelve pulses in $1/60$ sec with a ripple of only 3 percent, giving virtually constant potential.

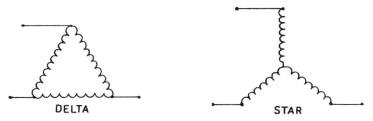

Figure 14.15. Types of transformer windings in three-phase circuits.

though the effective kVp at the transformer is about 95 percent with a 12-pulse, and 87 percent with a 6-pulse generator, the effective kVp during operation is actually 95 percent with either generator because of the voltage-smoothing effect of the capacitance in the high tension cables. The resulting ripple is about 3 to 5 percent in both instances, contrasted with about 70 percent in single-phase full-wave rectification.

Comparison of radiographic technics with the two systems shows that with the same mAs, 85 kVp three-phase is equivalent to 100 kVp single-phase generated x rays. On the other hand, if one uses 100 kVp with both types of equipment, the mAs for three-phase will be about 0.4 that of single-phase for equal radiographic density. Contrast will be less with the three-phase generator because of the greater half-value layer of the x-ray beam. When technique is adjusted to obtain the same density and con-

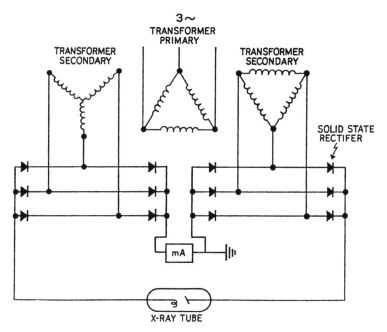

Figure 14.16. Simplified wiring diagram of a twelve-pulse-twelve-rectifier generator in a balanced circuit. Note the delta winding of the transformer primary, and the delta and star windings of the secondaries.

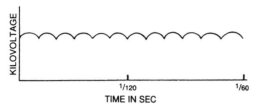

Figure 14.17. Rectified voltage curve with three-phase twelve-pulse x-ray generator. The voltage remains nearly constant throughout the alternating current cycle, the slight variation being called *ripple;* in this case, ripple is 3.5 percent. Note the twelve peaks or pulses per cycle of $1/60$ sec.

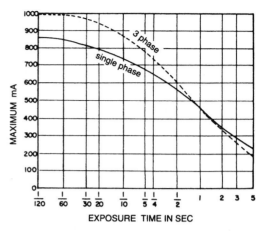

Figure 14.18. Comparison of tube rating curves with typical single-phase and three-phase x-ray generators. Note the higher rating with three-phase for short exposures, and the slightly higher rating with single phase for long exposures.

trast, the *radiation exposure of the patient is the same with both types of equipment.* Hence, from the standpoints of radiographic quality and patient exposure, there is no appreciable improvement with the three-phase generator.

Probably the single most important advantage of the three-phase system is the higher tube rating at *very short exposures* allowing the use of very high mA with exposure times as short as one millisecond. Figure 14.18 shows tube rating curves comparing the maximum mA that can be used with three-phase and single-phase equipment at various exposure times (Dynamax® "61" tube, 1 mm focus, 100 kVp). Note that at $1/120$ s the maximum ratings are 1000 mA with three-phase, and 860 mA with single-phase. On the other hand at five seconds the situation is reversed; now the maximum ratings are 190 mA with three-phase, and 230 mA with single-phase, because with very short exposures, heating occurs mainly at the anode surface and is readily dissipated. But with long exposures with three-phase equipment the large power input heats the anode to a greater depth, which increas-

es the danger of anode overheating and cracking, and therefore impairs tube rating.

Because of the complex circuitry of three-phase equipment, changeover from fluoroscopy to radiography is slow–1.5 to 2 s. This can be improved by a Q-stator in the circuit to reduce changeover time to 0.5 s.

The efficiency of three-phase operation is enhanced by conversion to a high-frequency system (see pages 165-166 and Figures 14.19 and 14.20).

In summary, then, the advantages of three-phase generation of x rays are as follows:

1. High mA with very short exposures, especially useful in angiography and spot-film radiography.

2. Nearly constant potential, more so with high-frequency conversion.

3. Higher effective kV.

HIGH-FREQUENCY GENERATION OF X-RAYS

On page 91, a brief description of high-frequency generation of x rays was given pertinent to kVp and mA control. A more detailed treatment of this subject will now be presented. Figure 14.19 is a block diagram that shows the basic steps in high-frequency generation of x rays. Although single-phase 115 or 230 volt alternating current or a storage battery can be used, the diagram starts with three-phase power supply. The AC immediately charges a capacitor bank to produce a steady DC which is less subject to line voltage fluctuations. A high-frequency converter then changes the DC back to AC that supplies the full-wave high-voltage transformer whose kV is controlled by an autotransformer. A feedback loop from the transformer to the kV control tracks kV output and continually regulates it to the selected value throughout the exposure. Finally, a filter capacitor changes kV AC to DC for application to the x-ray tube.

At the same time, the filament transformer feedback loop to the mA control tracks the filament current and closely adjusts it to maintain the selected mA throughout the exposure, in this case, by pulse width modulation labeled A and B in the diagram. Thus, if the filament current should fall during the exposure, pulse width is widened to compensate; and if filament current should rise, pulse width is narrowed. Other methods of current regulation may be used–frequency modulation or rheostat.

Advantages. There has been rapidly growing acceptance of high-frequency generators, not only for general radiography, but especially for mammography and angiography because of the following advantages:

1. Impedance of circuit (see page 76) is very low, so a much smaller transformer is possible than with ordinary single- or three-phase equipment.

2. Much shorter exposure times available than with single-phase.

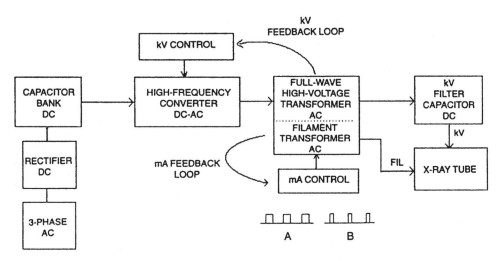

Figure 14.19. Block diagram of a high-frequency circuit for generation and regulation of kilovoltage and milliamperage. With a three-phase power source, this is the most efficient method of generating x rays for radiography today.

3. Higher kV and mA with very short exposure times are possible.

4. More accurate control of exposure time is provided for phototiming and serial radiography.

5. Waveform of kV virtually ripple-free, that is, constant potential (see Figure 14.20).

6. Conversion of AC line voltage to DC immediately, smoothing variations in line voltage and ultimate kVp.

7. Calibration of kV easier; also remains stable longer.

8. Real-time monitoring of kV, mA, and time by feedback control.

9. Error detection circuitry indicating, on control console, site of malfunction.

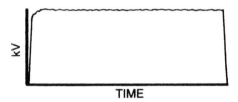

TIME

Figure 14.20. Kilovoltage waveform in a high-frequency circuit. Note the slight ripple, which represents virtually constant potential operation.

Microprocessor (computer) control further reduces equipment size. When technique is selected, a local microprocessor "speaks" to a remote microprocessor to select and monitor kV and mAs, with continuous regulation as explained above. This reduces the multitude of wires from the selector buttons to the activating circuitry, as only two wires are needed between the local and remote microprocessors. Automatic exposure control is also available for high-frequency equipment (also see pages 153-155).

Additional Control Features. A number of devices serve to protect the equipment from damage:

1. ***Tube Protection.*** Special circuits display a warning on the control panel, or abort tube activation when the selected technique exceeds the tube rating. Other circuitry displays the amount or percent of heat units accumulated in the anode, and/or the available remaining heat units (see pages 146-147).

2. ***Low and High Speed Rotor Controllers.*** These ensure rotor starting voltage, adjustable boost time, low running-voltage, and exposure circuit interlock to prevent energizing the tube before the anode is rotating.

3. ***Grounding of High-voltage Circuits.*** This is accomplished by grounding the mA meter and secondary circuit at the midpoint of the high voltage transformer secondary (also see pages 157-158).

Power Rating of X-ray Generators and Circuits

The power rating of x-ray generators and circuits ranges from about 30 to 80 kW (kilowatts), 1 watt being an energy expenditure of 1 joule/s. As mentioned earlier on page 77, electric power in AC circuits requires application of several modifying factors. However, for a steady direct current,

$$P = IV \; watts \qquad (1)$$

where P is power in watts, I current in A (amperes), and V voltage.

Since modern H-F generators produce a nearly constant potential waveform (ripple less than 2 percent), approximating direct current, we can apply equation (1) to express their power rating.

The power consumption in the ***secondary (high-voltage) circuit*** of an x-ray machine supplied by a H-F generator is expressed by

$$P = mA \times kV \; watts \qquad (2)$$

since mA = 0.001, or 10^{-3} A, and kV = 1000,

or 10^3 V. Substituting these values in equation (2),

$$P = 10^{-3} \, I \times 10^3 \, V$$
$$P = IV$$

because $-3 + 3 = 0$, and $10^0 = 1$ (recall that exponents are added). To change watts to kilowatts, divide the right side of equation (1) by 1000.

Power in the primary and secondary circuits must be the same according to the law of energy conservation, since P = energy-per second and therefore the energy will be the same in a given time interval.

Example 1. An exposure technique calls for 200 mA at 80 kV with an H-F unit. Calculate the energy expenditure.

Example 2. What is the power rating of an H-F generator when the maximum mA for 100 kV is 500? (Hint: convert mA to A, and the answer will be in kW.)

Example 3. Find the maximum power rating of an x-ray generator when the maximum exposure factors are 500 mA and 60 kVp.

FALLING-LOAD GENERATOR

An ingenious method of reducing exposure time and simplifying the selection of exposure techniques is provided by the *falling-load generator.** Unique circuitry, controlled by a microprocessor, automatically reduces mA in tiny steps closely following the selected kV curve of the pertinent tube rating chart (see Figure 14.21). In this process, the tube operates at near-maximum rating along the required initial segment of the kV curve, to produce the optimal mAs at each point.

Power supply is three-phase, converted to high-frequency, which permits a feedback loop from the secondary side of the full-wave rectified high-voltage transformer to

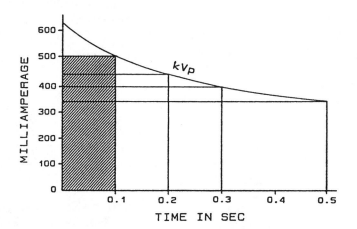

Figure 14.21. Tube rating chart shows the principle of the falling load generator. The shaded area includes the kVp and mA allowable for 0.1 sec. Following the course of the upper right-hand "corners" of the rectangles along the curve, the generator summates the mAs values at small intervals until the full mAs has been delivered.

---

* Medical Systems Division, North American Philips Corp.

the H-F converter to adjust for random voltage fluctuations in the primary circuit. A filter capacitor then "smooths" the kV before it is applied to the x-ray tube.

The falling-load generator permits either of two kinds of technique selection, depending on the circuitry of the machine:

1. *One-knob selection* allows the technologist to set *kV* by means of an "up-down" touch pad, like channel selection on a modern TV set. This requires a phototiming circuit, which automatically sets the optimum mAs for the desired examination.

2. *Two-knob selection* is available in the absence of phototiming. As before, kV is selected by means of an up-down touch pad, but now the mAs must be set by a separate up-down touch pad.

Demand for the falling-load generator has declined, mainly because it does not provide the very short exposure times available with three-phase high-frequency generators.

SPECIAL MOBILE X-RAY EQUIPMENT

Battery-Powered Mobile X-Ray Units

Self-contained mobile ("portable") apparatus energized by powerful storage batteries eliminates the need of an external power supply, except for recharging the batteries. An example of the updated version* of the original model utilizes a sealed lead-acid minimum-liquid battery that does not leak even if punctured. It actually comprises nine 12-volt packs connected in series, operating both the x-ray tube and the drive motor for the mobile unit.

A message center contains an *automatic voltage sensor,* which flashes a warning message indicating the need of a recharge when 90 percent of the stored mAs has been exhausted. A second message gives the warning that no charge remains; whereupon the unit shuts down automatically, but still allows a short interval in which the motor still operates to allow time to move the unit to a wall outlet. Recharging involves plugging the electric cord into a 110- or 220-V line. The charging process stops automatically when the battery attains full charge. This procedure must be conducted in an open area, never in a closet or operating room.

The circuitry is not too complex. An *inverter* changes the battery DC voltage to 1 kHz (1000 Hz) AC. The high-voltage *transformer* with full-wave rectification steps this up to 125 kV at nearly constant potential, and yields a usable storage of 20,000 mAs.

Microprocessor control provides highly accurate output as to kV and mAs, the kV being available as follows: 50-76 kV in 2 kV, and 80-125 kV in 5-kV steps. A touch panel allows two-point selection of kV and mAs; no separate timer is needed.*

The options for technique selection are as follows:

kV	mAs
50-90	0.4-320
95-105	0.4-250
110-125	0.4-200

This unit operates a rotating anode tube with a 0.75 mm nominal focal spot and a thermal capacity of 275,000 heat units.

Capacitor (Condenser)-Discharge Mobile X-ray Units

One of the early methods of generating x rays utilized the principle of storing electric

---

* GE Medical Systems, as of 1994.

charge in a capacitor (condenser) and then discharging it through an x-ray tube. At this point review the capacitor on pages 56-57. For our present purpose, we may say that the **capacitor stores mAs.** (Note that mA is a measure of charge per sec, so mA times sec should represent charge.)

The voltage and charge on a capacitor are related by the following equation:

$$Q = CV \qquad (1)$$

where Q = charge in coulombs
V = potential in volts
C = capacitance in farads (constant for a particular capacitor)

By selection of appropriate units, equation (1) is readily adapted to x-ray equipment; thus, with a capacitance of 1 microfarad (μF),

$$mAs = \mu F \times kV$$

In other words, if the capacitor is designed to have a capacitance of 1 μF, then 1 kV of potential is acquired for every mAs quantity of electric charge stored in the capacitor. For example, at a potential of 75 kV, it will have

a charge of 75 mAs, so a total of 75 mAs will be available to the x-ray tube. To charge the capacitor in ten seconds requires only 75 mA/10 sec = 7.5 mA, a small current, indeed.

The time it takes a **capacitor** to discharge to 37 percent of its initial charge is called the **time constant,** τ (Greek letter, *tau*), which is determined by the equation

$$\tau = RC \qquad (2)$$

where R and C are the resistance and capacitance of the circuit, respectively. By selecting the proper capacitance for a particular circuit we have a means of controlling the discharge time.

In practice, one should avoid too great a drop in kV during exposure, a situation that would exist if the capacitor were allowed to discharge completely. Instead, some type of interval timer must be used to control the duration of the exposure. The term **wave-tail cutoff** refers to the process of stopping the discharge of the capacitor at some preselected point on the discharge curve (see Figure 14.22). The first portion of the diagram shows the charging curve as the capacitor is charged to a desired peak kV, V_p, in

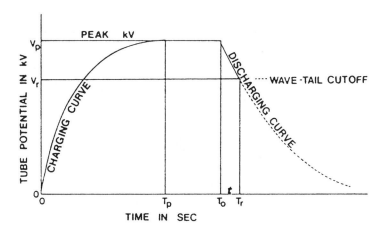

Figure 14.22. Charging and discharging characteristics of a capacitor discharge radiographic unit. V_p is peak kV; V_r is residual kV after capacitors have discharged during exposure time $(T_r - T_o)$; and T_p is time required to charge capacitors to peak kV.

time, T_p. *At time T_o* the exposure is initiated and the capacitor discharges along the discharging curve which, in this case, is cut off at time T_r. The tube current decreases along a similar curve. Wave-tail cutoff is represented by the dotted portion of the discharging curve.

If we were to start with 100 kV and a tube current of 500 mA, and used 30 mAs for an exposure, the exposure time, *t,* would be obtained from

$$mAt = mAs$$
$$500t = 30$$
$$t = 30/500 = 0.06 \text{ sec}$$

During this exposure time the tube potential will have dropped from 100 kV to 70 kV (a loss of 30 kV accompanying a loss of charge of 30 mAs), when the total capacitance is 1 μF.

How do we start and stop the discharge of the capacitors, thereby activating and deactivating the x-ray tube? This requires a ***grid-controlled x- ray tube*** (see Figure 14.23). An ordinary x-ray tube is a diode, having two electrodes. On the other hand, a grid-controlled tube has a third electrode or "grid," making this type of tube a triode. The grid, which in ordinary vacuum triodes consisted of a metal wire mesh, is actually the f***ocusing cup*** itself, completely insulated from the filament in a triode x-ray tube. When a ***negative potential*** or ***grid bias*** of 1 to 2 kV is applied to the cup relative to the tube kVp, it "breaks" the tube current (mA).

In the capacitor discharge unit, grid bias functions as a switch: activation of the expo-

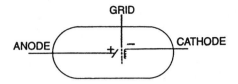

Figure 14.23. Simplified diagram of a grid-controlled x-ray tube. Exposure is started by removing the negative charge on the grid (nearly to zero); it is stopped by restoring a negative charge to the grid.

sure switch instantaneously reduces grid bias virtually to zero, permitting the applied kV to drive the space charge electrons to the anode as in a conventional x-ray tube. Once the selected mAs value has been attained (determined either by an mAs control device or an automatic exposure system), a grid bias of –1 to 2 kV is automatically applied and wave-tail cutoff achieved.

Figure 14.24 shows a block diagram of a capacitor discharge unit. For those who may be interested in greater detail, Figure 14.25 contains a wiring diagram of such equipment. As shown, one capacitor is charged during each half cycle of the alternating current to one-half the desired kV. The two capacitors in series discharge almost continuously through the tube during exposure, with a relatively small ripple in kV. Note the summation of the kV of the two capacitors in series.

A grid-controlled tube requires a cathode cable with a fourth connector or pin at its tube terminal, through which the negative bias voltage can be applied to the focusing

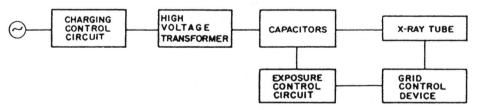

Figure 14.24. Block diagram of a capacitor discharge radiographic unit.

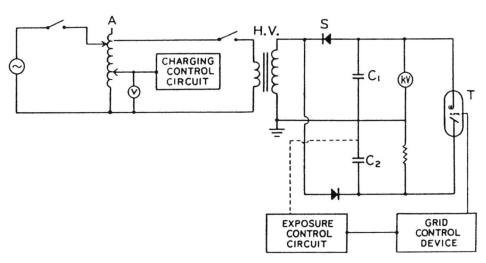

Figure 14.25. Simplified wiring diagram of a capacitor discharge radiographic unit. *A* is auto-transformer. *H.V.* is high voltage transformer. *S* is solid state diode (rectifier). C_1 and C_2 are capacitors.

cup (what are the other three connectors?)

In actual radiography with capacitor discharge apparatus, after the main switch is turned on, the line voltage is adjusted and the kV selected. Next, the capacitors are charged by depressing the charger button. When the preselected kV has been acquired by the capacitors, charging automatically stops. The mAs is selected by another control, although this step is omitted with phototiming (optional). Finally, the exposure button is activated, causing the capacitors to discharge through the x-ray tube.

Note that after completion of the exposure, kV may not return immediately to zero because of a residual charge on the capacitors. Because of electric shock hazard the high voltage cables cannot be disconnected while any charge remains. Another warning should be sounded here–if a charge remains on the capacitors, leakage may cause generation of x rays without activation of the exposure switch.

QUESTIONS AND PROBLEMS

1. Describe with the aid of a diagram the overload circuit breaker.
2. What is the function of a fuse and how does it work?
3. Discuss the principle of the prereading kilovoltmeter.
4. Why is a remote control switch necessary? How is it constructed?
5. Describe the three main types of exposure timers. Describe automatic exposure control.
6. What is the advantage of a circuit breaker over a fuse?
7. Show by diagram the filament circuit and its important components.
8. Why can a milliammeter be safely mounted on the control panel? What is a ballistic mAs meter and under

what conditions is it an essential part of the control apparatus?

9. Explain why the center of the transformer secondary coil is grounded.

10. Explain how the kilovoltage control varies the kilovoltage. Which device does it operate in the primary x-ray circuit?

11. Which device in the filament circuit is operated when the filament control is manipulated? What function does it serve?

12. Describe a shockproof cable with the aid of a diagram.

13. Prepare a simple wiring diagram of an x-ray machine, including four-diode rectification.

14. Of what does a three-phase current consist?

15. How does the design of a three-phase transformer differ from that of a single-phase? What type of rectifiers is used with three-phase equipment?

16. What is meant by ripple? Discuss its importance in radiography.

17. What are the advantages and disadvantages of the three-phase over the single-phase generator?

18. List the advantages of high-frequency generation of x rays. How is mA and kV controlled in HF circuits?

19. The minimum exposure time with an x-ray unit is set at 0.04 sec. A particular technique calls for 80 kVp and 3 mAs at 200 mA. Would the radiograph be under- or overexposed? Why?

20. A manual backup timer is set for 0.5 sec. If a technique calls calls for an exposure of 100 mAs at 500 mA, would the radiograph be over-, under-, or correctly exposed? Why?

21. Explain the principle of a capacitor discharge unit. How is exposure time controlled? What is wave-tail cutoff?

22. Describe the construction and operation of a grid-controlled x-ray tube.

23. Describe a storage-battery operated mobile (portable) x-ray unit and list its advantages.

Chapter 15

X-RAY FILMS, FILM HOLDERS, AND INTENSIFYING SCREENS

Topics Covered in This Chapter

- Composition of X-Ray Film
    Base
    Emulsion
        Speed
        Latitude
- Types of Films
- Film Exposure Holders
    Cardboard
    Cassettes
- Intensifying Screens
    Composition
    Principles

Fluorescence
Phosphorescence
Screen-Film Matching
Speed Factor
Efficiency
Screen-Film Contact
Graininess
Quantum Mottle
Crossover
Care of Screens
- Questions and Problems

ON PASSING THROUGH THE BODY, an x-ray beam interacts with the constituent atoms by photoelectric and Compton processes. As a result, the beam emerging from the body–the *exit beam*–consists of a pattern in which different areas have different numbers of photons corresponding to the pattern of tissue thicknesses, atomic numbers, and densities through which the beam has passed. The term *aerial image* refers to such an exit beam, since it contains information about the tissues through which the beam has passed. This information pattern can be revealed by various image receptors such as film, film-screen combinations, magnetic tapes and discs, and xerographic plates.

In the early days of roentgenography, glass photographic plates coated with an emulsion sensitive to light were used to record the radiologic image. The disadvantages of plates included the danger of breakage, the hazard of cutting one's hands, the difficulty in processing, and the inconvenience in filing these plates for future reference. With the introduction of flexible films, these disadvantages were eliminated. This chapter deals with the use of films and intensifying screens–one type of image receptor system whose purpose is to display the aerial image.

Composition of X-Ray Film

There are two essential components: the *base* and the *emulsion*. Figure 15.1 shows the

173

configuration of a typical double-emulsion x-ray film.

1. **The Base.** Modern film has a tough base measuring about 0.2 mm thick, consisting of a uniformly transparent sheet of *polyester plastic* that does not shrink or stretch and is usually tinted blue for eye comfort. It is called "safety" because it is no more flammable than an equal thickness of paper.

2. **The Emulsion.** This consists of microscopic crystals of **silver bromide** suspended in **gelatin.** A small amount of sodium iodide is also included. The gelatin, a transparent material similar to that used as a food, is obtained mainly from cattle skins and bones treated with mustard oil to improve the sensitivity of the emulsion by providing **traces of sulfur.** To the gelatin dissolved in hot water are added, in total darkness, silver nitrate and potassium bromide to produce **silver bromide;** and a much smaller amount of silver nitrate and potassium iodide to produce **silver iodide.** Silver bromide and silver iodide are examples of **silver halides.** Heating the mixture to a temperature of 50 to 80°C–a process called digestion–further improves the sensitivity of the emulsion. After cooling, it is shredded, washed, reheated, and mixed with additional gelatin. Not only does the gelatin keep the silver halides uniformly dispersed in the emulsion, but it is also readily penetrated by the processing chemicals and by the water used in the final wash.

Next, the emulsion is spread in a layer about 0.007 to 0.02 mm thick on both sides of the polyester base which has previously been coated with a special adhesive agent. Cooling requires strictly controlled conditions.

Finally comes the application of a gelatin protective layer as a **supercoat** on both sides of the film. Such double-coated film is routinely used in general radiography, but single-coated film is available for special purposes such as mammography.

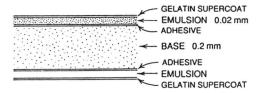

Figure 15.1. Cross section of a typical x-ray film.

Note that from the moment before the ingredients are brought together in the emulsion, until the finished sheet film has been packed in boxes, the entire process must be carried out in **total darkness.**

Film Emulsion Characteristics. Emulsions have been designed to respond to certain wavelengths of the ultraviolet, blue, or green light, as well as to x rays. Such adaptation of film sensitivity to specific colors is called **spectral sensitivity matching.** Moreover, an emulsion should display the radiologic image of a variety of tissues with optimum recorded detail, density, and contrast, qualities that will be discussed later.

An ordinary film contains the silver halide crystals clustered in the form of irregular microscopic pebble-shaped grains. The speed of such emulsions can be increased only by increasing the size of the grains and the thickness of the emulsion. This inevitably impairs recorded detail or image sharpness because of the coarser overall granular pattern in which the grains vary in size and shape, measuring about 1 to 1.5 micrometers in width.

A type of emulsion called **tabular** or T-grain (Eastman Kodak Company) contains **flat** silver halide grains (see Figure 15.2) whose size approximates that of conventional silver halide grains. Their flatness, as well as the addition of special sensitizers, allows them to capture more light without the need of thickening the emulsion. Consequently, the films display increased speed without sacrificing recorded detail. Since the T-grain

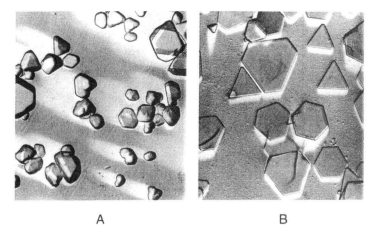

A   B

Figure 15.2. Microscopic appearance of silver halide crystals in a film emulsion. *A.* Irregular three-dimensional crystals in an *ordinary* emulsion. *B.* Flat hexagonal crystals in a *tabular grain* emulsion. They are thinner and more uniformly dispersed. (From Pizzutiello MS Jr., Cullinan JE, *Introduction to Medical Radiographic Imaging.* Courtesy of Eastman Kodak Company.)

emulsion is sensitive mainly to the light from green-emitting screens, a magenta dye is added to the emulsion to minimize crossover (see page 186, Figure 15.13) and further improve recorded detail.

To be suitable for radiography, an emulsion should have two important characteristics:

1. ***Speed or Sensitivity***–the relative ability of an emulsion to respond to radiation such as light and x rays. An emulsion is said to be fast or have high speed if a small exposure produces a radiograph of adequate optical density (darkening).

2. ***Latitude***–a term that encompasses three related properties of film emulsions:

    a. *Contrast Latitude*–the ability of an emulsion to display a radiographic image with a reasonably long tonal range, from white, through various shades of gray, to black. This is ***long scale*** or ***low contrast,*** to be explained later in more detail.

    b. *Exposure Latitude*–the range of exposure factors that will produce an acceptable radiograph. Obviously, the emulsion should have sufficient lati-

tude to allow a reasonable margin of error without appreciable impairment of radiographic quality. Both aspects of latitude–long scale and permissible margin of exposure error–are closely related. However, excessive film latitude may impair image visibility, that is, the ability to see fine recorded detail.

    c. *Subject Latitude*–the dependence on subject contrast (see page 179). This has to do with the range of densities resulting from contrast in the subject undergoing radiography. A high contrast subject requires a wide latitude film and a low contrast subject a narrow latitude film for optimum imaging in terms of tonal range.

Types of Films

There are three main types of x-ray films that are, or were, available for medical radiography.

1. **Screen Film** is the type most often used in radiography. Films are designed to

have maximum sensitivity (response) to the principal light-color emitted by the screen with which they will be used (spectral sensitivity matching). Such film-screen combinations have made possible the high speed and excellent quality of radiography. Manufacturers produce screens and films having a variety of speeds and contrasts to suit the fancy of any radiologist, both for general radiography and for special examinations.

2. **Nonscreen or Direct-exposure Film** had a thicker emulsion than screen-type film and was used without intensifying screens to obtain fine recorded detail. This type of film is no longer available, but ordinary film can be used in cardboard holders to obtain superb recorded detail of small anatomic parts such as fingers or toes, especially important in the early diagnosis of rheumatoid arthritis.

3. **Mammography film** is a single-coated, fine-grain film, designed to be used with a single intensifying screen. The combination must be fast enough to deliver a minimal dose to the glandular tissue of the breast. Double-coated film is available for dense breasts. Available now is a single-emulsion screen-type film—min-R 2000 (Eastman Kodak Company)—that provides increased contrast with narrower latitude, and greater visibility of detail. It requires precise selection of exposure factors, but this is facilitated by automatic exposure control. Breast glandular dose remains almost unchanged.

4. **Duplicating Film** is used to copy radiographs. The original radiograph is inserted into a cassette whose opaque front has been replaced by a pane of clear glass. A sheet of special duplicating film is then placed, with emulsion side down, onto the radiograph and the cassette lid closed. An exposure is made to an ordinary illuminator at a distance of 12 in. (30 cm) for 6 s, and the film processed as usual. When necessary, the density of the copy may be improved by adjusting the exposure time; a *longer* expo-

sure will *decrease* image density, and conversely. Duplicating films, when properly exposed, produce excellent copies of the original. Special copiers are available commercially to expedite the procedure.

Practical Suggestions in Handling Unexposed Film

1. Films deteriorate with age; therefore, the expiration date stamped on the box should be observed, the older films being used first.

2. Since moisture and heat hasten deterioration, films should be stored in a cool, dry place.

3. Films are sensitive to light and must be protected from it until processing has been completed.

4. Films are sensitive to x rays and other ionizing radiations, and should be protected by distance and by shielding with protective materials such as lead.

5. Films are marred by finger prints, scratches, and dirty intensifying screens, and by crink marks due to sharp bending.

6. Rough handling causes black marks due to static electricity. These appear as jagged lines, black spots, or tree-like images after development.

Film Exposure Holders

Each x-ray film must be carried to the radiographic room in a suitable container which not only protects it from outside light but also allows it to be exposed in radiography. Film holders are available in standard film sizes.

There are two types of film holders used in general radiography: *cardboard holders* and *cassettes.*

1. **Cardboard Film Holder**—an enclosed light-proof envelope into which the film is loaded *in the darkroom.* The film is

placed in the folder and the long flap of the envelope is folded over it, followed by the two shorter side and end flaps. After being closed and secured by the hinged clip, the holder is ready to be taken to the radiographic room for exposure. During exposure, the front of the film holder must face the tube, the back cardboard being lined with lead foil to prevent fogging of the film by x rays scattered back from the table. Cardboard holders are seldom used, except for extremely fine recorded detail of small parts.

2. **Cassette**–a case measuring about one-half inch in thickness and having an aluminum, stainless steel, or plastic *frame,* a hinged lid with several closure clamps, and a bakelite or light metal front that faces the x-ray source. The cassette front has a low average atomic number to minimize x-ray absorption. One pair of intensifying screens is mounted on the inside of the cassette front, and the second screen is mounted on the inside of the lid. As will be shown below, these screens convert the energy of x-ray photons to light photons, thereby amplifying the photographic effect on the film. This entails not only a marked reduction in the exposure needed for satisfactory radiographs but also a corresponding decrease in patient exposure. Moreover, it makes possible the use of grids for improved contrast.

The front of the cassette faces the x-ray tube during exposure. To load the cassette, raise the hinged lid *in the darkroom,* slip a film gently into the cassette of the same size,

and close the lid by means of the springs. The film is thus sandwiched between two screens and in close contact with them. Figure 15.3 shows a cassette in cross section. Lead foil cemented to the lid behind the back screen helps absorb radiation scattered back toward the screen, but this must be omitted in phototimed exposures.

Intensifying Screens

We have already mentioned the use of intensifying screens and shall now describe their structure and function.

1. **Composition.** An intensifying screen (see Figure 15.4) consists of microscopic crystals of a *phosphor* (luminescent material), incorporated in a binding material, and coated on one side of the white reflecting surface of a sheet of high grade cardboard or Mylar® plastic, 0.25 mm thick, called the *base.* The phosphor coating is the *active layer,* usually measuring from about 70 to 250 micrometers thick. It is, in turn, coated with a thin, smooth, abrasion-resistant material, its edges being sealed against moisture. For years *calcium tungstate* was the most widely used phosphor because it:

a. Emits blue and violet light to which ordinary x-ray film is particularly sensitive.

b. Responds well in the kV range ordinarily used in radiography.

c. Does not deteriorate appreciably with use or with age.

Figure 15.3. Cross section of a cassette with intensifying screens and film.

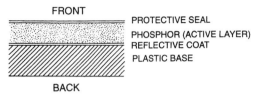

Figure 15.4. Magnified cross-sectional diagram of an intensifying screen.

d. Does not have significant afterglow or lag (see page 179).
e. Can be used to manufacture screens of uniform quality. As will be seen below, other phosphors (eg, rare earth types) have largely replaced calcium tungstate because of their greater efficiency.

2. **Principle.** Phosphors are crystalline materials that have the unique ability to *luminesce* or give off visible light when struck by x rays. This involves the transition of electrons between *energy bands* in the crystal lattice, shown diagrammatically in Figure 15.5 (Note the similarity to the concept of semiconductors, pages 101-105). The *valence band* contains atomic valence electrons and is usually filled, the *forbidden band* normally remains unoccupied by electrons except at certain points, and the *conduction band* allows free movement of electrons. An imperfect crystal results from faults or metal-lic impurities which serve as *traps* for electrons in the forbidden band; in fact, the minute amount of impurities added to the phosphor crystals in intensifying screens produce traps, many of which already contain electrons.

There are two kinds of luminescence–*fluorescence* and *phosphorescence*–classified, respectively, according to whether immediate or delayed emission of light occurs following exposure to x rays. A brief, simplified discussion of these processes will now be given.

a. *Fluorescence*–light emitted *promptly,* within 10^{-8} sec after absorption of an x-ray photon by photoelectric or Compton interaction. The resulting primary electron raises one or more valence electrons to the con-

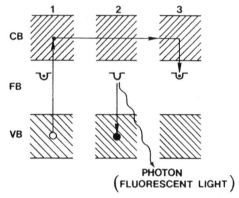

Figure 15.6. Process of fluorescence. An incoming photon (not shown) initiates a photoelectric or Compton interaction in the valence band *(VB),* releasing a primary electron which, in turn, imparts energy to a number of secondary electrons and raises them to the conduction band *(CB).* In *1,* a secondary electron has been raised to the *CB,* leaving a hole in the *VB.* In *2,* any electron previously trapped in a flaw in the forbidden band *(FB)* immediately falls into a hole in the *VB* and a photon is emitted as fluorescent light. In *3,* an electron from the *CB* finally drops into the empty trap, resetting it. Note that fluorescent light emission occurs *promptly,* in less than 10^{-8} sec.

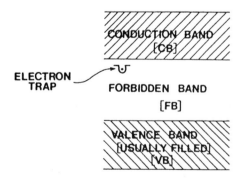

Figure 15.5. Schematic concept of a phosphor crystal. Note the similarity to a semiconductor crystal.

duction band, leaving a "hole" in the valence band (see Figure 15.6). Immediately, in less than 10^{-8} sec, a trapped electron drops from the forbidden band into the hole in the valence band, giving off visible light. The trap is "reset" when it is refilled by transition of an electron from the conduction band. A much less common event in radiographic phosphors is the direct transition of an electron from the conduction band to a hole in the valence band without preliminary trapping. As a result of fluorescence, such as that occurring in a pair of medium speed calcium tungstate screens, *one incident x-ray photon ultimately causes the emission of about 300 light photons,* about one-half of which leave the screen and reach the film (Ter-Pogossian); so the *screen efficiency* of this particular screen is 50 percent. The ability of a given phosphor to change x-ray photon energy to light energy is called its *conversion efficiency.*

b. *Phosphorescence*–light emitted *after* a delay of 10^{-8} sec or more after the absorption of an x-ray photon. Such delayed emission of light is also known as *afterglow.* Here, absorption of x-ray photons causes elevation of electrons to the conduction band, leaving many holes in the valence band (see Figure 15.7). However, in crystals subject to intense phosphorescence, such as zinc sulfide, the traps in the forbidden band are normally empty of electrons. When the newly arrived electrons drop from the conduction band into traps in the forbidden band, they remain there for a variable length of time. Heat from the kinetic energy of the atoms then lifts the trapped electrons back to the conduction band, whereupon they may drop into holes in the valence band, accompanied by the delayed emission of visible light.

Thus, there are two types of luminescence occurring in certain crystals following the absorption of x rays. One is the *prompt emission of light* by the process of *fluorescence.* The other is the *delayed emission of*

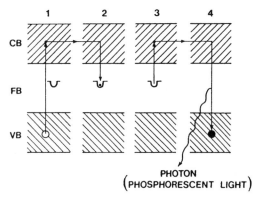

Figure 15.7. Process of phosphorescence. Here, traps are normally empty. In *1* we have the same sequence of events as in fluorescence (Figure 15.5). But in *2,* electrons drop from the conduction band *(CB)* into empty traps, one of which is shown here. Electrons remain in the traps for variable lengths of time before a small amount of energy, such as the heat present in the crystal, raises them back to the *CB* as in *3.* Finally, they drop into holes in the *VB.* Note that delayed emission of light occurs in phosphorescence.

light through *phosphorescence* (lag). In both types, the tiny phosphor crystals absorb energy in the form of x rays and convert this energy to visible light, which is emitted in all directions. Phosphorescence is virtually nil in intensifying screens and in cesium iodide image intensifier screens. The total amount of light emitted in a particular region of the active layer depends on the quantity of x rays striking that region, and is a summation of the light given off by the innumerable, closely packed crystals.

In the object being radiographed, those parts which are readily penetrated by x rays will appear on the *screen* as brighter areas than those parts which are poorly penetrated. Thus, the screen registers, although temporarily, the *aerial image* and includes zones of various degrees of brightness. These, in turn, ultimately produce corresponding differences in darkening of the radiograph.

Since the film emulsion is specially sensitive to the particular color of light emitted by the screens, this *photographic effect* is of major importance. In fact, *as much as 98 per cent of the recorded density (blackening) in a film exposed with intensifying screens is photographic in origin,* that is, due to light emitted by the screens; the remaining 2 percent resulting from direct x-ray exposure. Thus, screens greatly intensify the effect of x rays on the film emulsion, thereby reducing the exposure needed to obtain a particular degree of film blackening and, at the same time, substantially decreasing patient exposure. Intensifying screens made possible the use of grid radiography of thick anatomic parts such as the abdomen, skull, and spine, and led to the development of high speed radiography.

To summarize, we may say that the *exit* or *remnant radiation* (radiation that has passed through the patient and carries the aerial image) passes through the front of the cassette and impinges on the front intensifying screen. This emits light which varies in brightness depending on the amount of radiation reaching any particular area of the screen. In fact, there is a light and dark pattern on the screen corresponding to the relative transmission of x rays through the vari-

ous body structures in the path of the beam; the screen image represents the aerial image in the form of light. The image in light emitted by the screen photographically affects the film emulsion with which it is contact, recording the light and dark pattern, although this is reversed on the finished radiograph. X rays also pass on directly to the film causing a variable, though small, photographic effect. Some x rays reach the back screen which fluoresces and affects the film emulsion nearest it in the same manner as the front screen. Thus, a film is coated on both sides with a sensitive emulsion so that one side receives light from the front screen, and the other from the back screen, as shown in Figure 15.8.

3. **Screen Speed.** A screen is said to be fast or to have high speed when a relatively small x-ray exposure produces a given output of light and causes a certain degree of blackening of a film. Conversely, a screen is said to be slow when a relatively large exposure is required for a given amount of film blackening. *The speed or intensifying factor of a pair of intensifying screens may be defined as the ratio of the exposure required without screens to the exposure required with screens to get the same degree of blackening of x-ray films.* Another name for intensifying

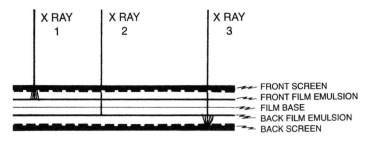

Figure 15.8. Principle of intensifying screen. X-ray photon *1* strikes the front screen causing it to fluoresce at that point. Another x-ray photon such as *2* may pass directly to the film and affect it at that point. Or a photon such as x ray *3* may pass through the front screen and the film, striking a crystal in the back screen and exciting fluorescence in it. Actually, the screens and film are in close contact. Not shown here is crossover from one screen to the emulsion on the opposite side of the film (see Figure 15.11).

factor is *speed factor.*

$$\text{intensifying factor} = \frac{\textbf{exposure without screens}}{\textbf{exposure with screens}}$$

Since the denominator is always less than the numerator, the intensifying factor of a pair of screens always exceeds unity, which means simply that the exposure with screens is less than that without screens for the same amount of film blackening.

At present, screen speed is rated according to **arbitrarily** assigned numbers as follows:

Screen Speed Number	Generic Name
less than 100	detail
100	medium speed
more than 100	high speed

On this basis, screens with a speed of 400 require about one-fourth the exposure of screens with a speed of 100, for equal radiographic density. It is expected that the generic names for screens will eventually be discontinued, and screen speed will be identified by number alone. Note again that these numbers are arbitrary and relative; they are not the same as the intensification factors.

But we must take into account the speed of the *film-screen system,* since this depends on speeds of both the screen and film being used. Thus, high-speed screens with slow films may be no faster than slower screens with high-speed films. So, choice of radiographic image receptors for a particular application should be based on a *film-screen system* that provides the requisite speed and image quality, rather than on screen speed alone.

The *speed* of intensifying screens depends on two main classes of factors—intrinsic and extrinsic.

a. **Intrinsic Factors** are those inherent in the composition of the screens and include the *phosphor,* the *thickness* of the active layer, the *packing density* of the phosphor, the size of the phosphor crystals, and the *reflectance* of the backing.

(1) *Phosphor.* The speed conferred on an intensifying screen by a particular phosphor depends on *its conversion efficiency* which is determined by two factors: (a) its ability to absorb x-ray photons, and (b) its ability to convert the absorbed x-ray energy to light energy, but this is a very inefficient process. Actually, only about 5 percent of x-ray energy in ordinary calcium tungstate, and 20 percent in rare earth phosphors, is converted to light energy (see Curry and others). Conversion efficiency can be enhanced by the addition of special activators; for example, activated calcium tungstate has twice the conversion efficiency of ordinary calcium tungstate.

Note that not all the light emitted by the phosphor reaches the film—about 50 percent is absorbed in the screen itself. *Screen efficiency* refers to the ability of light given off by a phosphor actually to leave a screen and expose a film. Thus, the intensifying factor of a screen depends on both phosphor conversion efficiency and screen efficiency.

Calcium tungstate screens emit blue and violet light to which ordinary x-ray film is especially sensitive. As a rule, the maximum color sensitivity of a particular film should be matched to the color of the light emitted by the screens with which the film is used.

Some of the newer phosphors with intrinsically higher conversion efficiency include *barium strontium sulfate* which fluoresces at *lower x-ray energy* (in radiographic range) than calcium tungstate and gives screen speed about 1.5 times that of medium speed screens. Another phosphor, *barium fluorochloride* provides screen speed about four times that of medium speed screens, but with moderate increase in mottle.

The rare earth phosphors, comprising salts of gadolinium, lanthanum, and yttrium tantalate, display much greater conversion

Table 15.1
PHOSPHOR COMPOSITION AND LIGHT EMISSION OF VARIOUS
INTENSIFYING SCREENS AND THEIR SPEED FACTORS
RELATIVE TO 100 FOR MEDIUM (PAR) SPEED SCREENS[a]

	Phosphor	*Principal Light Emission*	*Relative Speed*
Non-Rare Earth	Calcium tungstate[b]	Blue	60-400
	Barium lead sulfate[c]	Blue	20-50
	Barium strontium sulfate[c]	100-300	
Rare Earth Phosphors	Gadolinium oxysulfide[c]	Green	120-1200
	Lanthanum oxybromide and Gadolinium oxysulfide[d]	Blue/green	400-800
	Yttrium tantalate (high density)[d]	UV/Blue[e]	100-400
	Yttrium tantalate and Lanthanum oxybromide[d]	UV/Blue	400-800

[a] depends on film speed
[b] Kodak; Sterling
[c] Kodak
[d] Sterling
[e] UV = ultraviolet

efficiency than calcium tungstate while retaining excellent image quality. As shown in Table 15.1, these phosphors emit strongly at their characteristic light frequencies; for optimum efficiency, they should be used with spectrally matched films–maximum sensitivity to the colors emitted by the particular phosphor*. Screen-film combinations having a speed of 400 should be used in general radiography because they provide optimum speed and image quality while substantially reducing patient exposure. Although 800-speed systems reduce patient exposure even more, radiographic image quality suffers due to excessive quantum mottle (see page 211).

(2) *Thickness of Active Layer.* For a given phosphor, the thicker the active layer the greater will be the speed. In fact, this contributes more to screen speed than does crystal size.

(3) *Packing Density of Phosphor.* Screen speed increases with closer packing of the phosphor crystals.

(4) *Size of Phosphor Crystals.* Larger crystals impart greater speed to intensifying screens than do smaller ones for two reasons. First, more light is emitted by larger crystals because they receive more x-ray photons. Second, larger crystals scatter less light within the screen's active layer, increasing screen efficiency.

(5) *Reflectance of Backing.* The more light reflected back to the active layer by the cardboard or plastic backing, the greater will be the speed. However, this may be undesirable with thick active layers because of the detrimental effect on image sharpness caused by the diffusion (spread) of light in the screen.

A faster screen incorporates a more luminescent phosphor, such as specially activated calcium tungstate in a thicker layer, or some other phosphor that is inherently faster, such as one of the rare earths, or greater packing

* Rare earth phosphors fluoresce maximally at lower keV absorption peaks ("edge effect") than does tungstate phosphor, so they more closely match the 35 to 50 keV x rays that predominate in the radiographic range.

density. But note that *as the thickness of the active layer is increased the radiographic image becomes more blurred* (ie, *less sharp)* because an increase in average distance the emitted light has to travel to the film results in greater diffusion (spread) of the light rays (see Figure 15.9). Consequently, there is a practicable limit to the thickness of the active layer. However, denser packing of phosphor crystals increases screen speed without affecting screen thickness or image sharpness.

Theoretically, we should expect the size of the crystals to influence image sharpness because a larger crystal would produce a larger spot of light than a smaller crystal, but this turns out to be unimportant in actual practice. The crystals are extremely minute, averaging about 5 microns (0.005 mm) with a range of 4 microns (slow or "detail" screens) to 8 microns (faster calcium tungstate screens). This range of crystal size has a smaller effect on recorded detail than does active layer thickness. The resolving power of various screen-film combinations and a 1 mm focal spot appear in Table 15.2.

b. **Extrinsic Factors** include the conditions under which the screens are used, namely temperature and kV.

(1) *Temperature.* As room temperature rises, screen speed decreases and film speed increases. These two tendencies nearly cancel each other, so that no temperature correction is necessary for screen speed under ordinary conditions. However, at the high temperatures prevailing in the tropics, basic mAs values have to be increased about 40 percent at 50°C (120°F). On the other hand, in extreme cold such as may be encountered in industrial radiography, a decrease of about 25 percent in the basic mAs is required at −10°C (15°F).

(2) *Kilovoltage.* The speed of rare earth screens (except for yttrium tantalate*) does not remain constant over the useful kVp range, being maximal at about 100 kVp and gradually dropping to about 75 percent at 70 kVp. These values depend on the phosphor. On the other hand, calcium tungstate and yttrium tantalate screens show negligible change in speed with change in kVp.

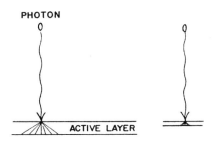

Figure 15.9. Effect of active layer thickness of an intensifying screen on image definition. Note that the light originating at points deep within the thinner active layer (right figure) spreads less before reaching the film, thereby producing a sharper image.

Table 15.2
RESOLVING POWER OF FILM-SCREEN COMBINATION (1mm Focal Spot)
AND MAMMOGRAPHY FILM

	Film-Screen Relative Speed				Mammography
	100	200	400	800*	Film
Resolution (line pairs/mm)	10	9	8	6	20**

* Sterling
** Kodak

* Ultra-Vision™ (DuPont)

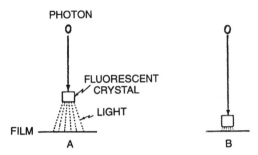

Figure 15.10. Poor screen contact causes a blurred image. In *A*, there is an appreciable space between the fluorescent crystal and the film (when screen contact is poor), so that a tiny but spreading bundle of light rays strikes the film, producing a blurred image instead of a fine point. In *B*, the crystal is in close contact with the film so that the image is about the same size as the crystal. (The crystal is magnified many times in the diagram.)

4. **Screen Contact.** The film must be sandwiched uniformly between, the two screens, with perfect contact throughout. Even the smallest space between the film and screen at any point will permit the light rays emerging from the screen at that point to spread over a wider area, producing a blurred image of that particular point (see Figure 15.10). You can easily ascertain the uniformity of contact of the screens with the film by placing a piece of hardware cloth with $1/8$ inch mesh over the front of the cassette and making a film exposure—40 kV, 10 mAs, and 100 cm (40 in.) focus-film distance, with a 3 mm Al filter. With satisfactory screen contact, the image of the wire mesh is sharp over the entire radiograph; but in areas of poor contact the image appears blurred and patchy (see Figure 15.11).

Screen contact is usually, though not universally, good in smaller cassettes. However, larger cassettes, especially 30×35 cm (11×14 in.) and 35×43 cm (14×17 in.) often exhibit poor contact even when new. Therefore, it may be necessary to cement a layer of felt, fiberglass, or foam rubber (preferably by an expert) as a cushion between the back screen and cassette lid. When the cassette is closed, the padding tends to equalize the pressure applied to the screens, squeezing them tightly against the

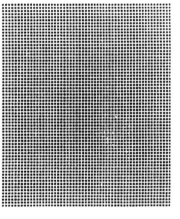

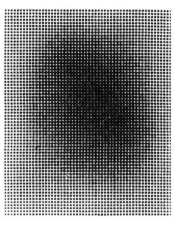

A   **B**

Figure 15.11. Tests for screen contact. In *A*, with excellent screen contact, the image of the wire screen is uniform. In *B*, with poor contact at the center, the image of the wire is blurred, producing a typical blotchy, dark area in the radiograph.

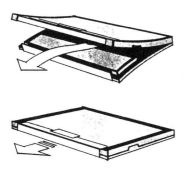

Figure 15.12. Cassette with front and back panels curved inward when open. Clamping them on closure squeezes the air out as they flatten, yielding excellent contact. (Courtesy of Eastman Kodak Company).

film.

Improved cassette design some years ago (Eastman Kodak) deliberately curved the front and back panels in opposite directions as shown in Figure 15.12 to squeeze the air out when the lid is closed. This contributes greatly to intimate film-screen contact.

Cassettes should be checked periodically for contact. Whenever a radiograph appears sharper in one area than another (for example, edges sharper than the center), poor screen contact should be suspected and the cassette tested without delay.

The *causes of poor screen contact* include (1) warped cassette front, (2) cracked or twisted cassette frame, (3) loose hinges or spring latches, and (4) local elevation of a screen by a foreign body underneath. Rough handling and dropping of cassettes are mainly responsible for damage leading to poor screen contact; also, this can result from warping of a cassette when placed under a bed-patient without a cassette tunnel.

5. **Recorded Detail with Screens.** In previous sections we described the relationship of intensifying screens to radiographic image sharpness. These data may be summarized under a single heading. In general,

recorded detail is never so sharp with screens as with direct radiographic exposure of x-ray film in a cardboard holder, provided the patient can be adequately immobilized. In actual practice, image sharpness with screens becomes worse as the thickness of the active layer increases. As noted before, a thicker active layer causes greater diffusion of light and blurring of the image. But remember that as the active layer thickness is decreased to improve recorded detail, the screen becomes slower. Another fact to bear in mind is that regardless of the type of screen, recorded detail suffers when film-screen contact is poor. In nongrid radiography of small parts, detail screens give much better recorded detail and less mottle than medium speed or fast screens (see also page 212).

Another cause of impaired recorded detail with intensifying screens, related to diffusion of the image, is a condition known as *crossover* or *punch-through,* shown in Figure 15.13. Note that crossover occurs only with the double-emulsion films and two screens prevailing in general radiography. Light from one screen expands in the form of a cone as it passes successively through this screen, the nearer film emulsion, the film base, and the farther film emulsion where a slightly enlarged, less sharp image is formed. Recorded detail is thereby worsened. However, special dyes incorporated in the emulsion minimize crossover.

6. **Care of Screens.** These must be kept scrupulously clean, since dust and other foreign matter absorb light from the adjacent screen and cast a white shadow on the film. Screen cleaners should be of the type recommended by the manufacturer. Some may be cleaned with cotton *dampened* with bland soap and water and wiped with additional pieces of cotton moistened with water, never dripping wet. They can then be air-dried by half-opening the cassette and standing it on its side. However, such washing must first be approved. Ultravision (yttrium tantalate)

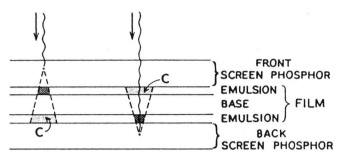

C = CROSSOVER RADIATION

Figure 15.13. Crossover or punch-through of light from one screen to the other in the form of a tiny expanding cone of light. This causes a blurring effect in the corresponding emulsion of the film, resulting in impaired recorded detail. Crossover is minimized by incorporating special dyes in a crossover control layer between each screen and its support.

screens require special instructions.

Care must be taken not to nick, scratch, or chip the screens as, for example, by carelessly digging the film out of the cassette with the fingernail, or by scratching the screen with a corner of the film. In loading the cassette, one carefully slips the film into the cassette with the lid elevated about 5 cm (2 in.). Doing this gently helps avoid static marks. To remove the film after exposure, one carefully raises a corner of the film, being certain that the fingernails do not slip across the surface of the screen.

To be avoided is the use of ultraviolet-blocking lotions on the hands before loading cassettes having ultraviolet-emitting yttrium tantalate screens; any lotion adhering to the films or screens will block the ultraviolet radiation and create white patches on the processed film.

The cassette must always be kept closed, except while being loaded or unloaded, to prevent accidental damage to the screen surface and help keep out dust. Scratch marks and dust block the light from the screen, leaving a *white spot* on the finished radiograph.

QUESTIONS AND PROBLEMS

1. Show by cross-section diagram the structure of x-ray film.
2. What is meant by safety film? Non-screen film?
3. State two differences between screen and nonscreen film. What is mammography film?
4. Describe the structure of an intensifying screen. How does it differ from a fluoroscopic screen?
5. Define the intensifying factor of an intensifying screen. State the speed *numbers* of slow, medium, and fast screens; how are they related to intensifying factor.
6. What are three differences in the composition of a slow screen and a fast screen? How do they differ in their effect on radiographic sharpness?
7. How does an intensifying screen intensify the x-ray image?
8. Why is an x-ray film coated on both

sides with sensitive emulsion?

9. How do rare earth screens differ from ordinary intensifying screens?

10. Explain screen lag. Luminescence. Phosphorescence. Fluorescence.

11. How does active layer thickness affect image sharpness? Explain. What other factors cause impairment of image sharpness by screens? Explain.

12. Discuss the four intrinsic factors affecting screen speed.

13. How do temperature and kilovoltage influence screen speed?

14. In general, is recorded detail better with direct exposure radiography or with intensifying screen radiography? Why?

15. What is meant by phosphor conversion efficiency of an intensifying screen? Screen efficiency?

16. Discuss rare earth phosphors on the basis of emitted color and conversion efficiency, as contrasted with calcium tungstate.

17. Explain crossover or punch-through related to film-screen systems. How can it be reduced?

18. Why should we avoid scratching or otherwise marring a screen surface?

19. What is the effect of poor screen contact on radiographic sharpness? Why?

20. Describe a test for film-screen contact.

21. What special precaution must be taken in handling Ultra-Vision™ film/screen systems?

22. A radiograph is correctly exposed with a 200-speed screen-film combination. How would you change the mAs for a 400-speed system? kVp?

Chapter 16

THE DARKROOM

Topics Covered in This Chapter

- Building Essentials
  - X-Ray Protection
  - Passboxes
  - Air Conditioning
- Entrance
  - Light-Tight Door
  - Maze
  - Entrance Hall with Electrically
      - Interlocked Doors
      - Revolving Door
- Lighting
  - Safelights
  - General Illumination
- Equipment for Manual Processing
- Questions and Problems

Introduction

THE IMPORTANCE of the darkroom in radiography cannot be exaggerated. Radiography unquestionably begins and ends in the darkroom, where the films are loaded into suitable light-proof holders in preparation for exposure, and where they are returned for processing into a finished radiograph. In general, a darkroom is a place where the necessary handling and processing of film can be carried out safely and efficiently, without the hazard of producing *film fog* by accidental exposure to light or x rays.

The term "darkroom" overstates the case, since complete blackout is unnecessary. In fact, as will be shown later, a great deal of safe illumination can be provided to facilitate darkroom procedures. Although the term "processing room" is more accurate, it is not widely used. We shall therefore continue to refer to it as the "darkroom," bear-

ing in mind that it is dark only insofar as *it must exclude all outside light and provide "safe" artificial light.*

Location of the Darkroom

Because of its overriding importance, the location of the darkroom should be determined during the planning stage of the Radiology Department. Convenient placement of the darkroom with relation to the radiographic rooms should save time and eliminate unnecessary steps. Walls adjacent to the radiographic rooms should be shielded with the correct thickness of lead–equivalent material (see pages 403-404). Stored radioactive materials should be located as remotely as possible from the darkroom because even as little as 5 mR total exposure from x or gamma rays causes detectable fog. As a general rule, protection of films approximates that of personnel.

Windows should be avoided because they are extremely difficult to render lightproof. Furthermore, they serve no useful purpose, since much more satisfactory methods of ventilation and lighting are available.

Finally, the darkroom should be readily accessible to plumbing and electrical service.

Building Essentials

Walls between the darkroom and adjoining x-ray rooms should contain enough lead thickness, or its equivalent in other building materials, for adequate protection of the films. This is especially important because efficiency demands that the darkroom be close to the radiographic rooms (see above).

In a busy department, the efficiency of work flow can be improved by having *pass-boxes* built into the walls at appropriate locations. Typical passboxes have two light-tight and x-ray proof doors that are so interlocked that both cannot be opened at the same time. The cassettes, after radiographic exposure, are placed in the passbox through the outside door, and then removed through the inside door by the darkroom technician. The most suitable location for the passbox is obviously near the film-loading bench although this is not always possible.

Darkroom walls should be covered with chemical-resistant materials, particularly near the processing tanks. Such materials include special paint, varnish, or lacquer. Ceramic tile or plastic wall covering provides a more durable finish.

Floor covering should consist of chemical-resistant and stainproof material such as asphalt tile. Porcelain or clay tile may be used, but a nonskid abrasive should be incorporated to minimize the danger of slipping. Ordinary linoleum and concrete are unsuitable because they are readily attacked by the processing solutions.

Entrance

Access should be conveniently located with relation to the darkroom equipment. The simplest type of entrance is a single door which must be made absolutely light-tight by weatherstripping.

Such an entrance generally prevails in offices and hospitals, but it should have an inside lock to prevent opening while films are being processed.

Another type of protective entrance is a small hall with two electrically interlocked doors, so designed that one door cannot be opened until the other is completely closed, thereby preventing entrance of light (see Figure 16.1). A separate door should be provided for emergency use and for moving equipment into and out of the darkroom.

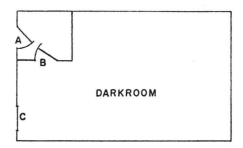

Figure 16.1. Darkroom floor plan with interlocked doors. When door *A* is opened, an electrical relay prevents anyone from opening door *B,* and vice versa. *C* is an emergency door which is usually kept closed.

A more elaborate type of entrance is the *maze,* shown in Figure 16.2. Note that one must execute a complete turn in going through the three doorways. Serving as a light trap, the maze requires no doors, especially if the walls are painted black. Because of the high cost of the required floor area, mazes are rarely used today.

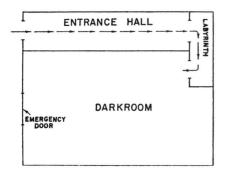

Figure 16.2. Darkroom with maze or labyrinth entrance.

For most installations, the preferred modern type of entrance is the revolving door, shown in Figure 16.3. Several sizes are available. For example, one measures about 3 feet in diameter and, built into the wall, it extends about 15 in. into each adjoining room so that it occupies relatively little floor area. At the same time, it offers convenient, lightproof access to the darkroom. Models with three-way and four-way doors are also

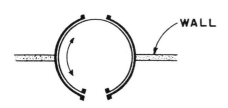

Figure 16.3. Diagram (top view) of a two-way revolving darkroom door. The outer cylindrical chamber, built into the wall, has two openings, one into the darkroom and the other into the lightroom. The inner chamber has one door which, suspended at its top by a central bearing, can easily be rotated by means of a handrail until its opening coincides with either of the openings in the outer chamber. The entire unit is completely lightproof and takes up very little space.

available, as are special modifications for handicapped employees.

Size

Darkroom size will vary, of course, with the size of the department, but it must be large enough to house conveniently all the necessary benches for loading and unloading films, a film storage bin, cupboards, and manual or automatic processing equipment. On the other hand, the darkroom should not be too large, since the excessive distances between various units of equipment result in wasted time and steps. Placement of the loading bench across the room from the tanks minimizes accidental splashing of water and solutions on the films, benches, cassettes, or other equipment that may come in contact with the films, but this is rarely a problem in the darkroom with an automatic processor.

Ventilation

Essential to efficient work and to the technologist's health are the adequate removal of stale humid air and the supply of fresh air. Air conditioning is definitely the preferred method of ventilating the darkroom. However, in a small office or department in areas where climate permits, an exhaust fan may provide adequate air circulation. The system should be absolutely light proof and should include a means of filtering out dust at the point of entry of the fresh air.

Lighting

A properly designed darkroom should have three types of illumination: safelight, general, and radiographic.

1. **Safelight.** We must have a source of light which will not fog films and still provide adequate illumination under processing

conditions. ***Safelight lamps*** serve this purpose; they must have filters of the proper color, such as the Wratten Series 6B® filter for ordinary blue-violet sensitive film, but a special filter such as the Kodak GS-1 must be used with green-sensitive Ortho-G film. The ***working distance*** from the safelight should be no less than 1 meter (3 ft), and bulb wattage should correspond to that specified on the lamp housing. If a brighter bulb is used, light transmitted even through the correct filter may cause film fog. Safelights are designed for either indirect ceiling illumination or for direct lighting.

The safety of darkroom lighting may be tested as follows: subject a film in a cassette to a very small x-ray exposure, just enough to cause slight graying; screen film is more sensitive to fogging by light after initial exposure to the fluorescence of intensifying screens. Then remove the film from the cassette in the darkroom, cover one-half the film with black paper, and leave it exposed under conditions simulating as closely as possible those normally existing when a film is being loaded and unloaded. Process the film as usual. If the uncovered portion appears darker than the covered, you may conclude that the darkroom lighting is unsafe, and must then make every effort to eliminate the source of light responsible for the fogging. This may result from a cracked or an incorrect filter, a light leak in the safelight housing, or a leak around the entrance door to the darkroom.

Note that the darkroom walls do not have to be painted black. A light color enhances safelight illumination without increasing the risk of film fogging.

2. **General Illumination.** A source of overhead lighting is needed in the darkroom for general purposes such as cleaning, changing solutions, and carrying out other procedures that do not require safelight illumination.

3. **Radiographic Illumination.** A wall-mounted fluorescent illuminator should be available in the light room at a convenient location for viewing, as needed, radiographs exiting the processor.

Film Storage Bin

A ***lightproof storage bin*** for unexposed film should be placed under the loading bench. Vertical partitions subdivide the bin to accommodate film boxes of different sizes. Counterweighting the drawer of the bin makes it close automatically when released. A warning light must be posted on the front of the bin to prevent its being opened in white light.

QUESTIONS AND PROBLEMS

1. Where is the best location for the darkroom? How should the walls be shielded?
2. Discuss the various kinds of darkroom entrances, including the advantages and disadvantages of each.
3. Describe interlocking doors. Passboxes.
4. Describe a darkroom safelight lamp. What precautions are necessary to assure that the safelight is really safe?
5. How do you test a darkroom for safe illumination?
6. Which colors should be used in painting the walls of the darkroom? Of the maze?
7. How can you prevent x rays from entering the darkroom, and why?

Chapter 17

CHEMISTRY OF RADIOGRAPHY
AND FILM PROCESSING

Topics Covered in This Chapter

- Introduction
- Silver Halide Crystals
- Aerial Image Formation
- Conversion of Latent to Manifest Image
- Manual Processing
    Development
    Rinsing
    Fixing
    Washing
    Drying

    Problems
- Automatic Processing
    Principle
    Processor Components
    Film Characteristics
    Care of Rollers
    Sensitometry
- Silver Recovery
- Questions and Problems

Introduction

THE BASIC PRINCIPLES of radiographic image formation in sensitive emulsions have been established for many years, evolving as they have from ordinary photography. However, as we shall see later, certain changes were required to adapt conventional film processing to automation, these changes being perhaps more *physical* than chemical. Therefore, a thorough understanding of radiographic image formation and photographic chemistry should precede the study of both manual and automatic processing.

Radiographic Photography

Fundamental to an understanding of the production of an image on an x-ray film, is the concept of the *latent image.* An ordinary radiographic film emulsion contains silver halides in the form of minute crystals that are invisible to the naked eye (1 to 1.5 μm, micrometers). These halides consist mainly of silver bromide (AgBr) but also include a small amount of silver iodide (AgI) to enhance sensitivity.

Figure 17.1 presents a diagram of the cubic lattice structure of a silver halide crystal. In it, the positive silver ions alternate with the negative bromide ions (and an occasional negative iodide ion), all being held together by ionic bonds (see page 35), although some silver ions drift freely through the lattice. During the ripening phase of emulsion manufacture, *silver sulfide particles,* serving as *sensitivity specks* or *development centers,* appear on the surface of the crystals. Only those crystals with sensitivity specks, acting as *electron traps,* can be affected by expo-

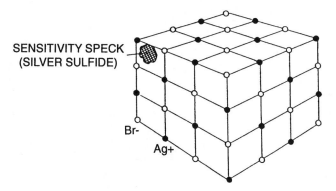

SENSITIVITY SPECK
(SILVER SULFIDE)

Br-

Ag+

Figure 17.1. Lattice diagram of a silver bromide (AgBr) crystal. The straight lines joining the Ag^+ and Br^- ions represent the electrovalence forces holding the ions together in the crystal. An occasional iodide ion (I^-) is also present in place of a Br^- ion. The sensitivity speck (silver sulfide) renders the crystal highly sensitive to light and x rays.

sure to light or x rays.

According to the ***Gurney-Mott hypothesis*** of latent image formation, four steps are involved:

1. A ***photon*** entering a sensitized silver halide crystal may interact with a bromine ion, liberating a loosely-bound valence electron and leaving a neutral bromine atom.

2. As the ***free electron*** drifts through the lattice, it may be trapped by a sensitivity speck to which it imparts a negative charge.

3. A migrant ***positive silver ion,*** attracted to the now negatively charged sensitivity speck, picks up the electron and becomes a neutral silver atom adhering to the speck. This process, called ***nucleation,*** is repeated a number of times within a very short time interval.

The silver atoms that have been deposited on the sensitivity speck are too few in number to be visible, so they constitute the ***latent image*** whose stability increases with the number of silver atoms it contains.

Radiographic Chemistry

When the latent image is acted upon by ***reducing agents*** known as ***developers,*** the

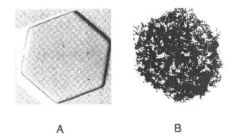

A   B

Figure 17.2. Grain development in a tabular (T-grain) silver halide crystal. *A.* Three nucleated undeveloped grains (tiny black specks). *B.* Fully developed grains that have undergone billions of times' amplification. (Adapted from Hass, AG, Ed., section by Dickerson, RE, Courtesy of Medical Physics Publishing.)

process initiated by photon action is greatly speeded up; ***the sensitivity speck serves as a development center for the entire crystal.*** In fact, development amplifies the latent image about 5×10^9 (5 billion) times (Dickerson). The speck rapidly traps electrons from the reducing agents and attracts more silver ions which become reduced to silver atoms and grow into thread-like or, in some film emulsions, tabular clusters of metallic silver (see Figure 17.2). Thus, the ***dark areas consist of metallic silver in a very fine state of subdivision,*** the amount of silver deposited in a

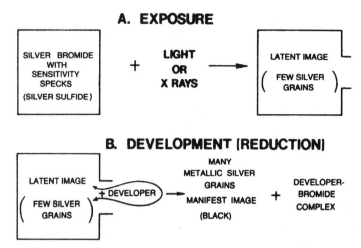

Figure 17.3. Scheme of the basic theory of photographic chemistry. *A.* Radiant energy in the form of light or x rays converts a silver bromide (or silver iodide) crystal containing a sensitivity speck (silver sulfide) into a *latent image center. B.* The developer reacts with the altered crystal and reduces it to metallic silver, which constitutes the *manifest image.*

given area increasing with the amount of radiation received by that area. The basic concepts of photographic chemistry are shown in Figure 17.3.

During development the bromine (and iodine) ions diffuse out of the developed crystals and into the solution. This, in addition to the gradual exhaustion of the reducing agents, eventually causes deterioration of the developer to the point where it must be discarded.

What happens to the portions of the film emulsion that are not affected by light or x-ray photons? Since the silver halides (AgBr and AgI) in these areas have not been altered, they are relatively unaffected by the developer. However, they must be removed in order to render the unexposed film areas *transparent,* and also to prevent *fogging* by subsequent exposure to light. The unexposed and undeveloped silver halides are eliminated from the emulsion by immersion in a *fixing agent,* ammonium thiosulfate, a process known as *fixation.* As a result, the areas from which the silver halides have

been removed become clear, while the black areas remain black since metallic silver is not dissolved by the fixing agent in the ordinary course of processing. However, prolonged immersion in the fixing solution will cause bleaching of the image; this may be appreciable even in twenty-four hours.

The amount of blackening of a particular area of a radiograph depends on the amount of radiation it has received. This is the radiation that has passed through the various thicknesses, types, and densities of tissue interposed between the tube and the film and is called *remnant* or *exit radiation.* It forms the *aerial image* (image in space) which, in cross section, consists of more or less closely spaced photons that will be displayed on the image receptor (film, intensifying screen, fluorescent screen) (see Figure 17.4). On this basis, a *radiograph* may be defined as a processed film, recording the aerial image or exit radiation of an anatomic part.

Consider, for example, the simple case in which a radiograph of the hand is made by

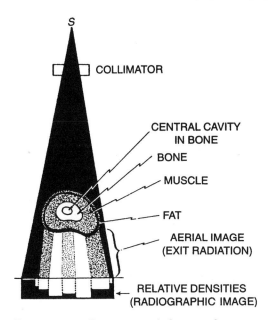

Figure 17.4. Formation of the aerial image. The x-ray beam, on passing through the body, undergoes absorption and scattering which depend on the kV and on the atomic number and density of any particular tissue. The beam emerging on the opposite side of the body contains information in terms of the number of photons per unit cross-sectional area of the beam; this comprises the *remnant* or *exit radiation,* which forms the *aerial image* or *image in space.* The resulting image on the film—the *radiographic image*—consists of the various densities corresponding to the aerial image.

direct exposure to x-rays (ie, without intensifying screens). The light areas represent the bones; because of their high calcium content, the bones absorb a large fraction of the incident x rays so that very few x-ray photons pass through them to the silver halide crystals in the film emulsion underlying them. Therefore, the reducing action of the developer on these crystals produces almost no darkening. The soft tissues, on the contrary, absorb only a relatively small fraction of the incident x rays and therefore the areas of film emulsion beneath the soft tissues receive a relatively large amount of remnant or exit radiation. As a result, more silver halide crystals are affected in these regions and the developer causes considerably greater blackening. Notice that the finished radiograph is really a negative, corresponding to a negative in ordinary photography. It is the recorded aerial image of the tissues of different thicknesses and densities, which vary in their attenuation of x rays (mainly by photoelectric absorption). As was shown earlier (see pages 177-178), the radiographic effect of the aerial image is greatly enhanced by intensifying screens.

MANUAL PROCESSING

This section deals with film processing by the **manual,** or **time-temperature** method, which involves conversion of the latent image to a visible image. Included are primarily development, rinsing, fixing, washing, and drying; and secondarily, fixer neutralization and detergent rinse. The basic principles also apply to automatic processing, to be described below. Cleanliness of hands, utensils, and darkroom are essential. Figure 17.5 depicts a typical arrangement of tanks for manual processing, which is now obsolete in radiology, but may be of historic interest.

Manual processing includes the following steps:

Step 1. **Development** in a solution containing an organic reducing agent, such as hydroquinone and metol, to convert ionic silver to metallic silver. An alkali, sodium hydroxide, serves as an accelerator; and a

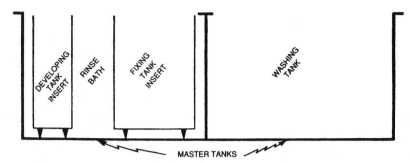

Figure 17.5. Arrangement of tanks for manual processing.

restrainer prevents chemical fogging. Finally, an antioxidant, sodium sulfite, slows oxidation of the developer by air.

Step 2. **Rinsing** in a solution of acetic acid (vinegar) to neutralize the alkali carried over from the developer, thereby stopping development.

Step 3. **Fixation** in a solution of "hypo" to remove the unexposed, undeveloped silver halides and so preserve the image. A hardener, alum, toughens the emulsion, actually tanning it.

Step 4. **Fixer neutralization** in a special solution to remove any residual fixer, to help prevent deterioration and discoloration of the image.

Step 5. **Washing** by immersion of the films in running water for a prescribed time to remove effectively any residual chemicals.

Step 6. **Drying** in a special cabinet, preferably with a source of forced heat.

The minimum time possible for manual processing is about 40 min. Figure 17.6 is a time-temperature curve, which shows how development time decreases as temperature of the developer is increased, hence the term *time-temperature* development. This curve is a composite of data from two American manufacturers of x-ray film.

Replenishment of developer and fixer is needed because the levels of these solutions drop as they adhere to the films and are carried into the next tank. Developer requires

a special replenisher, but fixer is replenished by ordinary fixer solution.

Film Fog, Stains, and Artifacts

It may be of some help to indicate the causes of the more common film defects.

1. **Fog.** There are many causes of film fogging, that is, a generalized darkening of the film.

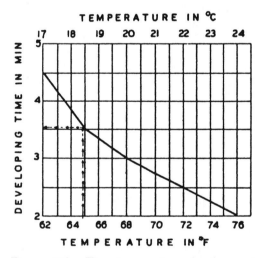

Figure 17.6. Time-temperature development curve for manual processing. This is a composite based on the data furnished by the two leading American manufacturers for medium speed film and rapid developer.

a. ***Exposure to Light.*** This may occur when the darkroom is not light proof; the safelight contains too large a bulb; the safelight housing or filter is cracked; the safelight filter series is incorrect; or the exposure of the film to the safelight is prolonged, especially at short distances.

b. ***Exposure to X Rays or Radionuclides.*** Films should be shielded from these sources of radiation by distance and sufficient thickness of lead.

c. ***Chemical Fog.*** The many causes include overdevelopment or development at excessively high temperatures; oxidized, deteriorated developer, which may also stain the film (oxidized developer is brown); prolonged or repeated inspection of films during development; and contamination from corroded tanks.

d. ***Age Fog.*** Either mottled or uniform fogging due to outdated films, or films stored under conditions of high temperature and excessive humidity.

2. **Stain.** Various types of discolorations may appear on films at different intervals after processing. These can generally be avoided by the use of reasonably fresh solutions and correct processing.

a. ***Brown.*** Oxidized developer.

b. ***Variegated Color Pattern.*** Inadequate rinsing.

c. ***Grayish yellow or Brown.*** Excessive fixation, or use of exhausted fixer.

d. ***Grayish-white Scum.*** Incomplete washing.

3. **Artifacts.** There are several kinds of extraneous shadows that appear when films are not handled gently.

a. ***Crinkle Marks*** are curved black or white lines about 1 cm in length which result from bending the film acutely over the end of the finger.

b. ***Static Marks*** are lightning or tree-like black marks on the film, caused by static electricity due to friction between the film and other objects such as intensifying screens and loading bench. To avoid this, films should always be handled gently. In addition, the loading bench should be grounded in order to prevent the buildup of static electricity.

c. ***Water Marks*** are caused by water droplets on the film surface, which leave round dark spots of various sizes because of migration of silver particles.

d. ***Cassette Marks*** are caused by foreign matter such as dust, hair, fragments of paper, etc, or by screen defects, which leave a corresponding white mark on the radiograph.

e. ***Air-bell Marks*** result from formation of air bubbles in the developer. A bubble prevents developer from reaching the underlying film, and so leaves a small, clear circular spot on the radiograph.

f. ***Streaking*** is caused by a variety of technical errors and is one of the most troublesome types of film defects. It usually results from: (1) failure to agitate the films in the developer; (2) failure to rinse the films adequately; (3) failure to agitate the films when first immersed in the fixer; and (4) failure to stir the processing solutions thoroughly after replenishment.

g. ***Horizontal bubbly line*** across top of radiograph, with clear film above, indicates low level of developer.

AUTOMATIC PROCESSING

A major improvement in radiography has been the perfection of automatic processing of films. Successful marketing of automated processors has brought about a reduction in cost, thereby making such equipment available even to the department with a limited budget. In fact, it has replaced manual processing in radiology.

There are many outstanding advantages of automatic processing. First, it *shortens total processing time to as little as one and one-half minutes,* in contrast to about one hour for the manual method. (Even with hypo neutralizer and wetting agent, manual processing could be reduced, at best, to about forty minutes.) Second, it *improves quality control* by more precise temperature regulation and replenishment in the automatic processor. Third, it increases the capacity of the Radiology Department or office by expediting work flow. One-minute processing is now available.

The manufacturers of the first automatic processors were beset by four main problems: transport mechanism, processing chemicals, temperature control, and film characteristics. A discussion of these problems and their solution is essential to the understanding of the principles of automatic processing.

1. **Transport Mechanism.** The basic mechanism of the automatic processor is a series of rollers which transport the films from the loader through the various sections—developing, fixing, washing, and drying (see Figure 17.7).

Special *crossover rollers* move the films from one roller rack to the next. Each film is gripped between rollers, which include the larger *master* and the smaller *planetary rollers;* they are arranged in racks that are removable for servicing and cleaning.

The speed of transport must be constant to assure correct sojourn of the films in each section. Spacing between the rollers must be accurate to an extremely small tolerance to avoid slipping or jamming of the films. Besides these operations, the transport system provides brisk *agitation* of the solutions over the film surfaces to assure uniform action of the developer. *Pump circulation* of the developer and fixer also contribute to their agitation.

The squeezing action of the last few rollers in each section removes a large amount of solution from the films before they enter the next section, thereby lessening contamination of the solutions. Finally, the terminal rollers in the wash section squeeze out most of the water and so hasten drying of the films. Figure 17.7 shows diagrammatically a typical automatic processor; its apparent simplicity belies the difficulties encountered during its early development.

2. **Processing Chemicals.** Although the basic principles of photographic chemistry apply to both manual and automated processing, conventional solutions were found to cause insurmountable difficulties in the first automated experimental models. Roller spacing was so critical that the swelling of the emulsion in the developer, and its subsequent shrinkage in the fixer, caused the films either to jam between the rollers or to wrap around them. Further complications arose when an attempt was made to speed up processing through the use of stronger solutions.

Figure 17.8 shows a variation of about eleven units of *gel swell* of the emulsion during manual processing. This renders conventional solutions completely unsuited to automated processing; in fact, it turns out that the problem is largely *physical.*

To adapt solutions to automation with its decreased processing time, certain changes had to be made. These may be summarized as follows:

a. *Filters in Water Supply Line* to remove small-particle impurities. The cold and hot water lines have separate porous filters, which must be replaced at regular intervals, depending on the purity of the water. Should the cold water filter become clogged and the hot water continue to enter the processor, the solutions would become *overheated,* resulting in *overdeveloped,* dark radiographs. Furthermore, the emulsion could soften, causing slippage or wraparound of the films as they proceed through the rollers.

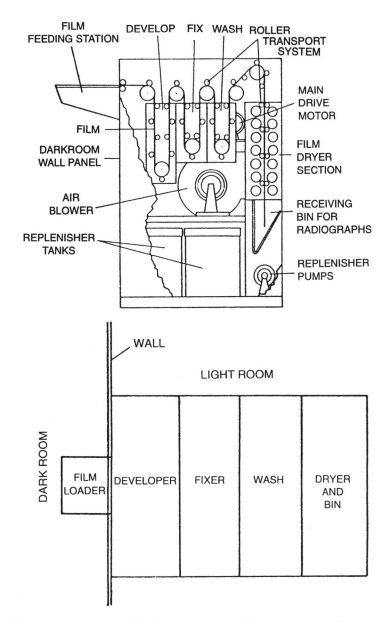

Figure 17.7. Automatic processing. In the upper part is shown a cutaway diagram of the processor (side view). In the lower part is shown a diagrammatic top view. *(Courtesy of Eastman Kodak Company.)*

On the other hand, should the hot water filter become clogged, the ***thermostat would cycle on and off*** more frequently to maintain processing temperature, increasing the likelihood of thermostat failure. Some processors operate on only a cold water supply requiring a single filter; in these units the thermostat and heater are designed to temper (warm) the incoming water without the need of a separate hot water supply.

b. ***Increased Concentration of Solutions*** to shorten greatly the processing times in the various sections. Hydroquinone and phenidone are the developing agents, and cycon is added as an antioxidant (retards oxidation). A fresh ***developer*** bath consists of replenisher, properly mixed, to which has been added a ***starter solution*** to prevent excess alkalinity. Thereafter, a pump automatically feeds properly mixed replenisher solution from a storage tank into the development section. The ***fixer,*** ammonium thiosulfate in high concentration, is automatically replenished from a storage tank by a pump.

c. ***Increased Processing Temperatures*** to help speed up processing; for example, 35°C (95°F) in developer and fixer, 31°C (88°F) in wash, and 57°C (135°F) in dryer are typical of ninety-second processors. Drying time is reduced to 20 seconds. To combat chemical fog at these high temperatures, antifogging agents such as aldehydes are added to the developer.

d. ***Hardening of Emulsion*** to prevent softening by the solutions, as well as sticking to the rollers. For this purpose, special hardening agents such as glutaraldehyde or sodium metaborate are included in both the developer and fixer.

e. ***Control of Emulsion Thickness*** to keep it as constant as possible throughout processing, to permit the use of constant spacing between the rollers. Compounds such as sulfates are added to the developer to minimize swelling of the emulsion.

f. ***Precise Replenishment*** of the developer and fixer to maintain the proper alkalinity of the developer, acidity of the fixer, and the chemical strength of both solutions. Replenishment must be adjusted to a constant rate for each film processed. Inserting each film in the feed tray with the ***narrower*** side against the film guide rail activates a ***microswitch,*** which turns on the replenisher

system. As the film moves through the rollers in the developer section, replenishment occurs at a constant rate that has been set initially according to the length of film travel. The same process takes place in the fixing section. Films should be fed alternately along both guide rails.

3. **Temperature Control.** As just noted, ***high temperatures*** prevail in automatic processing. This has resulted in marked shortening of processing time, but now extremely accurate temperature control has to be maintained in each section of the processor. This is accomplished by a pump, which circulates water at the proper temperature around the developer and fixer sections; the water finally enters the wash section. Incoming water temperature is usually lower than that of the processing solutions so that it can be heated to the correct temperature under thermostatic control. Temperature is even more critical than in manual processing because the technologist can no longer adjust development or fixing time for differences in temperature. Moreover, the very short developing time at high temperature makes the films much more susceptible to serious under- or overdevelopment when these factors differ even slightly from the normal.

4. ***Film Characteristics.*** The manufacture of films requires even more precise control than before to maintain constant thickness of the base and emulsion. Curling tendency must be eliminated to prevent wrapping around rollers or misdirection to the wrong rollers, causing jamming of the films. The emulsion characteristics must conform to the new types of processing solutions. Finally, stickiness of the emulsion has to be minimized to prevent adherence to the rollers and wrap-around.

The various modifications just described have made automatic processing so dependable that it has become practicable even in the private office and small hospital. Figure

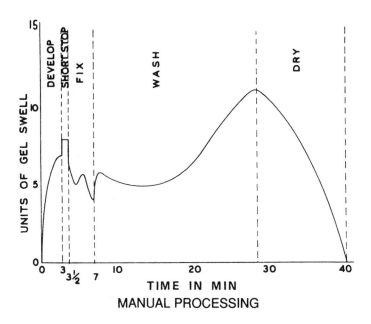

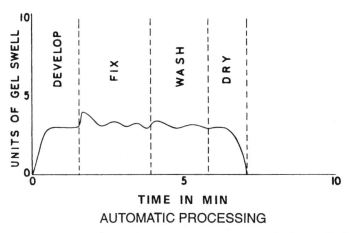

Figure 17.8. Comparison of manual and automatic processing on the basis of relative swelling of film emulsion during the various phases. The units of gel swell are arbitrary. Note the much smaller variation in the degree of emulsion swelling with automatic processing; this results from special additives in the solutions. *(Courtesy of R.E. Humphries, GAF.)*

17.8 shows how constant the thickness of the emulsion remains. The overall variation in gel swell from start to finish in automatic processing is about four units, but after the initial relatively slight swelling of the emulsion in the developer, the thickness remains

virtually constant well into the drying phase.

Table 17.1 contrasts the time frames in manual and automatic processing.

The question naturally arises as to the frequency of breakdown of automatic processing equipment. Experience has shown that

Table 17.1
COMPARISON OF MANUAL AND AUTOMATIC PROCESSING

	Manual	Automated*
		$1^1/2$-minute
Developing Temperature	20 C (68 F)	35 C (95 F)
Fixing Temperature	20 C (68 F)	35 C (95 F)
Washing Temperature	20 C (68 F)	35 C (95 F)
Drying Temperature	43 C (110 F)	57 C (135 F)
Developing Time	3-5 min	25 sec
Fixing Time	2-10	21
Washing Time	15-30	9
Drying Time	15-20	20
Surface Change	Manual Agitation	Transport Rollers
Replenishment	Manual	Automatic

* Data vary with type of equipment, films, and chemicals.
** Sixty-minute processors are now available.

the greatest cause of breakdown is ***failure to keep the rollers scrupulously clean,*** as recommended by the manufacturer. Another cause is improper rate of automatic replenishment which should be carefully checked at regular intervals.

Sensitometry is coming into wider use to check the correctness of developer replenishment. It is long overdue as a method of quality control. Sensitometric testing of the developer should be done routinely every morning before starting the day's work, to assure uniformity of development during the useful life of the solution.

One and one-half minute processing time is now standard. Occupying a relatively small space, these units can accommodate about 300 films of assorted sizes per hour. Newer models offer 1-minute processing with almost no loss of quality.

To assure optimum film speed, contrast, and latitude, proper film and chemical combinations must be used.

As a rule, two smaller units are preferable to one larger unit of the same total capacity

because (1) in the event of equipment failure, the second unit is still available; and (2) there is less waiting time in loading two processors during peak hours.

Several brands of automatic processors are available, and selection should be made carefully because they vary in performance. One should especially ***avoid*** those requiring more than ninety seconds because they lack important features and also demand an annoying increase in processing time. Furthermore, one should have ready access to competent repair service.

Automatic Daylight Film Loaders. Available today are light-tight units that attach to the automatic processor to allow loading and unloading cassettes in daylight. When the cassette is inserted into the unit, it automatically unloads in the dark section and the film is pulled automatically into the processor. The cassette then reloads with fresh film and returns to the daylight. This eliminates the need of an elaborate darkroom and saves technologists' time.

SUMMARY OF PROCESSOR CARE

Details of quality assurance are covered in Chapter 22. Here, a brief summary of processor maintenance will be given, but manufacturer's recommendations must be carefully followed. Records should be kept, including dates.

1. ***Daily.***
   a. Replenishment rate check.
   b. Developer temperature.
2. ***Weekly.***
   a. Sensitometry strip and densitometry test for density and contrast. The mid-range step on the strip is the ***speed index*** (normally 1.2 D), and two densities above this—2.2 D—is noted; this is the contrast index. Processor performance is thereby monitored.
   b. Gross film fog for base + fog. Since the first step on the sensitometric strip is unexposed, its densitometry reveals increased fogging, which may be due to a cassette or darkroom leak.
   c. Remove and clean roller racks.
3. ***General Maintenance.*** Regular inspection of rollers and gears, and replacement of defective parts on schedule of service personnel.

SILVER RECOVERY FROM FIXING SOLUTIONS

All radiology departments should consider reclaiming silver from exhausted fixing solutions as a large amount of unexposed silver halide is dissolved by the ammonium thiosulfate. In larger cities professional service is available, but some departments have the personnel to do their own thing.

Exhausted fixing solution is transferred from its section into the electrolytic recovery unit, of which a number can be obtained commercially. They all work on the same principle: two electrodes—a cathode and anode—are immersed in the transferred fixer. Turning on the electric current causes the positive silver ions (Ag^+) to migrate to the cathode and a variety of anions (negative ions) move to the anode. By this method, one obtains very pure metallic silver, which is sent to a commercial facility for conversion into bars.

Not only should silver be reclaimed for financial reasons, but also for environmental friendliness. Additionally, discarded radiographs can be sold for silver recovery, which can provide more financial return than one can imagine.

QUESTIONS AND PROBLEMS

1. Define latent image. How is it produced?
2. Of what do the black areas on a roentgenogram consist? What factors determine the various densities in a radiograph?
3. What is the purpose and theory of development? Fixation?
4. List the ingredients of a manual developing solution and describe the function of each one.
5. List the ingredients of a manual fixing

solution and describe the function of each one.

6. What is meant by time-temperature development?

7. Discuss replenishment of developer and fixer in manual and automated systems.

8. What is included in the term "film processing"?

9. Name five causes of film fogging and state how they can be avoided.

10. Describe four causes of streaking of radiographs.

11. Why should overfixing be avoided?

12. Describe the important differences in the composition of processing solutions in automated and manual systems.

13. What problems would arise if manual processing solutions were used in an automatic processor?

14. List the important advantages of automated processing over manual processing.

15. State and explain briefly the three functions of the transport system.

16. Why are automatic processors operated at high temperature? How critical is the developer temperature in automated processing in comparison with manual processing?

17. How can one decrease the frequency of automated processor breakdown?

18. Why are filters used in the water supply of automated processors? What is the result of clogging of these filters?

19. Compare manual and automatic processing times.

20. Summarize the care of an automatic processor.

21. How is silver recovered from fixing solution?

Chapter 18

RADIOGRAPHIC QUALITY

Topics Covered in This Chapter

- Recorded Detail
    - Factors Governing Focal Spot Blur
    - Focal Spot Evaluation
        - Size
        - Resolution Tests
    - Motion Blur
    - Screen Blur
    - Object Blur
- Optical Density (density)
    - Peak Kilovoltage (kVp)
    - Milliampere-Seconds (mAs)
    - Source-Image Receptor Distance (SID)
        - Inverse Square Law
- Contrast
    - Radiographic
    - Subject
    - Film–Characteristic Curve
- Distortion
    - Size–Magnification
    - Shape–Distortion
- Direct Magnification–Macroradiography
- Modulation Transfer Function (MTF)
- Questions and Problems

MEDICAL RADIOGRAPHY aims to provide maximum diagnostic information about a given anatomic structure by recording its x-ray image, usually on film, its success depending on the production of radiographs of superb quality. Poor quality may entail serious errors in diagnosis by inadequate recording of information which, for us, is a collection of images. Recent advances in digital imaging show great promise in the improvement of image quality, but the basic principles of radiography still apply.

Radiographic quality refers to the sharpness of structural image details such as bone trabeculae and small vessels. We use the term *recorded detail* to denote radiographic quality, otherwise known as definition, sharpness, and simply detail. The opposite of sharpness is *blur* or fuzziness of image

margins.

Resolution measures the ability of an *image receptor* (screen-film, fluoroscopic intensifier image, etc) to produce separate images of closely spaced small objects; whereas *sharpness* can be measured objectively by *acutance,* the abruptness of the boundary between an image detail and its surroundings.

Visibility of detail refers to the ability of the observer to see recorded detail, which can be obscured by *motion* of the part, or *noise:* light-fog, scattered x rays, or screen mottle.

Five principal factors influence recorded detail: blur, optical density (or simply, density), contrast, distortion, and noise. These will now be explained individually and how they relate to one another.

BLUR

There are four major causes of image blurring or unsharpness. One is geometric in origin, depending primarily on the measurable size of the focal spot. The other three have to do with motion of the object, the nature of the image receptor (eg, film-screen system), and the shape of the object being radiographed.

Geometric or Focal Spot Blur

The arrangement in space relating the x-ray source (focal spot), anatomic part, and film, controls the degree of geometrically produced blurring. We shall simply call this *geometric* or *focal spot blur.* The terms *penumbra* and *edge gradient* may be used as synonyms for geometric blur, but the latter is preferred.

Geometric blur depends on three factors: (1) effective focal spot size, (2) source-to-image receptor distance (SID) or focus-film distance (FFD), and (3) object-to-image receptor distance (OID) or object-film distance (OFD) (see Figure 18.2.) All these factors affect recorded detail (image sharpness), which improves as geometric blur decreases. The term focus-film distance has been replaced by *source-to-image receptor distance (SID);* this refers to any medium, like a fluorescent screen or x-ray film, that converts incident x-ray photons either to visible light or some other form that can, in turn, be changed to visible light.

1. **Effective Focal Spot Size.** According to elementary geometry, the effective or apparent size of the focal spot (see page 139) has a marked influence on recorded detail. Figure 18.1 shows an anatomic object situated between an x-ray tube and a film. If the focal spot were assumed to be a point (having no dimension), all x rays passing the edge of the object would produce a

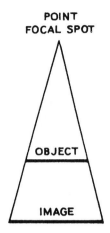

Figure 18.1. Image production by a hypothetical (imaginary) *point* focal spot. All the rays passing the edge of the object would produce a point-for-point sharp image of the object margin. The result would be nearly perfect geometrically recorded detail (no blur or penumbra).

point-for-point sharp image of the object margin on the film. But in practice the focal spot has a finite width, usually ranging from 0.3 to 2.0 mm. Therefore, as shown in Figure 18.2, x rays originate over innumerable points on the focal spot, spread as they pass the edge of the object and proceed toward the film. This produces a blurred margin, the width of the blur being proportional to the effective focal spot size (compare Figure 18.2A with Figure 18.2B). In other words, to obtain minimal blur we should use the smallest practicable focal spot. Since blur impairs recorded detail, we may conclude that *tubes with smaller focal spots provide better recorded detail, with improved image quality.*

2. **Source-to-Image Receptor Distance (SID).** The degree of blurring (ie, blur width) also depends on the distance between the source (focal spot) and the image recep-

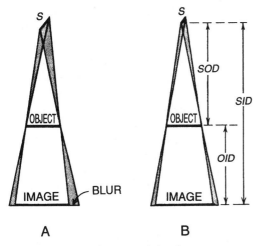

A   B

Figure 18.2. Influence of focal spot size on recorded detail. In *A*, the larger effective focal spot produces more blurring (wider penumbra) than the smaller effective focal spot in *B*. Focal spot blurring is caused by x rays, which are emitted from the entire focal spot area, converging toward the edge of the object and then diverging toward the edge of the image at the film or other image receptor. Thus, *a smaller focal spot produces better geometrically recorded detail.*

tor. As the SID (FFD) is increased while the OID (OFD) remains unchanged, the x rays arising from innumerable points on the focal spot undergo less spread after passing the edge of an object, on their way to the film. As shown in Figure 18.3A, at a shorter SID, the image margin is blurred more than in Figure 18.3B where the SID is longer. Thus, blurring decreases and recorded detail improves with increasing SID.

3. **Object-to-image Receptor Distance (OID).** The third geometric factor in blurring is the object-to-image receptor distance (OID) or (OFD). If the focal spot size and the source-to-image receptor distance remain constant, a decrease in the OID decreases blur (improves recorded detail). This is readily seen by comparison of Figures 18.4A and 18.4B.

The above three factors in geometric blurring (penumbra) may be conveniently summarized by a simple equation derived from Figure 18.3, where F_e is the *effective* focal spot size, *a* the source-to-object distance (SOD), *b* the object-to-image receptor distance (OID), and *B* the blur width, all in cm. According to the rule for similar triangles (shaded areas on right),

$$B/F_e = b/a$$
$$B = F_e\, b/a$$
$$B = \frac{\text{effective focal spot size} \times \text{OID}}{\text{SOD}} \text{ cm} \quad (1)$$

As will be shown later (see pages 225-226), recorded detail deteriorates with

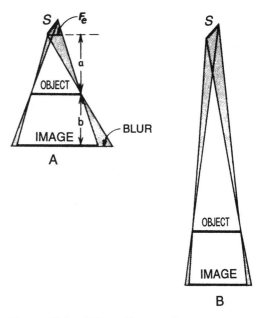

A   B

Figure 18.3. Effect of source-image receptor distance (SID) on recorded detail. The object-image receptor distance (OID) is the same in *A* and *B*, but the SID (FFD) is longer in *B* than in *A*. Therefore, there is less blur of the image in *B*. Thus, *increasing the* SID (FFD) *improves geometrically recorded detail.* Also, note the greater image magnification at the shorter SID when SOD remains constant.

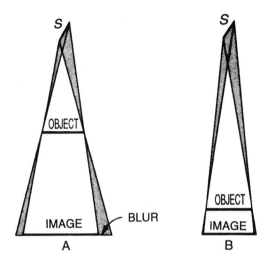

Figure 18.4. Effect of object-image receptor distance (OID) on recorded detail. The SID (FFD) is the same in *A* and *B*, but the OID is shorter in *B* than in *A*. Therefore, image blur is less in *B* than in *A*. Thus, *decreasing the OID (part-film distance) improves geometrically recorded detail.*

increased magnification, which results from an increase in OID and a decrease in SID (see Figure 18.4).

In *summary*, then, recorded detail (image sharpness) is enhanced by any factor that decreases geometric blur: namely, (1) small focal spot, (2) long source-to-image receptor distance (SID), and short object-to-image receptor distance (OID). Therefore, you should *place the part to be radiographed as close to the film as possible and use an appropriately long SID with the smallest applicable focal spot.*

Focal Spot Evaluation

As we have seen, the size of the focal spot strongly influences recorded detail. Unfortunately, there is no simple method of evaluating the focal spot, nor is there universal agreement, even among physicists, as to

how it should be done. At the present time there are three general ways of designating effective focal spot size:

1. **Nominal Size.** As quoted by the manufacturer, this is based on *computation* or *measurement* at very low mA (pinhole camera). However, the nominal size often understates the *effective* or *projected size* of the focal spot by a significant margin. Based on the ±50 percent tolerance customarily allowed in manufacture, a 0.3 mm focal spot may be as large as 0.45 mm, a 0.6 mm focal spot 0.9 mm, and a 1.0 mm focal spot 1.5 mm.

We shall now turn to the methods of practical evaluation of focal spot size.

2. **Pinhole Camera.** The *projected (effective) focal spot* dimensions can be measured by a *pinhole camera* as described in detail in *ICRU Report No. 10f.* A pinhole

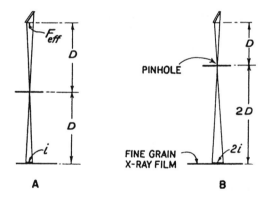

Figure 18.5. Principle of the pinhole camera for measurement of focal spot size. In *A* the pinhole is located midway between the tube focus and the film, so the image *i* of the focal spot has the same size as the effective or apparent focal spot F_{eff}. In *B*, with the pinhole located twice as far from the film as from the focus, the image *2i* is twice the size of the focal spot. Thus, proper correction must be made for the position of the pinhole in measuring focal spot size. In addition, the vertical axis of the pinhole must be perfectly aligned with the central ray.

camera consists of a tiny hole drilled in a plate made of a gold-platinum alloy. For example, hole diameter is 0.03 mm for focal spots smaller than 1 mm, and 0.075 mm for focal spots 1 to 2.5 mm. These very small holes minimize blur (penumbra), making for more accurate measurement of the focal spot image. At the same time, the small pinhole requires very long exposures, so that care must be taken not to damage the target of the x-ray tube. It is extremely important that a pinhole camera conform to rigid specifications and be purchased from a reputable dealer.

The pinhole is placed between the focal spot and a fine grain x-ray film, perfectly aligned with the central ray, perpendicular to it and parallel to the film. Figure 18.5 shows the arrangement. For focal spots up to 2.5 mm the pinhole-film distance is twice the focus-pinhole distance to give a magnification factor of two. Measurement of the pinhole image is made with a calibrated lens and corrected for magnification.

As shown in Figure 18.6, the x rays emitted by the focal spot have a nonuniform intensity distribution (ID), being concentrated along two edge bands. In fact, such an ID makes the focal spot behave as though it consisted of two narrow focal spots, the effect being to impair image resolution when compared with an ID that has a central peak of intensity. Doi and Rossman have shown that with a computerized simulated system in small vessel angiography, focal spot size is more important than ID.

Focal spot size increases with increasing mA, a condition known as *blooming,* but decreases with increasing kV. Although the focal spot is smaller when hot than when cold, blooming does not depend on temperature; rather it results from space charge effects associated with higher mA. In angiography, for example, the projected focal spot size increases about 50 percent in

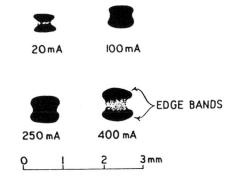

Figure 18.6. Effect of milliamperage (mA) on focal spot size. Note that x-ray emission is most intense along the edges of the focal spot—the so-called *edge* or *band effect.* Also, the size of the focal spot increases significantly with mA, a condition known as *blooming.* *(Adapted from Bernstein, Bergeron, and Klein, courtesy of Radiology.)*

going from 200 to 1000 mA. Because focal spots are measured at the factory at mA values much below those prevailing in radiography, their sizes are significantly understated. So the pinhole camera permits the evaluation of focal spots as to size, shape, and intensity distribution under actual operating conditions in the Radiology Department.

3. **Resolution Test Patterns.** Inasmuch as the pinhole camera does not show directly the resolution capability of a focal spot, another method has had to be devised—the *star pattern resolution object.* Resolution designates the smallest distance between two objects such that they appear as two separate images. Figure 18.7A shows a star pattern object. It provides a continuous change in spacing between thin radiopaque lines (usually lead), the limit of resolution being indicated by the blending of the line images. The theory underlying the use of resolution test objects is too complex to be considered here. However, it should be noted that there may be a ringed zone of apparently good resolution within the blurred zone, but this

A   **B**

Figure 18.7. Examples of resolution test patterns. *A.* Star type, useful in measuring focal spot resolution. *B.* Parallel line type for measuring resolution of intensifying screens. *(Courtesy of Nuclear Associates, Inc.).*

represents spurious resolution.

Another type of resolution test object consists of a series of grouped *parallel opaque lines* that differ in thickness and spacing (see Figure 18.7B). With this test object, resolution is related to the thinnest and most closely spaced line pairs resolved by the particular focal spot and is expressed as the maximum number of line pairs per mm. A *line pair consists of a line and one adjacent space.*

In summary, then, focal spot measurement is necessary to assure satisfactory recorded detail, since the true effective focal spot size may be appreciably larger than the nominal size. However, such measurements require experienced personnel using precision instruments. The two most widely used devices are the standard pinhole camera and some type of resolution test object; the former measures the physical size of the projected (effective) focal spot (also, shape and intensity distribution), whereas the latter measures its resolution capability. It behooves the purchaser of an x-ray tube to demand a suitably narrow tolerance range, especially for fractional focal spots such as 0.3 mm, and for angiography and magnification radiography. But in any case, measurement should be verified by one or both of the described methods.

Motion Blur

Motion of the part being radiographed may be regarded as the greatest enemy of recorded detail because it produces a blurred image and cannot be completely avoided. Motion can be minimized in three ways: (1) by careful immobilization of the part with sand bags or compression band, (2) by suspension of respiration when examining parts other than the limbs, and (3) by using exposures that are as short as possible, generally with intensifying screens of adequate speed, provided quantum mottle is not excessive (see page 211). Although screens contribute blur, this detrimental effect is usually more than compensated for by their reduction of motion blur.

Screen Blur

Recorded detail is better in radiography with cardboard holders than with intensifying screens, provided *immobilization is adequate.* However, screens are routinely used in general radiography. Under these circumstances, the gain in sharpness with screens and short exposures more than compensates for the increased motion blur with cardboard holders or slow screens and long exposures. In general, screen blur is greater than that caused by a 1 mm focal spot, but extremely short exposures using high speed screens with very low mAs can impair recorded detail because of excessive quantum mottle (see below).

The sources of screen blur include:

1. **Crystal Size.** Recall from Chapter 15 that intensifying screens produce a radiographic image because of *fluorescence of crystals* in the active layer. Since each crystal, despite its very small size, produces a spot of light of similar size on the film, image lines are broader than when they are formed directly by x-ray photons. Image blur increases with increasing crystal size, but this is of relatively minor importance in practice because it does not vary appreciably among general-purpose screens.

2. **Active Layer Thickness.** The diffusion (spreading) of light in the active layer due to its measurable thickness increases image blur (see pages 182-183).

3. **Film-screen Contact.** Cassettes are provided with clamps to secure close contact between the film and screens. Even very slight separation between them allows the light from any point in the screen to spread on its way to the film, thereby contributing to blur, with impaired recorded detail (see pages 183-185).

Because of these factors, slow screen-film systems (detail screens) should be used in the radiography of small parts (hands, feet, wrists), especially in infants and small children, to obtain adequate recorded detail. However, in the radiography of parts measuring 8 to 10 cm thick, high speed screens should be used routinely; and above 10 cm, a grid also.

4. **Noise.** In examining, with a magnifying glass, a radiograph that has been exposed by means of *intensifying screens,* you will find that it has a mottled or grainy appearance. Only a small part of this is contributed by radiographic mottle due to the granular structure of the screens and the resulting clumps of silver in the radiographic image, a condition known as *structure mottle.*

More important is *quantum mottle,* produced by the nonuniform intensity over the cross section of an x-ray beam as it leaves the tube port. Recall that the beam consists of photons or quanta having a random distribution in space. Thus, in a film that has been exposed directly to such a beam, different areas will have received different numbers of photons. Quantum mottle is a form of *noise,* which may be defined as any random audible or visible disturbances that obscure information. Other examples of visible noise on a radiograph are *fogging* of film by light or by scattered x rays.

Thus:

screen mottle = structure mottle + quantum mottle

With *slow image receptors* such as slow screen-film combinations, or direct exposure film, a *large number* of photons are needed for a particular degree of film density (darkening). Under these conditions the photons striking the image receptor are closely crowded, so the variation in the number of photons from area to area is small, mottle is minimal, and the image is relatively uniform. On the other hand, with *fast image receptors* such as high-speed screen-film combinations, a smaller number of photons can provide the same overall density, but now there is a larger variation in the number of photons from area to area (statistical fluctuation), so quantum mottle is greater. This is a simplified version of *quantum mottle.*

Quantum mottle increases with high contrast films because density differences are exaggerated. But increased diffusion of light in intensifying screens, such as those with thicker active layers, increases image blurring and at the same time makes quantum mottle less apparent. Quantum mottle, as a form of **noise**, impairs visibility of recorded detail.

Aside from the image receptors themselves, kVp is extremely important in the production of quantum mottle. As kVp is increased, x-ray photons give rise to more light photons, so the screen intensifying factor increases and fewer information-carrying x-ray photons are ultimately required for a given degree of film darkening with a particular film-screen combination. This explains why high kVp increases quantum mottle, especially with very low mA. Therefore, **low kVp** and **high mAs** should be used for small parts such as fingers and toes in infants. The *mAs favors information ("signal") over mottle ("noise")*, thereby improving the **signal/noise ratio** and image quality.

In summary, then, quantum mottle increases with fast image recording systems and with high energy photons. In any given instance, the decision has to be made as to whether high speed or minimal mottle is more important.

Screen mottle is not the only source of noise, which also includes scattered radiation, fogging from light or x rays, and anything else that impairs visibility of detail.

Object Blur

Most anatomic structures, which we here refer to as objects, have rounded borders. These introduce a blur factor, demonstrated in Figure 18.8. Note that in *A* the object has a shape conforming to that of the x-ray beam, so all the rays pass through the full thickness of tissue. As a result, the radiographic image presents a sharp boundary or **density gradient** as seen from the density trace at the bottom of the figure. In *B*, the rays pass through progressively thinner portions of the object toward its periphery so that image density fades off gradually at the border. In *C*, the round object also gives rise to an image with fading density at its boundary. Thus, in both B and C the image border has a long density gradient with increased blurring, its degree depending on the distance over which the density falls off. The effect of object blur is greater than is often appreciated, and it may even exceed geometric blur.

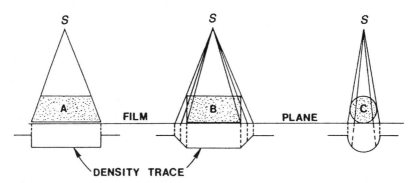

Figure 18.8. Object blur caused by object shape. In *A* the shape of the object conforms to that of the beam, so the density falls off sharply at the edge of the image which therefore has an abrupt or sharp border. In *B* a progressively smaller thickness of the object intercepts the beam toward the edge, so the density falls off gradually at the image boundary, which is therefore less sharp than in *A*. In *C* a rounded object also produces a less sharp image than in *A*. *(Adapted from Seemann, HE, Physical and Photographic Principles of Medical Radiography;* by permission of John Wiley & Sons. Inc.)

RADIOGRAPHIC DENSITY

The degree of darkening of a particular area of film is called its *radiographic density.* Some authors call it *optical density,* since it is the density we see, and to differentiate it from the scientific term density meaning mass per unit volume. To simplify matters, we shall adhere to the term *density* for film darkening.

It should be recalled that any region of a film emulsion exposed to light from intensifying screens, or x rays directly, becomes susceptible to the action of the developer. The silver halides that have been so affected are changed by the developing agent into tiny particles of *metallic silver* which appear black because of their finely divided state. The greater the total amount of radiation that reaches the film, the greater will be the final degree of blackening. Areas receiving only a small amount of radiation undergo little or no subsequent action by the developer, such underexposed areas appearing gray or translucent in the finished radiograph. Thus, in the final analysis, *radiographic density* (ie, degree of blackening or opacity) *depends on the amount of radiation reaching a particular area of the film and the resulting mass of metallic silver deposited per unit area during development.* Standardization of development should produce optimal film density for a correctly exposed film.

Density is measured by an instrument known as a *densitometer,* which indicates the relationship between the intensity of light falling upon one side of a given area of a radiograph (as from an illuminator) and the intensity of the light passing through (transmitted). This relationship is given by the following equation:

$$\text{density} = \log \frac{\text{incident light intensity}}{\text{transmitted light intensity}} \quad (2)$$

The equation can best be explained by numerical examples. If the incident light intensity is ten times the transmitted intensity, then the density is $\log 10 = \log 10^1 = 1$; if the incident light intensity is one hundred times the transmitted intensity, then the density is $\log 100 = \log 10^2 = 2$; etc (see pages 14-15). One need not be conversant with logarithms to use a densitometer, since it is calibrated to read density directly. With this method, clear film base has a density of 0.06 to 0.2, depending largely on the amount and shade of blue dye present. A diagnostic radiograph usually has densities varying from about 0.4 in the lightest areas to 3.0 in the darkest. Obviously, excellent radiographic quality requires optimal density for any particular anatomic structure; the correct exposure factors are selected automatically or from a proven technique chart.

Density is an extremely important factor in radiographic quality because it carries *information.* Without density there is no image and therefore no recorded detail. However, density must be optimal because, in excess, it may conceal information through loss of visibility of recorded detail.

Five factors govern the radiation exposure and resulting density of a radiograph: (1) kilovoltage, (2) milliamperage, (3) time, (4) distance, and (5) thickness and nature of part being radiographed.

1. **Kilovoltage.** An increase in kVp applied to the x-ray tube increases both the output and the percentage of higher energy (short wave length) photons. These more penetrating photons are not so readily absorbed by the structures being radiographed, and therefore a larger fraction of the primary radiation eventually reaches the intensifying screens. Thus, an increase in kVp increases the exposure rate at the film and the resulting radiographic density. An *increase of 15 percent in kVp approximately doubles the exposure.* For example, we can double the exposure at 40 kVp by adding 6

kVp, and at 70 kVp by adding 10 kVp.*

Example. A film exposed at 60 kVp is underexposed. What kVp would be needed to double the exposure?

Using the 15% rule, multiply $60 \times 1.15 = 69$ kVp.

Under ordinary conditions, with grids having a ratio of 8:1 or less, it is unwise to exceed 85 kVp because this produces an excess of *scattered radiation,* which fogs the radiograph and impairs contrast. On the other hand, the efficiency of x-ray tubes improves at higher kVp because of the considerably smaller heating load on the anode (see page 146) and the greater certainty of penetration. It has been found that *high voltage radiography* at 100 to 130 kV and a high ratio grid (12:1 or 16:1) offer advantages over conventional radiography, including greater latitude and a smaller dose to the patient.

2. **Milliamperage (mA).** This is a measure of electron flow per second from cathode to anode of the x-ray tube. As this flow rate is increased, more photons per sec are produced at the target. In fact, x-ray output (R/min) is proportional to mA: doubling mA doubles x-ray output; tripling mA triples output, etc. Note that mA determines output *only;* it has nothing to do with the penetrating ability of an x-ray beam, which is governed by kVp.

3. **Time.** An increase in exposure *time* causes a proportional increase in the *number* of photons emitted by the target. Thus, doubling exposure time doubles total exposure, and tripling exposure time triples total exposure—a longer exposure time allows the radiation at a given exposure rate to act longer, thereby affecting more silver halide crystals in the film emulsion and increasing radiographic density. In practice, mA and exposure time are multiplied: *mA × sec = mAs,*

called *milliampere-seconds.* This is a measure of the *charge* transferred from cathode to anode during an exposure. For example, if the selected technique calls for 100 mA and 0.1 sec, then multiplying $100 \times 0.1 = 10$ mAs. If the same exposure is to be given in a shorter time, so that radiographic density remains unchanged, mA can be increased to 400 and exposure time reduced to 0.025 (= $^1/_{40}$) sec: 400 mA × 0.025 sec = 10 mAs as before. We can generalize this relation by the simple equation

$$mA_2 s_2 = mA_1 s_1 \qquad (3)$$

where, if $mA_1 s_1$ is known and we select a particular time s_2, we can determine mA_2 from equation (3). This relation, known as the *reciprocity law,* holds true from about 1ms to 10s provided the x-ray generating equipment has been *properly designed* and *calibrated.* You should cultivate the habit of thinking in terms of mAs because it greatly facilitates the modification of established technique.

4. **Distance.** The effect of distance on exposure rate is not so simple as some of the other factors, but it is easily understood by keeping in mind the following elementary geometric rules:

a. X-ray photons actually originate from innumerable points on the focal spot of the x-ray tube, spreading in all directions from their points of origin.

b. In considering the effect of distance on exposure rate, we shall assume that the focal spot acts as a point source from which x rays spread in the form of a cone after leaving the circular port of the tube housing. But we invariably modify it with a beam-limiting device ("collimator"), which gives it a rectangular or square cross section.

Since the x-ray photons diverge (spread), *the width of the beam increases as the dis-*

---

* X-ray output is related to kVp^2. That is, output $\propto kVp_1^2 / kVp_2^2$. But density deviates from this because of increased penetration at higher kVp. The 15% rule applies much more closely in practice.

tance from the target increases, so the same amount of radiation is distributed over a larger area the farther this area is from the tube focus. Obviously, the same radiation spread over a larger area must spread itself "thinner." If, at a certain distance from the tube focus, the beam were to cover completely a film of a certain size, then at a greater distance it would cover a larger film. However, the radiographic density of the latter would be less in a given exposure time because each square centimeter of this film would have received less radiation than each square centimeter of the first film. We may therefore conclude that ***x-ray output decreases as SID increases.*** By how much is the next question.

We can determine by simple geometry *how much* output decreases with increasing SID. In Figure 18.9, the slanting lines represent the edges of a beam emerging from the x-ray tube at *S* and restricted by a square collimator. Two planes, *ABCD* and *EFGH*, are chosen perpendicular to the direction of the central ray of the beam (represented by the dotted line). Both planes are assumed to be squares. Plane *EFGH* is located twice as far from the target as plane *ABCD*. Therefore side *HE* of the lower plane is twice as long as a side of the upper plane, such as *DA,* because triangles *SHE* and *SDA* are similar and their corresponding sides are proportional. To simplify the discussion, let *X* equal a side of the upper plane. Then *2X* must equal a side of the lower plane. The area of the upper plane will then be $X \times X = X^2$, and the area of the lower plane will be $2X \times 2X = 4X^2$. Thus, the lower surface, *EFGH* has four times the area of the upper surface, *ABCD,* because $4X^2/X^2 = 4$.

It is evident that ***when the distance is doubled, the same radiation is spread over an area four times as great.*** Therefore, the radiation intensity or output must be $1/4$ as great. The ***inverse square law of radiation,*** which applies to both x rays and light may be stated as follows: ***the x-ray output or exposure***

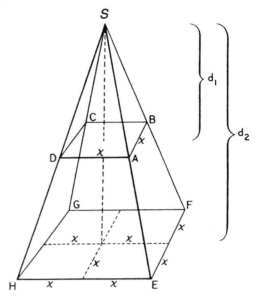

Figure 18.9. Inverse square law of radiation. The lower plane surface *(EFGH)* is selected at twice the distance from the point source of radiation (tube focus) than is the upper plane *(ABCD)*. Each side of the lower plane *(x + x)* is twice as long as each side of the upper plane *(x)*. It is evident from the diagram that the lower surface area is four times as large as the upper surface area, which means that at twice the distance from the target the x-ray beam covers four times the area and therefore the radiation intensity must be only one-fourth as great.

rate at a given distance from a point source is inversely proportional to the square of the distance.

Example 1. If the output of an x-ray beam at 20 in. from the focus is 100 R/min, what will it be at 40 in.? Set up the inverse square law in equation form:

$$I/i = d^2/D^2 \qquad (4)$$

where

I is the output at 40 in. = ?
i is the output at 20 in. = 100 R/min
d = 20 in.
D = 40 in.

Note again the inverse square proportion. Substituting the above values in equation (4):

$$I/100 = (20)^2/(40)^2$$
$$I = \frac{\cancel{20} \times \cancel{20} \times 100}{\cancel{40} \times \cancel{40}}$$
$$\frac{}{2 \times 2}$$
$$I = 25 \text{ R/min}$$

Thus, at twice the distance the output is one-fourth the initial value (25 as compared with 100). This means that in radiography, doubling the distance while the kVp and mAs are held constant reduces the output to one-fourth. Therefore, to keep the output constant the mAs has to be multiplied by four.

Example 2. If the exposure rate of radiation at 60 in. is 10 R/min, what will the output be at 20 in.? Again using equation (4), let *i* represent the unknown output at point *d* located 20 in. from the focus. Then, using the data given in the above problem,

$$i/10 = (60)^2/(20)^2$$
$$i = \frac{10 \times \overset{3}{\cancel{60}} \times \overset{3}{\cancel{60}}}{\cancel{20} \times \cancel{20}}$$
$$i = 90 \text{ R/min}$$

In other words, the distance has been reduced to $20 \div 60 = 1/3$ and the output has increased by $90 \div 10 = 9$ times.

There is a *much simpler method* of applying this law. Rearranging equation (4),

$$ID^2 = id^2 \qquad (5)$$

This means that for a given set of factors (kVp, mA, filter, etc) the output at a particular point in the beam, times the square of the distance of the point from the focal spot, is a *constant.* Thus, in equation (5), id^2 is a constant. In the problem on page 215, it is stated that $i = 100$ R/min at distance *d*, 20 in. Therefore,

$$id^2 = 100 \times 20 \times 20$$

The problem is now solved by substituting this constant and the *new* distance, 40 in., in equation (5).

$$I \times 40 \times 40 = 100 \times 20 \times 20$$
$$I = \frac{100 \times 20 \times 20}{40 \times 40}$$
$$I = 25 \text{ R/min}$$

In actual practice, a therapy machine must be calibrated for each treatment distance because of deviation from the inverse square law due to the finite source. However, the law is sufficiently accurate for approximation in diagnostic radiology and in protection problems.

In *radiography* we use a *direct square law* when changing source-film distances because we are *compensating* for the reduced or increased exposure as the distance is increased or decreased, respectively. Assuming that the kVp remains constant, we may obtain the mAs required to compensate for a change in distance by using the following equation:

new mAs/old mAs = (new distance)²/(old distance)² (6)

Example 3. If the factors 100 mAs, 100 kVp and 40-in. (100 cm) distance produce a radiograph having the proper density, what will be the mAs needed to maintain the same radiographic density at a focus-film distance of 60 in. (150 cm)?

$$\text{new mAs}/100 \text{ mAs} = (60)^2/(40)^2$$
$$\text{new mAs} = 100 \times 3600/1600 = 225 \text{ mAs}$$

5. **Radiographic Object.** The human body consists of tissues and organs that differ in thickness and density with resulting differences in radiolucency—the ability to transmit x rays (review pages 194-195). Thicker and denser anatomic parts, whether normal or pathological, attenuate (ie, absorb or scatter) x rays to a greater degree, leaving less exit or remnant radiation to reach the film. Ranging from the greatest to the least densi-

ty or radiopacity are: (1) dental enamel, (2) bone, (3) tissues of "water density" such as muscle, glands, and solid nonfatty organs, (4) fat, and (5) gas. Pathologic processes such as pneumonia increase the density of lung tissue. On the other hand, destructive bone diseases reduce the density of bone. Exposure factors–kV and mAs at a particular distance–are optimally selected for the thickness and density of the part being radiographed. Finally, opaque media rank high on the list of dense materials.

CONTRAST

A radiograph consists of light and dark areas; that is, it shows variations in density. The range of density variation among the light and dark areas is called **radiographic contrast.** While density represents the amount of silver deposited in a given area, contrast represents the relative distribution of silver in various areas of a radiograph. To be perceptible to the average human eye, the difference in density of adjacent areas must be at least 2 percent. **Optimum contrast enhances recorded detail.**

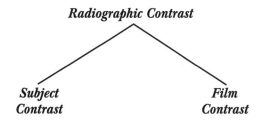

Radiographic Contrast

Subject Contrast   *Film Contrast*

There are three kinds of contrast. **Radiographic contrast**–the overall contrast of a radiograph–depends on **subject contrast** and **film contrast.** These will be discussed separately and interrelated. However, the unqualified term contrast usually means radiographic contrast.

Radiographic Contrast

We usually designate radiographic contrast as **long scale** (low) or **short scale** (high). This concept may be explained as follows: suppose a radiograph were cut into small squares representing each of the different densities present, from the lightest to the darkest, and these squares were then arranged in order of increasing density. If there should be many such squares with very little difference in the density of successive squares, then the original radiograph must have had **long scale** or **low contrast.** On the other hand, if there should be relatively few squares with a large difference in density between successive ones, then the radiograph must have been of **short scale** or **high contrast.** In other words, a long scale or low contrast radiograph has a long range of tonal gradation from white, through many shades of gray, to black; whereas a short scale or high contrast radiograph has a short range between white and black (see Figure 18.10 on page 219).

Contrast improves recorded detail and is optimal when it produces sufficient difference in density among the various details to make them distinctly visible in all areas of the radiograph. This is usually achieved by **medium scale contrast.** Excessively short scale contrast tends to impair detail; an example is the so-called "chalky" radiograph which may lead to error in interpretation, especially in the detection of fine fracture lines. However, short scale contrast is desirable under special circumstances such as studies with radiopaque media–urography, angiography, etc.

Subject Contrast

As indicated earlier, ***subject contrast*** is one of the factors in radiographic contrast, the other being film contrast. Before defining subject contrast, we must first introduce appropriate basic concepts.

An x-ray beam undergoes ***attenuation*** or loss of photons while passing through a patient, because of absorption, scatter, and inverse square law. But attenuation is not uniform throughout the cross section of the ***exit beam,*** that is, the beam leaving the patient. If we could somehow "see" the exit beam in cross section from the standpoint of the film, we would observe a variation in the number of photons per cm^2 at different locations in the beam. This results from unequal degrees of attenuation of x rays by various tissues. It is the spatial distribution of photons in the cross section of the exit beam that constitutes the ***aerial image*** (see Figure 17.4 page 195), which will eventually be recorded on the film (or some other image receptor) as the ***radiographic image.***

Thus, ***subject contrast may be defined as the contrast in the aerial image.*** It is the ratio of the number of photons in two or more equal small zones of the aerial image.

Note that the aerial image contains information derived from the patient. But in addition to the wanted information, the aerial image becomes contaminated with ***noise,*** this being ***unwanted*** radiation that impairs the quality of the information and ultimately degrades the quality of the resulting radiograph. Such noise factors, which include scattered radiation, quantum mottle, and fogging, obscure recorded detail.

We shall now explore the factors that affect subject contrast, namely, radiation quality, radiographic object, scattered radiation noise, and fog noise.

1. **Radiation Quality**–determines penetrating ability (transmission) of an x-ray beam. This is the inverse of attenuation. We found in Chapter 12 that an increase in the kV across the x-ray tube increases the penetrating ability of the resulting x-ray beam. A beam of high penetrating ability goes through tissues of various densities more uniformly than does a beam of low penetrating ability. For example, in radiography of a leg, the bony structures absorb a much greater fraction of the primary x rays than do the soft tissues. As the kVp and consequent penetrating ability of the primary photons increase, the difference in the absorption of the radiation, as between bone and soft tissues, becomes smaller. Consequently, the transmission of x rays through bone is more nearly like that through soft tissue and there is less difference in radiographic density between them; in other words, there is a reduction in subject contrast. Figure 18.10 shows the effect of various kilovoltages on contrast. You can see that an ***increase in kV produces longer scale or lower contrast*** in the radiograph of an aluminum step wedge, a bar made up of a series of increasing thicknesses of aluminum (see Figure 18.10).

An additional factor at higher kVp is the loss of contrast resulting from the increase in the fraction of scattered radiation, but this is largely removed by the use of high ratio grids.

2. **Radiographic Object**–atomic number, density, and thickness of the object being radiographed.

Atomic Number. As noted before, photoelectric absorption increases dramatically with increasing atomic number. Therefore, attenuation is higher per cm of bone (high atomic number of calcium and phosphorus) than per cm of soft tissue.

Density. An increase in density (g/cm^3), when expressed in terms of electrons/cm^3), increases x-ray attenuation per cm of tissue because with higher density, more electrons are present per cm^3 of tissue to engage in Compton scatter.

Thickness. With increased total thickness

ALUMINUM STEP WEDGE

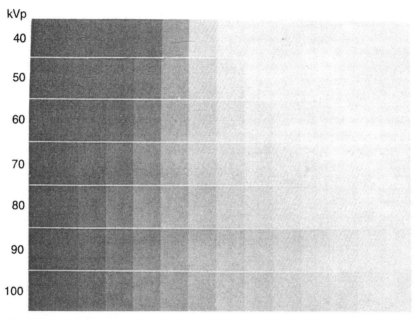

Figure 18.10. Effect of kilovoltage on contrast, using an aluminum step-wedge. The short range (high) contrast with 40 kVp, as compared with long range (low) contrast with 100 kVp is obvious.

of tissue traversed by the x-ray beam, there are more atoms available for interactions (absorption and scatter).

The degree of attenuation in a tissue is called its *radiopacity;* conversely, the degree of transmission is its *radiolucency.* Greater attenuation means that less radiation is transmitted through the patient; that is, less exit radiation is available for radiography (see pages 194-195).

As noted above (pages 216-217), the body contains tissues of various densities. For example, a radiograph of the abdomen has a light zone in the center representing the ver-

tebral column which attenuates more radiation than an equal mass of soft tissue. Various soft tissues are represented as shades of gray, depending on differential attenuation. The fat around the kidneys, along the psoas muscles, and in the abdominal wall, appears as dark gray or nearly black lines.

The differential attenuation of x-ray photons produces subject contrast in the aerial image, ultimately contributing radiographic contrast, which improves image quality. A number of organs and anatomic regions do not have sufficient inherent contrast for plain radiography. In such cases adequate contrast enhancement becomes possible by the use of ***contrast media.*** For example, dense media such as barium sulfate in the gastrointestinal tract, and iodinated compounds in the gall bladder, kidneys, vascular system, and bronchial tree. Note that the photoelectric effect produces a sharp peak in absorption of x-ray photons, approximately equivalent to a 65-kVp beam, because the binding energy of the K-shell electrons of iodine is 32 keV. Optimal contrast would therefore be obtained at about 65 kVp. However, we find in practice that adequate contrast between iodine and soft tissue occurs with x-ray beams generated at 65 to 80 kVp,

3. **Scattered Radiation**–a noise factor, obscuring information, or ***detail,*** in the radiographic image. Scattered radiation impairs contrast by a fogging effect mainly on the **lighter areas** of the radiograph, and may be controlled by means of stationary or moving grids and by limiting the size of the x-ray beam with a beam-limiting device, to be described in the next chapter.

Table 18.1
FACTORS AFFECTING
SUBJECT CONTRAST*

Factor	Change	Subject Contrast
Beam spectrum		
kVp	↑	↓
Generator % ripple	↑	↑
Target material	W to Mo*	↑
Scatter		
kVp	↑	↓
Beam collimation (field size)	↑	↓
Grid or air-gap technique	Yes	↑
Compression	Yes	↑
Nature of subject		
Differences in x-ray absorption	↑	↑
Patient thickness	↑	↓

Arrows indicate increase (↑) or decrease (↓)
*W-Tungsten, Mo-Molybdenum

* From Pizzutiello, R.J. Jr., & Cullinan, J.E.: *Introduction to Medical Radiographic Imaging. (By courtesy of Eastman Kodak Company.)*

4. **Fogging**–a form of noise that imparts an overall gray appearance to the radiograph with reduction in contrast and visibility of recorded detail.

Table 18.1 summarizes the factors affecting subject contrast.

Film Contrast

As we have already pointed out, film contrast and subject contrast make up radiographic contrast. However, in considering film contrast we must include the type of image receptor (in this instance, film), its use with or without screens, and the processing system. It is therefore more appropriate to speak of the ***film imaging system,*** a designation which includes film and processing, with or without screens.

Films themselves vary in their ***inherent contrast,*** depending on their emulsion characteristics. Thus, films are specifically designed for long, medium, or short scale contrast.

The ***development process*** also affects film contrast. With automated processing, optimum contrast is readily achieved by precise control of time, temperature, and replenishment rate.

Finally, you should recall that ***intensifying screens*** convert 98 percent or more of the radiologic image to light. In other words, the aerial image is changed almost completely to a light image, which is then recorded by the film. This conversion process enhances contrast because screen-type film has ***more inherent contrast*** for the light emitted by screens than for x rays directly. In general, contrast increases and latitude decreases with faster film-screen systems.

There is no simple definition of ***film*** contrast. It can be understood only by studying the so-called ***characteristic*** or ***sensitometric*** curve of a particular film recording system, shown in Figure 18.11. This type of curve was first used in photography by Hurter and

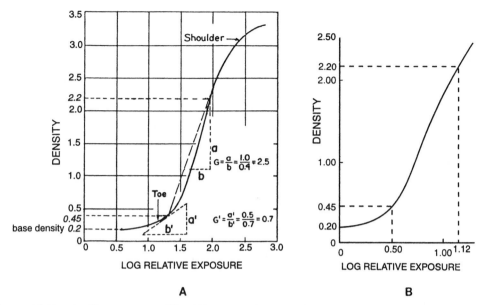

Figure 18.11. *A.* Characteristic (H & D) curve of a typical screen-type x-ray film. $G = a/b$, the slope (steepness) of the straight portion of the curve, represents the region of maximum film contrast. $G' = a'/b'$, the slope of one extremely short segment of the toe portion; note that here the slope varies along the curve and is less than that of the straight portion so that contrast is low. In the shoulder portion contrast is also low. Furthermore, the toe portion is a region of underexposure, and the shoulder a region of overexposure. The dashed line represents the *average gradient (average slope) and encompasses that portion of the curve corresponding to the useful exposure-density range of the film in question.* *B.* Simplified version of the H & D curve to show average gradient.

Driffield and is often referred to as an H & D curve. Film manufacturers use such curves to monitor film quality and consistency. H & D curves show the density response of film to various exposures under experimental conditions. Radiographic density is then plotted as a function of exposure (actually, density versus log relative exposure). Note that at a value of 0.5 on the horizontal axis the beginning of the curve already lies above this axis because of **inherent** or **base film density of 0.2.** As the exposure is increased, film density at first increases gradually with an upward curve in the **toe portion,** then more steeply along a **straight line,** and finally along a **shoulder portion.** The slope of the straight line joining the point on the curve near the toe corresponding to 0.2 (base density + 0.25 = **0.45,** and the point on the

curve near the shoulder corresponding to base density 0.2 + 2.0 = **2.2,** has been defined **arbitrarily** as the **average gradient** (see dashed line). The average gradient includes the useful exposure-density range of a film and varies with the type of film, **those films with greater average gradients having greater contrast for a given subject contrast** (see Figure 18.12). A film with an average gradient greater than 1 will heighten subject contrast.

The characteristic curve also shows that at low densities (toe portion) and at high densities (shoulder portion) contrast is poor. Maximum contrast occurs in the straight portion. An average radiograph displays poor contrast in both the lightest (toe) and darkest (shoulder) regions. Furthermore, underexposure exists in the toe region, and

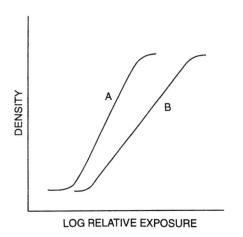

Figure 18.12. Characteristic curves reveal differences in radiographic contrast. Curve *A*, being steeper than curve *B*, indicates shorter scale, or higher contrast. Also, curve *B* represents more latitude.

the greater is the required exposure in terms of kVp, mA, and time. Some technic charts keep the mAs constant for a given region of the body and vary the kVp according to part thickness. Others show *optimum kilovoltage* for a given part and vary the mAs. If high kVp equipment is available, we can use 100 to 140 kVp with grids having a ratio of 12:1 or 16:1. High voltage radiography produces radiographs with good definition, wide latitude, and moderate contrast, provided suitable filters and grids are employed; besides, patient exposure is reduced with high kilovoltage exposure. No technique charts are required with automatic exposure.

Table 18.2 summarizes the factors affecting film contrast.

overexposure in the shoulder region. It is precisely in these regions where recorded detail is poor.

Film latitude has an inverse relation to film contrast—a low contrast film has wider latitude than a high contrast film. For example, a wide latitude film is preferred in chest radiography to avoid obscuring subtle detail of vascular markings. On the other hand, low latitude film is preferred when opaque media are of major interest as in angiography or urography.

Most radiographic *technique charts* are based on the different densities of various parts of the body, and on different thicknesses of the same part of the body in different individuals. The thicker or denser the part,

Table 18.2
FACTORS AFFECTING FILM CONTRAST*

Factor	Change	Film Contrast
Film type		
High contrast (short-scale)	Yes	↑
Wide latitude (long-scale	Yes	↓
Position on film characteristic curve (density level)		
Toe		↓
Linear position		↑
Shoulder		↓
Fog level	↑	↓
Film processing		
Suboptimal	Yes	↓

Arrows indicate increase (↑) or decrease (↓)

* From Pizzutiello, R.J. Jr., & Cullinan, J.E.: *Introduction to Medical Radiographic Imaging.* (By courtesy of Eastman Kodak Company.)

DISTORTION

A radiographic image does not faithfully represent the anatomic part, but differs from it in varying degrees of size and shape. Such *misrepresentation of the true size and shape of an object is called distortion.* The amount

of distortion depends on several factors, to be discussed in this section. While distortion generally has a degrading effect on radiographic quality, we cannot completely eliminate it because the image lies in a single plane and is two dimensional, whereas the object, being solid, is three dimensional. In fact, distortion was used deliberately to bring out a structure that would otherwise be hidden, for example, in oblique radiography of the gall bladder to separate it from the vertebral column.

The two kinds of distortion–*size* and *shape*–will now be described.

1. **Size Distortion–Magnification.** X rays emerge from the tube focus in all directions, but the tube port delimits them to form a beam that is circular in cross section. As the beam advances it broadens, becoming cone shaped. Finally as it passes through the collimator, the x-ray beam acquires a rectangular cross section.

Geometric rules prevail in the realm of x-ray image formation just as they do in photography with light. We can easily show that when an object is held between a source of light and a white surface, the size of the shadow enlarges as the *object* is moved nearer the light, and shrinks as it is moved closer to the white surface. This is due to the *divergence of image-forming light rays in* the beam. In Figure 18.13 the divergent beam of light *magnifies* the shadow more in *A* than in *B;*

that is, the shorter the distance between the object and the source of light the greater the magnification. *The law of image magnification states that the width of the image is to the width of the object, as the distance of the image from the light source is to the distance of the object from the light source.*

$$MF = \frac{\text{image width}}{\text{object width}} = \frac{\text{image distance (SID)}}{\text{object distance (SOD)}} \quad (6)$$

where MF is the magnification factor

Precisely the same law applies to radiographic image formation. For example, in cardiac radiography the standard procedure is to use a 2-meter (6-ft) source-image distance to minimize size distortion or magnification of the heart, so that measurement of the transverse cardiac diameter on the radiograph gives virtually the true diameter. On the other hand, contrary to what is often taught, the **cardiothoracic ratio,** or the ratio of the transverse cardiac diameter to the transverse thoracic diameter, undergoes no significant change as **the SID is varied between 100 cm (40 in.) and 1.8 m (72 in.), if the patient's position remains constant. This results from the fact that as**

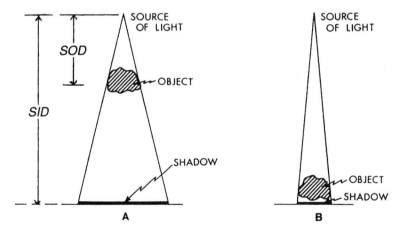

Figure 18.13. Image magnification. In *A* the shadow of the object is larger than in *B*, because in *A,* the object is nearer the source of light. The same principle applies to the formation of x-ray images.

the SID is changed, these two diameters change at virtually equal rates, as shown in Figure 18.14.

Magnification is unavoidable in radiography because of the geometric nature of image formation. As a rule, *magnification or size distortion can be decreased either by increasing the SID, or by reducing the OID.*

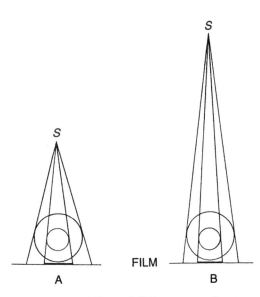

Figure 18.14. Effect of *SID* on magnification. In *A,* with a shorter *SID,* magnification is greater than in *B.* However, in this example, the *ratio* of the diameter of the *image* of the inner circle to that of the outer circle its essentially the same in both *A* and *B.*

Radiographic magnification is easily derived from the geometric relationship of similar triangles (see page 11). For example, suppose that an object measures 15 in. (37.5 cm) in diameter and lies 10 in. (25 cm) above the level of the film. If the SID is 40 in. (100 cm), what is the magnification of the radiographic image? We must first determine the width of the image by constructing a diagram as shown in Figure 18.15 and applying proportion (6):

$$\frac{\text{image width}}{\text{object width}} = \frac{SID}{SOD}$$

$$\frac{\text{image width}}{15} = \frac{40}{30}$$

$$\text{image width} = \frac{40 \times 15}{30} = 20 \text{ in.}$$

a. The *linear magnification of the image,* or the enlargement of its width relative to that of the object, may now be expressed in one of two ways:

(1). **Magnification Factor.** This is defined as the ratio of image width to object width:

$$\textbf{magnification factor} = \frac{\textbf{image width}}{\textbf{object width}} \quad (7)$$

In the above example,

$$\textbf{magnification factor} = \frac{20}{15} = 1^{1/3}$$

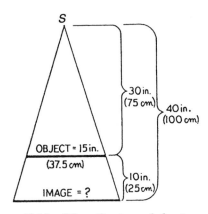

Figure 18.15. Magnification of the image in radiography. Image width/object width = *SID/SOD.*

(2). *Percentage Magnification (%M.)* We define this as the difference between image width and object width, relative to object width, expressed in percent.

$$\textbf{\%M} = \frac{\textbf{image width - object width}}{\textbf{object width}} \times \textbf{100} \quad (8)$$

In the above example,

$$\text{percentage magnification} = \frac{20 - 15}{15} \times 100$$

$$= {}^{1}/_{3} \times 100$$

$$= 33^{1}/_{3}\%$$

The same geometric law of image formation gives the diameter of the **object** being radiographed when the **image** diameter, the SID and the OID are known—draw a diagram as in Figure 18.15, insert the known values, and find the object diameter by using the proportion in equation (6).

2. **Shape Distortion.** This is caused by misalignment of the central ray, the anatomic part, and the film. In Figure 18.16, an oval piece of lead placed in the x-ray beam nonparallel to the film, casts a circular shadow. A rectangular piece of lead can be placed in a beam so that it casts a square shadow. With these objects placed perpendicular to the central ray and parallel to the film, the images will, of course, show no appreciable shape distortion and appear oval and rectan-

gular, respectively. Under certain conditions, deliberate distortion may bring out parts of the body that are obscured by overlying structures; for example, in radiography of the sternum you can rotate the patient enough to displace it from the image of the vertebral column.

We can summarize the data on distortion as follows: ***there are two types of distortion—size distortion and shape distortion. Size distortion*** is magnification caused by progressive divergence (spread) of the image-forming x rays in the beam: the shorter the OID and the longer the SOD, the less the degree of size distortion.

Shape distortion results from improper alignment of the central ray, object, and film. It is, of course, possible to have shape and size distortion occurring together if the causative factors of both conditions are present.

Since distortion (both size and shape) and recorded detail are influenced by the same factors—SID and OID—the greater the distortion of the radiographic image the poorer

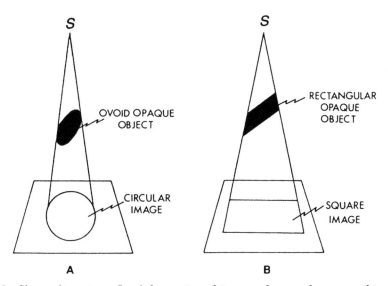

Figure 18.16. Shape distortion. In *A* the projected image of an oval opaque object is circular because the object is not parallel to the film. In *B* the image of a rectangular opaque object is almost square.

the recorded detail. You can readily demonstrate this principle by observing the shadow cast by an object such as a pencil on a white surface. As the pencil is moved toward the light (away from the white surface), the shadow not only becomes larger (magnification distortion) but also becomes more blurred (poorer recorded detail). As the pencil is moved nearer the white surface, the shadow becomes smaller and sharper.

Direct Magnification or Enlargement Radiography (Macroradiography)

We have already learned that as we move an object away from the film, toward the x-ray tube, image magnification increases if the SID remains unchanged (see Figure 18.14). At the same time, blurring increases and recorded detail worsens. If we could preserve recorded detail while achieving magnification, we should have a satisfactory method of *direct magnification radiography,* also known as *macroradiography.*

The introduction of tubes with *fractional focal spots,* that is, *microfocus tubes,* confirmed the feasibility of producing directly enlarged radiographic images. Although the 0.3 mm focal spot initially seemed to give reasonably good results insofar as image quality was concerned, subsequent work indicated that even smaller focal spots were needed to *minimize geometric blur and enhance recorded detail* during magnification to produce images with truly superb resolution. In fact, with two-diameter magnification a focal spot larger than an anatomic object (eg, tiny blood vessel) will generate blur wide enough to wipe out the image of the detail as shown in Figure 18.17. This explains why a focal spot no larger than 0.3 mm should be used in magnification angiography.

Tubes have been designed with focal

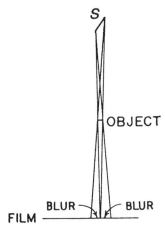

Figure 18.17. When the object is smaller than the effective focal spot, blur *B* dominates the image, so the true image or umbra disappears. This explains why, in two-diameter macroradiography, vessels smaller than the focal spot disappear in an angiogram and also emphasizes the importance of using the smallest practicable focal spot when utmost recorded detail is sought.

spots as small as 0.1 mm and, in fact, some experts believe that really high quality radiographic enlargement cannot be achieved with focal spots larger than 0.2 mm (Rao and Clark). However, their use in radiography is still severely restricted by the very limited heat loading capacities of these tubes.

In practice, two times linear (four times area) magnification is most often used. This requires placing the *anatomic part or object midway between the tube focus and the film* (see Figure 18.18). There we see that the magnification factor *(mf)* depends on the following relation:

$$mf = \frac{\text{source-to-image distance}}{\text{source-to-object distance}} \quad (9)$$

or

$$mf = \frac{SID}{SOD} \quad (10)$$

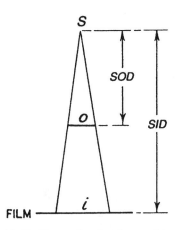

Figure 18.18. Principle of radiographic magnification. *Image size/object size = SID/SOD;* or, magnification factor $m = i/o = SID/SOD$.

Since, in Figure 18.19, SOD is equal to 1/2 SID, substituting the value 1/2 SID for SOD,

$$mf = \frac{SID}{1/2 SID} = 2$$

At the same time, Figure 18.19 shows that the geometric blur *B* (penumbra) at two times magnification equals the effective focal spot size.

The discussion thus far has dealt with *transverse magnification,* of an object lying in a plane parallel to the film. But *longitudinal magnification* (Doi & Rossman, 1975) refers to magnification of objects in a plane at some angle to the film plane. Longitudinal magnification increases the resolution of such objects by increasing their separation in the magnified image so that two objects lying above each other would show greater image separation (resolution) during magnification than they would in ordinary radiography (see Figure 18.20).

An appreciable *air gap* must exist between the object and the image receptor (film-screen system) in magnification radiography (see pages 226-227). For example, the SID is normally 40 in. (100 cm), so that with a magnification factor of two, an air gap of

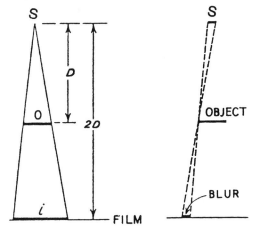

Figure 18.19. In the diagram on the left, the object is located midway between the tube focus and the film. Therefore, the magnification factor *mf* = 2, because *mf* = 2D/D = 2. In the diagram on the right, blur equals the effective focal spot size when *mf* = 2, *o* = object, *i* = image.

20 in. (50 cm) will be present between the object and the film (neglecting the thickness of the object itself). In fact, a 20 in. (50 cm)

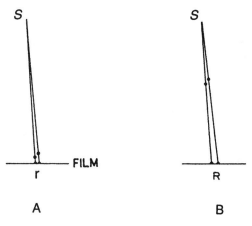

Figure 18.20. In *A,* transverse magnification is less than in *B.* At the same time, longitudinal magnification, or the resolution of two points lying at different distances above the film, is greater in *B.* (*R* in *B* is larger than *r* in *A.*)

air gap provides cleanup of scattered radiation as well as a 15 ratio grid. Therefore, we can omit the grid in magnification radiography. At the same time, close beam limitation and precise alignment to the anatomic area of interest are essential to high quality magnification radiography, just as in ordinary radiography.

Bear in mind that bringing the anatomic part closer to the tube increases the exposure at the level of the part. This is governed by the inverse square law; for example, with a magnification factor of two, the SOD is decreased by one-half relative to the usual SID in ordinary radiography. Reduction of the distance by one-half increases exposure four times = (magnification factor)^2 = 2^2. However, this applies only to the center of the object; the tube side of the patient will receive an even greater increase in exposure. But the increased dosage is partly offset by dosage reduction through elimination of the radiographic grid in the presence of the air gap.

As for *screen* and *film blur,* the latter is not significant. Much more important is the fact that *intensifying screens become more efficient when the same information in the enlarged image is spread over a larger area,* thereby enhancing recorded detail beyond that in conventional radiography.

Although the above discussion of macroradiography suggests that the last word has been written on the subject, you should know that experts differ as to its value in routine radiography. For example, convincing data by Rao and Clark favor the use of detail screens or nonscreen technique for radiography of well-immobilized small parts (eg, fingers and toes), the image being viewed with the aid of an excellent convex hand lens. This gives resolution approaching that in macroradiography. There is general consensus that tubes with focal spots larger than 0.3 mm should not be used in magnification radiography. Full advantage of the fractional focus tubes prevails only in close collimation to the area of interest, close apposition of the beam-limiting device to the tube aperture, and accurate centering of the beam.

In general, we can say that the advantages of magnification radiography, where applicable, depend on three factors:

1. *Sharpness factor*–the magnified image is spread over a larger area of the intensifying screens and therefore exhibits better recorded detail than the corresponding nonmagnified image.

2. *Noise factor*–while the image is magnified, screen mottle is not. Thus, recorded detail improves (greater resolution) but the noise factor (screen mottle) remains unchanged, so we say that the *signal-to-noise ratio is increased.*

3. *Visual factor*–visual recognition of images improves with magnification.

At present, macroradiography is seldom used in general radiography, but it does play an important role in *magnification mammography,* especially with breast compression. It is also helpful in angiography to detect small vessels, applications that have been aided by usable x-ray tubes with 0.1 mm focal spots.

MODULATION TRANSFER FUNCTION

The term *modulation transfer function (MTF)* often appears in textbooks and literature dealing with radiographic quality. The technologist should therefore be aware of its significance, at least in a general way. This has prompted a short, simplified treatment of the subject.

As noted earlier, the recording of a radiologic image by any type of image receptor (radiography, image intensification, TV, etc) entails the transfer of information from the object to the image. In the process there is inevitably some loss of information, such as recorded detail. The MTF represents a mathematical method of expressing the efficiency of any imaging system, that is, the part of the information put into the system, which becomes recognizable as the image. In a practical sense, then, ***MTF may be regarded as a measure of imaging system quality.***

It should be pointed out that we can obtain MTFs for the individual components of an imaging system; for example, in radiography we can derive the MTF separately for the focal spot, geometry, film-screen system, motion of the object, etc. Multiplication of the individual MTFs gives the total MTF of the system:

$$MTF_{total} = MTF_1 \times MTF_2 \times MTF_3 \times \ldots \ldots \times MTF_n$$

How is MTF obtained? It turns out that while this involves a complex mathematical procedure, we can still gain sufficient practical knowledge about it without necessarily understanding all of the involved mathematics. Basically, MTF is derived from what is known as ***line spread function (LSF).*** If we were to expose ***directly*** (without screens) a film to an extremely narrow beam of x rays collimated by a slit about 10 µm wide, we would obtain a sharp line image. However, a film-screen combination exposed to the same beam would record a line having unsharp edges due to screen and focal spot blurring. The LSF represents the width of this line as measured by a microdensitometer, with relative density plotted as a function of distance from the center of the line (see Figure 18.21).

By a complex mathematical process known as Fourier analysis, LSF may be converted to MTF which is then plotted (as in Figure 18.21). Here MTF on the vertical axis

receives a maximum value of one, since the ratio of the amount of information recorded in the image to the amount put into the system can never exceed unity, and is usually much less. In the examples shown in the figure, comparison of two different film-screen systems reveals that the superior film-screen system insofar as image quality (sharpness,

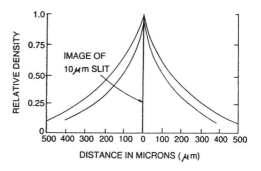

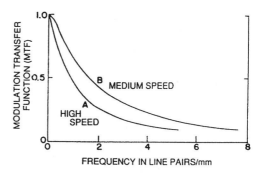

Figure 18.21. *Above.* Hypothetical example of line spread function for high speed (upper curve) and medium or par speed (lower curve) screens. The image of the slit with high speed screens is wider, giving poorer resolution, than with par speed screens. *Below.* Hypothetical modulation transfer function (MTF) for high speed *(A)* and medium or par speed *(B)* screens. Par speed screens have a larger MTF for a particular value of resolution (spatial frequency in terms of line pairs/mm) and are able to resolve more line pairs/mm than high speed screens (MTF curve *B* extends to higher values of resolution). (*Adapted from K Rossman, courtesy of Radiology.*)

resolution, and contrast) is concerned, has a narrower LSF, and an MTF that is higher at larger "spatial frequencies" (sine functions analogous to the number of line pairs resolved per mm). The MTF for other components of the imaging system can be obtained by mathematical transformation of the appropriate LSF.

At present, MTF is the best available means of evaluating the quality performance of an imaging system, derived from the MTFs of its constituent parts. As time goes on, anticipate a growing use of this concept in the specification of radiographic equipment and image receptors, both as a whole and with regard to their individual components.

QUESTIONS AND PROBLEMS

1. Define radiographic quality. How does it relate to recorded detail?

2. What is meant by sharpness; blur; acutance; resolving power; image receptor? How are they related?

3. List the five factors that affect recorded detail.

4. What are the three factors in geometric blur? Explain each one.

5. State the equation that relates the factors in geometric blur.

6. Calculate the image blur ("penumbra") produced by a 0.6 mm effective focal spot at a 100-cm SID when the OID is 5 cm.

7. Describe briefly the two main methods of evaluating focal spot size.

8. Define focal spot blooming. What exposure factor influences the degree of blooming?

9. How important is motion of the part being radiographed insofar as recorded detail is concerned? How can motion be controlled?

10. Discuss the factors of crystal size, active layer thickness, film-screen contact, and mottle as they affect blur.

11. How does object shape influence blur?

12. Define radiographic density. Explain the five factors that determine it.

13. State the inverse square law in your own words, and the two forms of the applicable equation. How is it modified in radiographic practice?

14. At a 50-cm SID the output is 20 R/min. What will the output be at 25 cm? 100 cm?

15. A certain technic calls for 50 mA and 0.1 sec at 65 kV. If we have an uncooperative patient and wish to reduce motion by using an exposure time of 1/60 sec, what will the new mA value be?

16. What is meant by the penetrating ability of an x-ray beam? How can the penetrating ability of a beam be changed?

17. A certain technique calls for an exposure of 70 kVp and 180 mAs at a distance of 60 in. (150 cm). If we wish to decrease the distance to 20 in. (50 cm), what new mAs value will be required to maintain the same radiographic density?

18. A radiograph exposed at 200 mAs and 75 kVp is judged to have been underexposed by about 50%. What kVp would produce about the desired density at the same mAs value?

19. How do intensifying screens affect density? contrast? recorded detail?

20. List the main materials in the human body in decreasing order of their radiographic density.

21. Define radiographic contrast. On what two kinds of contrast does it depend?
22. What is the function of radiographic contrast? Can there be too much contrast?
23. How are the following terms related: low contrast; high contrast; long scale contrast; short scale contrast?
24. Define subject contrast. How is it related to the aerial image? remnant (exit) radiation?
25. What is film contrast? Draw and explain the characteristic curve of a film. Discuss average slope (average gradient).
26. Name the factors that influence subject contrast; film contrast.
27. Define distortion. Name and explain the two kinds.
28. What effect does distortion have on recorded detail?
29. How can size and shape distortion be minimized?
30. An anatomic part is located 4 in. (10 cm) above the image receptor (film-screen system). The SID is 40 in. (100 cm), and the diameter of the anatomic part is 9 in. (23 cm). Find the size of the image percent magnification, and magnification factor.
31. What is quantum mottle? State its importance in radiography. Under what conditions is it accentuated?
32. Discuss the principle of direct magnification radiography (macroradiography). How do we compensate for the increased blurring inherent in this system?
33. How does magnification radiography affect patient exposure?
34. Why can the radiographic grid be eliminated in magnification radiography? Is this always desirable?
35. On what is modulation transfer function based? What does it measure?

Chapter 19

DEVICES FOR IMPROVING
RADIOGRAPHIC QUALITY

Topics Covered in This Chapter

- Scattered Radiation
- Removal by a Grid to Improve Contrast
    Stationary
    Moving
- Efficiency of Grids
    Grid Ratio
    Frequency
    Selectivity
    Contrast Improvement Factor
- Types of Grids
    Parallel

    Focused
- Practical Applications
- Air Gap Technique
- Reduction by Beam Limitation (collimation)
    Types of Collimators
- Other Methods of Improving Quality
    Moving Slit Radiography
    Anode Heel Effect
    Compensation Filters
- Questions and Problems

IN THE PRECEDING chapter we described the important factors affecting radiographic quality. We shall now show how the quality of a radiograph can be improved by the use of various special devices. Recall that *recorded detail* is the ultimate criterion of radiographic quality.

The rotating anode tube with its small focal spot provides excellent recorded detail. Furthermore, this tube has allowed the use of high mA with very short exposures, thereby minimizing motion, the greatest enemy of recorded detail. Thus, while the inherent construction of modern equipment should provide radiographs of superb quality, certain auxiliary devices are needed to reduce *scattered radiation,* a form of noise that impairs image quality by reducing contrast and obscuring recorded detail.

SCATTERED RADIATION

You should recall from Chapter 12 that the primary x-ray beam leaving the tube focus is polyenergetic in that it contains photons of various energies. The primary beam consists of *brems radiation* resulting from the conversion of the energy of the electrons as they are stopped by the target, and *characteristic radiation* emitted by the target metal due to excitation of its atoms. As the primary beam passes through the patient, some of the radiation is absorbed, while the rest is scattered in many directions. In the diagnostic range of 30 to 140 kVp the radiation *generated in the body* consists mainly of scattered photons produced by Compton interaction, but also includes characteristic radiation resulting from photoelectric interaction. Recall that the energy of the Compton scattered photons increases with increasing kVp. Characteristic photons have extremely low energy and are absorbed locally in the tissues, but *many of the Compton (scattered) photons have enough energy to pass through the body and approach the film from many directions.*

This multidirectional scattered radiation is a *noise factor* which seriously impairs radiographic quality by its fogging effect, diffusing x rays over the surface of the film and thereby *lessening contrast* (see Figure 19.1). You can see that the larger the percentage of scattered radiation relative to the primary radiation, the greater will be the loss of contrast of a detail such as *P*. In a radiograph of good quality, less than one-fourth the density should result from scattered radiation.

The ratio of scattered radiation to primary radiation is not uniform over the surface of the radiograph. In fact, with relatively opaque objects, the ratio increases toward the image border, so that scattered radiation tends to *degrade edge contrast* (Jaffe and Webster).

Of the total radiation reaching the film, the *ratio of scattered radiation relative to primary radiation increases with the following factors:*

a. An *increase in the area of the radiation field and the thickness of the part* traversed by the beam. As the beam increases in cross section it irradiates a larger volume of tissue and generates more scattered radiation.

b. An *increase in tube potential.* Increasing the kVp decreases the probability of Compton scatter, but photoelectric interactions decrease even more (see page 128). However at *higher kVp* the Compton scattered photons have more energy and are more likely to reach the film–the energy of the scattered photons is only slightly less than that of the primary photons as shown in Table 19.1; and an increase in the applied kV causes more and more of the radiation to scatter in a forward direction, that is, at an angle of 45° or less with relation to the direction of the primary beam. Thus, increasing kVp increases not only the energy of the scattered photons but also the number directed toward the film, so that a larger fraction of the scattered radiation reaches the film with increasing impairment of contrast.

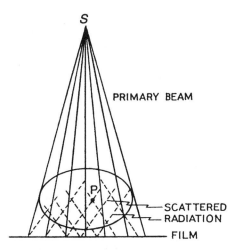

Figure 19.1. Deleterious effect of multidirectional scattered radiation on radiographic quality. This causes *loss of radiographic contrast* of a detail such as *P*. *F* is the focal spot.

Table 19.1
RELATIONSHIP BETWEEN PRIMARY
PHOTON ENERGY AND ENERGY
OF PHOTONS SCATTERED AT 45°

Primary Photon keV	45° Scattered Photon keV
40	39
60	58
80	76
100	94
120	112

c. An ***increase in the density of the tissues*** traversed by the primary beam; for example, as between air and soft tissue. In nongrid exposures of the chest and abdomen, the following are approximate relative values of the scattered and primary radiation reaching the film:

chest	50% scattered	50% primary
abdomen	90% scattered	10% primary

REMOVAL OF SCATTERED RADIATION BY A GRID

Since scattered radiation cannot be entirely eliminated because of the very nature of x rays and their interaction with matter, we must try to remove as much scatter as possible, once it has been produced, so as to improve contrast: the less the scatter, the better the contrast. The problem of scatter applies especially to radiography of large anatomic areas such as the abdomen because of the increased scattering by a large volume of tissue. Most effective in removing scattered radiation are the ***stationary grid*** and the ***moving grid.*** Although the latter is by far the more frequently used of the two, the stationary grid will be discussed first because of its historical priority and relative simplicity.

Principle of the Radiographic Grid

In 1913 Gustav Bucky introduced the ***stationary radiographic grid,*** a device placed between the patient and the cassette for the purpose of reducing the amount of scattered radiation reaching the film and so ***improving radiographic contrast.*** As shown in Figure 19.2, Bucky's original grid was of the ***cross-hatch type,*** consisting of wide strips of lead arranged in two parallel series, which are mutually perpendicular. Despite the coarseness of the early grids, with their 2 cm × 2 cm spacing, they removed a significant amount of scattered radiation, as shown by the frontal skull radiograph at the Smithsonian Institution.

A ***modern stationary grid*** has a very dif-

Figure 19.2. Photographic reproduction of Bucky's original radiographic grid. Note the cross-hatch pattern and the very coarse spacing. Yet, this grid achieved significant cleanup of scattered radiation, although a pattern of the lead strips appeared on the radiograph. *(Photograph* [No. 4 7022] *furnished by courtesy of Smithsonian Institution.)*

ferent appearance (see Figure 19.3). It consists of extremely thin, closely spaced lead strips measuring about 0.05 mm in width. They are separated by radiolucent material, usually plastic or aluminum, measuring about 0.33 mm wide. ***Aluminum*** is preferred for improved durability of the grid. and to absorb scattered radiation from the lead strips. A grid is manufactured by stacking and bonding together alternate sheets of lead and spacer material of appropriate thickness like the strips just mentioned. The

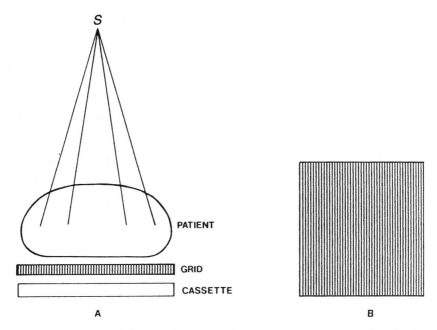

Figure 19.3. Stationary parallel or nonfocused grid, seen in cross section in *A* and top view in *B*. *S* is the x-ray source.

stack is then sliced crosswise, the thickness of each slice representing the height of the grid. Thus, each slice is a grid with alternating thin strips of lead and aluminum (or plastic).

Figure 19.4 shows an enlarged diagram of a grid. Only those x-ray photons passing directly through the narrow aluminum spacers will reach the film-screen system. Since the scattered photons proceed at various angles, most of them will strike the lead strips and undergo absorption without reaching the film. Such a grid may absorb as much as 90 percent of the scattered radiation and entail a notable improvement in contrast and recorded detail. Some ***primary radiation*** strikes the lead strips and is absorbed, a condition called ***grid cutoff.***

The stationary grid has one main disadvantage—as the lead strips absorb radiation, they cast shadows on the radiograph as thin white lines; but at the usual radiographic viewing distance, and especially with high quality grids, these lines are almost invisible. Besides, their presence is more than compensated for by the improved quality of the radiographic image. Because of the interposition of the grid in the x-ray beam, an appreciable part of ***both primary and scattered radiation is absorbed,*** and therefore the exposure must be increased according to the type of grid being used.

Figure 19.3B shows a conventional grid with lead strips parallel to the 17-in. edge. Also available is "decubitus" grid with its lead strips parallel to the 14-in. edge for cross-table radiography to minimize grid cutoff, since holding the cassette perpendicular the table is not so critical as with a conventional grid.

Efficiency of Grids

Certain criteria have been established to designate the quality or efficiency of radi-

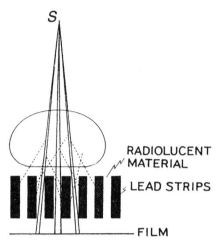

Figure 19.4. Principle of the radiographic grid. Scattered radiation, represented by the dotted lines, proceeds in various directions and is largely absorbed by the lead strips. Those primary rays that pass directly through the aluminum or carbon fiber spacers between the lead strips reach the image receptor (ie, film-screen system). Although the lead strips are very thin, any primary rays striking them are absorbed—an effect known as *cutoff.*

ographic grids; that is, *their ability to remove scattered radiation,* or *"cleanup."* These apply to both stationary and moving grids (to be described later), and include *physical* and *functional* aspects.

Physical Factors in Grid Efficiency. The three physical factors mainly responsible for grid efficiency are *grid ratio, grid frequency,* and *lead (Pb)* content.

1. *Grid Ratio* is defined as *the ratio of the height of the lead strips to the distance between the strips,* as expressed in the following equation:

$$r = \frac{h}{D} \tag{1}$$

where *r* is the grid ratio, *h* is the height of the lead strips, and *D* is the width of the spaces between the strips. These factors appear diagrammatically in Figure 19.5. As an exam-

ple, if the lead strips are 2 mm high and the space separating them is 0.4 mm, the grid ratio is 2/0.4 = 5:1.

Notice in Figure 19.5 that grids *A* and *B* have the same ratio, despite the difference in their heights. Grid *D* has the same height as *B,* but *D*'s ratio is greater because its lead strips are closer. Further comparison of the grids reveals that the paths of the most divergent scattered photons in grids *A* and *B* make the same angle with the lead strip, whereas they make half that angle in grid *D.* Thus, the higher the grid ratio, the "straighter" the rays have to be to get through the grid interspaces. It is like a line of cars passing through a tunnel; any car that strays too far off the road will hit the side of the tunnel. All other factors being equal, *the higher the grid ratio the better will be the "cleanup" of scattered radiation and the*

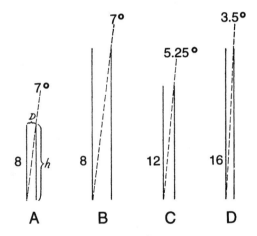

Figure 19.5. Effect of grid ratio on maximum angle of obliquity of a scattered ray that can get through the grid. In *A* and *B,* it is obvious that the ratio may be the same in grids of different heights, and 7° is the maximum angle that a ray can make with the vertical and still pass between the lead strips. In *D,* with the grid ratio twice that in *A,* the maximum obliquity of a transmitted ray, 3° 35' is half as large as that in *A.* Note the smaller frequency (lead strips farther apart) of grid *B* relative to grid *A* despite their identical grid ratio.

resulting radiographic contrast, with improved visibility of detail at the expense of increased patient exposure.

2. *Grid frequency,* defined as the number of lead strips per cm or per inch, has an important bearing on grid efficiency. Available grids range in frequency from 20 to 40 lines per cm (50 to 100 lines per in.). In general, the greater the grid frequency, the thinner the lead strips have to be, and the greater will be the likelihood of scattered photons passing *laterally* through the strips, especially at high kV. With grids of equal ratio, the one having fewer and thicker strips per cm possesses superior quality but at the same time the visibility of the grid lines becomes more objectionable (except in moving grids). Conversely, *as the grid frequency increases, the grid ratio must also be increased to maintain the same efficiency (cleanup).*

Functional Factors in Grid Efficiency. Knowing that a grid has a particular ratio and frequency (as well as lead content), how can we judge its ability to minimize scattered radiation? In other words, is there a quantitative measure of grid efficiency based on its actual function? Fortunately, there are several criteria of grid performance, the best ones being *selectivity* and *contrast improving ability,* as defined by the ICRU in *Handbook 89.*

1. *Selectivity.* Since the grid lies in the path of the x-ray beam, its lead strips absorb a fraction of *both* primary and scattered radiation. The absorption of primary radiation by a grid is known as *grid cutoff.* Obviously, a grid should transmit as large a fraction of the primary radiation as possible, to keep the exposure of the patient small. At the same time, it should absorb a maximum of scattered radiation to provide the best possible improvement in radiographic contrast. As specified by the ICRU, if Greek letter Σ (sigma) represents selectivity, T_p the fraction of *primary* radiation transmitted through the

grid, and T_s the fraction of *scattered* radiation transmitted through the grid, then

$$\Sigma = \frac{T_p}{T_s} \qquad (2)$$

Thus, the larger the ratio of the transmitted fraction of the primary radiation, to the transmitted fraction of the scattered radiation reaching the film, the greater will be the selectivity of the grid. At the same time, it must be remembered that the larger the fraction of the *total radiation* (primary plus scattered) that is transmitted and reaches the film, the less will be the exposure required for a given radiographic density.

Closely related to grid ratio and frequency is the *lead content* in terms of mass per unit area–g/cm²–of grid surface. For grids of *equal ratio,* the one with thicker strips (lower frequency) will be more efficient in removing scattered radiation and improving contrast, by virtue of its greater lead content (see Figure 19.6). But it will also remove

Figure 19.6. Two grids with the same ratio, but the one on the right has thicker lead strips. The latter provides better cleanup of scattered radiation, but also absorbs more primary radiation and is therefore less selective.

more primary radiation, thereby impairing selectivity. At the same time, a larger patient exposure is required for the same radiographic density. In general, lead content serves as an indicator of performance if the grid has an adequate ratio and frequency.

2. *Contrast Improvement Factor.* The ultimate criterion of the efficiency of a grid is its actual performance in improving radiographic contrast, as explained earlier in this chapter. The contrast improvement factor, *K,* is defined as the ratio of the radiographic contrast with grid divided by the contrast

without grid:

$$K = \frac{\text{radiographic contrast with grid}}{\text{radiographic contrast without grid}} \quad (3)$$

Unfortunately, this factor is so dependent on kV, field size, and thickness of part that it cannot be specified for a given grid under all conditions. (The ICRU has given special recommendations for measuring K in a 20-cm thick water phantom at 100 kV.)

We must emphasize that selectivity and contrast improvement factor have certain limitations because they are measured in a water phantom, which is inherently different from the human body. In making these measurements, certain conditions of standardization have to be laid down, and these may or may not hold during practical radiography. However, selectivity and contrast improvement serve as useful criteria for the intercomparison of grids.

Types of Grids

Stationary Grids. These thin *wafer grids* come in various sizes and have to be taped to the front of the cassette. Because they are easily damaged—a slight bend will impair their quality—these grids are now built into the cassette front. Moreover, because grid technique requires increased exposure factors, grids must be used with intensifying screens. The thinness of the grid does not appreciably increase OID, so it results in negligible blurring of the image. The cassette with a built-in grid is called a *grid cassette.*

A special *crossed grid* consists of two superimposed grids with the lead lines rotated 90° with relation to each other (see Figure 19.7). They are not often used in general radiography but may be advantageous in special procedures such as cerebral angiography with biplane equipment. A more versatile arrangement is the superimposition of

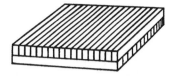

Figure 19.7. Crossed grids improve cleanup of scattered radiation. Note that the upper grid lines are perpendicular to the lower ones.

two lower ratio grids, one of which is turned 90° relative to the other; the grid ratio of such a combination of crossed grids equals the sum of the individual grid ratios. For example, crossing grids with ratios of 5:1 and 8:1 give a cross-grid combination with a ratio of 13:1, although centering of the x-ray beam need not be any more perfect than with an 8:1 grid (Cullinan). The grids can also be used separately when a lower grid ratio suffices. Crossed grids do not lend themselves to angled beams because of excessive cutoff.

Grids are used with intensifying screen cassettes because of the relatively heavy exposures that would otherwise be required. Because the grid is very thin and does not appreciably increase the object-film distance, it causes little or no distortion of the radiographic image. The *grid cassette,* whose front has a built-in grid, is much more convenient to use than an ordinary cassette with a separate grid.

1. *Parallel or Nonfocused Grid.* In this type of grid, used now mainly in mobile radiography, the lead strips are all oriented parallel to one another. When such a grid is used at too-short SIDs (see Figure 19.8A) the more slanting rays toward the periphery of the beam strike the sides of the more peripheral lead strips. This causes progressively greater absorption of primary radiation toward the edges of the grid, with corresponding underexposure of a radiograph toward its sides relative to its center an example of *peripheral cutoff,* caused by *dis-*

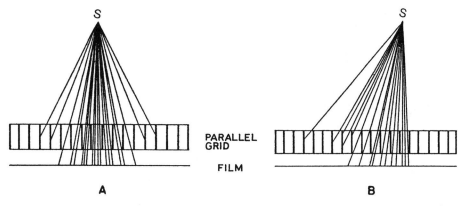

Figure 19.8. Cutoff with a parallel grid. *A*. Effect of distance decentering at short SID (less than 48 in. or 125 cm); the more divergent lateral rays undergo progressively greater absorption, resulting in *symmetrical cutoff*. *B*. Beam angulation across the lead strips causes *asymmetrical cutoff* on the side to which the beam is angled.

tance decentering of the grid. To minimize this problem, parallel grids should be used at a distance of 48 in. (125 cm) or more with films larger than 8 in. × 10 in., because at SIDs beyond 48 in., the central portion of the divergent beam contains nearly parallel rays. In principle, the greater the SID with a parallel grid, the more nearly will the rays be parallel (see Figure 19.8).

If the beam is not perpendicular to a parallel grid and is angled across the long axis of the lead strips, asymmetrical cutoff results. This is more severe on the side toward which the beam has been angled, especially at short SIDs (see Figure l9.8). In this case, we have *asymmetrical peripheral cutoff* due to *angulation decentering*. However, with the beam tilted parallel to the long axis of the grid strips, density will not vary across the radiograph, except for the heel effect (see pages 253-254).

2. *Focused Grid*. This most frequently used type comprises parallel strips that slant more and more toward the lateral edges of the grid as shown in Figure 19.9A. If one extends the planes of these tilted strips upward, these imaginary planes intersect along a line called the *convergence line*,

which is parallel to the grid surface. Naturally, when seen end-on as in Figure 19.9B the convergence line appears as a point. The vertical distance between the convergence line *C* and the center of the grid is called the *focusing distance f_o* a term that must *not* be confused with focus-film distance.

To eliminate the grid lines produced in a radiograph by a stationary grid, Dr. Hollis Potter, in 1920, conceived the idea of *moving the grid* between the patient and the film during x-ray exposure, in a direction perpendicular to the lead strips. The moving grid has come to be called, loosely, a Potter-Bucky diaphragm, but the ICRU recommends instead the term *moving grid,* reserving the name Potter-Bucky for the mechanism that activates the grid. However, the term Bucky for the grid and mechanism together has become so standardized in diagnostic radiology that it would seem futile to discard it.

The direction of grid travel is very important. While moving parallel to its surface, the grid must move perpendicular to the long axis of its lead strips. Thus, in Figure 19.10, during exposure the grid would move

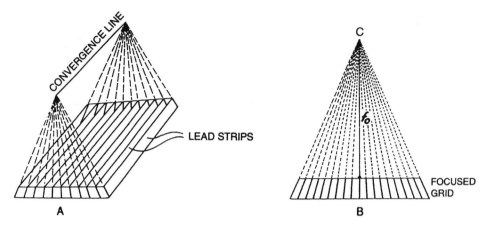

Figure 19.9. *A.* Three-dimensional view of a focused grid; imaginary extension of the inclined lead strips upward would intersect along the *convergence line*. *B.* Frontal view of a focused grid showing the convergence line end-on at point *C.* f_o is the *grid-focusing distance*–the perpendicular distance from the center of the convergence line to the center of the upper surface of the grid.

back and forth across the table.

The moving grid is of the focused type, but the lead strips are thicker than those in a stationary grid because its motion during the exposure blurs out the lead shadows ("grid lines") on the radiograph.

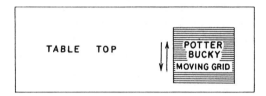

Figure 19.10. Top view of a radiographic table equipped with a moving grid. Note that the lead strips are parallel to the long axis of the table.

Precautions in the Use of Focused Grids

Care must be taken to use *focused* grids properly to take advantage of their maximum efficiency; that is, to achieve the least absorption of primary radiation and the greatest absorption of scattered radiation in the lead strips. In the following sections we shall deal with the factors governing *grid cutoff,* which may be defined as the increased absorption of primary radiation in the lead strips resulting from an incorrect SID or improper alignment or centering of the beam to the grid.

1. *Source-Image Receptor Distance.* A grid with a ratio of 8:1 or less tolerates, without appreciable loss of efficiency, a range of equal to the focusing distance ±25 percent. Thus, with a focusing distance of 40 in. (100 cm), SIDs of 30 to 50 in. (about 75 to 125 cm) give satisfactory results with an 8:1 grid. Below or above these limits, *peripheral cutoff* becomes significant as shown in Figure 19.11. With a grid ratio of 12:1, and even more so with a ratio of 16:1 centering tolerance is *extremely small,* so the SID must virtually coincide with the focusing distance to minimize peripheral cutoff.

2. *Angulation of the Beam.* With a *focused* grid and the source *level with the convergence line,* angulation of the beam across the long dimension of the lead strips causes cutoff with *uniformly* decreased den-

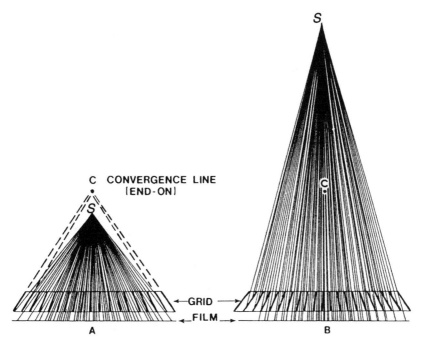

Figure 19.11. The effect of the SID on the efficiency of a focused grid. In *A*, the SID is less than the grid focusing distance minus 25 percent, and more and more rays are absorbed by the lead strips near the edge of the film. In *B*, the SID is longer than the grid focusing distance plus 25 percent, and again there is relatively greater absorption of the beam at the edges of the film. In either case, the periphery of the radiograph will show reduced density–*grid cutoff.*

sity of the entire radiograph. (Recall that with parallel grids such angulation causes asymmetrical peripheral cutoff.) This resembles the decentering of the beam relative to the center of the grid, as will be shown below. The tube, however, may be angled in a direction parallel to the strips without causing grid cutoff, provided the vertical SID remains within the correct limits of grid focusing distance. In practice, a moving grid is located under the table with the lead strips oriented parallel to the long axis of the table, as shown in Figure 19.10. Consequently, the tube may be **angled only in the direction of the long axis of the table** without entailing grid cutoff.

3. ***Centering of the Tube.*** The tube must be centered to the center of the grid to avoid ***off-axis decentering.*** With the tube decen-

tered, that is, off-centered across the long axis of the lead strips, ***the effect on radiographic density depends on the SID relative to the grid focusing distance.*** With the SID equal to the grid focusing distance, and the tube decentered across the lead strips, there will be uniform reduction in the density of the radiograph, that is, uniform cutoff, as in Figure 19.12A. Indeed, this resembles an underexposed or underdeveloped radiograph. Uniform cutoff intensifies with increasing grid ratio and lateral decentering distance, and diminishes with increasing grid focusing distance.

If the tube focus is at an appreciable distance ***above*** the convergence line, and is at the same time decentered, cutoff is asymmetrical with the area of the film directly

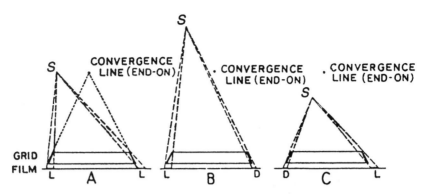

Figure 19.12. Effects of focus-film distance and lateral decentering on efficiency of a focused grid. *L* = lighter. *D* = darker. *A*. The tube focus is decentered but level with the grid convergence line. The projections of the lead strips on the film are equal in size, but are broader than they would be if the tube were not off-center. Therefore, the radiographic density is *uniformly decreased over the entire radiograph*. *B*. The tube focus is decentered but is *above* the level of the grid convergence line. The projections of the lead strips lying more directly under the tube are broader than those of the corresponding lead strips on the opposite side *(L is broader than D)*. Therefore, the radiograph will show less density on the side toward which the tube is shifted. *C*. The tube focus is decentered but at a level below the grid convergence line. The result is opposite that in *B*.

beneath the focus being lighter than the remainder of the film. If the tube focus is **below** the convergence line and decentered, then the area of the film directly below the

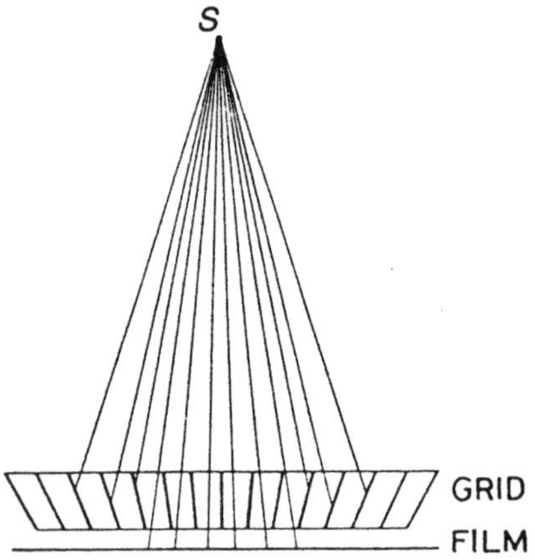

Figure 19.13. The focused grid is inverted, with the "tube side" away from the source. The outer regions of the radiograph will show symmetrical peripheral cutoff–dark band in the center, lateral zones clear.

focus will be darker than the rest of the film. These relationships are illustrated in Figure 19.12.

4. ***Tube Side versus Film Side.*** The focused grid has a "tube side" and a "film side." In all the above figures, the grid is shown correctly with the tube side toward the tube target. If the grid is inserted reversed, as in Figure 19.13, the tilted strips near the edges of the grid will absorb progressively more radiation, causing severe peripheral cutoff. However, with parallel grids either side may face the tube or film.

Practical Application of Grids

Bucky or Grid Factor. Absorption of both primary and scattered radiation by a grid demands an exposure increase above nongrid technique. Moving grids with a 12:1 ratio are standard in conventional radiography at or below 100 kV; here the **Bucky** or **grid factor** ranges from about four at 70 kV, to about 5 at 100 kV (Liebel-Flarsheim

Company). The Bucky or grid factor is defined as follows:

$$\text{grid factor} = \frac{\text{mAs with grid}}{\text{mAs without grid}}$$

or,

mAs with grid = grid factor × mAs without grid

for equal density at a given kVp. Table 19.2 summarizes the grid factors for various

Table 19.2
GRID or BUCKY mAs FACTORS AT TYPICAL
kVp VALUES FOR VARIOUS GRID RATIOS*

Grid Ratio	70 kV	95 kV	120 kV
No Grid	1	1	1
5:1	3	3	3
8:1	3.5	3.75	4
12:1	4	4.25	5
16:1 crossed	4.5	5	6
5:1 crossed	4.5	5	5.5
8:1 crossed	5	6	7

* Experimental data from Liebel-Flarsheim Company. In practice, these may have to be increased by 25 to 30 percent for grids with ratios of 12:1 or greater.

kVps. To compensate for grids by changing kVp instead of mAs, thereby reducing patient dose, refer to Table 19.3 adapted from Cullinan. Because of uncertainties in

Table 19.3
APPROXIMATE EXPOSURE COMPENSA-
TIONS FOR GRID CONVERSIONS

Grid ratio	mAs increase	kVp increase
5:1	2x	+ 8
8:1	4x	+15
12:1	5x	+20-25
16:1	6x	+20-25

*Adapted from J. E. Cullinan: *Illustrated Guide to X-ray Technics,* Philadelphia, Lippincott, 1980. *(By permission of the publisher.)*

radiation absorbed by a grid, the quoted grid factors should be confirmed or modified by trial exposures with a phantom. Grids should be used in radiographing parts measuring 10 cm or more (Cullinan).

At this point we should state the relative merits of grids with ratios of 12: or 16:1. In *routine radiography below 100 kV* a 12:1 ratio *is preferred* for the following reasons:

1. Its efficiency in cleaning up scattered radiation (contrast improvement) is not much less than that of a grid with a 16:1 ratio in routine radiography below 100 kV.

2. Centering is less critical, allowing a larger permissible margin of error.

3. Less exposure of the patient is required for a particular film exposure.

4. Lead strips can be thicker (more lead content), thereby absorbing more efficiently

Table 19.4
CHARACTERISTICS OF VARIOUS TYPES OF GRIDS

Ratio	Type	Recommended kV	Positioning Latitude	Cleanup
16:1	linear	above 100 kVp	very poor	superlative
6:1	crossed	above 100 kVp	good	
12:1	linear	up to 100 kVp	poor	excellent
5:1	crossed	up to 100 kVp	excellent	
8:1	linear	up to 90 kVp	fair	good
6:1	linear	up to 80 kVp	excellent	moderate
5:1				

the higher energy scattered radiation; although this yields better selectivity, it increases patient dose.

Above 100 kVp, a 12:1 or 16:1 grid must be used for optimum radiographic quality. Table 19.4 compares the practical features of various kinds of grids. Positioning latitude–permissible error in centering, angulation, and distance–is worse with parallel than with focused grids of like ratio at SIDs at or below 40 in. (100 cm). The same applies to 16:1 *vs* 12:1 grids, the latter having wider latitude.

Grid Specifications. In 1963 the ICRU, in *Handbook 89,* recommended that manufacturers include certain data to specify the important characteristics of focused grids:

1. *Grid ratio*–the height of the lead strips divided by the distance between successive pairs.
2. *Grid frequency*–the number of lead strips (lines) per in. or per cm; for example, 100 lines/in.
3. *Focusing distance*–the distance between the convergence line and the grid surface.
4. *Source-grid distance limits*–the maximum percent variation in the tube focus-to-grid surface distance within which no excess grid cutoff (ie, absorption of primary radiation) occurs. This is usually equal to the focusing distance ±25 percent.
5. *Contrast improvement factor*–under standard conditions.

Although the first four items are to be included in grid labeling, the fifth is also included in grid specification.

Patient Dose with Grids. Various combinations of kV and grid ratio affect the radiation exposure of the patient. Table 19.5 compares these factors obtained at the University of Rochester School of Medicine, in the radiography of the lumbar spine in the lateral projection. All the radiographs were diagnostically equivalent. Evidently, in radiography at 70 kVp, the radiation exposure of the patient doubles as the grid ratio increases from 8:1 to 16:1. Therefore, *grids with a ratio of 16:1 should not be used in conventional radiography*–below about 100 kV. On the other hand, at *120 kV* a 16:1 grid produces far better cleanup of scattered radiation than lower ratio grids without a significant increase in patient exposure. For general radiography *up to about 100 kV* a 12:1 grid serves as a satisfactory compromise both from the standpoint of grid efficiency and patient exposure, while less sensitive to cutoff.

Grid-Moving Mechanisms. As mentioned before, the Potter-Bucky grid moves during the x-ray exposure. In the obsolete

Table 19.5
EFFECT OF GRID RATIO AND CORRESPONDING kVp AND mAs
ON ENTRANCE SURFACE EXPOSURE OF PATIENT
IN LATERAL PROJECTION OF LUMBAR SPINE.
ALL RADIOGRAPHS OF COMPARABLE DIAGNOSTIC QUALITY*

Grid Ratio	kVp	mAs	Entrance Exposure	Entrance Exposure as % of Maximum
			R	%
16:1	70	1,000	18.9	100
8:1	70	500	9.5	50
16:1	100	160	5.9	31
8:1	100	120	4.5	25

*University of Rochester School of Medicine, Department of Radiology.

single-stroke type, one first "cocked the Bucky"; that is, pulled it to one side of the table by a lever, putting a spring under tension. When the grid was released (either by a string or by an electromagnetic tripping device), the spring pulled it across the table, its motion being smoothly cushioned by a piston acting against oil in a cylinder.

The *reciprocating* mechanism has no separate timer, since it oscillates continuously during the exposure. It is moved rapidly forward 2 to 3 cm in 0.3 s by a solenoid while tension is placed on a spring, an returns slowly (1.7 s) by action of this spring against oil in a chamber as in the single- stroke Bucky. Minimum exposure is $1/20$ s.

The *recipromatic* mechanism is activated entirely by an electric motor, the speed of both strokes being identical; exposure times may be as short as $1/60$ sec.

The *oscillating* or *trill* mechanism is the most advanced type available today and is relatively simple. The grid is supported on four blade springs near its corners (see

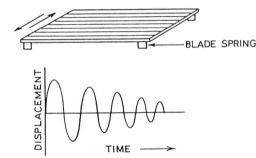

Figure 19.14. Oscillating "trill" mechanism for moving grids. The upper figure shows the grid mounted on four blade springs. A system of relays, synchronized with the exposure switch, starts the grid vibrating in the direction of the arrows just before the beginning of the exposure and stops the exposure just before the grid stops vibrating. The lower figure shows the "damped" (decreasing) oscillation or vibration of the grid. Very short exposures are possible with this type of moving grid mechanism, without the possibility of synchronism.

Figure 19.14). Just before the start of an exposure, the grid is pulled an appropriate distance across the table by a series of relays and then suddenly released, whereupon, it oscillates or vibrates over a rapidly decreasing distance until after the end of the exposure. The trill Bucky allows very short exposures without grid lines because of the continually changing position of the grid with relation to the bursts of x-ray photons in the beam throughout the exposure.

There are several causes of *grid lines* in a radiograph made with older types of moving grids. Since they have relatively thick lead strips, the appearance of these grid lines in the radiograph as alternating light and dark stripes–the so-called corduroy effect–is especially objectionable because it impairs recorded detail. The causes of grid lines with moving grids of various kinds include:

1. Synchronism. We found in Chapter 13 that an x-ray beam is not generated continuously, but rather in intermittent showers or bursts of photons corresponding to the peaks in the voltage applied to the tube. If the travel of the grid is such that a different lead strip always happens to be above a given point on the film at the same instant that a kV peak is reached, the images of different lead strips will be superimposed on the same point on

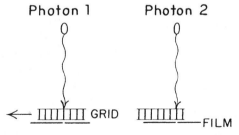

Figure 19.15. Grid synchronism. If the grid moves to the left at certain speeds, a different photon is intercepted by a different lead strip over the *same* point on the film. Hence, a shadow of the lead strip appears on the film just as though the grid were stationary.

the film, even though the grid is moving (see Figure 19.15). The net effect is the same as though the grid were stationary. This series of events cannot occur with an oscillating grid under normal operating conditions.

2. Exposure starting before grid has reached full speed, or continuing after grid travel has slowed down or stopped. Correct relationship between grid travel and exposure time prevents the appearance of grid lines. An electric contactor starts the exposure only when grid travel has reached the correct speed, and stops the exposure before the grid stops moving. Grid lines appear whenever exposure occurs with the grid stationary.

3. Grid moving too slowly may produce either clearly visible grid lines or density variations in the radiograph.

In recent years, **stationary grids** have been so perfected that the grid lines are virtually invisible at the usual viewing distance. Such ultrafine line grids may often be used in place of moving grids. Stationary grids are considerably less expensive because the Bucky mechanism is eliminated. Furthermore, there is less exposure of the patient with a stationary grid for the same radiographic density because no decentering occurs during the exposure. With a moving grid, as much as 3 cm of decentering occurs with reduction in density of almost 20 percent (Curry, Dowdey, and Murry).

REMOVAL OF SCATTERED RADIATION BY AN AIR GAP

Although the radiographic grid is the most effective and widely used device for removing scattered radiation arising in the patient, one can do this nearly as well by increasing the object-film distance. The space between the patient and the image receptor (eg, film-screen system) is commonly referred to as an *air gap*. Referring back to Figure 19.1, you can see that when an x-ray beam enters the body, some is scattered in all directions. That which reaches the film, being non-image forming, causes loss of image contrast and impairs recorded detail. Increasing the thickness of the air gap allows more and more of the scattered photons to move laterally outside the film area (see Figure 19.16). This is especially true of photons that have been scattered at large angles (ie, more oblique to the central ray). In other words, less scattered radiation reaches the film as we increase the object-film distance. This is *entirely a geometric effect;* there is virtually no absorption of scattered radiation in the air gap. Because of the loss of scattered photons, air gap technic

requires a compensatory increase in exposure factors to keep radiographic density

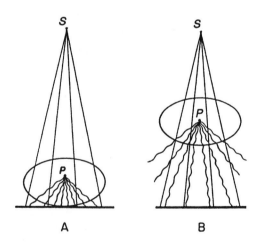

Figure 19.16. Effect of air gap thickness on reduction of the amount of scattered radiation reaching the image receptor. In *A,* scatter reaches all parts of the image receptor from all points such as *P* at short *OID*. In *B,* (longer *OID)* more scattered radiation misses the image receptor.

unchanged, provided the focus- film distance remains constant.

Gould and Hale, in comparing contrast improvement with an air gap *vs* a grid using a water phantom as a subject, found that:

1. A 10-in (2.5 cm) air gap provides about the same cleanup as a 15:1 grid in a thin "subject" measuring 10 cm.
2. The 15:1 grid is more efficient than a 10-in air gap in a 20-cm "subject."

The effectiveness of an air gap is not influenced by kVp. In the radiographic range of 40 to 120 kV, scattering occurs almost equally in all directions, although scatter is more penetrating at higher kVp.

Air gap technic has a place in chest radiography, magnification radiography, and in cerebral and renal angiography. Further improvement in recorded detail can be achieved, especially in cerebral and renal angiography, by the use of a low-ratio grid in addition to the air gap.

In chest radiography, a 10-in. (25-cm) air gap effectively reduces the amount of scattered radiation. The accompanying image magnification must be compensated by an increase in SID, from the customary 6 ft (2 m) to 10 ft (3.3 m). At the same time, this helps reduce the blurring associated with the increased OID. However, with nongrid technique at both distances, an increase in SID from 6 ft to 10 ft requires a much larger exposure by a factor of about $10^2/6^2 = 100/36 = 2.8$ times (according to inverse square law). Air gap technic in chest radiography therefore calls for 110 to 120 kVp.

Magnification technic (see pages 226-228) can aid in the detection of suspected hairline fractures. An air gap is inherent in this procedure. The increase in patient exposure that would occur with the part midway between the tube and image receptor is partly offset by elimination of the radiographic grid. Similar conditions prevail in cerebral and renal magnification angiography. In each instance, the loss of recorded detail associated with the increased object film-distance (air gap) must be compensated by use of a fractional focal spot.

REDUCTION OF SCATTERED RADIATION BY LIMITATION OF THE PRIMARY BEAM

By modifying the primary x-ray beam (ie, the beam coming from the focal spot) we can decrease the fraction of the photons that might be scattered in the body, depending on two well-known facts. First, *an increase in kVp increases the fraction of radiation reaching the film.* Therefore, the kVp for a given anatomic part should be just enough to penetrate it. Excessively high kVp should be avoided, unless a grid with higher grid ratio is used. Under ordinary conditions, with a 12:1 grid, 100 kVp should not be exceeded. Above this level, one should use a 16:1 grid.

Second, *the fraction of radiation scattered increases as the area (and volume) of irradiated tissue increases.* Not only would this impair radiographic quality through loss of contrast, but it would also increase patient exposure. In fact, halving the width of a square beam reduces the exposed *area* to $1/4$, and the *volume* to $1/8$. Thus, the cross-sectional area of the beam should be limited to encompass the smallest area that includes the selected anatomic region, by placing various *beam limiting devices (BLDs)* or *beam restrictors* in the path of the x-ray beam as *close to the tube port* as the housing permits.

Beam limitation is most effective when the field is less than 6 in. (15 cm) in diameter, especially in nongrid radiography. However, with larger areas such as the abdomen, the proper **grid** excels as a means of reducing scattered radiation; beam limitation further improves radiographic contrast when large areas are to be covered, but this is most effective if the collar of the beam-limiting device reaches and surrounds the tube port to remove as much off-focus radiation as possible. Such off-focus radiation arises when unfocused electrons strike parts of the anode other than the target area itself and may amount to as much as 25 percent of the on-target radiation (Ter-Pogossian).

About 50 to 90 percent of the density of a radiograph may result from scattered radiation, so that restriction of the primary beam requires an increase in exposure to compensate for the loss of density. The increase in exposure must be found by trial or in published tables. In setting up a technic chart, we must state the dimensions of the beam (or film size) for each technique. Any further increase or decrease in beam area may entail a compensatory decrease or increase in exposure, respectively.

1. ***Aperture Diaphragms.*** Figure 19.17 shows how an aperture diaphragm decreases scattered radiation, thereby enhancing contrast. The aperture diaphragm would be very efficient if the x rays originated at a true point source. Since the tube focal spot has a finite (measurable) area, x rays originate from it at innumerable points and proceed in all directions. Some of these photons blur the image of the edges of the diaphragm opening; they appear to "undercut" these edges and pass outside the desired confines of the beam, showing up on the radiograph as a gradually fading rather than a sharp border and also increasing patient exposure. Thus, a diaphragm fails to provide adequate beam limitation (see Figure 19.18), although it may be better than nothing under unusual circumstances.

2. ***Cones.*** A radiographic cone is a metal tube that flares outward, away from the tube aperture (see Figure 19.19). However, it deserves the same criticism as an aperture diaphragm in that it fails to provide adequate

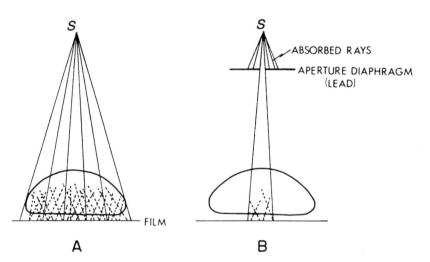

Figure 19.17. The effect of aperture diaphragm *and a point source* on the volume of tissue irradiated and the resulting scattered radiation. In *A* there is no restriction of the x-ray beam; a large volume of tissue is irradiated, with resulting abundance of scattered radiation. In *B* the aperture diaphragm narrows the beam, a small volume of tissue is irradiated, and there is less scattered radiation.

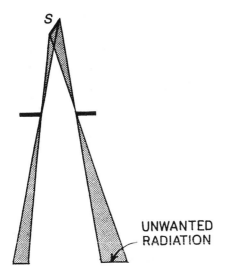

Figure 19.18. Inefficiency of an aperture diaphragm as a beam-limiting device or collimator. Since, in practice, a focal spot has a finite (measurable) size, the edge of the beam is not sharply defined–unwanted radiation beyond the margins of the diaphragm exposes an unnecessarily large volume of tissue.

beam limitation, as shown in Figure 19.19A. A cone limits a beam by virtue of its upper opening only, which acts as an aperture diaphragm. A modification of the cone, the extension cylinder, has openings of equal size at both ends, so the lower opening acts as an aperture diaphragm. Because of the longer distance of the lower aperture from the focal spot, beam limitation is better than that with a cone (see Figure 19.19B).

Another serious shortcoming of the flared cone is the excessive area of the circular field required to cover a rectangular film of any given size (see Figure 19.20). This is objectionable because not only does it lead to the production of more scattered radiation, but it also exposes the patient to radiation outside the area of interest. A partial solution might be to have cones with rectangular openings to cover each film size at various focus-film distances, but this would be impracticable. The variable aperture device does this much more efficiently, as described in the next section.

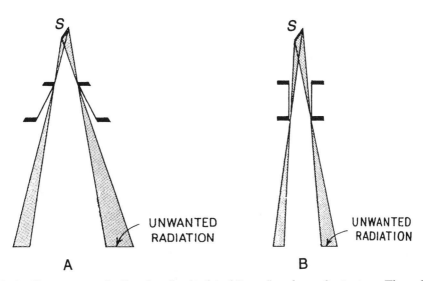

Figure 19.19. Comparison of a flared and cylindrical "cone" on beam limitation. The cylindrical cone in *B* provides better beam limitation than the flared cone in *A,* according to the difference in the unwanted radiation.

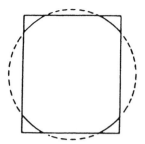

Figure 19.20. The x-ray beam cuts the corners of the film when limited by an aperture diaphragm or cone, but the dashed circle indicates the excessive volume of irradiated tissue.

3. ***Variable Aperture Beam-limiting Devices.*** These–also known as ***collimators***–combine the best features of the cone and aperture diaphragm. The term collimator, although generally used, is inaccurate because, strictly speaking, it is a device that is supposed to line up rays so as to make them parallel to one another. In radiograph-

ic usage, the collimator simply limits or shapes the beam to the desired area; the radiation continues to diverge after leaving the collimator. However, the term collimator has become so ingrained in radiology and is so much simpler than the term variable aperture beam-limiting device that we shall continue to use it. It should be mentioned that some are using the term ***BLD*** (for beam-limiting device) to designate a collimator.

A typical collimator comprises a box-like apparatus (see Figure 19.21B) equipped with two or more sets of adjustable diaphragms, stacked one above the other, which can be opened or closed to delimit accurately a beam of any desired rectangular configuration. In addition, the multiple diaphragms more effectively reduce the amount of radiation passing outside the edge of the beam, thereby giving it a sharper margin, provided they are properly aligned. The best collimators have a collar extending to and sur-

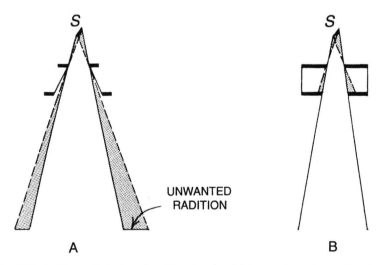

Figure 19.21. Effect of a well-designed collimator (variable aperture beam-limiting device) on beam restriction as compared with a cone. In *A,* a cone reduces the beam area, but, in effect, the upper opening of the cone serves as an aperture diaphragm so that the beam is not sharply defined (see Figure 19.18). In *B,* a double-diaphragm collimator markedly improves beam restriction by absorbing penumbral radiation in the lower diaphragm. Further improvement can be achieved with more than two levels of variable diaphragms.

rounding the tube port for the purpose of further improving recorded detail by removing off-focus radiation. As stated above, circular beams shaped by aperture diaphragms or cones allow radiation to pass beyond the zone of interest (see Figure 19.20), increasing the fraction of scattered radiation and unnecessarily exposing the patient. Thus, they do not adequately limit the beam (see Figure 19.21A).

When used with an extension cone, the collimator should be narrowed to the smallest square that encloses the circular field to improve further the efficiency of beam limitation. However, not all collimators can accept an extension cone.

Collimators are equipped with an ***illuminated field and beam-centering device,*** and also have calibrated scales to allow aperture adjustment for films of various sizes at different SIDs. According to federal regulations, collimators for all new equipment must be electrically interlocked with the clamps in the Bucky tray so that the apertures adjust automatically to the size of the cassette. Such a collimator is known as a ***positive beam-limiting device (PBLD).*** A manual override allows beam-shaping for special field sizes.

Film Coverage by Aperture Diaphragms and Cones. As just noted, collimators are calibrated, thus obviating the need of calculations. However, with aperture diaphragms and cones, we can compute film coverage by apertures of various sizes at the desired SID. Since film sizes are still generally stated in inches, we shall use these units in our example. Computation of film coverage requires the following information:

1. Distance from source to aperture.
2. Distance from source to film (SID).
3. Diameter of aperture.

The computation of film coverage by a beam limited by an aperture diaphragm is not difficult. In Figure 19.22, with an open-

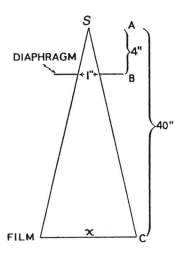

Figure 19.22. Method of calculating film coverage with an aperture diaphragm. In this example, x/1 = 40/4. Therefore, 4x = 40 and x = 10 in., disregarding geometric blur.

ing 1 in. in diameter and 4 in. from the focal spot, and an SID of 40 in., what will be the diameter of the beam at the surface of the film, *disregarding geometric blur?* Note that the x-ray beam continues to broaden after passing through the aperture. In longitudinal section, this forms a larger triangle with its apex at the tube focal spot and its base on the film, and a smaller triangle with its apex at the tube focus and its base at the aperture. These are similar triangles (see pages 11-12), so by the application of elementary geometry we obtain the proportion:

$$AB/AC = d/x$$

where AB = source-to-aperture distance
AC = SID
d = diameter of aperture
x = diameter of beam at film plane

Substituting the numerical values in the proportion,

$$4/40 = 1/x$$
$$4x = 40 \text{ in.}$$
$$x = 10 \text{in.}$$

An x-ray beam 10 in. in diameter is obviously too small to cover a 10 in. × 12 in. film but will cover adequately an 8 in. × 10 in. film. Since the diagonal of this film is 12.8 in., its corners will be clipped.

The same principle applies to the **f*lared cone*** whose ***upper*** opening limits the beam (see Figure 19.19A). Calculations must be made as though an aperture diaphragm were located at that level, the lower opening being ignored.

With a ***cylinder*** the ***lower*** aperture always limits the size of the beam (see Figure 19.23),

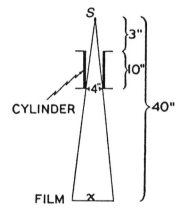

Figure 19.23. The effect of a cylinder in limiting the size of an x-ray beam.

and therefore the calculations are made as though an aperture diaphragm were present at the level of the lower opening. The following proportion may be used to find the film coverage diameter, that is, the diameter of the beam at the film:

$$\frac{\textbf{film coverage diameter}}{\textbf{cylinder diameter}} = \frac{\textbf{SID}}{\textbf{source-to-lower aperture distance}}$$

where the "source-to-lower aperture distance" is obtained simply by adding the distance between the focus and the top of the cylinder, to the length of the cylinder. Substituting the corresponding values from Figure 19.23 in the above equation.

$$\frac{film\ coverage\ diameter}{4} = \frac{40}{3 + 10}$$

$$= \frac{4 \times 40}{13} = \frac{160}{13} = 12.3 \text{ in}$$

Thus, an 8 × 10 film will have about 1/4 in. cut off all four corners under these conditions, provided the beam has been centered exactly to the film. Bringing the x-ray tube and cylinder closer the patient greatly improves beam limitation, but increases dose.

OTHER METHODS OF ENHANCING RADIOGRAPHIC QUALITY

Moving Slit Radiography

Introduced by Jaffe and Webster as a method of improving radiographic contrast, moving slit radiography extends the principle that a very narrow (slit-limited) moving x-ray beam engenders less scattered radiation than does an ordinary beam of the same area. A suggested prototype model (see Figure 19.24) consists of two lead sheets of appropriate thickness, in which have been cut a series of twelve parallel slits 3 mm in width spaced 2.5 cm apart. One lead sheet would be placed above, and the other below the patient, while corresponding upper and lower slit pairs must be coaxially aligned in the beam. Movement of the assembly during exposure wipes out the images of the slits in the same manner as a moving grid. With this system, radiographic contrast should be enhanced by a factor of two to three relative to that with a grid. However, this would require a tenfold increase in x-ray output by the tube. The ultimate practicability of this device must await further development.

equipment.

There is a type of transparent filter material composed of clear plastic incorporating 30 percent lead by weight. It is the Clear-Pb™ compensation filter (Victoreen Nuclear Associates), which can be mounted on the face 'of the collimator by a specially designed magnetic system. These filters are thin and lightweight. The simplicity of this system will undoubtedly find wide application in general radiography in such examinations as combined mediastinal-pulmonary tomography, aortic arch angiography, and anteroposterior projections of the foot.

Some dedicated chest radiographic units provide for a **trough filter**–one that is thinner at the center than at the edges–to permit heavier exposure of the mediastinum than of the lungs (see Figure 19.27).

A simple trick has been used to compensate for difference in density when one lung has been **partly** opacified by disease. One can use a sheet of black paper, cut to size, to block one-half of one screen behind the normal lung. It may then be possible to penetrate the diseased lung without overexposing

the normal lung. However, computed tomography has replaced this procedure.

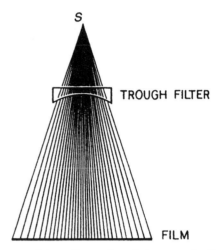

Figure 19.27. Principle of the trough filter. There is more attenuation of the beam as it passes through the lateral portions of the filter than in the center. Such a filter provides greater exposure of the mediastinal structures than of the aerated lungs, thereby compensating for inherent density differences.

SUMMARY OF RADIOGRAPHIC EXPOSURE

In Figure 19.28 a typical arrangement of x-ray equipment and patient in radiography is shown to summarize the important items in radiographic exposure.

QUESTIONS AND PROBLEMS

1. Discuss the three factors affecting the percentage of scattered radiation relative to primary radiation at the film surface.
2. Describe the principle and construction of a radiographic grid. What is the grid ratio when lead strips are 3 mm high and their spacing 0.2 mm?
3. Explain the physical factors in grid effi-

ciency, including grid ratio and grid efficiency.
4. If a correct radiographic technique with a 5:1 grid is 70 kVp and 50 mAs, what new mAs would be required with a 12:1 grid for the same radiographic density?
5. Discuss the functional characteristics of grid efficiency, including selectivity

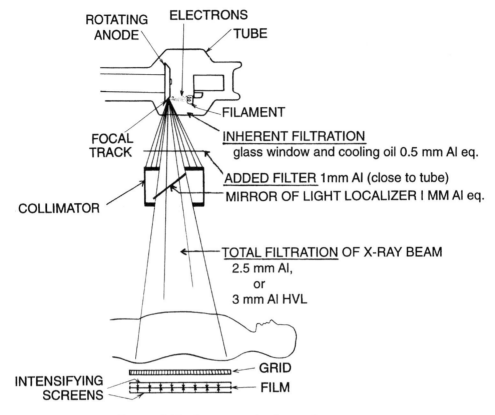

Figure 19.28. Summary of radiographic exposure.

and contrast improvement factor.

6. What is meant by the convergence line of a grid? The focusing distance of a grid?

7. Discuss four precautions in the use of a grid.

8. What is the preferred grid ratio in conventional radiography? In high voltage radiography?

9. Explain the decreased density at the edges of a radiograph, which may occur when a grid is used. What is it called?

10. Describe briefly the main types of moving grid mechanisms.

11. Name three causes of grid lines with a moving grid and state how each can be corrected.

12. Under what conditions may the x-ray beam be angled with respect to a radiographic stationary or moving grid?

13. Describe four causes of grid cutoff.

14. Define grid ratio. What is the relationship between grid ratio and grid efficiency?

15. Explain how air gap technic reduces the amount of scattered radiation reaching an image receptor?

16. How does the efficiency of an air gap compare with that of a grid in improving contrast (ie, "cleanup")?

17. Describe three examples in which air gap technic may be advantageous. How does one minimize blurring associated with an air gap, and what effect does it have on exposure?

18. How does an aperture diaphragm diminish scattered radiation?

19. An aperture diaphragm measuring 2 in. (5 cm) in diameter is located 5 in. (12.7 cm) below the source (focal spot). What will be the diameter of the beam at 40 in. (100 cm)?

20. In problem 19, if the effective focal spot diameter is 1 mm, find the width of the blur (penumbra) of the aperture edge on the radiograph.

21. Describe a typical radiographic collimator (variable beam-limiting device). Why is it the most desirable type of beam limiter?

22. A cylinder measuring 12 in. (30.5 cm) in length is used to limit an x-ray beam. The lower opening is 15 in. (38 cm) from the tube focus and measures 4 in. (10 cm) in diameter. What is the largest standard film that will be cov-ered (except for the corners) at an SID of 40 in. (100 cm)?

23. Calculate the width of the geometric (focal spot) blur with a 0.6 mm effective focal spot at a 100 cm SID when the OID is 5 cm.

24. Summarize the methods of minimizing scattered radiation in radiography.

25. Define and explain the cause of the anode heel effect. How is it influenced by the focus-film distance? By the anode angle? By the size of the film?

26. Draw a simple diagram summarizing radiography.

27. What is the purpose of a compensating filter? Discuss the available materials for compensation filters and give their relative advantages.

28. Describe the form, function, and application of a trough filter.

Chapter 20

FLUOROSCOPY

Topics Covered in This Chapter

- Human Eye
- Image Intensifier
    Input and Output Phosphors
    Electronics
    Brightness Gain
    Conversion Factor
- Multiple-Field Intensifier
- Viewing Fluoroscopic Image
    Optical Lens

    Television (closed circuit)
        Video Camera
        Television Monitor
        Charge-coupled Device (CCD)
- Television Image Quality
- Recording Fluoroscopic Image
- Recording Television Image
- Questions and Problems

THIS CHAPTER will deal with image intensification in fluoroscopy, and the various systems for viewing and recording the fluoroscopic image.

The Human Eye

Fluoroscopy, an essential procedure in diagnostic radiology, makes possible the visualization of organs in motion, positioning of the patient for spotfilming, instillation of opaque media into hollow organs, insertion of catheters into arteries, and a variety of other procedures. Unfortunately, *direct* or *dark* fluoroscopy with a zinc cadmium sulfide screen is an extremely inefficient process, both physically and visually. This results from the very low brightness level of the direct fluoroscopic image, about $1/30,000$ that of a radiograph on a fluorescent illuminator, but also from the nature of human vision which we shall explore briefly in

introducing the subject. Therefore, direct fluoroscopy has become obsolete.

The human eye is a living camera. Light reflected from an object enters the eye and passes through the lens which focuses it on the *retina,* a membrane lining the inside surface of the back of the eye (see Figure 20.1). Within the retina there are two types of light-sensitive cells, the **cones** and the **rods.** Present in other areas, but concentrated in a small spot at the center of the retina–the *fovea centralis*–the cones respond to light at ordinary brightness levels, and not to the dim light from a zinc cadmium sulfide fluoroscopic screen. Cone vision is *photopic.*

The rods, microscopic rod-shaped cells distributed throughout the periphery of the retina, respond to light at the low **brightness levels** prevailing in direct fluoroscopy, but they exhibit very poor visual acuity–the ability to discriminate small images. In fact, we know that the contrast (difference in bright-

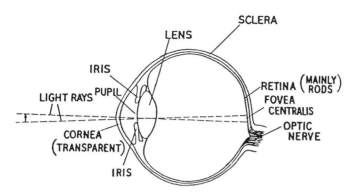

Figure 20.1. Diagram of midsagittal section of human eye.

ness) of two adjacent images must approximate 50 percent to be perceived by rod vision, but cones can discriminate images with a contrast of about 2 percent in high level radiographic illumination. Since fluoroscopic image contrast averages one- fourth to one-half that in a radiograph, the visibility of details is poor on the dim fluoroscopic screen. Another factor in rod vision is the prolonged dark adaptation, 20 to 30 min, required for the rods to attain adequate sensitivity, which then has to be preserved during fluoroscopy in an almost completely darkened room. Fortunately, the rods are particularly sensitive to the green light emitted by the zinc cadmium sulfide screen. Rod vision is *scotopic.*

In summary, then, direct fluoroscopy is an exceedingly inefficient process because of the low brightness level, the low contrast, the dependence on rod vision with its poor visual acuity, and the need of prolonged dark adaptation.

Fluoroscopic Image Intensification

In 1942 Dr. W. E. Chamberlain, in a classic paper, predicted that if the brightness of the fluoroscopic image could be increased about 1,000 times, cone vision would be brought into play and visual acuity enhanced by a factor of ten (ie, 10 times greater) relative to rod vision. Besides, dark adaptation and a completely darkened room would become unnecessary.

After extensive research by various manufacturers, marked brightening of the fluoroscopic image was realized through the development of the ***x-ray image intensifier.*** Its main component is the image tube, an electronic device that promptly converts, in several steps, an x-ray image pattern into a corresponding visible light pattern of significantly higher energy per square cm of viewing screen. Figure 20.2 shows the basic construction of a typical image intensifier. After passing through the patient, the x-ray beam enters the ***highly evacuated glass envelope*** of the image tube and produces a fluorescent image on the cesium iodide ***input screen (IS).*** It emits several thousand light photons per x-ray photon received from the fluoroscopic exit beam. A thin, x-ray transparent aluminum layer behind the input screen reflects back to it any light photons that have been emitted in the reverse direction, that is, toward the subject. The light from the input screen falls on the closely applied ***photocathode,*** which emits several thousand electrons per absorbed x-ray photon. The electrons form an electronic image that duplicates the

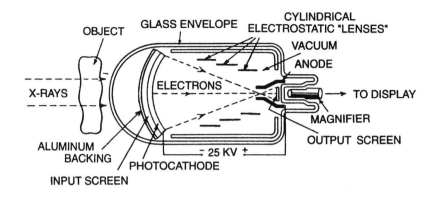

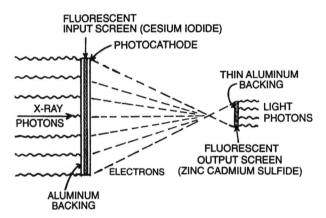

Figure 20.2. *Above.* Image intensifier. X rays produce an image on the *input screen,* whose light liberates *electrons* from the *photocathode.* The concentration of the electrons is proportional to the light intensity at any given point on the photocathode, so they form an *electron image* in space. The electrons are then accelerated at 25 kV and electrostatically focused on the *output screen,* where the final intensified image is formed. This image may be viewed by a system of lenses and a mirror, which has been largely replaced by closed circuit TV; it may also be recorded on cine film, video tape, or video disc. *Below.* Simplified diagram of an image intensifier, showing detail of the input and output screens.

light image but consists of various densities of electrons instead of photons. In other words, a light image has been converted to an electron "image." According to Hendee, about 15 to 20 electrons are emitted by the photocathode for every light photon absorbed.

Sharpness of the electron image is enhanced by the slight forward concavity of

the photocathode, in the direction of electron emission. These electrons are accelerated toward the anode by an applied high voltage–25 to 35 kVp–and are focused by a series of cylindrical electrostatic "lenses" onto the ***output screen (OS)*** whose phosphor is ***silver-activated zinc cadmium sulfide.*** A thin aluminum coating on the output screen, on the side toward the photocathode, is transparent to the electrons but prevents light from flowing back to the photocathode and creating noise. The aluminum is grounded to prevent accumulation of electric charge. A hole in the anode lets the electrons through to the output screen.

In the output screen the electrons' energy is converted to a corresponding, but inverted light pattern—an image whose brightness is about 4,000 to 6,000 times that of the light image on the input screen, a comparison designated as *brightness gain.* The image finally undergoes processing through an appropriate system for one or more of the following options: direct viewing with a mirror, closed circuit television, cine (motion picture) recording, videotape, or videodisc.

Brightness Gain (BG). This is derived as follows:

1. *Flux Gain (FG)*–acceleration of the electrons in the image tube causes multiplication of photons in the output screen, a process called *flux gain,* for a given exposure:

$$FG = \frac{\text{number of photons on OS}}{\text{number of photons on IS}} \quad (2)$$

2. *Minification Gain (MG)*–concentration of light photons as they are focused on the output screen, expressed by

$$MG = \frac{\text{area of IS}}{\text{area of OS}} \quad (3)$$

For example, with a 9-in. (23-cm) input screen and a 1-in. (2.5-cm) output screen

$$MG = \frac{9^2}{1^2} = 81$$

Since the light photons have been "crowded" on a screen approximately $1/80$ as large as the input screen, brightness increases 80 times simply from minification.

In general,

$$\textbf{BG} = \textbf{FG} \times \textbf{MG} \quad (4)$$

With a minification gain of 80 and a flux gain of 50, for example,

$$BG = 80 \times 50 = 4,000$$

which means that the image at the output screen is 4,000 times brighter than at the input screen. Image intensifiers have a brightness gain up to about 6,000.

On passing through the optical system, the image will have regained about 80 to 90 percent of its original size without appreciable loss of brightness.

Conversion Factor (CF). The ICRU has proposed a more useful standard for specifying the intensifying ability of an x-ray image intensifier, based on strictly defined conditions. This expressed as follows:

$$CF = \frac{\text{luminance of OS (candela/m}^2)}{\text{exposure rate at IS (mR/s)}} \quad (5)$$

The conversion factor is now the preferred way to specify image intensifier quality.

The main advantage of the image intensifier lies in its brightness gain, with activation of cone vision (also called photopic vision) and improved visual acuity–ability to see fine detail. Besides, it has done away with the inconvenience of dark adaptation.

Contrast in the intensified image may be lessened by the "fogging" effect (noise factor) of some x-ray photons passing directly from the input screen to the output screen, bypassing intermediate conversion to light and electron images. Moreover, some light photons may pass in the reverse direction, from the output screen to the input screen, eventually contributing to image fog and loss of contrast. But contrast can be significantly improved by a television system.

The introduction of cesium iodide (CsI) as the input screen phosphor has resulted in notable improvement in recorded detail by virtue of the increased image resolution and contrast. This depends on the fact that CsI crystals can be crowded much more closely (high packing density) and in a thinner layer than zinc cadmium sulfide crystals onto the input screen. Because CsI is a better absorber of x-ray photons, thereby improving recorded detail and contrast, it has become the preferred phosphor in image intensifiers.

Image detail is not quite so good with image intensification as with direct screen fluoroscopy. In fact, the latter provides a resolution of about 4 line pairs/mm, in comparison with 1.6 lp/mm for a 4.5-in and 0.8 lp/mm for a 9-in. image intensifier (Curry & others). The decreased resolution with image intensification results from lost information at each step of image formation–x rays to light, to electrons, to final viewing system. Also, there is poorer focusing of electrons toward the periphery of the output screen especially with large image intensifiers causing loss of detail, a condition called *vignetting.* Another problem is the distortion of lines, which curve inward at the edges of the output screen–the *pincushion effect.*

The effect of image intensifier size on image quality–recorded detail and its visibility–can be measured by the modulation transfer function of the entire imaging system. This process is beyond our scope, but image brightness of the input screen is an extremely important factor. A larger input screen requires a higher x-ray exposure rate than a smaller one for equal image quality. In fact, this is directly related to the square of the input screens' diameters. For example, a 9-in. screen, compared to a 6-in. screen requires

$$\frac{9^2}{6^2} = 2.25 \text{ times greater exposure rate.}$$

Image intensification has led to the demise of direct, or dark fluoroscopy because of the following advantages:

1. Elimination of dark adaptation of the viewer's eyes.

2. Reduction in exposure of the patient, although this may be less than initially expected because a certain irreducible minimum brightness is needed at the input screen to reduce quantum mottle and provide good recorded detail in the final image.

3. Introduction of special viewing and recording systems, such as television, cineradiography, cut film, digital, and laser recording.

Figure 20.3 shows some of the methods of

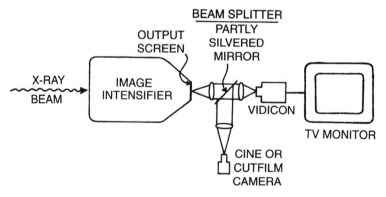

Figure 20.3. Options in viewing the output screen image in fluoroscopy. A half-silvered mirror placed at 45° to the light beam allows one-half the beam to be transmitted through the mirror to the vidicon, and the remainder to be reflected at a right angle toward the film camera. The vidicon image, as will be described later, can be viewed on a TV monitor and recorded on magnetic tape or disc, or on a laser disc.

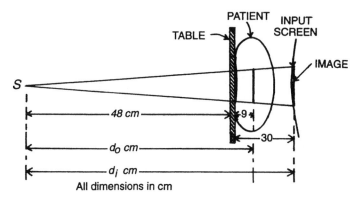

Figure 20.4. Image magnification by an image intensifier.

Magnification in Image Intensifier

As shown in Figure 20.4, anatomic structures in the patient undergo magnification on the input screen. The equation for the magnification factor M is expressed by

$$M = \frac{i}{o} = \frac{d_i}{d_o}$$

where d_o is the object-source distance, d_i the image-source distance, o the object size within the body, and i the image size on the input screen. In figure 20.4,

$$d_i = 48 + 30 = 78$$
$$d_o = 48 + 9 = 57$$

so

$$M = \frac{78}{57} = 1.37$$

Multiple-Field Intensifiers

Special image intensifiers can produce images of more than one size on the output screen, and are therefore called **multifield** or **multiple-field** intensifiers. As shown in Figure 20.5, a dual mode unit typically has a 9 in. (22.5 cm) input screen with an optional 6 in. (15 cm) format. The smaller mode is achieved by increasing the voltage applied to the electrostatic focusing lenses; an increase in voltage brings the focal point (where the electrons in the beam cross) closer to the *input* screen. Stated another way, the focal point is farther from the output screen in the 6 in. mode so the electron beam spreads farther onto the output screen than does the beam in the 9 in. mode. As a result, only the central 6 in. of the input screen appears in this mode, so the **field of view** is reduced and the image magnified, but brightness decreases. One can obtain a triple-mode intensifier, which includes a 4.5 in. option.

A special circuit automatically increases

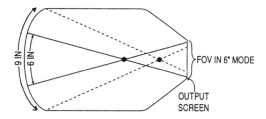

Figure 20.5. Dual-field image intensifier showing the fields of view in the 6 in. and 9 in. modes.

mA to compensate for the loss of brightness, but patient dose obviously increases in the ratio of the square of full field to that of the smaller field (Hendee). In our example, this can be simplified without introducing an error by using the squares of the diameters:

$$\text{dose increase} = \frac{9^2}{6^2} = \frac{81}{36} = 2.3 \text{ times}$$

Since information increases as mA is raised, the signal/noise ratio increases and image quality improves in the smaller mode.

VIEWING THE FLUOROSCOPIC IMAGE

Optical Lens System

Before the advent of television in radiology, the output screen (OS) of the image intensifier was observed through a lens-and-mirror system. Included were three lenses to enlarge the image, which was projected onto a mirror attached to a swivel arm. Manipulation of the mirror allowed the fluoroscopist to adjust the *field of view,* that is, the area on which an image is produced by a lens system. This unwieldy arrangement demanded frequent manual adjustment of the mirror. Moreover, the system gathered light inefficiently.

Television Viewing System

A TV camera (vidicon or plumbicon) and a TV monitor, coupled by an electric cable, serve to image the OS of the image intensifier, both being cathode ray devices designed specifically for their particular function. Although described separately, bear in mind that they operate together as a unit, in complete synchrony, to create the final TV image. It is simply a closed circuit TV system (see Figure 20.3).

Video Cameras

There are two kinds of video camera tubes, *vidicon* and *plumbicon.* The vidicon is preferred in general fluoroscopy because it

yields a brighter image, although it has greater lag, or image persistence. The plumbicon is not so bright, but has very little lag and is therefore favored for digital radiography. Both are cathode ray tubes.

The vidicon "sees" the image on the output screen and converts it to an electronic "image," which is reconverted to a light image by the TV monitor. By way of summary,

light image →electron image →light image
(output screen)   *(vidicon)*   *(TV tube)*

Although the process is shown in steps, the vidicon and TV tube operate *simultaneously.* However, for descriptive purposes, they will be presented separately.

A schematic diagram of a vidicon appears in Figure 20.6. It has the basic features of other cathode ray tubes: a highly evacuated glass envelope containing at one end an *electron gun* (grid-controlled cathode); a bottle-shaped *anode* occupying almost the entire length of the tube, closed at the broad end by a fine wire mesh; and a *target* at the opposite end facing the output screen of the image intensifier.

The *target* has a unique composition. Closely applied to the glass (see Figure 20.6), it consists of two layers: (1) a thin transparent electrically conducting *signal layer* made of graphite (a form of carbon called "lead" in a pencil) plated onto the glass, and (2) a second closely applied *photoconductive layer* consisting of antimony trisulfide in the *vidicon* tube, or lead oxide in the *plumbicon*

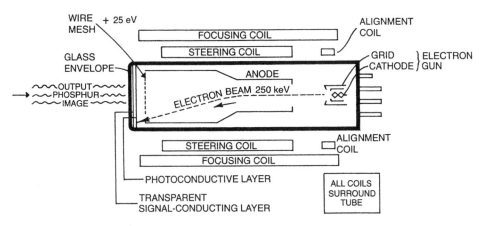

Figure 20.6. Details of a TV vidicon camera. The output screen image is converted to electric signals by a rapidly scanning electron beam. These signals are converted to a light image in the TV monitor tube, by a scanning electron beam synchronous with the one in the vidicon.

tube. The compounds in the photoconductive layer are in the form of tiny granules coated with mica, measuring 25 μm* in diameter (ie, about three times the diameter of a red blood cell). Photoconductivity is the property of certain substances to become electrically conductive when exposed to light.

At the opposite end of the tube is enclosed a ***cathode*** with a surrounding grid whose function is to expel electrons when they are released by the heated cathode. Together, the cathode and grid make up the ***electron gun.***

Outside the tube are the surrounding cylindrical focusing, steering, and alignment coils that control the direction of the ***pencil*** cathode ray beam, vertically and horizontally toward the target. As expected, the beam originates with thermionic emission by the electron gun and acceleration of the electrons by moderately high voltage (insufficient to generate x rays); the electrons are deliberately slowed down before they strike the target.

Summary of Vidicon Function. This involves the process of converting the visual image in light from the output screen (OS) to electric signals that carry the information to the TV monitor.

1. Light from the image on the OS, linked preferably by a fiberoptic cable, passes through the vidicon's signal layer to the photoconductive layer. Here the light photons interact with the photoconductive particles (antimony trisulfide) which emit photoelectrons and remain positively charged. The photoelectrons are "clipped," that is removed by the anode. The positive particles form a finely granular image that resembles the OS image, and attracts and holds electrons in a similar pattern on the signal plate.

2. In the meanwhile, an electron beam is being generated in the vidicon: the electron gun shoots out electrons which undergo acceleration to 250 volts, but are slowed down to 25 volts by the wire mesh grid at the end of the anode. The electron beam is focused, and the steering coil causes the beam to scan, or sweep repeatedly across the target in a definite pattern, to be described

---

* μm = micrometer = 10^{-6} or 1 millionth of a meter

later. In doing so, the electron beam neutralizes the positive particles in the photoconductive layer so the electrons just held on the signal plate are now free to move, creating electric signals.

3. The signals fluctuate in intensity corresponding to the light intensities in various areas of the initial OS image, and also reproduce the granular structure of the photoconductive layer. Leaving the signal plate through a resistor, the signals are amplified and conducted to the cathode of the TV monitor.

Television Monitor

The final step in TV viewing of the output screen (OS) converts the vidicon electric signals to a visible image, which copies the OS image by reversing the vidicon process. Although they are both cathode ray tubes, the TV tube is much larger and has a flared viewing end coated internally with a fluorescent material (see Figure 20.7).

In the TV tube, amplified signals from the vidicon cause the grid- controlled cathode to vary, from instant to instant, the number and distribution of electrons in the beam according to the strength of the incoming signals. Recall that these signals in the signal plate of the vidicon depended initially on the brightness distribution of light in the image intensifier's OS. An applied potential of 10 kV accelerates the electrons in the TV tube while the focusing and steering coils not only create a pencil beam, but cause it to scan the TV screen *very rapidly* from side to side and from top to bottom, *synchronously with the scanning beam in the vidicon.*

Scanning "sweeps" follow a prescribed linear pattern duplicating that on the vidicon, in which each line is a series of signals that represent the light intensities in the original OS image. Sweeping starts in the upper left corner of the screen and moves across almost horizontally as the ***active sweep.*** Immediately, the beam returns to the left side of the screen to a point slightly below the first line as the ***horizontal retrace sweep,*** which contains no signals. This process continues until $262^{1}/_{2}$ active traces have occurred in a 525-line system (see Figure

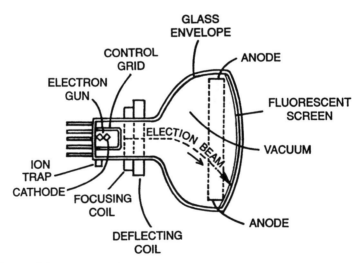

Figure 20.7. Essential features of a TV monitor tube in longitudinal section. The observer watches the real-time image on the broad face of the monitor, located on the right end of the diagram.

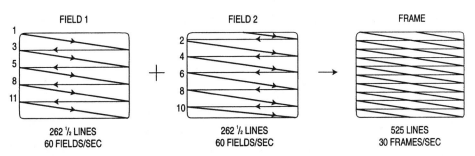

Figure 20.8. Explanation of image formation on the TV monitor screen. The number of lines have been reduced for simplicity. Each *field* actually consists of 262.5 lines produced by the pencil beam, rapidly scanning the phosphor in the face of the TV monitor tube. The two sets of lines, or *fields*, are *interlaced* to produce a *frame*.

20.8), thereby making up a *field*. The scanning beam promptly does a *vertical retrace* and repeats the entire sweeping pattern, but this time places the lines *between* the first set of lines, a process called *interlacing*. Two fields constitute a *frame*, the final TV image consisting of many frames. Occasionally one may see a horizontal linear pattern called a *raster*.

Now to explain interlacing: Figure 20.8 has been simplified by letting 10 lines represent $262^{1/2}$. A frame (two fields) has a frequency of 30 light flashes per sec, but these comprise two interlaced fields that each produce 60 flashes per sec, close enough to take advantage of retinal lag, or persistence, so that the light flashes appear to be continuous and flicker is minimized. Image flicker occurs when light flash frequency is about 50 per sec or less.

To assure complete synchronization of scanning between the vidicon and TV tubes a camera control unit automatically adds or subtracts pulses during the retrace sweep when the field momentarily goes blank. Thus, the vidicon and TV images remain in "sync," while the viewer is unaware of the synchronization process.

Charge-Coupled Device (CCD) TV Camera

Used now in home video cameras, the CCD is a specially designed semiconductor chip. It contains, in columns and rows, pixel sites, each of which acquires a charge proportional to the intensity of the light that strikes it. A timing system rapidly moves successive rows of pixels to the top edge of the chip, where a wire picks up the charges, row-by-row. The charge impulses along the wire resemble those along the active traces of a vidicon, and produce an image on a fluorescent screen in the rear of the camera; the image can be reproduced on a TV monitor or magnetic tape in real time. A standard CCD is a 525-line system, which involves interlacing to minimize flicker. It is all that one chip can handle; two-chip models for a 1050-line system are much more expensive, but a vidicon still gives best results in digital fluoroscopy.

Quality of the TV Image

Television image quality depends on resolution, contrast, and brightness. The vidicon must be of top quality, since it is potentially the weakest link in the system from the standpoint of quality loss due to noise in cheap vidicons.

Resolution. This has two components because of the manner in which the image is produced–horizontal active sweeps and electric information-bearing signals along the lines.

1. *Vertical Resolution.* The actual number of horizontal lines and their interspaces determines vertical resolution. Simply stated, a 525-line system would have a 525-line resolution, were it not for blanking during retrace. According to Hendee and Ritenour, in a 525-line system, there are 490 active sweeps, which must be multiplied by the Kell factor 0.7 as follows:

$$\frac{\text{vertical}}{\text{resolution}} = \text{active traces/frame} \times 0.7$$

$$= 490 \times 0.7 = 343 \text{ lines, or 171-line pairs}$$

In practice, this must be further modified by the size of the image intensifier input screen and the size of the TV monitor screen: larger input and monitor screens entail poorer resolution. For digital imaging, a 1025- line system (really 700-line resolution) becomes mandatory for acceptable image quality.

2. *Horizontal Resolution.* As already noted, this is determined by the amount of information that can be transferred through the system, from the OS of the image intensifier to the final TV image, *along* the scan lines. Since we are dealing with an analog image (ie, realtime or continually varying), resolution is expressed in terms of frequency–the number of times per sec that the sweeping electron beam can be modulated (varied) to conform to variations in brightness of the original image. The highest frequency of which the system is capable is called *bandpass* of that TV system. According to Curry, Dowdey, and Murry, at about 4.5 MHz* a 525-line system will have

equal vertical and horizontal resolution, a desirable goal.

Contrast. A vidicon tube reduces contrast to about 20 percent, whereas the TV monitor doubles it, so there is a net gain in the system. A plumbicon does *not* impair contrast. A great advantage of a TV system is its provision for adjusting image contrast by two manual controls, one for contrast and the other for brightness. As one turns up contrast, noise level increases. Upon reducing contrast slightly, one can then turn the brightness control slightly to achieve the desired contrast. Adjusting the brightness control alone, reduces contrast as brightness is increased.

Brightness Stabilization. The brightness or luminance of the image has to be continuously adjustable as the image intensifier is moved from radiopaque to radiolucent areas (eg, from abdomen to lungs) because the latter blank out. Manual brightness adjustment is highly impractical so the preferred method of solving this problem uses automatic adjustment of kVp and mA–instantaneous feedback of video signal strength to the kVp and mA control stabilizes brightness. In this process, kVp and mAs combinations are optimized not only for brightness but also for minimal patient exposure. According to van der Plaats, exposure rates may range from about 0.6 to 6 mR/min.

Recording the Fluoroscopic Image

Thus far, we have dealt with *real-time* imaging in fluoroscopy with image intensifiers and TV systems wherein we obtain transient images during their actual formation. Auxiliary systems come into play when *hard copy* is needed, that is, recording of images on film, paper, magnetic tape, magnetic discs, or laser discs.

---

* 1 MHz = 10^6 or 1 million Hz

1. **Spotfilm Devices.** In dark fluoroscopy, and then with image intensifiers, mechanically driven spotfilm devices permitted the exposure of an entire film or of multiple sectional exposures on the same film in a cassette with intensifying screens. A high-tension changeover switch shifted the operation from fluoroscopy to the radiographic **spotfilm device.** This system was continued when the image intensifier was introduced, the spotfilm device being positioned between the patient and the image intensifier. Phototiming improved exposure accuracy. Image quality was excellent but patient exposure was higher than it is today.

2. **Film Strip and Cut Film Cameras.** These receive light from the OS of the image intensifier. As shown in Figure 20.3, to allow spotfilming without interrupting fluoroscopy, a half-silvered beam-splitting mirror is placed in the light from the OS at an angle of 45°. Part of the light is reflected at an angle of 90° toward the film camera (cut or strip film). The remainder of the light beam passes straight through to the TV camera (vidicon). The advantages of this system include:

a. Viewing and spotfilming simultaneously.
b. Larger field of view, especially with large image intensifiers.
c. Shorter film exposure time, allowing more *individual* exposures/s.
d. Lower x-ray exposure of patient (about 30%)
e. Shorter time between exposures.
f. Large saving in film cost (as much as 80%).
g. Viewing cut film (100 mm) or roll film (70 or 105 mm) for individual exposures only, not cine.

There are certain disadvantages, not of major importance:

(1) Small films; require special processing.
(2) Small images, but one can get used to 100 or 105 mm.
(3) Square format requires larger exposure area to encompass circular image (see Figure 19.20 page 250, for analogy with cone beam restriction).

3. **Cinefluorography.** Another method of recording the image on the output screen of the image intensifier uses a modified conventional movie camera. It records the fluoroscopic image from the output screen of the image intensifier by being substituted for the film camera, as in Figure 20.3, so that viewing of the fluoroscopic image and cinegraphic recording can be accomplished at the same time. By far, cinefluorography has its major application in cardiac and angiographic procedures, for which 35 mm film is used (70 mm also available). To limit patient dose, an automatic control allows individual cine frame exposures only during instances of x-ray tube activation. Still, patient exposure is high in cinefluorography.

Recording the Video Image

Magnetic Tape and Disc

The basic components include a **magnetic tape** (or disc), a **recorder** to receive the electric signals from the vidicon or plumbicon via cable and imprint this information on the rapidly moving tape, and a **playback** device that sends the information to the TV monitor, as summarized in Figure 20.9.

A magnetic tape is made of polyester, coated with tiny particles of iron or chromic oxide that serve as magnetic dipoles. Normally, the particles are scrambled as in nonmagnetized iron.

As the tape is pulled through a magnetic recording head by the takeup reel, the head orients the magnetic dipoles on the tape according to the varying intensities of the video signals. To display the information, the tape is drawn through a playback elec-

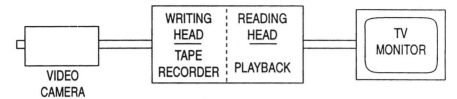

Figure 20.9. Block diagram of a TV tape recording and viewing system in fluoroscopy.

tromagnetic head whose core becomes magnetized with various intensities from the magnetized tape and sets up, by electromagnetic induction, fluctuating current signals in its windings. These signals go to the TV tube where images are produced as described on pages 266-267. Figure 20.10 is

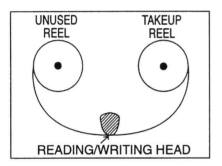

Figure 20.10. Principle of magnetic tape: writing (recording) and reading (viewing of recorded image).

a simple diagram showing that the takeup reel pulls the magnetic tape through the writing head in the recording mode, whereas the unused reel pulls the tape past the reading head in the reading mode. The magnetic "image" is stored on the tape and can be played back at will. But we cannot be certain about the permanence, or archival quality of magnetic tapes.

Magnetic discs operate on the same principle as tapes except for the configuration of the recording and playback equipment. But they yield static images, analogous to conventional spotfilms. They record along closely spaced, concentric grooves and are not touched by the recording or reading heads so they are less subject to wear than tapes.

Laser Discs

Also known as **optical discs,** laser discs record video information in **laser light,** which consists of **coherent** radiation—light waves that move in step (ie, in phase), as shown in Figure 20.11A. Also, a laser beam can be very tightly collimated and can travel great distances. For example, in the moon landing in 1969, the astronauts left behind a reflector so that a laser beam was subse-

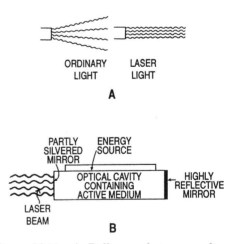

Figure 20.11. *A.* Difference between ordinary light and a laser beam. *B.* Production of a laser beam.

quently directed to it from earth to allow extremely accurate measurement of the distance between the two bodies.

How is laser light generated? Three things must first be available: (1) an active medium (eg, specific solids, liquids, or gases), (2) an energy source (eg, flash lamp, electric signal, or another laser beam), and (3) an optical cavity (a resonator enclosing the active medium) as shown in Figure 20.11B.

Atoms normally exist in ground or unexcited state. To produce a laser beam, the *majority* of the atoms in the active medium must be "pumped up" to an excited state, a condition known as *population inversion,* by the introduction of outside energy. Now, when *a* photon enters and stimulates an already excited atom, *two* photons are emitted with wavelength equal to that of the entering photon. This creates an amplifying chain reaction by rapid doublings of the number of photons, further amplified by extremely fast back-and-forth reflection between the mirrors at the ends of the opti-

cal cavity. Soon they burst through the partially silvered mirror at one end as an extremely narrow laser beam. The word laser is an acronym of *L*ight *A*mplification by *S*timulated *E*mission of *R*adiation.

Videodiscs are actually *laser discs.* In the recording mode, electric signals from the TV camera serve to generate the tightly focused laser beam, which burns pits in the photoactive coating on the metal disc; the depth of the pits depends on the strength of the electric signals. In playback mode, a low-intensity laser beam is reflected off the disc surface and read by a light-sensitive diode—the amount of light scattered depends on the depth of the pits. The diode circuit converts the light to digital (numeric) electric signals, which are changed back to light images by the TV monitor tube. With the ordinary compact disc, the electric signals are generated by sound and subsequently played back as sound.

Laser beams have found wide applications in surgery, dentistry, radiology, music, and industry in general.

QUESTIONS AND PROBLEMS

1. Define photopic and scotopic vision. Which of these applies to the image intensifier?
2. Describe, with the aid of a diagram, the construction and operation of an image intensifier (I I).
3. Define brightness gain; flux gain; minification gain.
4. What is the minification gain of an I I with a 15-cm input and a 2.5 cm output screen?
5. Name the phosphors on the input and output screens of an I I.
6. Define conversion factor.
7. What is the resolution, in lines/mm, of the average I I? Why is it not higher?
8. Find how many times greater the expo-

sure rate for a 9-in. input screen must be to equal that of a 6-in. input screen.
9. What is the brightness gain of an image intensifier that has a minification gain of 100 and a flux gain of 50?
10. Summarize the function of a vidicon tube.
11. State the methods of capturing the image on the output screen of an I I.
12. Define television frame; field; interlacing.
13. What is the most common use of a CCD?
14. What does the quality of a vidicon TV camera have, to possess outstanding quality?
15. Briefly explain the two kinds of resolu-

tion of a TV image.

16. List the main methods of recording the I I fluoroscopic image.

17. In what ways can the TV image be recorded?

18. Describe the principles of magnetic tape/disc recording.

19. What is a laser beam? How does it record an image?

Chapter 21

VANISHING EQUIPMENT

Topics Covered in This Chapter

- Stereoscopic Radiography
- Tomography
- Xeroradiography
- Questions and Problems

STEREOSCOPIC RADIOGRAPHY

STEREOSCOPY means "seeing solid," that is, seeing in three dimensions. An ordinary photograph appears flat, lacking depth, because the camera has one "eye" and projects the image of an object on the film in only two dimensions. But when a pair of human eyes sees the same object, the brain perceives it as a three-dimensional solid body.

How is this brought about? Each eye sees the object from a slightly different angle, depending on the distance of the object, its size and shape, and the distance between the pupils of the eyes (see Figure 21.1). As a result, slightly different images are formed on the retinas of the respective eyes and are carried separately over the optic nerves to the brain. Here the two images are fused and perceived as one having three dimensions–height, width, and depth.

Stereoscopic vision also permits us to judge the relative distances of various objects. You can easily verify this by trying to touch something with one eye closed, and then with both eyes open. Still, it is possible for a one-eyed person to learn to judge distance fairly well, although stereoscopy is

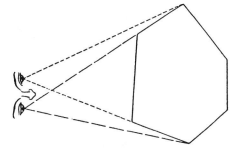

Figure 21.1. Stereoscopic vision. Each eye sees a slightly different view of the solid object. These two images are blended in the brain to give the impression of one image having a solid appearance; that is, height, width, and depth.

impossible.

We are not born with stereoscopic vision but develop it during infancy. Many years ago Dr. O. V. Batson showed that we differ in our ability to fuse stereoscopic images, and that in many instances this ability can be improved by corrective glasses, exercise, and practice.

Principle of Stereoradiography. We can produce a stereoscopic image radi-

273

ographically by applying the principle of stereoscopic vision. First, we make *two separate exposures of the same object from different points of view;* the x-ray tube takes the place of the eyes, exposing two films from slightly different positions (see Figure 21.2). The part being examined is immobilized while the tube is shifted to obtain the two stereoradiographs. Second, we must *view properly* the pair of stereoradiographs if we are to achieve stereoscopic or depth perception.

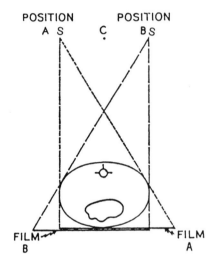

Figure 21.2. Stereoradiography. Two films are exposed separately, one with the tube focus at *A,* and the other with the focus at *B*. Note that the object is not moved. The focus positions are at equal distances from the center *C*.

Tube Shift in Stereoradiography. This is related to the interpupillary distance–the distance between the pupils of the eyes, which averages about $2^{1/2}$ in. (6 cm), whereas the average viewing distance is 25 in. (64 cm). These two distances are in the ratio of $2.5/25 = {}^1/{}_{10}$, and the same ratio governs the relationship between tube shift and SID:

$$\frac{\text{tube shift}}{\text{SID}} = \frac{1}{10} \qquad (1)$$

According to this equation, stereoradiography requires the following typical shifts:

SID 40 in. (100 cm)   tube shift 4 in. (10 cm)
SID 72 in. (180 cm)   tube shift 7 in. (18 cm)

We may conclude that the *tube shift is theoretically one-tenth the SID for a viewing distance of 25 in.* (64 cm). While, in most instances, this ratio lays down the conditions for excellent stereoradiographs, some observers find it difficult to fuse the images unless the ratio is reduced to $^1/{}_{13}$ or even $^1/{}_{16}$. However, the resulting apparent depth of the stereoscopic image will be less than that with the larger ratio of $^1/{}_{10}$.

The *direction* of the tube shift relative to the anatomic area should be perpendicular to the dominant lines; thus, chest stereoradiography requires a tube shift across the ribs. With low-ratio grids, tube shift perpendicular to the grid strips should entail no significant difference in the density of the two radiographs if the tube has been correctly centered to the grid, but with a *high ratio grid such as 12:1 or 16:1,* because of the criticality of centering, the shift *must be parallel to the long axis of the table* to avoid off-center cutoff. One, must be careful to open the collimator so as to cover the films in both positions of the tube.

The steps in stereoradiography may now be summarized:

1. Center accurately the part to be radiographed.
2. Determine the total shift of the tube for the given focus-film distance. Suppose this shift is to be 4 in. (that is, $^1/{}_{10}$ of 40 in.).
3. Determine the correct direction of shift.
4. Now shift the tube one-half the total required distance, away from the center. In this case, $^1/{}_2 \times 4 = 2$ in. Make the first exposure.
5. With the patient immobilized in the same position, change the cassette.

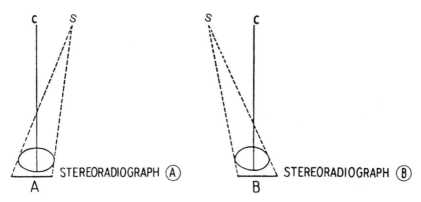

Figure 21.3. Method of shifting the tube in stereoradiography. In *A,* the tube has been shifted one-half the total distance to one side of the center, *C,* for the first exposure. In *B,* it has been shifted the full distance in the opposite direction for the second exposure.

(For chest stereoradiography, patient should suspend respiration.)

6. Now shift the tube the total distance, in this case 4 in., in a direction exactly opposite the first shift, and make the second exposure. (This is the same as shifting the tube one-half the total distance to the opposite side from the center point.) The exact method of shifting is shown in Figure 21.3.

7. Process the films.

8. The stereoradiographs are now ready for viewing.

Viewing Stereoradiographs. There are various methods of viewing a pair of stereoradiographs so that they are perceived three-dimensionally. These will now be described.

A number of optical devices facilitate stereoscopic viewing. They all shift the image of one of the stereoradiographs onto its mate. The simplest device is a 30° or 40° *glass prism.* Figure 21.4 shows how a prism bends the light rays coming from one radiograph so that they seem to come from the other one, thereby superimposing the images. For correct viewing the stereoradiographic pair is held up with the tube side

Figure 21.4. Basic principle of stereoradiographic viewing. The glass prism bends the light rays from stereoradiographic *2* in such a way as to superimpose its image on stereoradiograph *1.* This facilitates the fusion of the images to provide depth perception.

facing the observer; in other words, the eyes see the radiograph just as the tube did. At the same time, the stereoradiographs are positioned so as to represent a *horizontal shift of the tube.* With a vertical tube shift, as in radiography of the chest, the radiographs would have to be turned 90°, the chest then appearing to lie on its side. On one radiograph, the image appears *farther from the left hand edge;* this radiograph was made with the tube in the left position, and is therefore placed in the *left illuminator.* The other radiograph, exposed with the tube in the right position, is placed in the right illuminator. The observer now holds the prism between the thumb and index finger of the right hand in front of the right eye, with the ***thinnest portion near the nose,*** as shown in Figure 21.4. (If preferred, the prism can be held with the left hand in front of the left eye, again with the thinnest edge near the nose.) By slightly adjusting the position of the prism and stepping forward and backward, the observer soon succeeds in blending the images. This method requires very little practice, is inexpensive, and provides a convenient ***monocular stereoscope.*** If preferred, two 20° prisms may be used, one being held before each eye, again with the thinnest part near the nose; this is the simplest type of ***binocular stereoscope.***

Elaborate stereoscopes (Wheatstone; Stanford) were modifications of those just described, making use of prisms or mirrors to displace the images. Because of their bulk and high cost, they have been replaced by various types of small ***hand-held binocular stereoscopes.***

One must be careful to place the stereoradiographs in the illuminators correctly to achieve a genuine stereoscopic effect, so that an anteroposterior radiographic projection appears to be anteroposterior, and conversely. For example, with a chest stereoradiographed in a posteroanterior position, the anterior ribs should appear to curve away from the observer. Such verification, which is important in the relative localization of lesions or foreign bodies, can be simplified by taping an identifying marker—"stereo"—to the table top or cassette holder, to indicate the part that was farthest from the tube during radiography.

There is wide variation in the frequency with which stereoradiography is performed today, depending on the individual radiologist. However, stereoradiography helps not only in localizing a lesion or a foreign body but also in improving the perception of its shape, structure, and spatial relationship to other anatomic parts.

TOMOGRAPHY

In ordinary radiography a three-dimensional object—the body—is represented on a two-dimensional film. In effect, the object has been compressed virtually to zero thickness in the direction of the x-ray beam so that the images of a number of structures are superimposed on each other in the radiograph. Not infrequently, this makes radiographic interpretation difficult, if not impossible.

Before the introduction of tomography and computed tomography, radiographic separation of superimposed anatomic structures was achieved by rotation of the patient, as in cholecystography in a lanky patient whose gallbladder often overlies the lumbar vertebrae. Another trick had the patient chew rapidly to blur out the mandible dur-

ing anteroposterior radiography of the upper cervical vertebrae. Although such maneuvers often succeeded, they failed when a cavity was "buried" in a consolidated lung.

The invention of tomography in the 1920s was a landmark development in radiography. It is curious that a number of individuals developed tomographic machines *independently* in France: Bocage (1921), and Portes and Chasse (1922); in the Netherlands: Ziedes des Plantes (1921); and in the United States, the mathematical principles by Kieffer (1929), a radiographer, while recovering from tuberculosis in a sanatorium. A variety of designs and just as many names were applied to the process, but the official name is now ***tomography,*** and the product, ***tomogram.*** Tomography provides a reasonably sharp image of a particular structure, while images of structures lying in front of or behind it are blurred beyond recognition. Ideally, the desired image stands out sharply in a hazy sea of blurred shadows. *Tomography simply consists of motion of the tube and film in opposite directions in a prescribed manner during the exposure.*

The Three "Musts" in Tomography. To achieve true tomography, the following three conditions must prevail:

1. *Reciprocal motion of the tube and film,* that is, always in opposite directions.

2. *Parallelism of the film plane and the objective plane* (central plane of tomographic section).

3. *Proportional displacement of the tube and film.* The ratio of the SID to the source-objective plane distance must be constant throughout the excursion of the tube to keep magnification constant during tube travel.

To satisfy these conditions, the tube and the Bucky tray must be connected by a rigid rod that allows them to rotate about a pivot or *fulcrum.* There must also be movable *joints* or *linkages* between the rod and the tube carriage and between the rod and the Bucky tray. A simple drawing of a *linear* tomograph appears in Figure 21.5. Only images of points in the plane of the fulcrum—the *objective plane,* parallel to the film—remain in constant position on the film and display good recorded detail and contrast. Images of points lying in planes above or below the objective plane are blurred in proportion to their distance from the objective plane. Such blurred images are less visible than the sharper image of the objective plane. Figure 21.6 shows how the radiographic image of a point, *P,* outside the objective plane shifts on the film relative to the position of the image of a point, *E,* inside the objective plane.

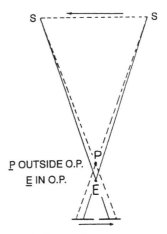

Figure 21.6. Relative shift of images of points lying outside *(P)* and within *(E)* the objective plane during tomographic excursion. Note that the image of point *E* retains the same location on the film, while the image of point *P* moves across the film and blurs. *O.P.* = objective plane.

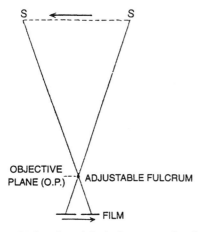

Figure 21.5. Simplified diagram of a linear tomograph.

In Figure 21.7 you see a schematic diagram of a simple linear tomograph with the standard terminology and symbols used by the International Commission on Radiologic Units and Measurements (ICRU).

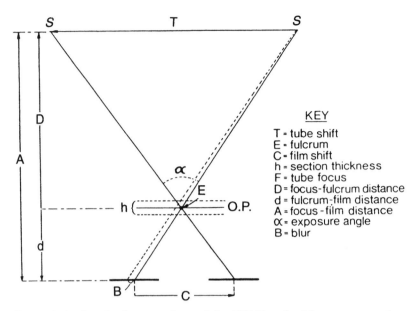

KEY

T = tube shift
E = fulcrum
C = film shift
h = section thickness
F = tube focus
D = focus-fulcrum distance
d = fulcrum-film distance
A = focus-film distance
α = exposure angle
B = blur

Figure 21.7. Standard terminology of the *ICRU* applicable to tomography.

Thickness of Section. A tomogram is really a radiograph of a slice or *slab* of tissue rather than of a plane of zero thickness. Since blurring gradually increases at greater distances from the objective plane (ie, level of fulcrum) there is no distinct boundary between the tomographic section (slab) and its surroundings. Thus, we define *the tomographic section as the slab between two selected planes of maximum permissible blur.* Obviously, the tomographic images of all points in the tomographic section will be sharper than the images of points beyond the boundary planes. If the maximum permissible blur has been properly chosen, say 0.5 mm, the images of all points within the slab will have acceptable recorded detail. Note that blur refers to the actual width of the blurred boundary of an image.

Two factors affect the thickness of section *h* for a given maximum permissible blur *B:*

1. *Exposure Angle*–the angle through which the central ray moves during the actual exposure; that is, the angle, at the fulcrum, made by the central ray between the begin-

ning and the end of the exposure. This definition applies regardless of whether the motion is rectilinear (straight line) or pluridirectional. It is to be emphasized that *the*

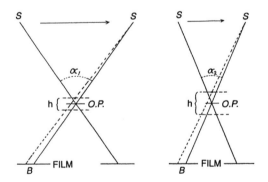

Figure 21.8. Thickness of tomographic section *h* increases with a decrease in exposure angle. The blur *B* of the limit of the thicker section in the right-hand diagram is the same as that of the thinner section in the left-hand diagram; thus, the recorded detail is equal at the upper and lower boundaries of the thick and thin section.

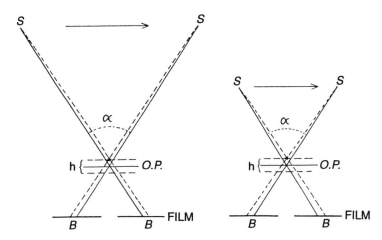

Figure 21.9. Tomographic section thickness is *not* affected by a change in the SID, provided the *exposure angle* remains constant. At the upper and lower boundaries of the tomographic sections, recorded detail-is the same in both cases.

larger the exposure angle the thinner the tomographic section, and conversely (see Figure 21.8). Note also that the angle through which the **tube** moves exceeds the exposure angle because tube motion starts before the exposure begins, and stops after the exposure ends.

2. *Source-Image Receptor Distance*–with a constant *exposure angle,* a change in SID does *not* affect section thickness (see Figure 21.9). But when tube excursion is stated as amplitude (*distance* tube moves), section thickness does depend on SID; amplitude is no longer used. SID and OID influence recorded detail just as in ordinary radiography (see pages 206-207).

Figure 21.10 gives the approximate section thickness at various exposure angles to achieve a maximum blur of 0.5 mm.

Tomographic Image. A tomogram consists of two parts. First, it presents good recorded detail of the tomographic section, which we call the **tomographic image.** Second, it contains the blurred images of points outside the tomographic section, called **redundant shadows,** which greatly influence the quality of the tomographic

image. Since the redundant shadows are generated by a moving x-ray tube, tomographic image quality ultimately depends on the **path of the tube during the exposure.** The more uniform the blurring of the redundant shadows, the better the recorded detail will

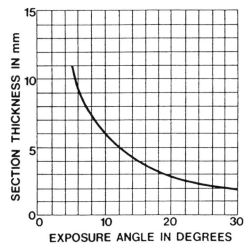

Figure 21.10. Relationship of tomographic section thickness to exposure angle. Note that the *section thickness increases as the exposure angle decreases.*

be. An example of nonuniform blurring is the streaking of the image in linear tomography.

Thus, tomographic image quality depends ultimately on (1) the recorded detail and contrast of the tomographic section and (2) the uniformity of blurring of the redundant shadows. More complex types of tube motion were developed to minimize streaking and thereby improve image quality. These included circular, elliptical, random, and hypocycloidal motion, but their increasing complexity and high cost discouraged wide acceptance, especially when the much superior CT scanners appeared on the market.

Linear tomography is the simplest and most commonly used system. Here, ***maximum blurring of redundant shadows and most effective tomography occur when the dominant axis of the tomographed structure lies perpendicular to the tube shift*** (see Figure 21.11). Conversely, ***minimum blurring and least effective tomography occur when the dominant axis is parallel to the tube shift.*** For example, in anteroposterior tomography of the vertebral column, the *lateral* vertebral margins, being parallel to the tube shift, will appear sharp in ***all*** sections without blurring of structures outside each section; therefore,

these margins are not truly tomographed. But recorded detail of the vertebral *plates,* oriented as they are perpendicular to the tube shift, will be adequately imaged.

Applications of Tomography. As described at the outset tomography is indicated whenever the desired radiographic image would be obscured by the images of structures lying in front or behind, or whenever the desired structure lies within a denser structure. Included are lesions of the chest, sternum, dens, temporomandibular joints, upper thoracic spine (lateral), thoracic spine pedicles, cholecystography and cholangiography, nephrotomography, orbits, middle and inner ears, and others. Figure 21.12. shows a typical setup in rectilinear tomography. A metal rod (behind the table) connects the tube and Bucky tray through a movable linkage at each end. The rod is provided with a pivot or ***fulcrum*** which is adjustable at various heights above the table and determines the ***objective plane.*** Because the rigid rod pivots at the fulcrum, horizontal motion of the tube in one direction causes the Bucky tray to move in the opposite direction with proportional displacement of the tube and film during the entire range of motion. Tube travel is activated by a motor through a switch located in

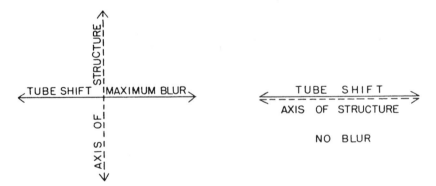

Figure 21.11. Relation of degree of blurring of redundant shadows to the direction of tube shift. Note that maximum blurring occurs when the tube is shifted perpendicular to the long axis of the object, and that no blurring occurs when the tube is shifted parallel to the axis.

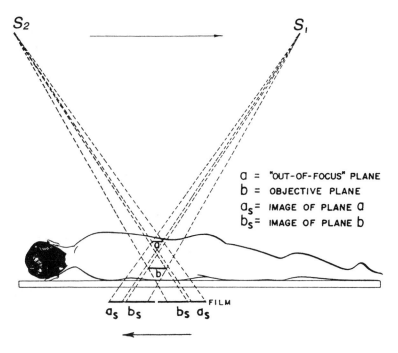

Figure 21.12. Details of linear tomography. The tube is shifted during the exposure in the direction of the upper arrow, while the film moves in the opposite direction The fulcrum level is selected at the level of the desired objective plane, *b*. The image of *b*, b_2, moves with the film, always striking the same spot. Its image is therefore sharp. But the image of *a*, a_s, constantly moving over the film during the exposure, is blurred out.

the control booth.

To obtain a tomographic series, we place the patient in the required position and localize the desired structure by appropriate radiographs. Next, we adjust the fulcrum for various heights above the table, corresponding to the series of objective planes needed to include the volume of interest. At each level of the fulcrum we expose a tomogram— commonly called a ***cut***. Usually, the separation of successive fulcrum heights (and objective planes) is 0.5 or 1.0 cm, with an exposure angle of 30°, but there are certain exceptions to be described later. We obtain a series of tomograms in which at least one should clearly reveal the desired structure.

Multitomography (multisection radiography). First proposed by Ziedes des Plantes

in 1931, multitomography yields a simultaneous set of tomograms, with one sweep of the x-ray tube (see Figure 21.13). This requires a special multisection "book" cassette containing a series of paired intensifying screens; a film is placed between each pair of screens. Moreover, each pair of screens is separated from the next pair by 0.5 or 1.0 cm of polyester foam or balsa wood spacers, depending on the desired separation. Because beam intensity decreases as the x rays penetrate the cassette, mainly by absorption in its contents, screen-pair speed is graded so that the slowest screens are on top and the fastest on the bottom. A disadvantage of multitomography is the progressive deterioration of image quality (detail and contrast) as the x-ray beam traverses the

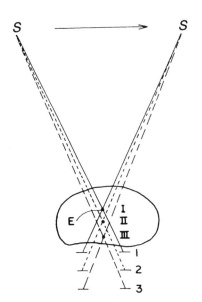

Figure 21.13. Multitomography. Simultaneous cuts of objective planes I, II, and III.

films and screens, although this was not usually a serious problem.

Zonography. First proposed by Ziedes des Plantes in 1931, zonography is simply tomography with a *small exposure angle*–5° to 10°–producing a thick section. For example, with a tomographic angle of 10° section thickness is 6 mm, and with 5°, 11 mm. Such a thick section, called a *zonogram,* superficially resembles an ordinary radiograph. It has better contrast and recorded detail than a conventional tomogram made with a tomographic angle of, say, 30°. Zonography is helpful when (1) the obscuring structures are far from the objective plane and (2) thin sections are not needed. Examples of struc-

tures usually suitable for zonography include the sternum (thoracic spine far removed), temporomandibular joint (opposite joint far removed), and nephrotomography (thin sections usually not needed). Zonography is ordinarily ill-advised in studies of the lungs because small cavities may be missed.

Patient Exposure in Tomography. In general, the surface dose for a given tomographic exposure is of the same order of magnitude as that for ordinary radiography, but is distributed over a larger area of the body. Patient exposure with multitomography is not much more than that with a series of individual tomograms with the same number of films. Precautions during tomography are the same as for conventional radiography–adequate filtration (HVL 3 mm Al), high kV, collimation, shielding of gonads, and optimum processing of films. Under these conditions, and because its use is relatively limited, tomography adds an insignificant amount of radiation to the patient.

Present Status of Tomography. Traditional tomography as described in this section can provide valuable information under limited conditions, especially in small hospitals where computed tomography is not available. However, in critical situations tomography may not be diagnostic and one must obtain computed tomographic (CT) studies as soon as possible. In the vast majority of hospitals traditional tomography has been replaced by CT. Still, radiographic-fluoroscopic machines are often equipped with a built-in linear tomograph, mainly for excretory urography and occasionally for chest radiography.

XERORADIOGRAPHY (XEROGRAPHY)

In 1937, C. F. Carlson invented *xerography* (pronounced "zero-"), a radically different kind of photocopying process. Not long afterward the xerographic method was

applied as a substitute for film in radiography where it became known as *xeroradiography,* which means "dry radiography."

The xerographic *principle* depends on the

fact that semiconductors such as selenium, while ordinarily behaving as insulators, become conductors when struck by light or x rays. A typical xerographic plate consists of *selenium* coated on a sheet of metal, usually aluminum. In use, the aluminum is electrically grounded.

To make a *xeroradiograph* or, in short, a *xerogram,* a positive electrostatic charge is first applied evenly to the selenium surface. Next, the charged plate is placed under the anatomic part and the proper x-ray exposure made, just as in film radiography, although the exposure factors are different. In any area of the plate struck by x-ray photons, the selenium loses positive charges through the grounded aluminum backing, the amount of charge neutralization depending on the x-ray exposure. Thus, the selenium loses much of its charge in areas beneath soft tissues which transmit large numbers of photons, whereas it loses relatively little charge in areas underlying bone, which transmits few photons.

This system yields an *electrostatic latent image* consisting of areas varying in the density of positive charges remaining on the plate after exposure. These densities are then converted to a visible or manifest image by spraying the plate with a toner—a fine cloud of negatively charged blue plastic powder that sticks to the various areas on the plate in amounts depending on the size of the residual positive charge. A permanent image is obtained by pressing a sheet of clear plastic or plastic-coated paper against the plate, thereby transferring the blue powder to the plastic, which is then peeled off the plate.

After it has been cleaned and subjected to 45 seconds of heating or *relaxation,* the plate can be re-used. Heating eliminates residual images that may appear as "ghosts" on the next xerogram.

Here are some of the advantages of xerography by comparison with film radiography:

1. *Recorded Detail Enhancement.* Xerograms have a unique property called *edge effect,* an accentuation of the borders between images of different densities. This produces an apparent increase in local contrast and recorded detail. The edge effect is caused by the "theft" of powder by the denser area from the less dense area along the border (electrical "fringe" phenomenon). As a result, details stand out strikingly, resembling bas relief. The edge effect becomes more pronounced as overall contrast is increased; therefore, low or medium kV technics are preferred. It is especially important in accentuating the image of tiny calcium deposits or *microcalcifications,* present in about 30 percent of breast cancers.

2. *Low Over-all Contrast.* Despite the apparently high contrast caused by the edge effect, the over-all contrast of xerograms is *low* in comparison with ordinary radiographs. The low contrast and the edge effect together produce xeromammograms in which details stand out boldly and all structures, from fat to bony ribs, are clearly seen with a single exposure.

3. *Wide Latitude.* Xeroradiography is characterized by wide exposure latitude, so that variation of about 10 kV at 80 kV in examinations has only a slight effect on xerogram quality. At the same time, a wider range of densities can be depicted on a single plate, facilitating interpretation.

4. *Magnification Effect.* Not entirely explained is the phenomenon of magnification of microcalcifications and other fine structures, making them much easier to detect.

5. *Rapid Process with no Liquid Solutions.* Processing is complete in ninety seconds in the xerographic processor in a light room. No plumbing is required, since this is entirely a dry process.

6. *Positive or Negative Image Mode.* By proper selection of charge, we can obtain either positive or negative xerograms.

Furthermore, these can be transferred to film or paper, depending on the preference for viewing by transmitted or reflected light respectively.

Xerography has several limitations. One is the artifacts caused by dust or imperfections in the plate, which may resemble microcalcifications in mammograms. Another is that xerography cannot be used in the examination of large parts because of the large exposures that are needed as compared with film-screen systems. Another problem is the specialized service required for proper operation of the xerographic unit, although minor repairs can be made by the radiographer after a short course in maintenance of the equipment.

In any event, xerographic equipment is no longer being manufactured, although still in use in a few facilities for mammography. It has been superseded by low-dose film-screen-grid combinations and dedicated mammographic x-ray machines.

QUESTIONS AND PROBLEMS

1. What is meant by stereoscopic vision?
2. How do we determine the correct tube shift in stereoradiography?
3. List the steps in making and viewing stereoradiographs.
4. Describe the principle of tomography.
5. Define tomographic section; fulcrum; objective plane; tomographic image; redundant shadows; exposure angle.
6. Discuss the various types of tomographic motion. Compare their advantages and disadvantages.
7. Explain the main modifications of tomography: zonography; multitomography.
8. What is the comparative exposure with multitomography and with a series consisting of an equal number of individual tomograms?
9. How should tomographic equipment be activated, for maximum safety of personnel?
10. How is tomographic section thickness defined? Explain the factors that influence section thickness.
11. Explain the xerographic process and list the advantages and disadvantages of xeroradiography as compared with ordinary film radiography.

Chapter 22

MAMMOGRAPHY

Topics Covered in This Chapter

- Soft Tissue Radiography
- Dedicated Mammography Unit
  X-Ray Tube Target and Filter
- Special Mammographic Film-Screen
  Combination
- Dedicated Film Processor
- Optimum Technique
- Quality Assurance (QA) and Quality

Control (QC)
  Qualifications of Personnel
  Equipment Acceptance Testing
  Scheduled QC Tests and Inspections
  Reports of Mammography Findings
  Patient Follow-up
- Mobile Mammography
- Questions and Problems

IN SOFT TISSUE RADIOGRAPHY, of which mammography is a specialty, we are dealing with essentially two types of structures–those of *water density* (eg, muscle, glands, fibrous tissue, blood vessels) and fat. With conventional exposure factors, these tissues exhibit poor subject contrast. However, by selection of suitable exposure factors we can enhance radiographic contrast between fat and any tissue of water density. But there is no way to distinguish, by ordinary radiography, the various tissues of water density unless they happen to be surrounded by fat, short of using contrast agents. This important principle applies directly in mammography in that lesions such as tumors (water density) can be separated radiographically from the surrounding fat which is often present in sufficient amount to act as a natural contrast agent, especially in older or obese women. Furthermore, since about one third of breast cancers contain microcalcifications (0.05

mm or less) that require viewing with a hand lens, superb recorded detail is essential.

The detection and analysis of soft tissue lesions in the breast, with minimal dose, had to await the development of appropriate x-ray equipment and film-screen-grid combinations that could function consistently at low kVp and with very small focal spots. The essential requirement for differentiation of soft tissue changes is *very low energy* x rays, generated at 24 to 28 kVp.

An extremely important consideration is the *dose* to the breast glandular tissue during mammography. Based on studies of breast cancer incidence in survivors of the atom bomb attack on the Japanese cities (Hiroshima and Nagasaki) by the U.S.A. at the end of World War II, experts *estimate* that approximately 6 excess breast cancers appear per million women per year per cGy (rad) exposure. This incidence is extrapolated (extended) from large doses, downward to vanishingly small doses (ie, non threshold response, according to the Beir Report of 1986). With the small doses used now in mammography, the benefit/risk ratio is very high (see page 384). This is especially important because breast cancer is high on the list of causes of death in women from age 40 and up.

Detecting the subtle changes in the soft tissues or the tell-tale cancer calcifications, while minimizing radiation exposure, became possible when, in 1966, a new type of radiographic tube was invented in France by J. R. Bens and J. C. Delarue. Knowing that optimum contrast is achieved in soft tissue radiography with x rays ranging in wavelength from 0.6 to 0.9 Å or average energy of about 17.5 kV, they designed an x-ray tube

with a ***molybdenum (Mo) target.*** They selected Mo because at about 30 kVp it emits characteristic radiation with energies of 17.14 and 19.5 kV (average 19 kV). Note that the energy of this radiation matches fairly closely the optimum for soft tissue radiography (in fact, it is slightly higher than the optimum 17.5 kV, and this is desirable). Furthermore, they introduced a 0.03 mm ***Mo primary filter*** into the beam to remove much of the radiation with energy above and below the desired range, thereby producing a ***nearly monoenergetic beam*** (see Figure 22.1). Using the ***same*** metal–Mo–for ***both*** the target and the filter utilizes the principle of the ***spectral window;*** that is, a filter readily transmits the characteristic radiation produced initially by

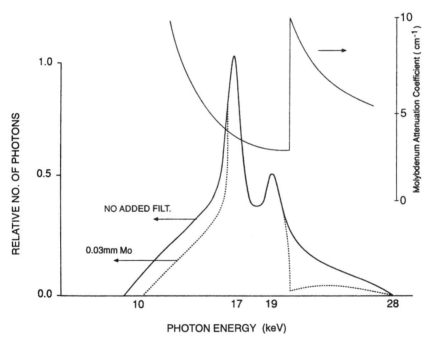

PHOTON ENERGY (keV)

Figure 22.1. X-ray spectrum from a molybdenum ("moly") target at 28 kVp, represented by the vertical axis on the *left*. The attenuation coefficient (degree of x-ray attenuation) in the moly filter is plotted on the vertical axis on the right–an increase in attenuation means greater removal of photons by the filter. Note that the spectral peaks at 17 and 19 keV pass through the filter, but the peak in the *attenuation curve* shows desired removal of higher energy photons, resulting in improved mammographic contrast. (From Hendee WR, Ritenour ER, permission of Mosby Year Book, ed 3, Chicago, 1992.)

the same element, while absorbing much of the general radiation (bremsstrahlen or continuous spectrum). Some units have a 0.025 mm rhodium (Rh) filter in the molybdenum x-ray beam.

Mammographic x-ray tubes have a beryllium window to minimize inherent filtration and permit low kVp for optimal low contrast required in soft tissue radiography. Mammographic rotating anode tubes have two focal spots: 0.3 mm for superb recorded detail at full-view imaging, and 0.1 mm for magnification radiography (see Figure 22.2). The anode rotates at high speed–9600 RPM–and a high heat-loading capacity–300,000 HU, or more.

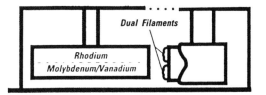

Figure 22.3. Dual anode mammographic x-ray tube, a detail extracted from the complete figure. (Courtesy of GE Medical Systems.)

The General Electric Senographe DMR x-ray tube has a rotating anode x-ray tube with a **bi-metal anode** and dual filaments (see Figure 22.3), one of the target tracks being a molybdenum/vanadium (Mo/Va) alloy, usable with an Mo, Va, or Al filter. Vanadium slows the rate of track pitting. The other track is rhodium (Rh) with an Rh or Al filter. At 20 kVp, the Rh track provides about 25 percent reduction in dose with thicker, glandular breasts, than does Mo at 17 kVp. Contrast is better with Mo for fatty or thinner breasts, whereas it is slightly better with the Rh track for thicker or more glandular breasts. An internal cone further improves contrast by reducing the exit of off-focus radiation.

All mammography units must be dedicated, that is, designed for mammography only (see Figure 22.4). A number of manufacturers sell FDA (Food and Drug Administration) approved mammography units: Trex LORAD and Bennett; General Electric Senographe; Picker International; and Siemens Medical Systems, Inc.; among others.

Dedicated H-F (high-frequency) mammographic units have the advantages of compactness, increased mobility, and reliability. In fact, they are being used not only for fixed installations but also in special vans with an on-board technologist for breast screening in outlying communities where no such equipment exists. Film processing may be done on-board, or the exposed films taken to

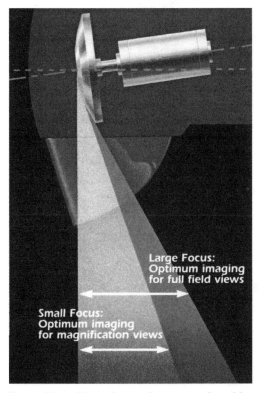

Figure 22.2. Typical x-ray beams produced by a double focal spot mammographic tube. (Courtesy of LORAD, Trex Medical Corporation.)

Figure 22.4. Example of a mammographic unit and control stand. (Courtesy of LORAD, Trex Medical Corporation.)

home base for processing. Because of developer and fixer cross-contamination during transport, the preferred method at present is to return the exposed films to home base where a dedicated mammography processor should be available. This should be done each evening after the day's run because saving the films for several days leads to deterioration of the image. Mobile mammography has unquestionably extended the reach of breast screening with its attendant early detection of more breast cancers.

From the standpoint of exposure, H-F generators entail reduction of breast tissue dose by about 25 percent relative to 60-Hz generators. Moreover, they greatly reduce exposure time to as low as 1 ms, thereby minimizing the effect of voluntary and involuntary motion and improving recorded detail. Finally, tube life increases because the very small ripple (less than 1 percent) distributes the heat in the anode more evenly than 60-Hz operation.

High-frequency generators for mammographic units are available with power ratings of about 4 to 8 kW at 200 to 240 V and 20 to 30 A. Some units still operate on ordinary house current–120 V–using an AC-DC converter to charge a battery or capacitor bank, and then a DC-AC converter to energize the full-wave rectified high-voltage transformer. In some models, an oscillator, or H-F converter, changes the battery output to H-F AC to supply the transformer.

Special low-dose mammographic film-screen and grid combinations provide the necessary speed and contrast. The film typically has a single emulsion and the grid a ratio of 4:1 or 5:1 with 30 lines/cm (80 lines/in.). Equally important is correct positioning to avoid repeats, thereby minimizing breast dose as well as contributing to efficiency of the entire process.

Breast thickness must be reduced by ***compression*** with an approved paddle device attached to the mammographic unit. Compression entails four beneficial effects:

1. Reduces image blur from motion.
2. Reduces geometric (focal spot) blur by squeezing the breast structures closer to the image receptor.
3. Increases subject contrast by decreasing breast thickness, thereby reducing the fraction of scattered radiation, an important noise factor.
4. Reduces breast exposure because a thinner breast requires less exposure for a high-quality mammogram.

Given all the above qualifications for a dedicated mammography facility, optimal technique with careful attention to detail is also essential for superb mammography, not the least being correct positioning and the selection of optimal exposure factors, which require a radiologic technologist well-trained in mammography.

QUALITY STANDARDS IN MAMMOGRAPHY

In the field of medical radiology, mammography stands out as one of its most demanding and meticulous branches. It should be obvious why; the American Cancer Society estimates that in 1998 breast cancer will comprise 30 percent of all cancers in women, representing 180,300 new cases and 140,000 deaths. Moreover, and this is very important, the staging, treatment plan, and prognosis depend intimately on the size of the lesion at the time of diagnosis. Missing a small lesion on mammography can make the difference between cure and failure, so it is imperative that everything contributing to a mammogram of superb diagnostic quality must be tightly controlled: training and experience of the interpreting radiologist, radiologic technologist (mammographer), medical physicist; dedicated high-quality mammographic x-ray units; special cassettes, films and grids; dedicated mammogram processor; properly designed breast compression system; and checking and testing of all these items at prescribed intervals.

To this end, the American College of Radiology (ACR) developed a detailed plan for systematic quality assurance (QA) and quality control (QC), publishing it as the *Mammographic Quality Control Manual* in 1990, with revisions in 1994 and 1999. Originally, the U.S. Mammographic Quality Standards Act (MSQA) was passed by Congress in 1992, published in the Federal Register by the Food and Drug Administration (FDA), and a revision scheduled to take effect on 4/28/99. These regulations have been published in great detail and are available to mammography facilities from the FDA department of HHS (Health and Human Services) Center for Radiological Health, 1350 Picard Drive, Rockville, MD 20850. The ACR's complete *Mammography Control Manual** covering sections on radiology, radiologic technology, and medical physics can be obtained from the American College of Radiology.

Quality assurance and quality control, described in minute detail in the compliance guides must, with few exceptions, be followed to the letter, for a mammography center to be approved initially and on subsequent periodic reexaminations. This involves the interpreting radiologist, the radiologic technologist, and the medical physicist, spelling out their qualifications and experience.

The present chapter will be limited mainly to the technologist's responsibilities and to equipment specifications. However, for complete details, the official guide *must* be followed closely and be readily available in the mammography facility to ensure the highest quality mammograms.

To avoid confusion, you should be aware of the difference between quality assurance (QA) and quality control (QC). *QA* is the general process of (1) testing and (2) providing the highest quality mammograms and their interpretation. *QC* is more specific, dealing with (1) testing and (2) maintaining the highest possible quality in both equipment performance and mammography.

A QA program requires a dedicated QA Committee, consisting of one or more of the following: radiologist, medical physicist, supervising radiologic technologist (mammographer), and others in the facility. Since QA is ongoing, it is often referred to as *CQA*–continuing QA.

Quality control, as an integral part of QA, includes a number of specific technical procedures that ensure acquisition of the highest possible quality screening and diagnostic mammograms. Four steps are involved, usually in the domain of the *medical physicist:*

1. Acceptance testing, to detect defects in new or major repaired equipment.
2. Establishment of baseline performance of equipment.
3. Detection of defects in performance of equipment before they become radiologically apparent.
4. Verification and correction of defects in equipment performance. Full details appear in the Compliance Guide.

The interpreting radiologist must assume ultimate responsibility for the implementation of QA and QC, but usually delegates control tests to a *designated mammographer.* A second mammographer should be trained to serve in the absence the designated mammographer-in-charge. The radiologist reviews test results periodically and provides direction when problems arise; he or she must provide adequate time for the QC program incorporated in the daily schedule– preferably at the start of the day's work. Finally, as a team, the radiologist, QC technologist, and medical physicist must develop and implement a QA program, in which the ACR/FDA Compliance guide must play a central role. An on-site or readily available medical physicist should supervise the QC program, perform those tests for which he or she is responsible, and oversee the QC technologist. Otherwise, the radiologist should supervise the program. To reiterate, the radiologist bears the ultimate responsibility for ensuring mammogram quality, QC testing, and QA procedures, and followup of patients.

*Prepared with support from the American Cancer Society, Cancer Control Grant #267A.

Qualifications of Radiologic Technologist (Mammographer)

A mammographer must be licensed by a State to perform general radiography, and be certified by the American Registry of Radiologic Technology (ARRT) or by the American Registry of Clinical Radiologic Technologists (ARCRT). In addition, a qualified mammographer must have completed at least 40 contact hours of documented training specifically in mammography under a qualified instructor, including breast anatomy and physiology, positioning and compression of the breast for mammography, QA and QC control techniques, and imaging of breasts containing implants. Later, at least 8 hours of training are required for any new modality. Stringent conditions for continuing education must be followed according to regulations.

Equipment

The basic requirement is a *dedicated,* single-purpose mammography unit. For FDA or designated State Agency approval, the facility must follow the specified guidelines for all accessories, such as compression paddle, film-screen-grid combinations, cassettes, film processor, and film viewing devices.

Quality Assurance

The duties of the leading individuals in the facility include the following responsibilities for QA:

Interpreting Physician (Radiologist)–followup on corrective action when mammograms are of unsatisfactory quality, and participation in the facility's medical outcome audit.

Medical Physicist–performance of annual mammographic equipment survey and supervision of equipment-related QA practices, including mandated reports.

QA Technologist (designated by facility)–responsible for all QC duties not assigned to medical physicist or radiologist. QA tasks may be assigned to other qualified personnel, but QC technologist must ensure that tasks are completed as required.

Quality Assurance Records

The three individuals in the preceding section must ensure that records be kept on:

1. Employee qualifications to meet assigned QC tasks.
2. Mammographic techniques and procedures.
3. QA, including monitoring data, problems detected in monitoring, corrective action, and effectiveness of corrective action.
4. Safety and protection.

QC records should include the results of each required test, specified below, until the next inspection and the facility has been declared in compliance with regulations; or until the test has been performed two additional times at the required frequency, whichever is longer.

Radiologic Technologist Responsibilities

The radiologic technologist–mammographer–is responsible for image quality, patient's breast positioning, breast compression, image production (mammography), and film processing. The mammographer assigned to quality control (QC) must perform the designated tests at the required intervals.

QUALITY CONTROL TESTS

The facility must adhere to scheduled quality control tests as required on a daily, weekly, monthly, quarterly, semiannual, and annual basis (see Table 22.1). Again note that these tests will be covered in general terms–specific details appear in the latest edition of the *Mammography Control Manual,* 1999; its QC instructions must be followed to the letter, in cookbook fashion. *Standard charts are available* to facilitate recording test results.

Table 22.1
MINIMUM TEST FREQUENCIES
FOR QUALITY CONTROL*

Minimum Frequency	Test
Daily	Darkroom Cleanliness Processor Quality Control
Weekly	Screen Cleanliness Phantom Images Viewboxes and Viewing Conditions
Monthly	Visual Checklist
Quarterly	Repeat Film Analysis Fixer Retention in Film
Semiannually	Darkroom Fog Screen-Film Contact Breast Compression

*Duties of QC control mammographer, according to the *Mammography Control Manual,* by the American College of Radiology (ACR).

Daily

The designated QC mammographer or a qualified substitute shall be responsible for:

1. *Darkroom Cleanliness*–includes damp-wiping the processor tray and counter top, damp-mopping the floor, *after* the processor water and power have been turned on to allow stabilization of the developer temperature.

2. *Processor Performance*–evaluated daily *after* darkroom cleaning. This includes check of developer temperature with a digital thermometer (fever type acceptable). A control film strip is exposed in a sensitometer containing a light and a step wedge, and processed just like a mammogram; and read with a densitometer (see Figure 22.5). The results are charted and evaluated; if necessary, corrective action must be taken according to the manual. Problems may be incorrect temperature, faulty replenishment, or incorrect development time.

Weekly

1. *Screen Cleanliness*–dusting or washing screens weekly, or when necessary. Only approved cleaner should be used.

2. *View Boxes and Viewing Conditions*–assuring high luminance (brightness), close masking of mammograms, readily available hot lights, and good quality magnifier lens. Extraneous room light must be minimized.

3. *Phantom Image Evaluation*–determination of imaging system's ability to detect imbedded simulated breast lesions in a tissue-equivalent breast phantom. Also, evaluation of image contrast by means of an acrylic disc. Results must be entered into a standard chart.

Monthly

A *visual checklist* involves monthly review of all items by simple inspection and manual operation. Some items pertain to patient safety; others concern operator convenience. Additional items may be included as needed. Missing items should be replaced and repairs made as soon as possible.

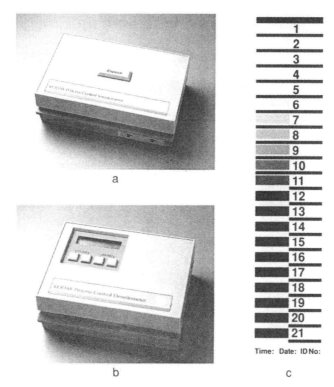

Figure 22.5. Typical sensitometer (a), densitometer (b), and sensitometric film strip (c). (From Pizzutiello RJ, Cullinan JE: *Introduction to Medical Radiographic Imaging,* by permission of Eastman Kodak Co.)

Quarterly

Mammography requires constant surveillance of repeat examinations and fixer retention in films.

1. *Repeat Examinations*–all unsatisfactory mammograms must be shown to the radiologist, as they may reveal a lesion not found on acceptable mammograms. All rejects must be tallied, collected, and analyzed quarterly. The percentage rejects are compared with previous month's rejects, and if a 3 percent increase, evaluate. Corrective action must be recorded.

2. *Fixer Retention*–must be measured in completely processed mammograms, and if excessive, must be corrected promptly.

Semiannual

1. *Darkroom Fog*–exposure of a control film on the counter top, for 2 minutes with room lights off, and safelights on. Fog density must not exceed 0.05 as measured with a densitometer. Requires immediate corrective action.

2. *Screen-Film Contact*–tested with exposure through a 40-mesh copper screen on the front of the cassette (refer back to Figure 15.11, page 184, for appearance of poor contact). The cassette should be removed, and repaired later.

3. *Breast Compression*–checked for prescribed compression according to the manual.

Annual control tests are in the province of the medical physicist, and cover all aspects of equipment performance according to the requirements in the manual. You should be aware that the mean glandular breast dose ***per view,*** as measured with a 4.2-cm thick phantom should not exceed 3 mGy (0.3 cGy, or rad) with a screen-film-grid combination.

Mobile Mammography Units

Quality assurance standards are the same as those required for fixed installations, except that at each location, performance standards of the system must be verified before starting mammography. For example, the mammographer can prove adequacy of image quality by evaluating the image of a breast phantom. The phantom image should accompany the clinical mammograms.

QUESTIONS AND PROBLEMS

1. What basic principle underlies soft tissue radiography?
2. Why is low-dose technique so important in mammography?
3. Name some typical target metals in mammography tubes.
4. State several target-filter combinations, and explain their advantages.
5. Give an example of a bi-metal anode. What advantage does it have?
6. Give an example of a double focal spot anode, and state when each spot is used.
7. What is a typical screen-film-grid combination, including grid ratio?
8. Give the four important benefits of breast compression during mammography.
9. Why are quality standards so high in mammography?
10. Who set the standards for mammography?

11. State the qualifications of a mammographer.
12. List the requirements for approved mammographic equipment and supplies.
13. Who is responsible for overall supervision of quality assurance?
14. Who is responsible for QC and QA duties not assigned to the radiologist and physicist?
15. At what intervals does the designated mammographer perform her/his QC tests?
16. How do standards for mobile mammography compare with those for fixed installations?
17. What is quality control? Quality assurance?
18. State the maximum mean glandular (ie, mid-glandular) dose as measured in a standard breast phantom.

Chapter 23

BASIC COMPUTER SCIENCE

Topics Covered in This Chapter

- History
- Data *vs* Information
- Computer Operations
- Components
    Hardware
    Software
- Input Devices
- Output Devices

- Primary Memory (RAM)
- Central Processing Unit (CPU) or
    Microprocessor
- Secondary Memory
- Computer Language
- Binary Numbers
- Questions and Problems

Introduction

A COMPUTER is a programmable electronic device that rapidly performs mathematical and logical operations in response to specific instructions. Since computers now play a major role in radiology for data processing (scheduling, record keeping, quality control, etc) and for control of equipment, the student radiographer should understand the basic principles and terminology of computer science.

History

Primitive humans instinctively realized the need to count–animals, people,and things in general–so early on they used their fingers or other objects such as sticks and stones. In fact, the root word of *calculate* is the Latin word *calculus,* meaning stone. The Roman numerals were represented by capital letters, a very cumbersome system that made mathematical operations extremely difficult. Although our numerals are called *Arabic,* their present form evolved over many centuries in Asia and Europe. Not until about 700 years ago did our decimal system, based on powers of ten and numeral position, develop in Europe after having originated in Greece in the third century A.D.

The ***abacus,*** an early calculating device (see Figure 23.1) probably invented in ancient Babylon, is still being used with some modifications in emerging nations. An abacus simply consists of a wooden or plastic frame strung vertically with wires on which movable beads are mounted. The wires represent, from right to left, the positions for units, tens, hundreds, thousands, and so on, with each wire holding nine beads.

Surprisingly, the abacus turned out to be a rapid method of doing addition and subtraction.

In 1642 Blaise Pascal, the great French mathematician and scientist, and the father

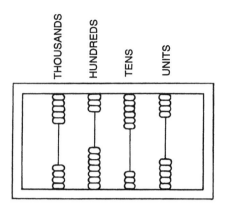

Figure 23.1. Simplified version of an abacus to show how it is used. In counting, one moves the beads successively up each wire, starting from the right: in this case it reads 5,364. It can also be used for addition and subtraction.

of probability theory, mechanized the abacus by inventing a cog-and-wheel calculator in which turning one cogwheel engaged adjacent cogs, rotating them in sequence. Thus, after 9 was reached on the first right wheel, the next rotation engaged the 10s wheel, and so on, allowing rapid addition and subtraction. Later, G. W. von Leibniz in Germany improved this device to perform multiplication and extraction of square roots.

In 1801, Joseph Marie Jacquard (France) developed a loom that was "computerized" to weave fabric in an intricate pattern by means of punch cards and a chain that picked up specific threads in a predetermined sequence; it was the first known use of punch cards in a mechanical computer. In 1833 Charles Babbage, an English mathematician, used stacked punch cards that could theoretically be fed into an analytic engine to generate navigational tables, but he never produced a working model.

For the 1890 U.S.A. census Herman Hillerith, an American statistician, invented an electromechanical computer that could "read" the position of holes in file punch cards and tabulate data such as age, sex, occupation, etc—in fact, an early spreadsheet. This so streamlined the census process that he went on to found an electromechanical computer company.

The market for faster and faster computers led to the use of vacuum tubes in the 1930s by Howard Aiken and others, but with the growth in size of these computers heat production increased to the point where air conditioning became necessary. In 1945 Mautchly and Eckert, at the University of Pennsylvania, built the ENIAC (*E*lectrical and *N*umerical *I*ntegrator and *C*alculator), which contained about 18,000 vacuum tubes, in a room of about 1800 square feet, the size of a house for the average family of four! And air conditioning posed a major problem for this first-generation, all-purpose electronic digital computer.

The introduction of the ***integrated circuit*** in the 1960s, the so-called ***chip,*** has led to ever faster, smaller, and more powerful computers, as evidenced by the progress of major players in computer manufacturing:

IBM System/36, 1964–first compatible computer family.

Intel 4004, Marian Hoff, 1971–first microprocessor, with arithmetic and logic function, and control serving the central processing unit (CPU).

Xerox Start System, 1981–first windows, icons, menus, and pointing device.

Water-scale silicon memory chip, 1987–stores 200 million characters.

Microsoft, IBM, and Apple-Macintosh–leading manufacturers of personal computers (PCs) and/or components

At present, thumbnail-size chips, with hundreds of thousands of tiny transistors mounted on a thin semiconductor wafer make up an integrated circuit which constitutes a microprocessor–the "brain" of a modern computer.

DATA *vs* INFORMATION

Although often used interchangeably, data and information are not the same. **Data** comprises words and numbers–raw material–that are entered into a computer for processing, whereas **information** is processed data in a meaningful, useful form. In other words, data *inputted* is processed and becomes *output* by means of a microprocessor, the central processing unit (CPU).

COMPUTER OPERATIONS

A computer is an extremely complex digital electronic processor that performs **arithmetic** and **logical operations** with great speed. There are two types of processors, analog and digital.

An **analog computer** processes data on unfolding events, or **real time.** For example, the hands of a clock move continuously, indicating time as they go. A **digital** watch or clock indicates the time at defined intervals–you can see the actual numbers of seconds, minutes, hours, days, and months "flip" successively with the passage of time.

A **digital computer,** with which we shall be concerned, processes alphanumeric data (words, numbers, punctuation marks, and mathematical and other common symbols), producing useful information. This takes two forms, **arithmetic** and **logical.** Arithmetic operations include addition, subtraction, multiplication, and division, while logical operations classify, compare, match, and record data, and state outcome. They include solution of equations, word processing (typing and editing), database (data about various entities assembled and organized as files and folders, for ready access), and spreadsheet (bookkeeping and accounting tables) in which arithmetic and logical functions overlap.

A computer requires an **operating system**– a master control program that manages the computer's functions such as accepting data and controlling its operations. In this chapter, discussion will be limited to Windows 98 (Microsoft) personal computer, although the basic principles apply also to the Macintosh and IBM units.

Personal computers, such as those found in most radiology departments, can perform the following basic operations: data input, memory, processing control, and output.

1. **Input**–includes two types of input: (a) **programs** that govern acquisition, manipulation, and output of data and (b) **data** inputted for processing by the computer. For example, data may have to be arranged in the form of a table or spreadsheet for accounting or bookkeeping with columns and rows, each under the desired heading.

2. **Memory**–includes three types of memory: (a) ROM or read-only memory, bearing basic unchangeable instructions for the computer; (b) RAM or random-access memory, functioning only when the computer is on and vanishing when the computer is off, except when information is deliberately saved to the hard drive; and (c) storage memory, including the hard drive, floppy disk(s), optical disk(s), tapes, etc. When data is input it goes to RAM, which has copied the relevant instructions from the hard drive.

3. **Processing**–includes the accessing and processing, by the CPU (microprocessor), of the data and program instructions temporarily stored in RAM. Rapid access by the CPU occurs through a high-speed **bus,** the pathways for transmission of electric signals

between computer components.

4. *Control*–involves the oversight of steps in processing to preclude errors and coordinate the opening and closing of microswitches.

5. *Output*–transfers the information resulting from data processing to appropriate output devices to make it available to the user.

COMPUTER COMPONENTS

We turn now to the parts of a computer that initiate and perform the operations described in the preceding section. Computers in the average radiology department are *microcomputers,* in contradistinction to medium-sized *minicomputers* and very large *main-frame computers.* Microcomputers are usually called *personal computers* or *PCs* (see Figure 23.2).

The two main types of components include:

1. *Hardware*–all components that can be touched (see Figure 23.1), including the console; monitor; keyboard; mouse (see Figure 23.2); various disks and their drives–hard, RAM, ROM, floppy; cables; and fax/modem. Speakers are optional. All

these items are lumped together as *peripherals*–any hardware that is plugged into the console for a special purpose. Some are input, while others are output devices, to be described below.

2. *Software*–"untouchables" that include the *operating system*–processes information stored on disk whose function is to tell the computer what to do. Additional programs include word processing, spreadsheet, and database; and contents of floppy disks, CD-ROM disks, and miscellaneous storage devices. The most prevalent operating systems in radiology departments are Microsoft Windows 95 and 98; or MacOS and iMac.

Input devices. These feed data into primary memory and comprise (1) keyboard,

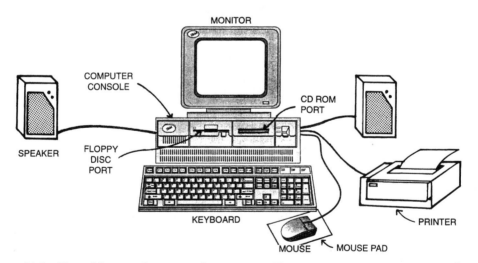

Figure 23.2. Typical layout of a personal computer. The items connected to the CPU (microprocessor) constitute *hardware,* also called *peripherals.*

(2) mouse, (3) disks (floppy and ROM), (4) modem, (5) graphics tablet, and (6) voice recognition.

1. *Keyboard*–closely resembles a typewriter keyboard in having the same basic alphabetical and numerical layout known as QWERTY, the first six letters on the upper row of letter keys. The next row up has the numbers keys and symbols keys like a typewriter, while the topmost row has a series of numbered F, or *function keys,* which provide special instructions. There are other keys with special functions, such as the numeric keypad. An *enhanced* keyboard has 101 keys, and three more have been added for Windows. It should be noted that the SHIFT key works just like the one on a typewriter, but the Caps Lock key produces only capital letters, and not punctuation marks. Other differences include separate keys for number 1 and small l, and for 0 and capital O. Touch typing, called "keyboarding" in computerese, helps greatly in developing speed and accuracy; programs like the *Mavis Beacon Teaches Typing* software are readily available. Incidentally, the symbol for multiplication is the asterisk "*", and for division the slash mark "/".

2. *Mouse*–invented by Douglas Engelbart in the 1960s, the mouse is indispensable as

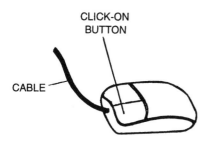

CLICK-ON
BUTTON

CABLE

Figure 23.3. A computer "mouse." The left button is most often used; it "clicks on" the computer-screen arrow, which follows the motion of the mouse as it is moved on its pad to select an icon, a menu, or an item from the menu.

an electromechanical input device that fits the palm and fingers (see Figure 23.3) and has one or more control keys on top and a rolling ball on its underside. As the mouse is moved around on its rolling ball, usually on a friction pad, it activates a *cursor* or arrow on the monitor screen so the cursor exactly follows the motion of the mouse. Thus, the mouse is a pointing device. In Windows it selects the desired *icon* (symbol for a program or other item) and, when the menu appears, allows the user to select the desired option and submenus. It can also be used to "drag" text and images from place to place on the screen, and to draw lines on the screen. Each time the cursor indicates the desired option, the proper button (usually the left one on the mouse) is depressed causing a click, called "clicking on," to activate the internal circuitry, which sends the appropriate signal to the computer. Another type, the optical mouse, uses a light beam to activate the circuitry to the computer.

3. *Disks*–floppy and ROM, floppy disks are for input or output; but ROM, input only. For input, a floppy contains a program of instructions or data to be entered into secondary memory (hard disk). ROM (**R**ead **O**nly **M**emory) ordinarily contains unerasable instructions for computer startup and basic information; but newer versions provide additional options (see below).

a. *Floppy disks* consist of thin, flexible mylar measuring $3^1/_2$ in. in diameter, capable of holding up to 1.44 MB (megabytes) of memory that is stored and read magnetically. These disks come in individual rigid protective envelopes. In use, a floppy is inserted into its special disk drive through a slot in the front of the console, so it is both a *storage* and *entry* device. In other words, an unused floppy disk can "read" processed information for storage, and "write" information to secondary (hard disk) storage. It is often used as overflow storage from the hard disk. Floppies must be protected from magnets of all kinds, heating or freezing, touching, excessive pressure (must not be placed under heavy objects), excessive bending, and mailing in anything but a special disk mailer.

b. *CD-ROM disks* come in a form resembling a music-CD. CD means *compact disk*. Such disks are inserted into a special slot in the front of the console and contain permanently recorded special programs that can be read by the computer. They have a capacity of about 600 MB (megabytes). CD-ROMs contain programs for "booting up" (starting up) a computer, as well as programs for graphics, sound, video, word processing, and so on. ROM *chips* are available to store a part of the operating system. Programmable ROM (PROM) chips can be reprogrammed once. Another type, the EPROM chip, which can be erased by ultraviolet light and reprogrammed more than once, has the advantage of allowing "debugging," the location and elimination of program errors.

4. *Modem*–device that carries information to or from the computer by *telephone*, and will be further described under *output devices*.

5. *Graphics Tablet*–consists of an electric pen and tablet. Writing or drawing on this tablet with the pen appears identically on the monitor screen, from which the graphics can be sent to the computer printer.

6. *Voice Recognition*–now in the process of development, it will eventually allow the user to dictate directly to the computer.

Primary or Main Memory. *RAM* (random access memory), a chip comprising many tiny transistors soldered together on a fiberglass string, is mounted near the CPU on the motherboard on which are located a variety of other chips having special functions. The motherboard is attached to the base inside the console–the housing for the processing devices. Input data can be processed only when RAM has copied the

appropriate program from secondary storage (hard disk or ROM); the result of processing data vanishes unless the computer has been instructed to save it to the hard disk. Obviously, the now blanked RAM chip remains, ready for the next input.

Note that input is temporarily deposited in RAM at specific sites called **addresses** to facilitate quick location for processing. RAM capacity is rated in **bytes** (see below) and should be at least 32 MB (megabytes = millions of bytes), but 64 to 128 is required for high-volume processing. Note again that, unless commanded to save, primary memory–RAM–disappears when the computer is turned off. One can expand memory by adding RAM chips.

Central Processing Unit (CPU). This term is synonymous with **microprocessor.** The CPU is the main processing chip, comprising thousands of tiny etched transistors and on/off switches on a semiconductor base. Mounted on the motherboard, this chip, about the size of a **thumbnail** (see Figure 23.4), can process an extraordinary volume of data in an extremely short time. Its arithmetic/logic unit (ALU) processes data arithmetically or logically while temporarily stored in RAM according to the instructions copied from secondary memory–hard disk or ROM. The CPU is entirely passive, since it cannot process anything unless it has available data and instructions in memory–RAM.

Since input and output are much slower than processing, it makes possible the processing, on one computer, of input from sev-

Figure 23.4. A microprocessor (CPU) chip, the "brain" of the computer, about life-size!

eral workstations that do not have a CPU. One CPU can serve them all virtually simultaneously because the speed of the main CPU, with extremely rapid switching among the workstations, is not noticeable to the user. This requires a ***network operating system (NOS),*** to provide a large number of hardware components connecting up to 50 workstations, and sharing of programs, database, and files.

A rapidly-growing, invaluable computer modality is national and international ***networking,*** which has revolutionized the spread of information from innumerable sources. Such information is downloadable on a PC, a process that can be greatly speeded up by connection to a community cable system.

Secondary Memory. This involves *permanent* maintenance of instructions and programs on three kinds of media: hard, floppy, and CD-ROM disks. Such stored information *cannot* be deleted by simply turning off the computer, which must be done deliberately.

The hard disk consists of a rigid metal platter coated with iron oxide particles that allow it to be magnetized and serve as a magnetic recording device (see pages 269-270). It is mounted on a spindle rotated by a motor, the assembly being sealed against dirt and moisture within the console. When activated, it rotates at high speed, about 3600 RPM (rotations per min). The disk drive includes the hard disk, motor, recording and reading heads and their motors, and controller to synchronize read/write operations and move information to and from the CPU. Information is recorded on a series of tracks within specified sectors and when the reading head moves over the disk, it detects the magnetized particles and sends this information to RAM. When the head is "writing" the reverse takes place— the head now transfers information electronically to the hard disk for permanent storage.

Each hard disk records on both sides, and there may be more than one disk in series. Hard disk capacity ranges from about 1 to 8 gigabytes (1 GB = 10^9 or 1 billion bytes!) Also available today are laser disks, CDs and DVDs. The latter, *digital versatile disks* can store up to 17 GB of data, ie, about two hours; this type of disk promises to replace present CD-ROMs, as well as laser disks, and will ultimately permit full read-write capability.

Output Devices. These are plugged into ***ports,*** or recessed sockets, on the back of the console and include mainly (1) the CRT monitor, (2) the printer, (3) the modem or fax modem. and (4) other peripheral devices such as supplemental storage devices, scanners, and others.

1. ***Monitor****–*a cathode ray tube (CRT) resembling a TV monitor, is not a conventional TV set. The monitor requires a special display adaptor called a ***video card,*** which is mounted on the motherboard. The monitor, together with the video card, constitute the ***computer video system*** that has special circuitry to display on the monitor whatever is taking place in the computer– text and graphics (drawings and pictures). Note that the card is one type of expansion unit added to the computer through posterior slots in the motherboard to increase the computer's capability. Video cards also generate color images with a tremendous range of hues, and also determine image resolution in conjunction with the monitor itself.

Resolution is usually stated in terms like ***VGA*** and ***SVGA.*** The latter is the acronym for **S**uper **V**ideo **G**raphics **A**rray with a "dot pitch" of 0.28. Since the image is made up of tiny dots or pixels (see pages 308-310), resolution depends on the closeness of the dots, so dot pitch 0.28 means that the space between adjacent dots is 0.28 mm. Depending on the amount of memory, such a monitor can resolve 800 pixels horizontally, and 600 lines vertically, and display 16 colors simultaneously. With powerful memory, literally millions of colors can be displayed, but there is some loss of resolution that is not readily perceptible.

It should be mentioned that video memory resides in special chips to which the CPU sends display information, read by the video controller and sent to the monitor. Finally, a super-VGA with at least 2 MB of memory is recommended for general use in the radiology department. However, viewing digital images may require much higher capacities– 4 to 8 MB.

2. *Printer*–on command, receives the information output of the computer and prints it on paper, known as ***hard copy.*** The three types of printers include the impact, and the nonimpact inkjet and laser.

a. *Impact Printers*–the least expensive called ***dot matrix printers*** create text and graphics by a vertical column of closely spaced pins which hammer the paper through a carbon ribbon as they move toward the right. This produces a pattern of dots in the form of a matrix within which the text character is formed (see Figure 23.5).

Figure 23.5. A letter formed by a dot matrix printer. The letter is actually much smaller, so the dots coalesce and become visually inseparable.

Print quality depends on the size of the matrix generated by the number of vertical times the number of horizontal dots. In common use are 5×7 and 7×9 matrices, the latter producing higher quality text. The main disadvantage of impact printers is the noise they produce, but they can make carbon copies for multipart forms. Font cartridges are available for various type sizes and styles.

b. *Inkjet Printers*–rapidly squirt a fine jet of black or colored ink onto paper to form both text and graphics, with excellent quality. They operate at relatively high speed–five to eight pages per minute. One can obtain font cartridges that plug into the printer to provide a variety of type styles and sizes, but most fonts are now managed by operation software (special programs) in Windows or MAC operating systems.

c. *Laser Printers*–speed about the same as inkjet printers, but produce the highest quality, similar to electronic typing, in black only. Font choice as in (b). Laser printing resembles output of copying machines–powdered ink fused to paper, a page at a time. Although more expensive than the others, many, if not most, users eventually buy laser printers for their high quality and speed.

Computers vary in the command to "print," but in any event, one should carefully review the monitor screen before printing to correct errors in advance.

3. *Modem*–an output device that converts the digital information output from a computer into audio signals, which it sends over an ordinary telephone system. Modems can be *internal,* located within the computer consoles; or *external,* within a plastic case connected to a serial port by a cable. The word modem is derived from "*mo*dulator-*dem*odulator." An internal modem is left on permanently, whereas an *external* modem must be turned off when not in use. One advantage of an external modem is that it can be connected to an existing microcomputer that does not have an internal modem. If one makes frequent use of a modem, obtain a *separate phone line,* or a *cable modem* if available.

A ***fax modem*** is a modification which, on command, takes a file that has been changed to a textual image (seen on the screen) and sends it to the designated fax number via the phone line. The fax modem can also receive faxes, storing them on the hard disk for viewing and printing. However, it cannot be entered directly into a word processing pro-gram without special software. A very rapidly growing means of communication is via the *internet.*

COMPUTER LANGUAGE

A computer does not understand ordinary language; in computerese, it cannot "read" what you and I can read. Therefore, data cannot be accepted until the computer has been programmed by a specialist called a *programmer.* He or she uses a unique *high-level language,* such as one of the following:

FORTRAN (*FOR*mula *TRAN*slation), 1954, science, math, engineering.

COBOL (*CO*mmon *B*usiness *O*riented *L*anguage), 1954, business.

BASIC (*B*eginner's *A*ll-purpose *S*ymbolic *I*nspection *C*ode), student instruction.

C and *C*++ now widely used by professional programmers; this language runs faster than other programs.

These are called *source codes,* typed instructions by programmers for translation into *machine code* by a *compiler* (special software) in the computer. Machine code is in the *binary system* (see below) that the CPU can read. So we have a very rapid translation of "languages" program: high-level language-to-compiler-to-binary machine code-to-microprocessor.

BINARY NUMBER SYSTEM

Digital systems have become increasingly prevalent in the radiology department for imaging and control of equipment, and more recently, for teleradiology. These all rely on *binary code* based on the binary number system.

Machine language uses binary code because of its simplicity and economy. It comprises only two digits, 0 and 1 (zero and one), which can dictate on or off, or high or low; this is extremely important in *microprocessor control of equipment* used in radiology.

In the binary system, 0 and 1 are each called a *bit,* acronym for *b*inary *d*igit. A *byte* consists of 8 neighboring bits, one byte being needed to represent one alphanumeric character in computer processing and storage. Such information in the form of strings of *0s* and *1s* is converted to electrical impulses carried over wires or fiberoptic cables and directed along the various circuits by extremely rapid opening and closing of microswitches, a process called *gating.* In computer operation this all occurs during data input, processing, and output.

We are accustomed to the *decimal system,* which consists of ten digits–0 to 9. Historically, humans in ancient times counted on their ten fingers (digits!), and it remains to this day as the best numbers system for general and scientific purposes. In any numbers system, of which there can be many, the value of a digit changes with its *position* in a number. For example, take the number *444:* the first 4 on the right means 4, the next one over is 40, and the one on the left is 400; 400 + 40 + 4 = 444. We can generalize this by arranging powers of 10 in a row, heading up a series of columns:

...... thousands hundreds tens   units
        10^3 10^2 10^1 10^0

Breaking down the number **1024** into its true components, we get

$$\underset{\text{thousand}}{\text{one}} + \underset{\text{hundreds}}{\text{no}} + \underset{\text{tens}}{\text{two}} + \underset{\text{units}}{\text{four}} = 1{,}024$$

Thus, every digit has an ***intrinsic*** value in and of itself, and a ***relative*** value by virtue of its ***position*** in a number consisting of more than one digit.

The ***binary system*** deals with only 2 digits, 0 and 1. By analogy with the decimal system,

2^3 2^2 2^1 2^0

or

8   4   2   1

which means

$$\underset{\text{eight}}{\text{one}} + \underset{\text{four}}{\text{one}} + \underset{\text{two}}{\text{one}} + \underset{\text{one}}{\text{one}} = 15 \text{ (decimal)}$$

In the binary system, this number would be represented as

1   1   1   1   = 1111 (binary)

If one number is missing, it is replaced by a 0 in that column.
Thus

8   0   2   1

would be

$$\underset{\text{eight}}{\text{one}} + \underset{\text{four}}{\text{no}} + \underset{\text{two}}{\text{one}} + \underset{\text{one}}{\text{one}} = 11 \text{ (decimal)}$$

or

1   0   1   1   = 1011 (binary)

Note that in this example, 11 in the decimal system is exactly the same as 1011 in the binary system; the numbers are just expressed differently. Table 23.1 shows the relation between the decimal and binary systems for the first ten decimal numbers.

One can convert from a decimal number to its equivalent binary number by a series of simple steps. For example, starting with 165,

1. Find the largest *power* of 2 that is less than 165.

Table 23.1
DIRECT COMPARISON BETWEEN
DECIMAL AND BINARY SYSTEM

Decimal	Binary	
zero	0	
one	1	(2^0)
two	10	$(2^1 + 0)$
three	11	$(2^1 + 2^0)$
four	100	$(2^2 + 0 + 0)$
five	101	$(2^2 + 0 + 2^0)$
six	110	$(2^2 + 2^1 + 0)$
seven	111	$(2^2 + 2^1 + 2^0)$
eight	1000	$(2^3 + 0 + 0 + 0)$
nine	1001	$(2^3 + 0 + 0 + 2^0)$
ten	1010	$(2^3 + 0 + 2^1 + 0)$

This is $2 \times 2 \times 2 \times 2 \times 2 \times 2 \times 2 = 128 = \mathbf{2^7}$
Now subtract 128 from the original number:
$165 - 128 = 37$

2. Find the largest power of 2 that is less than 37.
This is $2 \times 2 \times 2 \times 2 \times 2 = 32 = \mathbf{2^5}$
Subtract 32 from the last remainder 37: $37 - 32 = 5$

3. Find the largest power of 2 that is less than 5.
This is $2 \times 2 = 4 = \mathbf{2^2}$
Finally, subtract 4 from the last remainder 5:
$5 - 4 = 1 = \mathbf{2^0}$

4. We stop the process when the last remainder is 0 or 1.

The binary number is thus
one 2^7 + no 2^6 + one 2^5 + no 2^4 + no 2^3 + one 2^2 + no 2^1 + one 2^0

1   0   1   0   0   1   0   1

or 10100101
Check this by $128 + 32 + 4 + 1 = 165$

Summary of Applications in Radiology

The binary system underlies a host of equipment and devices in radiology:

Pushbutton controls: radiographic control panels
Digital radiography

Digital fluoroscopy
Digital ultrasound
Nuclear medicine equipment
CT imaging
MRI imaging
Ultrasound imaging
Subtraction

Ordinary **analog radiographs** can be digitized by a complex system using a laser beam. The digital image, consisting of numbers that represent various densities, can be displayed by conversion to analog form on a TV monitor and then manipulated electronically to vary density and contrast. However, resolution is not yet quite so good as that of the original radiograph.

A digital image consists of an image **matrix**–a collection of picture elements arranged arbitrarily in horizontal rows and vertical columns. These picture elements are termed **pixels,** shown in Figure 24.3, page 308, where you can see that the smaller the individual pixels, the greater will be the image resolution and the image recognition. As it appears to the observer, the width of the digitally reconstructed image is called the **field of view** (FOV). For a given FOV, the greater the number of pixels the smaller will be the individual pixels and the better the resolution. If you look through a magnifying glass at a photo in a newspaper, you will see that it is composed of tiny dots, which are pixels that disappear to the naked eye at the usual reading distance.

We can use an analog-digital converter to reconstruct radiographs digitally and store them in this form in a large computer for easy retrieval on demand, transmission over a network, and viewing on a monitor, with the option of density and contrast enhancement. Moreover, such a storage system obviates the recurring expense of filing, storage, and retrieval of original radiographs. Digital images can be reconstructed to analog by a digital-analog converter.

QUESTIONS AND PROBLEMS

1. Name some of the early methods of calculating or counting.
2. What is a computer "chip?" Of what does it consist?
3. How does input data become output information?
4. Define a modern computer.
5. State the difference between an analog and a digital device.
6. What is an operating system?
7. Describe three kinds of computer memory.
8. What is computer processing?
9. Define software and hardware, and give examples of each.
10. List and briefly describe four kinds of input devices.
11. Define RAM; ROM; modem; floppy disk; hard disk; CPU.
12. What is the main synonym for central processing unit (CPU)?
13. How do primary and secondary memory differ?
14. List the three main output devices, and describe each one.
15. Describe the three main kinds of printers.
16. What is meant by "dot pitch?"
17. Why is "computer language" necessary?
18. Define the binary system of numbers. Compare with the decimal system.
19. Change 231 to the binary system.
20. Convert 101101 to the decimal system.

Chapter 24

DIGITAL X-RAY IMAGING

Topics Covered in This Chapter

- Conventional Subtraction Technique
- Digital Fluoroscopy
- Analog Image
- Digital Image
  - Pixel
  - Matrix
- Subtraction
- Image Quality
- Noise Factor
- X-Ray Generator
- Digital Radiography
- Questions and Problems

Introduction

MARKED IMPROVEMENT in computer capability and appropriate ancillary equipment have led to more sophisticated radiologic imaging, as will be described later for computed tomography. Now we shall turn to a method of imaging opacified arteries with simultaneous creation of subtraction images.

Subtraction technique has been used for many years in conventional arteriography to enhance the images of opacified arteries when they are concealed by superimposed bone; for example, carotid and vertebral arteries overlying cervical vertebrae. *Manual subtraction* must be done after pro-cessing the arteriographic series by a *delayed* subtraction technique, according to the following steps:

1. *Copy* the best early **nonopacified** radiograph onto a special, high-contrast, single-emulsion subtraction film by exposure to light in a modified glass-front cassette or subtraction printer (see Figure 24.1). This copy film, after processing, serves as a **mask;** it is a **positive,** the original being a **negative.** The tones on the mask are reversed from the original: black to clear, and clear to black.

2. *Superimpose* in this order in the printer: the **mask** (positive), the selected **arteriogram** (negative), and an **unexposed** sheet of subtraction-type film. Close the printer.

3. *Expose* to a source of light such as a flu-

Figure 24.1. Cross-sectional diagram of a manual subtraction printer.

orescent illuminator for an appropriate time interval, and process the subtraction film in an automatic processor.

The subtraction radiograph will show the artery in black, while the bone has been selectively obscured by the mask; the positive mask and the negative arteriogram, when exactly superimposed, compensate for their opposite tonal scales, producing a subdued background for the opacified artery (positive). If desired, one can print the subtracted arteriogram on another sheet of film to reverse the image to a negative (artery white). The subtracted arteriogram has reasonably good image quality, especially since short-scale contrast is desirable.

Manual subtraction with conventional rapid-sequence arteriography has been virtually replaced by digital subtraction arteriography, to be described in the next section.

Digital Fluoroscopy

Automatic subtraction during arteriography becomes possible by the use of *digital fluoroscopy*. This is a sequential process starting with the image in light from the output phosphor of the image intensifier, an *analog image* having a continuous brightness range—it is a *real-time image*.

The essential sequence is as follows: the output screen's analog image is picked up by the video camera whose analog image is digitized, processed, and converted to a gray scale analog image. A digital image consists of *discrete image bits* (eg, a digital watch shows a sequence of numbers as discrete symbols—numerals—with abrupt transition from one number to the next).

Now let us examine in some detail how an image in light can be digitized to form a *digital image* (see Figure 24.2). The analog image at the output phosphor of a high-quality image intensifier enters a video camera, where various degrees of image brightness undergo a change to video electronic signals whose intensities correspond to the brightness range of the original image. These analog video signals are then fed via cable to an *analog-digital converter*, which digitizes the image and stores it in the computer's main memory as a matrix, that is, in rows and columns (see Figure 24.3). Next, a *digital-*

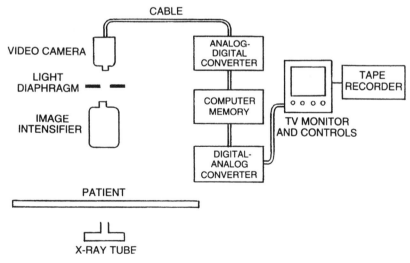

Figure 24.2. Block diagram of a digital photofluoroscopic unit.

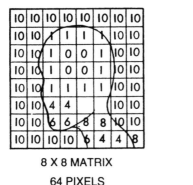
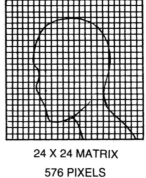

8 X 8 MATRIX

64 PIXELS

24 X 24 MATRIX

576 PIXELS

Figure 24.3. Examples of an 8×8 and a 24×24 matrix. Note that matrix overall size remains constant, but as the number of pixels increases, individual pixel size decreases. In practice, 256 $\times$ 256 or, better still, 512×512 matrices are used; their pixels are dot-sized, resembling the dots on a halftone print. Also see Figure 24.5.

analog converter changes the digital image to an analog image, following appropriate noise abatement and contrast optimization. Finally, the image is displayed on a TV monitor for direct viewing, and to a videodisc for later viewing and study.

A matrix is a square or rectangular array of numeric data, each in its own box or *pixel.* Recall that a pixel is simply a picture element from which the picture is constructed. The matrix contains the pixels arranged in horizontal rows and vertical columns (see Figure 24.3). Each pixel receives an assigned number, representing the sum of the exponents of 2, which depends on the size of the video signal (output phosphor brightness) assigned to that pixel. So a pixel with number 6 has the sum of the *logs* or exponents of 2^0, 2^1, 2^2, and 2^3 (ie, $1 + 2 + 3 = 6$), since computers use the *binary system,* with 2 as the log base (review logarithms, pages 14-15).

Pixel numbers are *relative,* not absolute; their values relate to a pixel containing zero, which represents the absence of a signal and is therefore "black." The larger the pixel number, the brighter the image (white is maximum); and the smaller the pixel number, the darker the pixel (black is minimum). The intermediate numbers indicate various shade of gray, so the range of signal brightness translates, in the end, into a gray-scale image—the number of shades of gray between white and black. It must be emphasized that the computer also assigns the *position* of each video signal to the proper pixel, corresponding to the source of the signal in the original analog image.

Matrix size equals the total number of pixels in a row times the number in a column, and *not* the dimensions of the matrix. This means that a "larger" matrix has more pixels than a "smaller" matrix, for the same area. Commonly used in digital fluorography is the 512×512 matrix, although a 1024 $\times$ 1024 matrix has smaller and more numerous pixels (see below).

Subtraction in Digital Fluoroscopy. Subtraction is performed as follows during image acquisition in the course of arteriography. An early negative image, without arterial opacification, is stored in computer memory as a positive *mask.* As the digital images with arterial opacification are obtained, the mask is automatically subtracted from each image, displayed on the TV

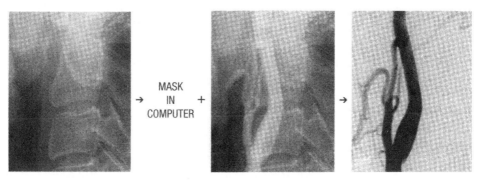

Figure 24.4. Digital subtraction radiographic series at peak opacification. From left to right, the first image is a preinjection digital radiograph (i.e., a "negative"). The computer instantly changes it to a positive image (vertebrae dark), storing it in memory as a "mask." Upon injection of contrast medium to opacify the cervical carotid arteries, images are generated in rapid succession, each being automatically masked in turn. The final image in this series has been "subtracted"– the bony structures have been eliminated by the mask. A notch in the common carotid artery just proximal to its bifurcation is due to turbulence of the blood and is not an abnormal defect.

monitor, and printed on film (see Figure 24.4). The images are also stored on a videodisc for later viewing and study. A variety of subtractions can be made by manipulation of the controls (see Table 24.1 for a comparison of manual and digital subtraction).

Image Quality. As in ordinary radiography, image quality with digital fluoroscopy is of paramount importance, since this dictates the amount of available information. Here are the factors in image quality with the dig-

Table 24.1
COMPARISON OF MANUAL AND
DIGITAL SUBTRACTION ANGIOGRAPHY

Manual	*Digital*
Advantages	
Not so subject to patient motion	Improved contrast resolution
Lower cost	Rapid process
	Programmed automatic immediate subtraction
	More flexible control of image quality
	Less radiopaque medium needed; eg, selective examination of carotid circulation by injection of aortic arch
Disadvantages	
Delayed process	Very dependent on patient motion, especially between mask and arteriograms
Limited manipulation of image quality (trial and error)	Artefacts possible by excessive contrast

ital system.

1. ***Pixel Size.*** As already noted, a matrix is specified according to the ***number of pixels*** it contains. The pixels are *not* really stored in boxes, but rather in locations in computer memory corresponding to points in the analog image. If a matrix has 8 columns and 8 rows, it is designated an 8 × 8 matrix and contains 64 pixels. A 512 × 512 matrix contains 262,144 pixels. Since these occupy the same overall space as the 64 pixels in the 8 × 8 matrix, they would have to be about 41,000 times smaller! In other words, ***the larger the matrix, the larger the number of pixels and the smaller their size,*** resulting in greater image resolution.

The ***width*** of a pixel can be easily determined by dividing the width of the image on the intensifier's ***input*** phosphor by the number of pixels in the horizontal row of the matrix:

$$\frac{\textbf{Pixel}}{\textbf{width}} = \frac{\textbf{image width on IS}}{\textbf{number of pixels in row}} = \textbf{mm}$$

Thus, with a 6-in. (152 mm) input phosphor and 512-pixels,

$$152 \text{ mm}/512 = 0.3 \text{ mm pixel width}$$

which is the theoretical limit of resolution of the system. With a 1024 × 1024 matrix, pixel width is 0.15 mm. To put this into perspective, ordinary radiography with a film-screen combination can image an object as small as 0.1 mm, nearly the same as the digital image. Pixels are actually about the *same size as the dots in a half-tone print.*

Look at pixel size this way with regard to image resolution. The larger the pixel size, the larger will be the fragments into which the image will be "broken," so the image will appear as though made up of square blocks, (see Figure 24.5) with few shades of gray

Figure 24.5. On the left is an *analog* image, a conventional photograph, in which the shades of gray (from white to black) vary continuously, since the film grains are invisible. The image on the right contains a finite number of shades of gray, the visible size of each pixel having been exaggerated to show the *digitized* version of the same image. (From Pizzutiello RJ, Jr. Cullinan JE. *Introduction to Medical Radiographic Imaging,* courtesy of Eastman Kodak Company.)

between white and black. In fact, the image becomes unrecognizable. But with a very small pixel size, the image is broken into smaller fragments which become less distinct, while the image becomes clear, with good recorded detail and resolution. In fact, it resembles a conventional radiograph. The computer must be designed to be able to handle the increased information presented to it by a large matrix, such as 1024×1024.

2. *Contrast.* Just as in ordinary radiography, contrast plays a major role in image quality. You have learned earlier that contrast scale should be neither too short or too long, although at times short scale contrast is desirable, as in arteriography. Excessive contrast may lose information on account of the absence of enough intermediate shades of gray, whereas extremely long-scale contrast may impair visibility of recorded detail. In digital fluorography, the range of pixel numbers, or numeric range, expresses the spread of signal intensities and determines the contrast scale. The numeric range increases with matrix size; in a 512×512 matrix the numeric range is 2^9 ($= 512$), this being designated a 9-bit range. Similarly, a 1024×1024 matrix has a numeric range of 2^{10}, or a 10-bit range. A 2^1 numeric range, with a 1-bit range, would have only two tones, white and black—an extreme case. By computer manipulation, one can modify the numeric range to shorten or lengthen contrast scale as needed.

3. *Noise.* As is true of any system that plays on our senses, digital fluoroscopy is afflicted by *noise.* Here, the *main source* of noise is spurious electric signals in the *video camera,* which must therefore be of high quality. These non-information-bearing signals obscure informational signals and are difficult to eliminate. The relation between the intensities of signal and noise is called the *signal-to-noise (S/N) ratio.* The higher the S/N ratio, the more the signal drowns out the noise; and the lower the S/N ratio,

the more the noise obscures the signal. Also, the higher the S/N ratio, the higher the fidelity of the system, which means that the received information corresponds closely to the transmitted information. For example, static in radio reception is noise, which is virtually eliminated by a high fidelity (hi- fi) system.

Noise has a severely detrimental effect on image quality in digital fluoroscopy. Anything that increases the signal without also increasing noise to the same degree, improves the ratio. The following factors improve the S/N ratio:

1. *Increase Signal Intensity.* High mA yields a brighter image, which results in enhanced signal intensity. Recall that mA is the most important carrier of information in radiography. Moreover, high mA reduces quantum mottle (see page 211).

2. *Reduce Scattered Radiation.* The factors in limiting scattered radiation were considered in detail under *radiographic quality;* they include beam limitation, grids, and lower range of kVp. Scattered radiation on the input phosphor eventually shows up as noise in the digital images. Thus, a decrease in scatter increases the S/N ratio.

3. *High Fidelity Video Camera.* As noted above, the video camera is the main contributor of noise to digital imaging. It is therefore imperative that the video camera have a high S/N ratio of no less than about 1000:1, an ordinary one having a ratio of about 200:1.

A method of reducing detectable noise involves a process called *smoothing.* Computer manipulation can be used to average the values of groups of adjoining pixels in a regular sequence throughout the matrix. Noise becomes less perceptible, but there is some decrease in image resolution.

Digital subtraction image quality is especially subject to the detrimental effect of noise. The mask and opacified images must have similar high fidelity. Since noise is a random process, the two digital images have a high probability of possessing different noise distributions and S/N ratios in various pixels. This again shows the need for high mA and a hi-fi video camera, as well as optional smoothing.

X-Ray Generator. Digital fluoroscopy starts with an under-the-table x-ray tube (see Figure 20.28). However, it must have a high enough rating to operate at several hundred mA, like an angiographic tube, capable of firing short bursts of x rays in rapid sequence. This requires on-and-off switching at the rate of 1 or a few per sec. The high mA produces a bright image without appreciable quantum mottle (see pages 211-212), to permit digitization yielding high-quality images during the repeated short exposures. A three-phase power supply provides the requisite high-intensity, short, repeated exposures with rapid on-off switching.

Future Prospects. A number of methods of digitizing radiography are in the process of development; they may properly be called *digital radiography.* One of these uses a closely restricted fan-shaped x-ray beam similar to that in moving-slit radiography (see pages 252-253) and in CT scanning (see pages 314-322). The beam is narrowed to a thickness of about 1 cm and an angle of about 45 degrees by two coaxial (in tandem) beam restrictors, one being located between

the x-ray tube and patient and the other on the exit side of the patient. As a result, scattered radiation is markedly reduced, thereby enhancing contrast. During radiography the fan beam moves broad-side (at a right angle to its flat side) across the anatomic part. The image receptor consists of an array of closely-spaced, tiny electronic detectors whose output signals are digitized and then processed by a computer into an analog image. Such images are stored in computer memory in digital form, to be retrieved later as needed for comparison with later studies, or for transmission over a network.

Other methods of digital radiography use a conventional x-ray beam and a special *photostimulable phosphor* as the image receptor which, when subjected to intense laser energy, emits a blue light; this acts on a photomultiplier tube whose output light is digitized and stored in a computer memory bank, later retrievable as an analog image.

It is believed that film radiography as we know it will eventually become obsolete and give way to digital radiography, with the images stored in computer memory banks in digital form. These can be retrieved for immediate or later use in analog form. The digital images lend themselves to easy manipulation for contrast enhancement, and can also be transmitted over telephone lines by means of a modem, or sent over a com-

QUESTIONS AND PROBLEMS

1. Describe, with a diagram, conventional (film) subtraction radiography. Compare with digital subtraction.
2. What are the steps in digital fluoroscopy? Why is this called "real time?"
3. Of what does a digital image consist?
4. Compare analog and digital watches.
5. Define pixel; matrix; matrix size.

6. State the two most common matrix sizes. How do their pixel sizes compare?
7. What is the relation of pixel size to image resolution?
8. State the factors in the quality of a digital image.
9. Define pixel width.
10. How do pixel width and matrix size

affect resolution?

11. Define "numeric range."

12. Define "signal/noise ratio" and state the factors that improve it.

13. What is meant by "smoothing?"

14. Why is 3-phase equipment needed in digital radiography?

15. Describe digital radiography.

Chapter 25

COMPUTED TOMOGRAPHY

Topics Covered in This Chapter

- History
- Basic Principles
- Types of Single-Plane Scanning Motion
- Detectors
- CT Numbers
- Image Reconstruction
    Pixels
    Matrix
- Spiral Computed Tomography
    Motion
    Detectors
    Interpolation
    Advantages
- Medical Applications
- Questions and Problems

CONVENTIONAL CT SCANNING

IN ORDINARY RADIOGRAPHY soft tissues are classified as having "water" density. Except for fat, soft tissues do not differ sufficiently in density to provide adequate contrast. In fact, anatomic structures must have a density difference of at least 2 percent to be shown in ordinary radiography (Hendee and Ritenour), and so we cannot differentiate between such structures as liver and muscle, nor can we see soft-tissue detail within the kidneys or brain. Conventional tomography also has limited usefulness in this regard. We can, of course, use contrast media to expand the diagnostic capabilities of routine radiography.

In 1961 a neurologist, William H. Oldendorf, suggested a method of detecting small density differences among soft tissues (ie, low contrast) by elaboration of tomography, but digital computers were not available to put his ideas into practice. Then, in 1971, Godfrey Hounsfield, a research physicist at EMI, Ltd, in Middlesex, England, installed in a London hospital a *computerized tomographic unit* which he had designed. This proved to be the first major advance in radiology since the discovery of x rays.

Computed tomography (CT) is a complex application of the law of tangents which states that in conventional tomography image quality improves with the number of image-forming rays passing tangential to an anatomic object over its entire border. However, this applies only to structures whose density differs sufficiently from their surroundings to provide adequate subject contrast. But with, modern CT each slice (ie, tomographic section) is scanned by a collimated x-ray beam in a circular path *around* the body over 360°.

A high thermal capacity x-ray tube and

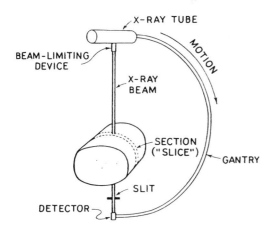

Figure 25.1. Basic arrangement of a computed tomography (CT) scanner. As the x-ray tube moves in an arc around the patient during the exposure, its narrowly collimated beam produces a tomographic section or slice. The exit or remnant radiation strikes a detector from many thousands of directions, and the resulting information is processed by a computer which reconstructs an image of the section. The tube and detector are mounted opposite each other on a rigid gantry so that the detector always remains directly opposite the tube focus. (Adapted from *The Fundamentals of Radiography*, 12th ed, Eastman Kodak Company.)

generator are mounted on a rigid circular support called a **gantry,** which rotates around the patient (see Figure 25.1). The exit beam for each position of the tube is received by one or more very small radiation detectors mounted on the gantry, directly opposite the tube. There are two kinds of detectors: one is solid, such as a sodium iodide or bismuth germanate crystal coupled to a photomultiplier tube (see pages 338-339); the other is a gas-filled (eg, xenon) ionization chamber. The total attenuation ("absorption") by structures that differ almost imperceptibly in density (ie, extremely low contrast) is measured for each position of the tube and detector. In other words, an extremely large amount of data–approaching 100,000 counts–is accumulated for each

tomographic section and then rapidly **computerized** in a number of stages to yield a tomogram. Such data processing requires a **digital computer.** The resulting tomogram reveals exquisite recorded detail of internal structures of each section, whereas with ordinary tomography they would appear as water density without visible internal structure.

We may summarize the steps in the scanning process as follows:

1. Numerous x-ray projections are made of each of a series of successive contiguous tomographic sections at intervals over an arc of 180° or 360°, depending on equipment design.

2. Images are received by appropriate detectors in the form of "counts."

3. The tremendous amount of accumulated data is reconstructed by the computer to yield a tomogram of internal structures with superb recorded detail, impossible with conventional tomography.

4. The computer stores the images in digital form, and retrieves them in analog form as needed. It also allows manipulation of the density and contrast of the image.

5. Ultimately, the *computer* controls the entire scanning process, selecting the various scanning factors, starting the scan, activating the x-ray tube and detectors during actual scanning, accrual of data, and monitoring the overall operation of the system.

Types of Scanning Motion. CT scanners have undergone rapid evolution in design over four "generations" or stages, in a period of about ten years. This description will be limited to two major types of motion.

1. *Rotation and Translation.* The EMI scanner originally designed by Hounsfield combined two kinds of motion: rotation around the body, alternating with translation (ie, rectilinear motion) across it, as shown in Figure 25.2; in other words, the x-ray tube moves over an arc of a specified number of degrees (part of a circle) and then moves in

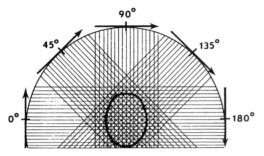

Figure 25.2. Simplified diagram of the scan motion of the first-generation EMI CT scanner, which used rotation-translation motion. At each designated angle, at one-degree intervals, a pencil beam of x rays enters the patient and the exit or remnant radiation is picked up by a pair of collimated detectors arranged side-by-side directly opposite the x-ray tube, on the other side of the patient. Scanning x-ray exposure occurs only during the linear motion of the tube as indicated by the arrows. After each scan, the gantry rotates one degree without x-ray exposure, and exposure is again made during linear motion in the new position. With 160 measurements of transmitted radiation, multiplied by 180 degrees of rotation, the total number of measurements is 28,800. These are assigned CT numbers and processed by the computer to a final reconstructed image. In the *second-generation scanner*, the pencil beam was replaced by a fan beam and multiple detectors to speed up scanning time.

a straight line. This sequence is repeated over the entire arc of 180° (one-half circle), with the x-ray tube activated **only during translation.** The closely collimated x-ray tube and detectors are maintained directly opposite each other at all times by the rigid gantry, which is free to rotate. A pair of solid detectors is mounted side by side so that with each rotation-translation of the tube and detectors, two tomographic sections are obtained simultaneously. The original EMI scanner used a narrow "pencil" x-ray beam and required about 5 minutes for each section pair, or 25 minutes for a whole brain

scan. This scanner used an oil-immersed 120-kV x-ray tube with a stationary anode and a large focal spot.

The **second generation scanner** still used rotate-translate motion, but the number of detectors was increased to reduce scan time. As before, exposures were made only during translation, so the detectors remained in a flat array.

2. **Rotation Only.** In the **third generation scanner,** a **fan-shaped beam** was introduced with a curved array of detectors that matched the exit width of the beam (see Figure 25.3). The x-ray tube and opposed

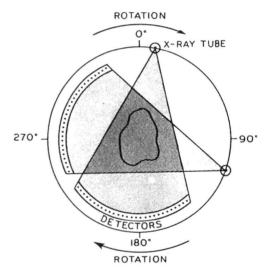

Figure 25.3. Third-generation CT rotational scanner with a fan beam and an array of thirty detectors. Mounted on a gantry, the tube and detectors rotate together around the patient. Note that the detectors always remain directly opposite the x-ray tube focus. Readings of the transmitted radiation (exit or remnant) are made continuously and processed by the computer during rotation of the gantry through a complete circle (360°). The x-ray beam is restricted by a beam-limiting device to a thin fan shape, and the detectors are also well collimated. This type of scanner markedly shortened scanning time.

detectors rotated together around the patient, and there was no translation of the tube. Because of slowness of the first two types of scanners, and the presence of artifacts, improvement soon occurred.

A full-circle array of stationary detectors marks *the fourth generation scanner,* together with a fan-shaped beam (see Figure 25.4). This brings much greater speed to the scanning process: the tube and detectors can now be moved rapidly all around the patient. Exposure is interrupted between the rotations while the patient is advanced within the gantry. Scan time per section has been reduced to as low as one second. Although the fan beam increases the amount of scattered radiation in comparison with the pencil beam, each detector has its own collimator (as does the x-ray tube itself) to minimize the amount of scatter reaching the detector. These scanners use rotating anode tubes with 0.6 or 1.0 mm focal spots. They are programmed to deliver x rays intermittently in 300 bursts or pulses for each tomographic

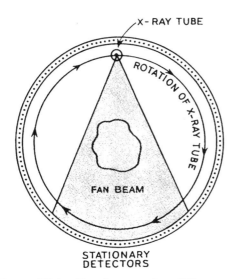

Figure 25.4. Fourth-generation CT scanner. The x-ray tube rotates full circle (360°) while about 600 detectors remain in a *stationary* array. Note that the beam is fan shaped.

section. With each pulse lasting about 3 ms (0.003 s), the tube is only activated over a total period of 1 s per section. This prolongs tube life, allows the use of a small focal spot for improved recorded detail, and shortens scan time by permitting a larger total exposure per scan.

The detectors in these later generation scanners are of two types. One is a bismuth germanate crystal coupled to a photomultiplier tube, which counts the x-ray photons reaching it in much the same way as a radionuclide scintillation detector. The bismuth germanate crystal has greater efficiency in absorbing photons and also has less afterglow than a sodium iodide crystal. The other type of detector is an ionization chamber filled with the heavy inert gas, xenon, under pressure; the gas molecules are ionized in proportion to the intensity of the detected x rays (proportional region, Figure 26.11, p. 337). Gas-filled detectors have virtually no afterglow and can be miniaturized, but they are slightly less efficient and stable than a crystal detector. The exit radiation from each tomographic section is absorbed by the detector and the resulting pulses in the photomultiplier tube counted automatically.

The ability of CT to detect a lesion depends on *subject contrast.* Further improvement is achieved by intravenous injection of iodinated contrast media or by ingestion of opaque media, a process known as *contrast enhancement.* Contrast may be enhanced even further by delayed scanning, as with hematomas.

Image Reconstruction. This is a complex mathematical process carried out by the computer in which the image is first reconstructed as a digital image (ie, expressed in attenuation or CT numbers) and then changed to a conventional gray-scale image. The image of the section is broken up into a number of tiny picture elements called *pixels,* which are projections of solid cores of tis-

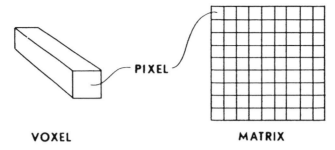

VOXEL   MATRIX

Figure 25.5. Explanation of imaging terms. A *voxel* is a volume element–a geometrically defined portion of the tissue section through which the beam has passed. The projection of a voxel, corresponding to its smallest side, is called a *pixel* or picture element in the *matrix,* which represents the tomographic image of the tomographic section. Pixels are actually tiny (ie, dot-sized).

sue known as **voxels** (volume elements) as shown in Figure 25.5; these are actually dot-sized. As we have just explained, the computer counts the light flashes in solid detectors, or ions in ionization detectors, as the radiation that has traversed each voxel reaches the corresponding detector. The computer instantaneously assigns a **CT number** to each pixel, depending on the linear attenuation coefficient (a measure of x-ray attenuation) in the corresponding voxel. CT numbers are not absolute values but are relative to the attenuation by water, which is assigned CT number 0. CT numbers range from +1000 for bone, to –1000 for gas, and are often called Hounsfield numbers in honor of the CT inventor. Thus, the computer yields a digital (numerical) image, that is, one in which the attenuation by tissue in the tomographic section is represented by numbers. Figure 25.6 shows how a single projection (tube and detector in a particular position) of a rectangular object might appear in terms of CT numbers. Note that the image pixels consist of higher CT numbers than the neighboring ones because we have chosen an object that is denser than its surroundings. But if the object were less dense than its background, its pixels would be represented by lower CT numbers.

We have just seen how a single projection

appears in CT numbers. Reconstruction of the whole image of the section from **multiple projections** can be accomplished in several ways that have become extremely complex, so we shall not try to describe them. The simplest method, **back-projection,** is no longer used in modern CT scanners, but it does provide a relatively simple concept of the principle. Image reconstruction requires, first of all, the handling of an immense amount of data by the computer.

In Figure 25.6, the CT number for each square pixel is proportional to the sum of the object densities (or attenuation) in the voxel through which the radiation has passed. In back-projection, a "magic square" consisting of pixels in, for example, nine rows across and nine columns down (81 pixels altogether) constitutes the image matrix, which includes a great number of pixels–eg, 80 × 80 = 6400–for satisfactory recorded detail. In Figure 25.6, the row of CT numbers from one projection is repeated in each succeeding row. Now let us assume that the gantry has been rotated 90° and another exposure made–the second matrix resulting from this back projection is shown on the left. Now the numbers in the corresponding squares (ie, corresponding pixels) are added to give the summated **matrix** shown on the far right of the figure, a procedure that is repeated for

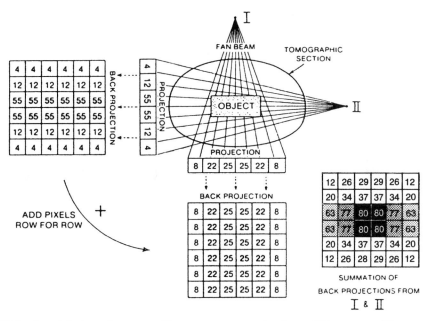

Figure 25.6. Simplified explanation of image reconstruction by a CT scanner. The corresponding pixel CT numbers (ie, numbers assigned by the computer to the radiation attenuation in each voxel) from each back projection are added as shown. The larger the number of projections and back projections resulting from rotation of the gantry around the volume of interest, the better will be the recorded detail in the reconstructed image. This is consistent with the principle in ordinary radiography that recorded detail improves with increased information, provided that noise does not increase as well. When the computer changes the CT numbers to a *gray scale* (analog image), we then have a recognizable tomogram.

each position of the tube as the gantry rotates. As more such projections are added to the matrix, recorded detail improves. The final matrix summating the four projections reveals the digital image of the rectangular solid. It should be mentioned that image quality is further improved by programming the computer to make a number of important corrections.

Finally, the computer converts the digital image to a gray-scale image to resemble a conventional tomogram, except that it has incomparable recorded detail. Figure 25.6 shows the gray-scale image of the tomographic section obtained by the four projections. In practice, the computer controls allow the operator to increase or decrease image contrast at will.

Figure 25.7 shows an axial projection of the brain.

SPIRAL (HELICAL) CT SCANNING

A newer method of computerized tomography scanning, ***spiral*** or ***helical*** CT, has become increasingly available in radiology departments because it has a number of advantages over the conventional fourth generation scanner. The equipment had to

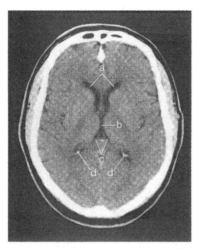

Figure 25.7. CT "slice" of the brain at one level, with labels indicating these cerebrospinal (CSF)-filled chambers:

    a. Anterior horns of lateral ventricles.
    b. Third ventricle.
    c. Ambient cistern.
    d. Calcified choroid plexus in posterior horns of lateral ventricles.

Irregular dark lines over surface of brain represent CSF in cortical sulci. The bodies of the lateral ventricles would appear at higher slice levels.

be redesigned to permit *continuous rotation* of the gantry without tangling the cables. This required the use of *slip-ring technology* in which an outer metal ring remains stationary while an inner rotating ring, holding the x-ray tube, has brushes that remain in electrical contact with the inside of the outer ring. As a result, the inner ring is free to rotate during its circular motion. Recall that with conventional CT, after each rotation of the gantry it returns to its initial position while the patient is advanced cephalad.

Principle of Spiral CT. The x-ray tube *continually* circles the patient while the couch moves the patient cephalad at a constant rate through the gantry. Thus, the motion of the tube generates a virtual *spiral* or *helix* around the patient during exposure, as

shown in Figure 25.8. In conventional CT scanning, the tube encircles the stationary patient once, following which the couch carries the patient cephalad over a preselected distance. The tube then describes another circle, and so on, "resting" in the intervals. Moreover, in conventional CT, the x-ray tube is energized intermittently for about a second at a time. Because motion is continuous with spiral CT, the tube may be energized for as much as 30 sec at a time, which demands a very high thermal anode capacity—at least 5 MHU (million heat units); by comparison, an x-ray tube for conventional CT scanning dissipates about 1 MHU. A typical x-ray tube (Varian) for spiral CT has a 5-cm thick anode, 15 cm in diameter (2 in. and 6 in., respectively). Tube life is rated for about 50,000 exposures, much like that of an ordinary CT tube. If the beam with the larger tube is pulsed (regularly intermittent) it can tolerate 1000 mA, but for continuous exposure, the maximum is 50 mA.

The detectors are solid state and stationary, with the x-ray beam rotating inside the detector array. These detectors belong to the class of scintillation photomultiplier tubes that have been miniaturized and modified as *crystal photodiodes.* At present, cadmium tungstate is the preferred crystal because of its high detection efficiency of about 80 percent.

Image Reconstruction. Since spiral scanning does not generate individual "slices," interpolation has to be done between successive spiral-generated images by a special computer algorithm. *Interpolation* means the estimate or calculation of intermediate values in a series according to the rules in an *algorithm*–a numerical-logical process that solves a difficult problem; in this case, to yield image slices in exquisite detail.

It has been found that the clearest images result from interpolation between 180° of successive spiral rotations; a full rotation is 360° (like a circle), 180° is one-half a rota-

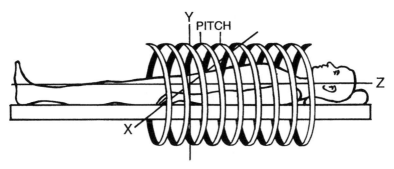

Figure 25.8. Spiral (or helical) CT scanner. The patient is advanced cephalad *during* gantry rotation.

tion. High resolution is achieved along the Z-axis by appropriate image reconstruction. Another option is reconstruction in transverse *(XY)* planes; obviously, this should occur within a slice because if it should occur between slices, a small lesion could be missed. It is also possible to overlap successive images to increase the chance of revealing a small lesion.

The higher speed of spiral scanning allows imaging of the patient within a single breath (about 30 s) while reducing the effect of involuntary motion. Should a patient be unable to hold a breath long enough, a series of shorter skip-scans can be obtained.

Finally, spiral CT allows *three-dimensional image reconstruction,* a real asset in radiotherapy planning.

The following list summarizes the advantages of spiral over conventional CT imaging:

1. Shorter scanning time.
2. Reconstruction at selected intervals along the Z-axis, and overlapping of successive images to render the scan more sensitive to the detection of small lesions.
3. Fewer motion artifacts.
4. Capability of three-dimensional reconstruction, especially valuable in radiotherapy planning.
5. Images obtainable at peak *angiograph-ic* and *nephrotomographic* opacification.

Disadvantages are minor and include longer acquisition time and image processing, and the need for very heavy-duty x-ray tubes that are more costly than conventional CT tubes. Still, spiral CT is more efficient than conventional CT in its ability to detect smaller lesions, such as hepatic metastases, among others.

Medical Applications of CT Scanning. Although CT scanning was first applied to the examination of the brain, it has also revolutionized the diagnosis of lesions in virtually all parts of the body; to name a few: head and neck, chest, abdomen, and vertebral column. In the brain, normal structures are readily depicted–gray and white matter, ventricles, and subarachnoid and subdural spaces. Lesions in the brain are also well shown–tumors, infarcts, abscesses, hematomas, edema, hydrocephalus, cerebral atrophy, and calcifications. Contrast enhancement brings out structural details that would otherwise remain invisible. While there is still an important place for radionuclide imaging and conventional angiography, CT has become a major noninvasive or slightly invasive diagnostic procedure. In fact it can demonstrate lesions by guided needle biopsy that might otherwise be undiagnosable, short of major surgery.

Lesions in the abdomen and chest that are particularly amenable to detection and evaluation by CT scanning include tumors of various kinds. In the abdomen, liver, biliary tree, and kidneys can be studied by CT, but ultrasound may be superior in some instances. An important application of CT in the abdomen at present is in the detection of pancreatic tumors and enlarged retroperitoneal lymph nodes. In fact, CT has assumed a prominent role in staging Hodgkin's and non-Hodgkin's lymphoma, as well as other malignant tumors, by virtue of its ability to discover enlarged lymph nodes.

QUESTIONS AND PROBLEMS

1. Describe the principle of computed tomography (CT).
2. What does CT accomplish that is impossible with linear tomography?
3. Summarize the steps in CT scanning.
4. What are the two major types of rotation? Which one is used today?
5. Define pencil beam; fan beam.
6. What are the two main types of detectors?
7. Define pixel; voxel.
8. How is CT (Hounsfield) number derived?
9. What is image reconstruction, and how is it achieved?
10. Is the CT process analog or digital? Explain.
11. State the principle of spiral (helical) scanning.
12. How does the x-ray tube in spiral scanning differ from the one in ordinary CT, and why?
13. Define "interpolation" and state why it is required in spiral scanning.
14. Define "algorithm."
15. List five advantages of spiral over conventional CT scanning.
16. What are the drawbacks in spiral scanning?

Chapter 26

RADIOACTIVITY AND DIAGNOSTIC NUCLEAR MEDICINE

Topics Covered in This Chapter

- Introduction
- Radioactivity
- Natural
  Radium
    Types of Decay Products
    Decay Constant
    Half-Life
    Average Life
  Radon
- Nuclear Reactor
- Artificial Radioactivity
  Isotopes
  Nuclear Transformations

  Properties of Radionuclides
  Radioactive Decay
  Application in Diagnosis
  Instrumentation
  Sources of Error
    Geometric Factors
    Counting Errors
    Important Medical Radionuclides
- Gamma Camera
  Collimators
    Available Procedures
    Image Quality
- Questions and Problems

Introduction

ONE OF THE OUTSTANDING ACHIEVEMENTS of the late nineteenth century was the discovery that certain elements had the unique property of giving off penetrating, ionizing radiation. In 1896, Henri Becquerel found that the heavy metal *uranium* emitted rays that passed through paper and darkened a photographic plate in a darkened room. Marie Curie and her husband Pierre then began looking for other elements that might behave in a similar way; in 1898, after two years of extraordinary hardship and personal sacrifice, they announced their discovery of *polonium* and *radium,* elements that also emitted radiation. Not only did this discovery open new therapeutic opportunities in medicine, but it also expanded immeasurably our knowledge about the structure of matter. The term *radioactivity* was applied by Marie Curie to the special behavior of these elements.

NATURAL RADIOACTIVITY

Unstable Atoms

The accepted designation of nuclear structure is *nuclide, defined as a nuclear species with a particular number of protons and neutrons, and having a particular energy state.* Radioactive nuclides, known also as *radionuclides,* are exceptional in that they have *unstable nuclei,* which means that their nuclear structure is inherently changeable. In the process of undergoing such spontaneous change, called *radioactive decay* or *disintegration,* they give off *ionizing radiation.*

Why are certain nuclides naturally radioactive? Nuclear particles exist in energy levels, analogous to those of orbital electrons that occupy various energy levels outside the nucleus. Transition of particles within the nucleus to lower energy levels in radionuclides are also accompanied by the release of radiation. In naturally occurring nuclides with a high atomic number (above 81), nuclear energy is large enough to allow random escape of nuclear particles, usually accompanied by electromagnetic radiation (gamma rays). Such unstable nuclides, as they give off radiation, change into other nuclides. It is uncertain which particular nucleus of a sample of nuclide will decay at any particular instant, but a *constant fraction of those present will decay in a given time interval* (eg, per sec). This fraction, called the *decay constant* (other terms: *deexcitation, disintegration,* or *transformation constant), is the same for all samples of a given radionuclide but is different for different radionuclides.* The decay constant of a radionuclide cannot be changed without extreme difficulty.

Radioactivity, then, may be defined as the ability of certain nuclides–radionuclides–to (1) emit ionizing, penetrating radiation while (2) undergoing spontaneous decay at a constant, uncontrollable rate.

There are two kinds of radioactivity–natural and artificial–but, as we shall see, they both obey the same laws. *Natural radioactivity* is a property of naturally occurring radionuclides such as radium. *Artificial radioactivity* pertains to artificial radionuclides such as radiocobalt; these are nuclides that have been made radioactive by irradiation of the stable form with subatomic particles (neutrons, deuterons, etc) in special high energy devices such as the nuclear reactor or cyclotron.

The *artificial radionuclides* make up an impressive array of material that has important application in biologic research, clinical medicine, and industry.

Radioactive Series

The naturally occurring radionuclides can be grouped in three series or families, based on their parent elements: (1) *thorium,* (2) *uranium,* and (3) *actinium.* Each of these gives rise, by spontaneous decay, to a series of nuclides characteristic of that series. The chain of breakdown of the successive nuclides always occurs in exactly the same way and at a fixed rate. Some daughter nuclides, such as thorium A (^{216}Po), decay at extremely rapid rates and are therefore difficult to isolate; others, such as radium, in the uranium series, decay very slowly and can more readily be obtained in pure form. Radium formerly held an important place in the treatment of some types of cancer. Radium and its offspring comprise the last descendants of the uranium series.

RADIUM

Introduction

Although radium (Ra) has been virtually abandoned in radiotherapy and replaced by artificial radionuclides such as cesium 137 and iridium 192, it serves as a useful model for the behavior of radionuclides in general. Therefore, it will be discussed briefly as an introduction to the subject of radioactivity and radioactive substances.

Properties

Radium, a heavy metal having a silvery-white appearance when pure, behaves chemically like stable barium and calcium. It belongs to the uranium radioactive series and has atomic number 88 and mass number 226. Because its radioactivity persists unchanged when combined with other elements, it is available only in the form of one of its salts, usually *radium chloride* ($RaCl_2$), which is far more easily prepared than the pure metal.

Types of Radiation

There are three different kinds of radiation emitted by radionuclides: (1) alpha particles, (2) beta particles, and (3) gamma rays. A given radionuclide may emit all of these radiations or only one or two of them. Various members of the radium series, as will be shown below, give off one or more of these radiations. Alpha and beta particles ionize matter directly, whereas gamma rays do so indirectly, like x rays.

1. **Alpha particles** are simply *helium nuclei* (ie, helium ions) consisting of two protons and two neutrons. Since this gives them atomic number 2 and mass number 4, they can be represented by the symbol $_2^4He$. Alpha particles, having a large mass and two positive charges, strongly ionize the atoms of matter through which they pass. Upon accepting two electrons, an alpha particle becomes a neutral helium atom:

alpha particle + 2 electrons → helium atom
$$He^{++} + 2e^- \rightarrow He$$

Alpha particles leave the nucleus at high speeds, 1.5×10^7 to 3.0×10^7 meters per sec (9,000 to 18,000 miles per sec) but do not penetrate matter for any appreciable depth because of their strong ionizing ability; in fact, they are absorbed by an ordinary sheet of paper.

2. **Beta Particles,** often called beta rays, consist of high speed *electrons* represented by the symbol e^- or β^-. Ejected from the nucleus at speeds approaching that of light rays, they are able to penetrate matter to a greater depth than alpha particles, their maximum range in tissue being 1.5 cm. In fact, 0.5 mm of platinum or 1 mm of lead is required to stop these beta particles. They have been used in therapy, mainly for superficial skin lesions in small areas. Beta particles are less strongly ionizing than alpha particles.

3. **Gamma rays** are high energy electromagnetic waves (photons). For example, the gamma rays of radium have an average energy of 0.83 MeV, equivalent to 1.5 million volt (MV) x rays. They travel with the speed of light and are much more penetrating than alpha or beta radiation. Gamma rays are *physically identical to x rays,* the difference being that gamma rays originate spontaneously within *nuclei,* whereas x rays arise from electronic interactions with or away from nuclei. Since gamma rays are electromagnetic waves (and photons), they are not deflected by a magnetic field. They ionize matter in the same way as x rays by first releasing fast electrons.

When an atom of radium decays, it does

	URANIUM→	RADIUM→	RADON→	RADIUM A→	RADIUM B→	RADIUM C	LEAD
				POLONIUM	LEAD	BISMUTH	
Atomic No.	92	88	86	84	82	83	82
Mass No.	238	226	222	218	214	214	206
Half Life	4.5×10⁹yr	1622 yr	3.8 da	3 min	26.8 min	19.7 min	stable

Figure 26.1. Successive decay products of radium, shown from left to right. These make up the radium series which is the last part of the uranium series. Data on radium D, E, and F have been omitted.

not disappear but undergoes **radioactive transformation** to a different radionuclide. Let us "look into" a radium atom and see what happens during this process. The atom ejects an alpha particle, which carries a charge of +2 because of its two protons. Since the atomic number of radium is 88, that of the daughter atom **radon** remaining after the emission of an alpha particle is 88 − 2 = 86. The mass number of radium is 226 and that of the ejected alpha particle is 4, so the mass number of radon is 226 − 4 = 222.

1 radium atom → 1 radon atom + 1 alpha particle

$$^{226}_{88}\text{Ra} \rightarrow {}^{222}_{86}\text{Rn} + {}^{4}_{2}\text{He}$$

Note that the sum of the mass numbers (superscripts) of the products radon and alpha particle equals the mass number of radium. The same holds for the atomic numbers (subscripts), so the equation is balanced.

The radon atom is radioactive, decaying to a different radionuclide, which in turn decays to still another radionuclide, and so on. This chain of decay goes on until a stable atom is left (in this case a metal, **lead**). Figure 26.1, showing the radium series, is included mainly to indicate pertinent data and not to tax your memory.

Radioactive Decay

As already indicated, radionuclides such as radium decay, emitting one or more types of radiation while changing to another nuclide. The radioactive **decay scheme** is a diagrammatic method of representing the process (see Figure 26.2). The parent

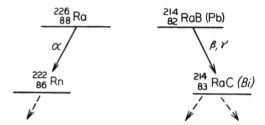

Figure 26.2. Decay schemes of radium 226 and radium B (lead 214). Only the first decay product is shown in each instance. By convention, when the product nuclide has a lower atomic number than the parent, the arrow points downward and to the left, as is the case when radium decays to radon. When the product nuclide has a larger atomic number than the parent, the arrow points downward and to the right, as occurs with the decay of radium B (lead 214) to radium C (bismuth 214). Note the two arrows below radium C–this is an example of branching decay by alternate routes.

radionuclide radium (Ra) should be drawn at the top. An arrow points downward and to the *left*, because the daughter nuclide radon (Rn) has a smaller atomic number than its parent. Each nuclide symbol has a superscript and a subscript indicating its mass number and atomic number, respectively. Finally, the type of radiation should be indicated by its symbol near the arrow, in this case by an α for the alpha particle.

In the case of *beta decay*, the daughter nuclide has a larger atomic number than the parent, so the arrow points downward and to the *right*. With *gamma-ray emission* the atomic number of the daughter nuclide remains unchanged, so the arrow would point *vertically downward*.

Decay Constant

Radioactive transformation occurs as a random process, that is, we cannot predict which particular atom will undergo nuclear decay at any instant. Starting with a given quantity of radium (or other radionuclide), the activity (number of disintegrations per second) decreases with time, *but the rate at which the activity decreases remains constant*. There are two ways of specifying the decay rate. One is the *decay constant* which is derived by calculus; to find the amount of activity remaining after a particular time interval requires the use of a special (exponential) equation.

The other way of indicating decay rate is *fractional decay,* that is, the fractional decrease in activity per given time interval. Thus if one gram of a radionuclide decreases 10 percent in one day, 0.9 g remains. At the end of the second day the 0.9 g will also decrease by 10 percent, leaving 90 percent of 0.9 g or 0.81 g, etc.

The decay constant is *not* the same as fractional decay unless fractional decay happens to be very small, about 0.2 or less (Pfalzner). Both are constant and unique for any particular radionuclide (see page 324). For example, the decay constant λ (Greek lambda) is 4.27×10^{-4} per yr for Ra, and 0.181 per day for Rn. But note that fractional decay for Rn is 16.6 percent per day, but its decay con-

stant 0.181 per day can be used only in the equation for exponential decay.

Half-Life

The caption *half-life* appears in Figure 26.1. This is a term which *denotes the length of time required for one-half the initial amount of a radionuclide to decay.* For example, suppose we have 1 gram of radium at the start. Figure 26.1 shows its half-life to be 1622 years (half-lives of artificial radionuclides used in medicine are much shorter). Thus, by the end of 1622 years the 1 g of Ra will have slowly decayed to 0.5 g. In another 1622 years it will continue to have decayed to one-half; that is, it will be one-fourth the original amount, or 0.25 g. Notice that the half-life is constant for any given radionuclide, and that the half-lives of different radionuclides show wide differences. Thus, the half-life of uranium is 4.5 billion years, while that of radium A is three minutes. Half-life is really a *statistical* concept indicating the time it takes for radioactive decay of one-half the atoms on the basis of pure chance.

The decay constant λ of Ra (or of any radionuclide) can be derived from its half-life T by the following equation:

$$\lambda = 0.693/T$$

Substituting 1622 yr, the half-life of Ra,

$$T = 0.693/1622 = 4.27 \times 10^{-4}/yr$$

This is the value of Ra half-life quoted above.

Average Life

There is a simple way of computing the probable survival time or life expectancy of an individual atom, similar to the determination of life expectancy by life insurance companies. This, the average life T_a, is simply the reciprocal of the decay constant

$$T_a = 1/\lambda$$

For example, the decay constant of Rn is 0.181/day, so the average life is $1/0.181 = 5.5$ days.

Radon

When an atom of radium decays and emits an alpha particle, an atom of radon remains (see page 326). Radon is a radionuclide that exists as a colorless heavy gas. Figure 26.3 shows the decay curve of radon, as well as the derivation of its half-life. As already noted, fractional decay is constant with time.

Because of its convenient half-life of 3.8 days and the availability of gamma rays from its daughter radionuclides, radon was at one time used in implant therapy, encapsulated in gold "seeds." However, because of problems with radiation safety during manufac-

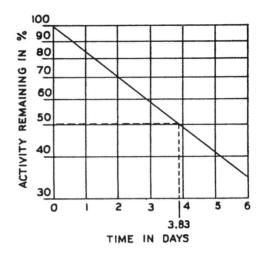

Figure 26.3. Decay curve of radon, semilogarithmic plot. The half-life is 3.83 days.

ture and possible leakage of the seeds, radon is no longer available in the United States, having been replaced by artificial radionuclides such as gold 198, iridium 192, and tantalum 182.

RADIOACTIVE EQUILIBRIUM*

A radionuclide decay series is said to be in *radioactive equilibrium* when, in a sealed source, the rate of appearance of each daughter equals its rate of decay; that is, the amount being formed per second equals the amount decaying per second. However, a distinction is made between *secular* and *transient equilibrium.*

In *secular equilibrium* the parent nuclide radium has a much longer half-life than any of its daughters. Because of its long half-life, sealed radium furnishes a virtually inexhaustible supply of daughter nuclides, which accumulate until secular equilibrium has been reached (after about thirty days); now

the daughters are present in constant amounts.

Radon is an example of a radionuclide which reaches *transient equilibrium* with its descendants because it has a relatively *short half-life* of 3.83 days. Upon reaching transient equilibrium at about five hours after being sealed, radon and its decay series continue to undergo rapid decay, with the same half-life as the parent radon. Therefore, the gamma-ray output decreases appreciably in a period of days, whereas the gamma-ray output of a radium source remains constant because of its extremely long half-life.

ARTIFICIAL RADIOACTIVITY

In chapter 4 we learned that matter is made up of extremely small particles called **atoms.** The concept of atomic structure has become extraordinarily complex, but we now review it in much simplified form—a central core or **nucleus** containing almost the entire mass of the atom and, surrounding the nucleus, the **orbits** or paths in which the negative **electrons** are in continual motion. The nucleus has two main constituents, **protons** and **neutrons,** collectively referred to as **nucleons.** A proton is a positively charged particle whose mass is nearly 2000 times that of an electron. A neutron has nearly the same mass as a proton but carries no charge.

All matter is made up of one or more simple entities called **elements.** The atoms of any particular element have a characteristic and distinct number of positive charges or **protons** in the nucleus and an equal number of negative charges or **electrons** in the orbits around the nucleus. The number of nuclear protons is called the **atomic number.** Notice that the atomic number is specific for a given element; all atoms of any one element have the same atomic number. Atoms of different elements have different atomic numbers.

Since the nucleons are responsible for the mass of an atom, the **mass number** of a particular atom has been defined as the total number of nuclear protons **and** neutrons. For example, hydrogen with one nuclear proton has a mass number of 1, and helium with two protons and two neutrons has a mass number of 4.

Isotopes

As stated before, all the atoms in a given sample of a particular element have the same atomic number, that is, the same number of nuclear protons (their chemical behavior is also identical). However, not all the atoms in this sample necessarily have the same *mass* number–total number of protons **and** neutrons. Since the number of protons is the same, the difference in mass number must be due to a difference in the number of **neutrons.** Such atoms whose nuclei contain the same number of protons, but different numbers of neutrons, are called **isotopes.**

The following equations show how a radionuclide differs from its isotope(s) only in the number of nuclear neutrons:

$$\begin{aligned} \text{nuclide} &= x \text{ protons} + y \text{ neutrons} \\ -\ (\text{isotope} &= x \text{ protons} + z \text{ neutrons}) \\ \hline \text{nuclide} - \text{isotope} &= y - z \text{ neutrons} \end{aligned}$$

Note that there is not an unlimited variety of isotopes of any particular element; as they occur in nature, there may be only two, or three, or several. However, the numerical ratio of the different isotopes is nearly constant for each element.

Since all the isotopes of an element have the same atomic number, they have the same chemical properties. The gas hydrogen, for example, has three isotopes (see Figure 4.5): ordinary hydrogen or **protium** with mass number 1, **deuterium** (heavy hydrogen) with mass number 2, and **tritium** with mass number 3. As would be anticipated, these three forms of hydrogen have the same atomic number 1 and identical chemical properties.

The term **nuclide** refers to a species of atom having in its nucleus a particular number of protons and neutrons. Accordingly, we can say there are three distinct hydrogen nuclides.

Artificial Radionuclides

Many artificial nuclides have been produced, a number of which happen to be radioactive and decay to entirely different elements while emitting ionizing radiation, as described above. Radioactive isotopes have the same *chemical* properties as their stable counterparts, differing only in their mass number (number of nuclear neutrons) and radioactive nature. We prefer to speak of radioactive nuclides or *radionuclides* rather than radioisotopes, since they are nuclides that happen to be radioactive, although these terms are often used interchangeably.

Artificial radionuclides appear during irradiation of stable nuclides by subatomic particles such as neutrons or deuterons in an atomic reactor or cyclotron, respectively. A neutron is an uncharged particle having mass number 1. A deuteron is a heavy hydrogen nucleus having a single positive charge and mass number 2. The irradiated nucleus captures a neutron or deuteron, becoming unstable, that is, radioactive.

When the daughter element differs from the parent, the process is called *transmutation.* An example of the transmutation of an ordinary, stable element to a different radioactive element will make this process more easily understandable. Upon irradiation by neutrons, ordinary sulfur changes (transmutes) to radioactive phosphorus:

sulfur + neutron → phosphorus + hydrogen

$$_{16}^{32}\text{S} + _{0}^{1}\text{n} \rightarrow _{15}^{32}\text{P} + _{1}^{1}\text{H}$$
(stable)   *(radioactive)*

In each term of the equation, the lower number or *subscript* represents the atomic number, and the upper number or *superscript* represents the mass number. The sum of the superscripts on one side of the equation equals that on the other side. The same holds true for the subscripts. Knowing that the atomic number of phosphorus is always 15, we can express radioactive phosphorus as ^{32}P. Since it has a specific number of protons (15) and neutrons (17) in its nucleus and therefore a particular species of atom, it is an example of a *radionuclide* or *radioactive*

nuclide. ^{32}P is unstable because its nucleus contains a surplus neutron (ordinary nonradioactive phosphorus is ^{31}P) and it therefore decays to ordinary sulfur, emitting a beta particle (negative electron) in the process:

$$_{15}^{32}\text{P} \rightarrow _{16}^{32}\text{S} + _{-1}^{0}\text{e} + \bar{v}$$
                            beta particle   *antineutrino*

Neutrinos and antineutrinos are uncharged low-mass highly-penetrating subatomic particles. At the present time, there is no way of altering the type of radiation emitted by a given radionuclide, natural or artificial.

Another method of inducing artificial radioactivity is to irradiate stable atoms with deuterons (heavy hydrogen nuclei) in a cyclotron.

Nuclear Reactor

As mentioned above, radionuclides can be produced by exposing stable nuclides to a flow of neutrons–*neutron flux.* In fact, this is how artificial radioactivity is often induced both for medical and industrial use. One type of nuclear reactor (see Figure 26.4) depends on the fission (splitting) of nuclei of uranium 235 (^{235}U), an isotope of ordinary

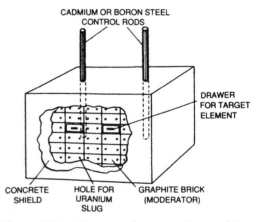

Figure 26.4. Cutaway diagram of a graphite-moderated fission reactor.

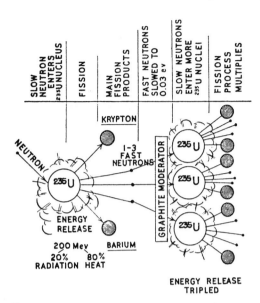

Figure 26.5. Fission of uranium 235 by slow (thermal) neutrons, resulting in a nuclear chain reaction. This differs from radioactive decay in the production of large nuclear fragments, as well as radiation, during fission.

uranium (^{237}U), by extremely slow, environmental neutrons (see Figure 26.5). These neutrons have a very small average kinetic energy, about the same as that of the molecules of air at the prevailing temperature and are therefore called *thermal neutrons.* When a thermal neutron is captured by a ^{235}U nucleus, the nucleus splits into fragments of various sizes while emitting one to three neutrons plus 200 million electron volts (MeV) of energy:

$$^{235}\text{U} + {}^1_0\text{n} \rightarrow \frac{\text{sum of masses of}}{\text{nuclear fragments}} + 1 \text{ to } 3 \ {}^1_0\text{n} + 200 \text{ MeV}$$

The splitting of the ^{235}U atom in this manner is termed *fission. Fast neutrons* emitted during fission must be slowed to thermal energy to facilitate their capture by other ^{235}U atoms which, in turn, undergo fission. In the first generation reactors graphite blocks served as moderators to slow the fast neutrons to ther-

mal energy, but later, graphite was replaced by heavy water, that is, water in which the hydrogen is in the form of *deuterium.* Both moderators are efficient in that they have a small capability of absorbing neutrons. Since a great deal of heat is generated during reactor operation, coolants must be used, ranging from air, to water, to heavy water.

The liberation of one to three neutrons each time a neutron undergoes capture, and the rapid multiplication of this process, constitutes a *chain reaction,* which is accompanied by virtually an instantaneous release of a fantastic amount of energy.

If a piece of ^{235}U is large enough to permit more neutrons to be emitted than can escape from the surface, the mass will explode. Thus, for the reactor to operate properly, the mass of ^{235}U must be *critical,* that is, just large enough for the fission process to continue at a steady rate, but still not so large as to go out of control or *go supercritical.* Fission rate can be varied by the use of boron or cadmium *control rods;* these have a strong affinity for neutrons, and adjustment of the depth to which the rods are inserted into the reactor regulates the size of the *neutron flux* (flow). Normally, the reactor's operation is adjusted by the cadmium control rods to maintain the fission process at the desired level without letting the reactor go out of control. Note that the "fuel" of the reactor is not pure ^{235}U, but rather natural uranium, of which only about $^1/_{140}$ is ^{235}U, enriched by the addition of ^{235}U.

Since the neutron carries no charge, it is not repelled by the positively charged nucleus; in fact, as a neutron approaches a nucleus there is actually an attractive force between them. Neutrons entering stable nuclei render them unstable. Thus, when placed in drawers and inserted into the reactor (see Figure 26.4), stable nuclides are exposed to neutron flux and changed to radionuclides through *neutron capture.*

Nuclear Transformations

The use of the nuclear reactor for the production of radionuclides will now be exemplified by typical equations.

1. ***Transmutation Reaction***–a different element is produced from the one entering the reaction, the latter being called the ***target element*** (see Figure 26.5). This reaction is designated as an *(n, p)* reaction because a

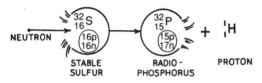

NEUTRON → STABLE SULFUR → RADIO-PHOSPHORUS + PROTON

Figure 26.6. Transmutation of sulfur to radiophosphorus by neutron capture.

neutron enters the nucleus and a proton (hydrogen nucleus) leaves it. The ^{32}P can be separated chemically from sulfur in ***carrier-free*** (pure) form:

$$\ ^{32}_{16}S \ + \ ^{1}_{0}n \ \rightarrow \ ^{32}_{15}P \ + \ ^{1}_{1}H$$
stable sulfur   neutron   radioactive phosphorus   proton

2. ***Radiative Capture***–the product nuclide is a radioactive isotope of the target nuclide and can be separated from it only with great difficulty. Contaminated by its stable isotope, the product radionuclide resulting from neutron capture is not carrier-free.

$$\ ^{59}_{27}Co \ + \ ^{1}_{0}n \ \rightarrow \ ^{60}_{27}Co \ + \ \gamma$$
stable   neutron   radioactive   gamma ray
cobalt   cobalt

This is an (n, γ) reaction because a neutron enters the nucleus and a gamma ray is emitted. On decaying, ^{60}Co changes to ^{60}Ni, while gamma rays and beta particles are given off.

3. ***Fission Products***–various radionuclides representing the fragments of the ^{235}U

nuclei that have undergone fission. They are actually "waste" products of the nuclear reactor, some of which can be separated chemically and used in medicine. In fact, they provide the main source of iodine 131 (^{131}I), cesium 137 (^{137}Cs), and strontium 90 (^{90}Sr), all three being applicable in radiotherapy.

Properties of Artificial Radionuclides

Some of the more important properties of artificial radionuclides will now be discussed. They decay in much the same manner as radium and differ widely in their decay rates.

Types Of Radiation. The radiations emitted by radionuclides have been described on pages 325-326. But those artificial radionuclides that are of medical interest emit beta particles, alone or with gamma rays.

Beta particles are fast-moving electrons ejected from the nuclei of certain radionuclides, actually arising from neutron decay:

$$\ ^{1}_{0}n \ \rightarrow \ ^{1}_{1}H \ + \ ^{0}_{-1}e \ + \ \bar{v}$$
neutron   proton   electron   antineutrino

For example, ^{32}P is a pure beta emitter, decaying as follows:

$$\ ^{32}_{15}P \ \rightarrow \ ^{32}_{16}S \ + \ ^{0}_{-1}e \ + \ \bar{v}$$
beta particle   antineutrino

A few radionuclides emit ***positrons***–positively charged electrons. Beta particles cause-significant ionization in tissues where they are for the most part absorbed. They cannot be detected directly by special external counting devices, although the brems radiation resulting from interaction with atomic nuclei can be so detected.

Gamma rays are electromagnetic in nature and are identical to x rays. Carrying no charge, they do ***not*** ionize directly but release primary electrons in the tissues by interaction with atoms. Gamma-ray emission is frequently accompanied by beta particles; thus, ^{60}Co and ^{131}I give off both gamma and beta radiation. External radionuclide detection by Geiger or scintillation counters depends on gamma radiation which usually has sufficient penetrating ability to pass completely out of the body.

Units of Activity. The rate at which a radioactive sample decays is its *activity,* measured in units called *curies* (Ci). *One curie is an activity of 3.7 (10)^{10} disintegrations per second (dps).* Smaller units representing decimal fractions of the curie are usually more convenient. One millicurie (mCi) is an activity of 3.7 (10)^7 dps, and one microcurie (μCi) is 3.7 (10)^4 dps.

The S.I. unit of radioactivity is the *becquerel (Bq),* defined as 1 dps; thus, 1 Ci = 3.7 (10)^{10} Bq. In this country, the S.I. system has not yet met with wide acceptance except for the gray, which is the unit of absorbed dose.

Specific Activity. A given sample of a radionuclide may not be carrier-free; that is, it may also contain the stable isotope. The *specific activity* of such a sample is defined as its activity (due to the radionuclide present) divided by its total weight. For example, if 10 g of stable cobalt irradiated in a nuclear reactor for a specific time acquires an activity of 55 Ci, then the specific activity of the resulting ^{60}Co sample is 55/10 = 5.5 Ci/g. In actual practice, the specific activity of ^{60}Co teletherapy sources may reach about 200 Ci/g.

Radioactive Decay

All radionuclides undergo a decrease in activity with time. The rate of decrease is exponential and is specific for any given radionuclide. So, regardless of the number of such atoms initially present in a sample, a certain fixed fraction will decay in a particular time interval. For example, if the initial number of atoms is 1 million, and the decay rate for the radionuclide is 0.5 per s, then 500,000 will have decayed in 1 s. In the next second, 0.5 of the remaining 500,000 atoms, or 250,000, will have decayed, etc. At present, there is no practicable method of changing the decay rate of radionuclides (see page 327 for a discussion of fractional decay vs decay constant).

Half-lives. In nuclear medicine there are three kinds of half-lives: physical, biologic, and effective. These depend on the method of measuring a radioactive sample over a period of time; representative curves for ^{131}I are shown in Figure 26.7.

1. *Physical Half-life (T)* is defined as the time required for a given radioactive sample to decay to one-half its initial activity. For example, with an initial activity of 10 mCi (3.7×10^8 Bq) the physical half-life is the *time* it takes for the activity to decrease to 5 mCi (1.85×10^8 Bq) as measured outside the body *(in vitro).* Examples of physical half-lives include: ^{131}I 8.1 days; ^{32}P 14.3 days; ^{60}Co 5.2 years.

Let us see how *physical half-life* is actually measured. First measure the activity of a sample of a selected radionuclide, such as ^{131}I by counting the number of decay events (disintegrations) per min by means of a suitable counter (see pages 336-339) under standard conditions, and subtract the background counting rate. This is the initial net counting rate. Repeat the procedure several times at suitable intervals over a period longer than the half-life. Finally, plot the data, with counting rate as a function of time elapsed since the initial count; the resulting curve is shown in Figure 26.8. To find the half-life, locate the point on the vertical axis corresponding to one-half the initial counting rate, then trace a horizontal line to the curve and, from the point of intersection, a vertical line to the horizontal axis where the final point of intersection locates the half-life. The physical half-life is constant for a partic-

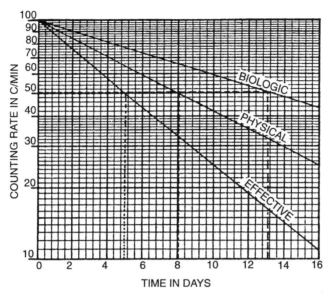

Figure 26.7. The three types of decay curves and corresponding half-lives for [131]I. In this example, the effective half-life is about five days and the biologic half-life about thirteen days. The physical half-life is constant—8.05 days.

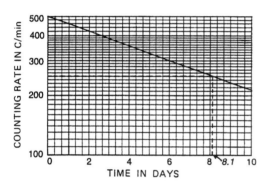

Figure 26.8. Physical decay of [131]I (semilogarithmic plot to obtain a straight-line curve.) With an initial net counting rate of 500 counts per min the half-life is the time required for the counting rate to decrease to 250 c/m. For ^{131}I this is 8.1 days.

ular radionuclide and is related to the decay constant; in fact, half-life T can be derived from the decay constant λ by the following equation:

$$T = 0.693/\lambda \qquad (1)$$

Or, knowing the half-life, we can derive the decay constant by rearranging equation (1) as follows:

$$\lambda = 0.693/T \qquad (2)$$

There is also a useful formula for the fraction of a radioactive sample *remaining* after n half-lives: $1/2^n$. Thus, if a sample of ^{131}I has an activity of 100 mCi, what fraction remains after twenty-four days? Since the half-life of ^{131}I is 8.1 days, the stated time approximates $24/8 = 3$ half-lives. The fraction remaining at the end of this time is

$$1/2^n = 1/2^3 = 1/8$$

The activity remaining is then 100 mCi $\times$ $1/8$ = 12.5 mCi.

2. *Effective Half-life* (T_e) is that due to both physical and biologic decay and is always less than the physical half-life. It is obtained from the curve plot of a series of absolute counting rates over a particular part of the body at suitable intervals after administration of a radionuclide. Effective half-life

relates to physical half-life T and biologic half-life T_b as follows:

$$1/T_e = 1/T + 1/T_b \qquad (2)$$

This can be rearranged as follows:

$$1/T_b = 1/T_e - 1/T \qquad (3)$$

and in words

$$\frac{\text{effective}}{\text{half-life}} = \frac{\text{physical half-life} \times \text{biologic half-life}}{\text{physical half-life} + \text{biologic half-life}}$$

3. **Biologic half-life** (T_b) is the time required for a given deposit of radionuclide in the body or a particular organ to decrease to one-half its initial maximum activity, due only to biologic utilization and excretion, and disregarding physical decay. It can be computed from equation (2), which may be rearranged as follows:

$$1/T_b = 1/T_e - 1/T \qquad (4)$$

For example, we know that ^{131}I has a physical half-life of 8.1 days. Suppose we find by absolute counting that the effective half-life of sodium iodide 131 in the thyroid gland is six days. Substituting these values in equation (3),

$$1/T_b = 1/6 - 1/8.1 = 0.167 - 0.123 = 0.044$$
$$T_b = 1/0.044 = 23 \text{ days, } \textit{the biologic half-life}$$

APPLICATION OF RADIONUCLIDES IN MEDICINE

A number of artificial radionuclides, when prepared in suitable compounds called **radiopharmaceuticals,** react in the body just like their stable isotopes. For example, the body cannot tell the difference between radioactive iodine and stable iodine. But once a radionuclide has entered the body, it acts as a radioactive tag that allows it to be detected, and its course through the body traced externally by special instruments; or, alternatively, it can be detected in the blood, secretions, or excretions within or outside the body. Furthermore, it can be measured with reasonable precision in most instances. In this chapter, only basic principles will be covered. We shall also touch upon the use of radionuclides in therapy, which depends on the fact that sometimes, as with iodine in the thyroid gland, a radionuclide will concentrate selectively in certain organs. Under such circumstances, a sufficiently large dose can be given to irradiate effectively the target organ.

The following outline summarizes the medical uses of radionuclides according to various indications:

1. **Diagnosis**–general procedure
   a. Compound tagged with radionuclide (radioactive isotope of a normal element substituted for it in the compound).
   b. Injected or ingested in tracer amount–small enough to be diagnostic, but not large enough to deliver excessive radiation exposure.
   c. Body cannot distinguish between radioactively tagged compound and its stable counterpart.
   d. Fate of tagged element determined by counting with special instruments (GM counter; scintillation counter; gamma camera)
      (1) Externally as ^{123}I, ^{131}I, or ^{99m}Tc, thyroid gland; ^{99m}Tc in various compounds: brain, lung, skeletal system, liver, kidneys, heart.
      (2) Excretions as in counting urine in Schilling test for pernicious anemia.
      (3) Secretions as in measuring ^{131}I output in saliva.

2. ***Therapy***–methods
  a. *Chemical*
    (1) Selective absorption as ^{131}I in thyroid gland: ^{131}I given orally in therapeutic amount; selectively absorbed in target organ–thyroid–which is irradiated by deposited radionuclide, in this instance, 90 percent of therapeutic dose from beta particles, 10 percent from gamma rays.
  b. *Physical*
    (1) External as ^{60}Co and ^{137}Cs in teletherapy; ^{90}Sr in contact therapy.

(2) Interstitial as ^{137}Cs needles; ^{192}Ir seeds; ^{198}Au grains; ^{125}I seeds; ^{182}Ta wires.

(3) Intracavitary as ^{137}Cs tubes; ^{32}P-chromic phosphate in pleural or peritoneal malignant effusions.

In recent years diagnostic nuclear medicine has evolved into two separate branches, ***laboratory tests*** and ***organ*** imaging. The remainder of this chapter will deal mainly with imaging, but first it will introduce basic instrumentation and a few examples of radionuclide counting procedures.

RADIONUCLIDE INSTRUMENTATION

The basic procedure in any quantitative diagnostic radionuclide study is to count, with an appropriate detecting or counting instrument, the number of photons or ionizing particles it receives and detects from a radioactive source in a specified interval of time and with a defined geometric arrangement. The source may be either an organ in the body or a specimen outside the body.

A general formula for radionuclide counting may be expressed as follows, ***provided the sample and the standard are counted under identical conditions*** (see pages 341-344):

$$\frac{\text{mCi in sample}}{\text{mCi in standard}} = \frac{\text{net counting rate in sample}}{\text{net counting rate in standard}}$$

The net sample counting rate, R_S, is the difference between the total counting rate, R_T, and the background counting rate, R_B.

$$R_S = R_T - R_B \qquad (5)$$

We obtain the background counting rate after having removed the sample and all other radioactive sources from the counting range of the detector. The counting rate is the total count divided by the elapsed time.

In any radionuclide procedure, the background counting rate must be subtracted from the total to obtain the net counting rate, unless the background is negligibly small.

We must have a method of counting the individual radiations emitted by a radioactive source. The same devices, with modifications, can be used to count both photons and charged particles. These devices consist of (1) the radiation detector plus its related electric circuitry and power supply and (2) the instrument for registering the number of counts.

Radiation Detectors. The two main types of detectors, which differ completely in principle, are the ***Geiger-Muller*** or ***GM tube*** and the ***scintillation detector.*** A detector with its related electrical circuitry is often referred to as a ***counter;*** thus, GM counter; scintillation counter.

1. ***GM Tube.*** An electronic pulse-type device that can detect individual ionizing particles or photons, it consists of a central wire ***anode*** and a cylindrical ***cathode*** enclosed in glass (see Figure 26.9). Within the tube is a suitable gas, often argon and alcohol under reduced pressure of about

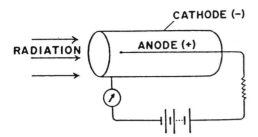

Figure 26.9. Simplified diagram of an end-window GM counter.

one-eighth atmosphere. The GM tube operates within a range of about 1,000 to 1,500 volts. There are two general types of GM tubes—end-window and side-window (see Figure 26.10). Radiation enters the end-window tube at one end and is therefore said to be *directional.* On the other hand, radiation enters a side-window GM tube at any point around its circumference and is therefore *nondirectional.*

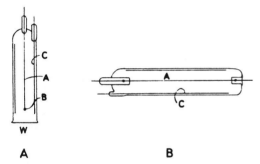

Figure 26.10. *A.* End-window type of GM tube. *B.* Side-window type. *A* is anode, *C* is cathode, *B* is bead, *W* is window.

The characteristic curve of a GM tube is shown in Figure 26.11. Although this curve is typical of all GM tubes, it may vary in numerical detail from one tube to another.

Photons or beta particles entering the tube cause ionization of the contained gas. With an applied potential of 800 to 1100

volts, the ions undergo sufficient acceleration to cause further *ionization by collision* with other atoms, resulting in an electrical discharge along the central wire anode. This discharge is manifested by a "pop" in the audio circuit. In this voltage range (800 to 1100 for this particular GM tube), the number of pops or counts per second is independent of the applied voltage but increases with the number of photons or beta particles entering the tube—this is the *GM region* of the characteristic curve (see Figure 26.11). Here, each entering photon or beta particle is detected as an individual count.

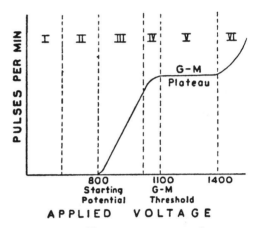

Figure 26.11. Characteristic curve of one type of GM tube. The voltage may be different for other GM tubes. The operating voltage should correspond to about the middle of the plateau. Zone III is the proportional region.

Such a counter can be used to detect the presence of photons or beta particles and, under suitable conditions, determine their quantity. A GM counter displays high efficiency in detecting beta particles; but low efficiency (about 1 to 2 percent) in detecting gamma radiation, which can be improved by coating the cathode with bismuth.

While the GM counter is still used in certain types of survey meters (see pages 418-

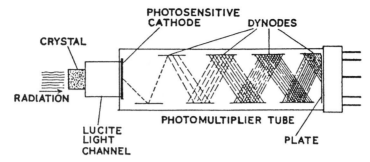

Figure 26.12. Diagram of a scintillation counter. Incoming radiation produces flashes of light (scintillations) in the detector crystal, one flash per photon. The light flashes are conducted by the Lucite channel to the light-sensitive cathode of the photomultiplier where photoelectrons are released. These are then accelerated by a series of dynodes (anodes) and are detected as pulses by the counting circuit after they strike the plate.

419, it has been replaced by the much more efficient and dependable scintillation counter in nuclear medicine.

2. *Scintillation Detector.* When x or gamma rays enter certain crystals, they interact photoelectrically with crystal atoms; the characteristic radiation is in the form of visible and ultraviolet light, the latter appearing as *scintillations* (light-flashes). Application of this principle gave us the scintillation counter and, as will be shown later, the *gamma camera* (also see pages 348-351).

In a scintillation detector (see Figure 26.12) all sides are encased except one; this clear side is placed in close contact with the photosensitive cathode, or *photocathode* of a *photomultiplier,* which also contains a series of about ten or eleven accelerating anodes called *dynodes.* The light-flashes from the crystal interact with the photocathode where they liberate electrons; these undergo serial acceleration and multiplication by the dynodes, to about one million times their original number. The electrons are collected to produce a voltage pulse in the circuit. One pulse occurs for each entering gamma photon, the size of the pulse being proportional to the energy of the photon entering the crystal. Since the scintillation counter has a dead time of only about 5 microsec, it is

capable of tremendous counting rates. Also, its high degree of sensitivity permits the counting of sources having extremely small activities.

Most scintillation counters have a thallium-activated sodium iodide detector crystal. However, anthracene crystals are available, especially for beta particles. The larger the volume of the crystal, the greater is the sensitivity or counting rate for a given activity. Crystals vary in diameter from a few mm for the measurement of exposure rates near sealed radioactive sources, to 22 in. (45 cm) or more in gamma cameras.

Scintillation counters are incorporated into a digital counting system that displays the number of counts in a pre-set time. In addition, selected "windows" confine the detected counts to a specified energy range, to count the photons from a particular radionuclide in a mixture containing those with energies other than the desired one.

The *well counter* is a scintillation counter specially modified for measuring samples of low activity. A hole bored in the detector crystal of a well counter can accommodate a small bottle containing the radioactive sample (see Figure 26.13). Because the crystal surrounds the sample on the sides and bottom, the well counter is extremely sensitive;

in fact, it can count a source with an activity as low as 10^{-5} μCi (microcurie). At the same time, the high sensitivity requires heavy lead shielding (usually about 5 cm Pb) to minimize background effects. Modern well counters are enclosed in a cabinet.

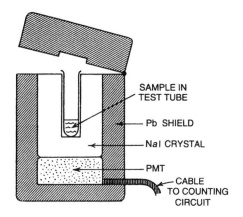

SAMPLE IN TEST TUBE

Pb SHIELD

NaI CRYSTAL

PMT

CABLE TO COUNTING CIRCUIT

Figure 26.13. Basic well counter, using a well crystal in a scintillation counter. Thick lead shielding is required to reduce the background counting rate.

Sources of Error in Counting

Various errors are inherent in the counting process, the most important ones being (1) statistical errors and (2) recovery, or dead time of the counter.

Statistical Errors. Since radioactive decay–a purely random process–obeys the laws of chance, statistical errors affect the accuracy of a given counting rate which, after all, reflects the rate of decay of a radioactive source. As a general rule, it is well to remember that the **reliability** (or reproducibility) *of a given count depends on the total number of counts recorded.* If the same sample is counted repeatedly under identical conditions, the counts will differ from each other because of chance variations (statistical fluctuation). How, then, can

we tell which count is correct and how much it differs from the ideal or "true" count? The answer lies in specifying a certain statistical reliability of a given count, but only the important elementary concepts will be included here.

1. **Standard Deviation,** σ (Greek letter "sigma"), is defined as follows:

$$\sigma = \pm \sqrt{N} \text{ (approx.)} \qquad (6)$$

where N is the total number of counts recorded in a given run. The standard deviation means that if we take the range $N \pm \sqrt{N}$, or the range of values from $N - \sigma$ to $N + \sigma$, there is about a 67 percent chance that the true count would lie somewhere within this range. For example, if the total count is 6400,

$$\sigma = \pm \sqrt{6400} = \pm 80$$

There is about a 67 percent chance that the true count falls between 6400 – 80 and 6400 + 80, or between 6320 and 6480.

2. **Percent Standard Deviation** (%σ). It is often more convenient to express σ as a percentage of the total count:

$$\%\sigma = \frac{\pm \sqrt{N}}{N} \times 100$$

$$\%\sigma = \pm \frac{100}{\sqrt{N}} \qquad (7)$$

or,

$$\%\sigma = \pm \frac{100}{\sigma}$$

In the preceding numerical example,

$$\%\sigma = \pm \frac{100}{\sqrt{6400}} = \pm \frac{100}{80} = \pm 1.3\%$$

The final count may then be stated as 6400 ± 1.3%; there is, again, a 67 percent chance that the true count lies between these limits. Figure 26.14 shows a curve plot of the percent standard deviation as a function of the total number of counts.

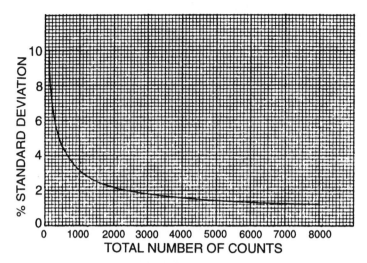

Figure 26.14. Variation in percent standard deviation relative to the total number of counts.

We can easily demonstrate that the percent standard deviation becomes smaller (ie, reliability becomes greater) as the total number of recorded counts is increased. Suppose we count a given sample long enough to accumulate 2500 counts. Then,

$$\%\sigma = \pm\,\frac{100}{\sqrt{2500}} = \pm\,\frac{100}{50} = \pm\,2\%$$

If the same sample is counted long enough to accumulate only 100 counts, then,

$$\%\sigma = \pm\,\frac{100}{\sqrt{100}} = \pm\,\frac{100}{10} = \pm\,10\%$$

Thus, the percent standard deviation is simply 100 divided by the square root of the number of counts. The percent standard deviation for a given *total count* also applies to the *counting rate* based on that total count.

In practice, with preset timing, we count a sample for a predetermined interval of time; and we also count the background, but not necessarily for the same interval. We then compute the counting rate by simply dividing each count by its corresponding time. The true sample counting rate R (without background), equals the gross sample counting rate (including background) R_T, minus the background counting rate R_B.

Recovery Time of Counter. As already mentioned in the discussion of GM and scintillation counters, a high counting rate may incur an error owing to the inability of the counter to detect radiation during a tiny interval of time following a pulse. This interval is called the ***dead time*** or ***recovery time*** of the counter–the higher the counting rate, the greater the likelihood that two ionizing events may occur so closely in time that only one is detected. A simple formula gives the approximate percentage correction for the dead time at extremely high counting rates:

$$\%C = 100\,\frac{R_t}{1 - R_t} \qquad (11)$$

where $\%C$ is the percentage correction that must be added to the observed counting rate R, to obtain the corrected counting rate, and τ (Greek letter "tau") is the dead time of the counter in question. For example, with a GM counter having a τ of $2(10)^{-4}$s, and observed counting rate of 200 cps should be corrected as follows:

$$\%C = 100 \ \frac{200 \times 2 \times 10^{-4}}{1 - 200 \times 2 \times 10^{-4}}$$

Performing the indicated multiplication first,

$$\%C = \frac{0\ 04}{1 - 0.04} = \frac{0.04}{0.96} = 4\%$$

The corrected counting rate is then

$$200 + 200(0.04) = 208 \text{ cps}$$

With a scintillation counter having a dead time of $(10)^{-6}$ s the correction factor is only about 1 percent even with an extremely high counting rate of 500 cps (the corresponding correction with a GM counter would be approximately 11 percent).

Efficiency and Sensitivity of Counters

The ability of a particular counter to detect radiation is generally known as its *efficiency*. But as we have seen, radioactive decay is a random process accompanied by the emission of radiation. In order to detect all of the radiation, the source would have to be placed completely within a special type of counter.

Sensitivity of Counters. In practice, a counter is placed at some distance from the source, which may be an organ or sample having sufficient activity. Because the radiation is emitted in all directions, the counter cannot receive it all, but can "see" only some of the radiation coming toward it, as in Figure 26.15 Furthermore, of the radiation actually entering the counter, only a fraction produces ionizing events and is detected. For example, an ordinary GM counter will detect about 1 to 2 percent of the entering gamma rays, although this can be improved by coating the cathode with bismuth. These two negative aspects of efficiency–the inability of a counter to "see" all the radiation and its failure to detect all the radiation that enters it–are embodied in *counter sensitivity*

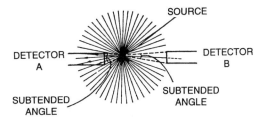

Figure 26.15. Angle of acceptance. The closer the counter is to the source, the greater is the number of photons entering it. Thus, detector *A* would record a higher counting rate than detector *B* with the same radioactive sample.

f defined by the following equation:

$$f = \frac{\textbf{counting rate}}{\textbf{activity}} \qquad (12)$$

Bear in mind that sensitivity is related to a given geometric arrangement of the source and the counter. Note that sensitivity may be expressed as counts per s per μCi, or counts per min per mCi, or any other convenient combination of proper units. Sensitivity can be used to convert counting rate to activity of a sample, by rearrangement of equation (8),

$$\textbf{activity} = \frac{\textbf{counting rate}}{f} \qquad (13)$$

provided the same geometric setup, counter, and radionuclide are used as in the initial determination of sensitivity. It also permits the computation of the sample activity needed to obtain a desired counting rate.

The preceding discussion applies primarily to the determination of the sensitivity of a particular counting system. In practice, there are two additional factors that influence the counting rate when counting is done over the body. One is attenuation of the radiation as it passes through tissues or other matter, on its way to the detecting instrument. The other is the scattering of radiation from nearby objects, with entrance of this scattered radiation into the detector.

We can minimize these factors by using radionuclides whose radiation is not appreciably attenuated, and by keeping the counting system isolated from scattering material. The influence of scattering can be further controlled by narrowing the energy window of the detecting system to match the energy of the primary radiation.

Minimum Detectable Activity. We sometimes need to know the smallest quantity of a radionuclide that a counter can detect to a reasonable degree of reliability, with a given geometric arrangement and type of counter. According to the ICRU*, the minimum detectable activity is that activity which, in a given counting time, records a number of counts on the instrument equal to three times the standard deviation of the background count in the same time interval. The equation specifying minimum detectable activity, A_{min}, is

$$A_{min} = \frac{3}{f} \sqrt{R_B/t_B} \qquad (14)$$

where $\sqrt{R_B/t_B}$ is the standard deviation of the background counting rate, and f is the sensitivity of the counter. Since $R_B = N_B/t_B$, we can rearrange as $t_B = N_B/R_B$. Substituting this value of t_B in equation (10),

$$A_{min} = \frac{3}{f} \frac{R_B}{\sqrt{N_B}} \qquad (15)$$

N_B is the number of background counts recorded in a preselected interval of time, usually about ten minutes. For example, if, with a particular geometric arrangement, f is 2000 cps per μCi, R_B 4 CPS, and N_B 2500 counts in 10 min, then the minimum detectable activity is

$$A_{min} = \frac{3 \times 4}{2000 \times \sqrt{2500}} = \frac{3}{500 \times 50} = 0.0001 \ \mu Ci$$

Note that the sensitivity and minimum detectable activity both depend on the geo-

metric arrangement of the source and counter, the type of counter, and the radionuclide being counted. Therefore, calibration of the equipment should be carried out separately for each counting setup and each radionuclide in use in the department.

Geometric Factors in Counting

In counting a radionuclide sample, we must take into account (1) *the distance between the source and the counter,* (2) *the presence of scattering material* in the vicinity of the source, (3) *the collimation of the counter,* and (4) *the size of the source.* Figure 26.14 showed how a source gives off radiation in all directions. The inverse square law of radiation applies (approximately because not a point source), so that the nearer the counter is to the source, the more radiation it "sees." Matter near the source gives rise to scattered radiation, which may be detected by the counter and give a falsely high counting rate. Collimation affects the fraction of emitted radiation reaching the counter. Figure 26.16 shows the response of the same counter with two different degrees of collimation. As the collimator is lengthened, the counting rate decreases, but the area of maximum sensitivity depends *only* on the diameter of the collimator hole. Therefore, a counter can be so collimated that it can detect, in the thyroid gland, small nodules whose activity is different from that of the surrounding gland. But collimation must not be so severe that the counter fails to see the entire organ, as in measuring radioiodine uptake by the thyroid gland.

A *focused collimator* (see Figure 26.17) provides maximum sensitivity for a particular degree of collimation, making it much more efficient than a single hole collimator.

Because of the fact that sensitivity depends so intimately on the geometric

---

* International Commission on Radiation Units and Measurements.

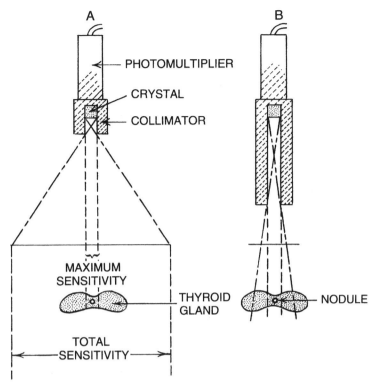

Figure 26.16. Relation between collimator length and field of view. In *A,* the field of view is too wide, allowing excess scatter (noise) from tissues surrounding the thyroid gland. In *B,* with a longer collimator, the field of view is too restricted to include the entire gland, but scatter is reduced and maximum sensitivity is the same as in *A.* Thus, image resolution of the nodule improves because of reduced scatter.

arrangement of the source and counter, equipment should be standardized and used under an identical geometric setup.

Methods of Counting

Two methods of counting, whether this is to be done on specimens outside the body, or over certain areas of the body, include (1) absolute counting and (2) comparative counting, both of which we shall now describe. Moreover health authorities require calibration of all test doses before they are given to the patient.

1. **Absolute Counting.** In principle, this is similar to the method of determining the sensitivity, f, of a counter. First we measure sensitivity with a standard known radioactive source (similar to the one to be counted in the patient or specimen) in the same geometric setup to be used with the clinical material. Suppose the standard source under these geometric conditions has an activity of 1 μCi and a net counting rate of 3000 cps. Then $f = 3000$ cps per μCi. If, now, we count the specimen under identical conditions (same radionuclide and geometric setup), and find a net counting rate of 1000 cps, then according to equation (13),

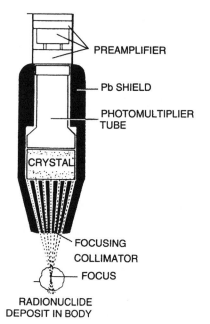

Figure 26.17. A photomultiplier tube (PMT) scintillation counter, with converging focused collimator.

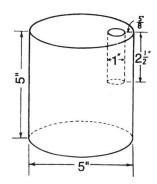

Figure 26.18. Diagram of a standard neck phantom, made of Lucite. The standard radioiodine capsule is placed in the 1-inch well.

$$\text{activity of sample} = \frac{\text{counting rate}}{f} \, \mu\text{Ci}$$

Therefore,

$$\text{activity of sample} = \frac{1000}{3000} = 0.33 \, \mu\text{Ci}$$

2. **Comparative Counting.** In clinical radionuclide studies utilizing the tracer method, we usually count the sample and standard separately under identical conditions. These net counting rates are then compared directly, the specimen counting rate being expressed as a percentage of the standard. Comparative counting may be exemplified by tracer studies of the radioiodine uptake by the thyroid gland: a predetermined dose of radioiodine is given to the patient and, after the appropriate time has elapsed, the counting rate over the thyroid gland is measured at a fixed distance (for example, 20 cm) with the "window" set for

the proper energy range. The same dose as that given the patient is now measured separately in a neck "phantom" (see Figure 26.18) at the same distance (see pages 346-347 for details).

Important Medical Radionuclides

The properties of some of the more important medical radionuclides will now be summarized.

Radioactive Iodine. At one time, iodine 131 (^{131}I) was the only isotope available for the diagnosis of thyroid disease. With a half-life of 8.1 days, ^{131}I emits 364-keV gamma rays and 183-keV (av.) beta particles. Obtained mainly by separation from the fission products of a nuclear reactor, ^{131}I as sodium iodide was widely used in the diagnosis and treatment of thyroid gland disease. The therapeutic effect of this isotope in hyperthyroidism and certain thyroid cancers depends principally on the beta particles, whereas tracer and scanning studies rely on the gamma rays, which are much more penetrating and can be detected outside the body with a GM or scintillation counter, or a gamma camera.

Another radioisotope of iodine—^{123}I—is preferred for uptake and imaging studies of the thyroid gland. Produced in the cyclotron by deuteron irradiation of tellurium, it has a half-life of 13.2 hr. Iodine 123 decays by electron capture and then emits 159-keV gamma rays, but no particles. Its main advantage, aside from the favorable energy of its gamma rays, lies in its short half-life which permits the administration of larger activities than with ^{131}I, but with a substantially smaller dose to the thyroid gland. For example, in imaging, 400-μCi of ^{123}I gives an absorbed dose of about 20 cGy to the normal thyroid gland, as contrasted with about 130 cGy from a 100-μCi of ^{131}I, a really significant difference. Ideally, thyroid uptake and imaging studies are now performed with ^{123}I, ^{131}I being reserved for the treatment of certain thyroid cancers and detection of their metastases. It should be mentioned here that [99m]Tc ("technetium"), to be described below, is a cheap and readily available pharmaceutical for thyroid imaging.

Radioactive Phosphorus. Phosphorus 32 has a half-life of 14.3 days and emits 700-keV (av.) beta particles. It is prepared mainly by neutron irradiation of stable sulfur in a nuclear reactor (see page 330). After administration in suitable form orally or intravenously, ^{32}P tends to concentrate in the bone marrow, spleen, liver, and lymph nodes. It can therefore be used in the treatment of certain blood disorders, mainly polycythemia rubra vera, a condition in which the bone marrow produces an excess of red blood cells. Phosphorus 32 suppresses the overproduction of these cells. Intravenous ^{32}P as phosphate delivers a whole body dose of about 1.5 cGy per mCi during the first three days, and thereafter, about 2.5 cGy per mCi to soft tissues and 25 cGy to bone (Shapiro).

Technetium 99m. The radionuclide ^{99m}Tc has a short, though convenient, half-life of six hours. Its gamma radiation, having a relatively low energy–140 keV–is easily collimated and at the same time has adequate penetrating ability to pass through the body and reach the detector. Technetium 99m is at present used mainly as the pertechnetate salt in aqueous solution. Because of the short half-life and the absence of beta radiation, large doses can be given with relatively little exposure to the patient. As a result we can obtain high count rates with excellent image resolution, using [99m]*Tc-tagged compounds* that have a more or less specific affinity for particular organs. Examples of ^{99m}Tc-tagged compounds and target organs include:

Diphosphonates (phosphorus compounds)– bones
Sulfur colloid–liver and spleen
Albumin aggregate–lungs
Cardiolite®–myocardium
*MAG3R** (mertiatide) and DTPA–kidneys
*Ceretec*** *–brain perfusion
Pertechnetate–thyroid gland

We should mention the ***technetium "cow,"*** a molybdenum ([99]Mo) generator which decays to [99m]Tc. Each day 0.9 percent sodium chloride solution is passed through the generator, dissolving out the ^{99m}Tc; this is called "milking" the "cow." The resulting ^{99m}Tc is filtered during the milking process and, after calibration, is ready for use.

Radioactive Cobalt. Cobalt 60 (^{60}Co), a radioisotope of cobalt 59, is produced in a nuclear reactor. It has a half-life of 5.2 years and emits gamma rays of two energies–1.17 and 1.33 MeV (average 1.25 MeV)–and a low energy beta particle (av. 96 keV). Cobalt 60 was once available in a variety of forms for brachytherapy (short distance therapy)–needles, beads, and wires–as a substitute for radium. More recently, cesium 137

---

* Mallinckrodt Medical
** Amersham Corporation

has been found to surpass ^{60}Co for this purpose. Large ^{60}Co teletherapy units designed for use in external irradiation have become virtually obsolete in this country.

Vitamin B_{12} or cyanocobalamin contains cobalt as a part of its molecular structure; when this compound is tagged with the radionuclide [57]Co, it is symbolized by [57]CoB$_{12}$ and is the agent that is widely used in the Schilling test for pernicious anemia (see page 347).

Radioactive Chromium. The radionuclide chromium 51 (^{51}Cr) is used in the form of chromate to determine red cell volume in polycythemia and in patients undergoing extensive surgery and to determine red cell survival in hemolytic anemias. Chromium 51 has a physical half-life of 27.8 days, and emits a 320-keV gamma ray and 242-keV (av.) beta rays. In the chromate form, this nuclide tags red cell, whereas in the chromic form it labels blood plasma.

More and more local commercial laboratories are supplying unit doses of various radiopharmaceuticals, including ^{99m}Tc compounds.

Examples of the Use of Radionuclides in Medical Diagnosis

It is beyond the scope of this book to give a detailed account of the numerous diagnostic procedures in the Nuclear Medicine Department. Instead, we shall include examples of the more important procedures based on various principles; these will include (1) *uptake by an organ* and (2) *excretion.* Organ imaging will be discussed in the next section.

1. **Uptake or Tracer Studies.** As mentioned earlier, the thyroid gland has a strong affinity for iodine. If radioactive iodine is taken it will normally be trapped in the thyroid gland over a period of twenty-four hours, its concentration depending on the

functional capacity of the gland. Obviously, before such a test can be run, the patient must avoid the intake of ordinary stable iodine ([127]I), as well as drugs that are known to suppress the activity of the thyroid gland. Such restriction may require weeks or months, depending on the type of drug; radiopaque media containing iodine may interfere with further iodine uptake for several months.

Depending on the type of equipment, a tracer activity of 2 μCi to 5 μCi of ^{131}I is taken orally in the morning by the fasting patient. A phantom simulating the neck should be available, made of Lucite or hard paraffin according to ICRU specifications (see Figure 26.18). A plastic test tube containing a standard of the same activity as that given the patient is placed in the hole in the phantom. At prescribed intervals, preferably two, six, and twenty-four hours, counting rates are taken at the same fixed distance—20 cm from the thyroid and from the standard phantom (comparative counting). The window is set for a range of 354 to 375 keV. Each time, a body background count should also be made over the lower end of the thigh, which resembles the tissues of the neck without the thyroid; the thigh counting rate should be subtracted from the thyroid counting rate each time. The percent uptakes at each counting session (two, six, twenty-four hr) are computed from the following equation.

$$\% \text{ thyroid uptake} = \frac{100 \ (\text{neck } R_T - \text{thigh } R_T)}{\text{phantom } R_T - R_B} \quad (16)$$

where R_T is the total counting rate of each item, and R_B is the background counting rate.

Figure 26.19 shows typical uptake curves obtained in various functional states of the thyroid gland. There is significant overlap of normal and abnormal values at both ends, this being more serious in the hypothyroid (low) range (see Table 26.1). Still, the thy-

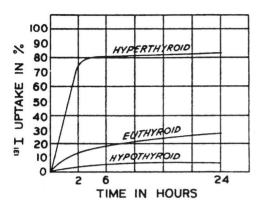

Figure 26.19. Representative iodine 131 uptake curves in the three functional states of the thyroid gland. Actually, there is no sharp dividing line between the normal (euthyroid) and abnormal (hypo- and hyperthyroid) states. The upper curve represents severe hyperthyroidism.

Table 26.1
RANGE OF 24-HR UPTAKES OF [131]I
BY THE THYROID GLAND
IN VARIOUS FUNCTIONAL STATES*

Euthyroid (normal)	5%30%
Hyperthyroid	more than 30%
Hypothyroid	less than 5%

* These values are typical; there may be much more overlap of the normal and abnormal than that shown here. Other tests are usually necessary to establish the diagnosis.

roid uptake of radioiodine is an extremely useful measure of thyroid gland function, especially when combined with supplementary chemical and radiochemical studies, and radionuclide imaging.

When available, ***123I is preferred for uptake and imaging studies*** of the thyroid gland because of the much smaller absorbed dose as compared with [131]I (see page 345).

2. **Excretion of a Radionuclide by the Body.** An example is the Schilling test for ***pernicious anemia,*** performed with cyanocobalamin (vitamin B_{12}) labeled with the radionuclide cobalt 57 (^{57}Co). Normal persons readily absorb this vitamin from the gastrointestinal tract, storing it in the liver within four to seven days. However, patients with pernicious anemia cannot absorb it because they lack ***intrinsic factor,*** which is normally secreted by the stomach. The fasting patient is first asked to void, the urine specimen being saved as a pretest control. The patient then swallows a capsule containing 0.5 μCi of $^{57}CoB_{12}$ with one-half glass of water (patient must not have received ordinary B_{12} orally or parenterally during the preceding two days). One hour later, a large dose of ordinary, stable B_{12}–1000 μg–is given intramuscularly to flush any absorbed radioactive B_{12} through the kidneys and into the urine, which is collected for 24 hours in a one-gallon plastic bottle with a screw cap, and should include the last bladder contents. After measurement of the total urine volume, a 4 ml sample of the pre-test urine (from previous day) and of the collected urine, and of a ^{57}Co standard should each be counted for 2 min in a well counter. A 2-min background count should also be made. Calculations are as follows after subtracting the pre-test urine count from the test sample count to obtain the net urine count:

$$\% \ ^{57}Co \text{ in urine} = \frac{\text{net urine count} \times \text{urine volume}}{4 \times \text{net standard count}}$$

Values greater than 6 percent are normal, and 0-3 percent are indicative of pernicious anemia. Intermediate values are questionable and require further study, as do low values. They may be confirmed by repeating the test after an oral dose of hog intrinsic factor together with the radiocobalt. However, the result may be equivocal if the patient has developed antibodies to the hog intrinsic factor during a previous test or from certain multivitamin pills.

IMAGING WITH RADIONUCLIDES: THE GAMMA CAMERA

A diagnostic imaging system called a *gamma camera* or *scintiscanner* can display the distribution of gamma-ray emitting radiopharmaceuticals (RPCs) that have been deposited in the selected target organ. By target organ is meant the one that selectively takes up a particular RPC. The gamma camera is a greatly expanded and modified version of the basic hand-held scintillation detector described on page 338; it is capable of using large numbers of radioactive decay events (counts) to create an image of the target organ. In a sense, imaging with a camera is analogous to radiography, if you can imagine that the RPC in the target organ is a miniature x-ray machine that produces an image from within outward with the gamma camera serving as the image receptor.

In a typical procedure, an appropriate RPC is first administered by injection, ingestion, or inhalation, depending on the designated examination. After an appropriate delay to allow distribution of the RPC, the patient is placed on the table directly under the camera at a specified distance. Figure 26.20 presents a block diagram of a gamma camera and its auxiliary devices, which will be described in the following sections.

Collimators

Just as in radiography, collimators absorb scattered radiation arising in the body by Compton interactions with atoms in the tissues. As will be explained shortly, the collimator acts like a grid. Radiation arising in the target organ must pass up through holes in the collimator toward the scintillation crystal (see Figure 26.21, page 349); scattered photons approach the collimator at various angles and are absorbed in the septa which separate the holes. As you already know, scattered radiation is a form of *noise* that obscures the final image.

Several types of collimators are available (see Figure 26.21). One must be aware of their *sensitivity*–the fraction of gamma photons passing through the collimator holes, relative to the number incident on its face; and their *spatial resolution*–the ability of the collimator to enhance contrast by absorbing

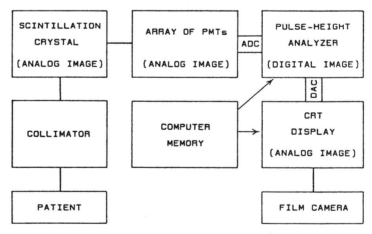

Figure 26.20. Block diagram of an Anger gamma camera. ADC = analog-digital converter. DAC = digital-analog converter. PMT = Photomultiplier tube.

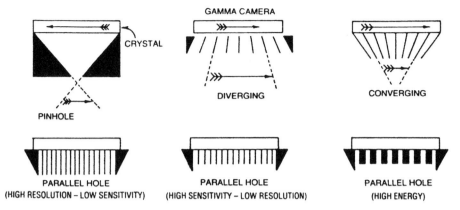

Figure 26.21. Varieties of gamma-camera collimators. (Modified from Mettler FA, Guiberteau MJ. *Essentials of Nuclear Medicine Imaging,* 3rd ed. New York, Grune & Stratton, 1990. By permission.)

as many of the Compton-scattered photons as possible. The collimator is situated between the patient and the camera's scintillation crystal.

1. ***Pinhole Collimator.*** This, the simplest type of collimator, consists of a small hole in the center of a lead (Pb) plate of required thickness. The hole serves as a conduit for gamma photons emitted by the radionuclide in the target organ. It provides high resolution in imaging small organs like the thyroid gland, but sensitivity is very low–relatively few photons pass through the hole, so the examination has to be prolonged to allow the camera to accumulate a sufficient number of counts. The use of the pinhole camera has gone into decline since the advent of the more modern gamma cameras with which parallel-hole collimators are used.

2. ***Multihole Collimators.*** Several forms are available. They all consist of lead plates in which multiple holes have been drilled. Unavoidable in such collimators is the penetration of photons through the septal walls between the holes; this should be kept well below 25 percent because septal-penetrating photons create noise, like scattered radiation in radiography. Newer types of collimators

have ***hexagonal holes*** to improve sensitivity, but this entails some loss of spatial resolution because their having thinner septa than round-hole collimators permits increased septal penetration.

Three kinds of multihole collimators include: (a) parallel-hole, (b) converging, and (c) diverging.

a. *Parallel-Hole Collimator.* The holes (and septa) are parallel to each other (perpendicular to the crystal face), resembling tubes. Sensitivity decreases markedly with increasing septal length. In fact, sensitivity is inversely proportional to the square of the septal length:

$$S \propto \frac{1}{L^2}$$

where S is sensitivity and L septal length.

The parallel hole collimator is the most popular today, especially in its low-energy high-resolution form. Low energy refers to its use with radionuclides whose gamma energy is less than 150 keV; this fits its application to technetium 99m (^{99m}Tc), by far the most widely used radionuclide.

The next two types of collimators are no longer important in imaging, but will be mentioned for the sake of historical interest.

b. *Converging Collimator.* Here the directions of the holes come together at a focus located at a distance beyond the collimator (ie, toward the patient). Figure 26.17 showed a hand-held version of a converging collimator to explain its principle. In use, the distance of the collimator's focus should be such that it lies within the target organ at the site of interest (eg, thyroid nodule). Note that maximum resolution prevails at the focal point and that there is a zone of acceptable image resolution or *focal depth* within a short distance above and below the focus. The converging collimator produces a slightly enlarged image.

c. *Diverging Collimator.* The directions of the holes move farther apart at increasing distances (ie, away from the camera). This type of collimator has been used in imaging large organs such as the liver or lungs. But sensitivity and resolution both decrease as distance increases. With the larger rectangular crystals available today, the diverging collimator is becoming obsolete.

Crystal-Photomultiplier Complex

As already mentioned, the collimator is in close contact with a special *scintillation crystal* of thallium-activated sodium iodide; this has the property of emitting flashes of light when gamma photons interact with its atoms by *photoelectric absorption.* Such interactions resemble those with x-ray photons, but in this case, the interactions occur in the outer shells of the crystal's atoms so the characteristic photon energy is in the visible rather than the x-ray range.

The circular crystals range from 10 to 22 in. (about 25 to 45 cm) in diameter, and $1/4$ to 1 in. (about 7 to 13 mm) thick. The newer cameras come with rectangular crystals measuring 20×25 in. (about 51×64 cm). Thicker crystals possess increased sensitivity (more light emitted because more atoms pre-

sent), but decreased resolution (more light scattered within crystal). If you could observe a crystal in the dark, you would see a light-image corresponding to the gamma-photon image emitted by the radionuclide in the target organ. This is equivalent to the light-image created by x rays in an intensifying or fluoroscopic screen.

Next, an array of many (about 40 to 90 or more) small, closely spaced PMTs (photomultiplier tubes), upon receiving the light flashes from the closely applied crystal, converts them to a pattern of voltage signals (see pages 307-308) as an analog image. An analog-digital converter changes this to a digital image. Next, the pulse-height analyzer, set to the selected energy "window," accepts a narrow range about the peak energy of the radionuclide in use, and rejects random pulses as well as those generated by scattered radiation that may have penetrated the collimator. Finally, the digital image is converted to an analog image for display on a CRT (cathode ray tube), from which a film can be obtained. At the same time, the digital image is stored on the computer's hard disc for retrieval later.

Single-Photon Emission Computed Tomography (SPECT)

A major advance took place in the early 1980s when single-photon emission computed tomography (SPECT) was introduced as a modification of the conventional gamma camera. A number of advances in engineering design have made SPECT a powerful tool in radionuclide imaging. One can set the window to accept photons of only a *single energy* according to the radionuclide being used. As a result, resolution is improved because extraneous signals such as scattered radiation and random noise in the system are rejected. By tying the camera into a self-contained CT system, images are created in single "slices" as with conventional CT in various planes, and reconstructed into three-dimensional images! An important factor in enhancing resolution is use of a digitizer for each individual PMT.

Another innovation has been the introduction of the **dual camera,** one above and the other under the patient. These produce simultaneous images that are combined into one by computer processing, which results in both improved resolution and reduced procedure time. For example, the usual whole-body skeletal imaging in which the patient's table moves caudad during the imaging process requires about one hour during which the patient must lie still; this time is reduced to about 15 minutes with the dual head unit. Also, a small organ like the thyroid gland can be imaged life-sized. Other types of examinations, previously impossible with a gamma camera, can now be accomplished with the dual-camera SPECT system.

Available Radionuclide Imaging

The process of gamma camera imaging is often called **scintigraphy,** and the images **scintigrams** or **scintiphotos,** although these are sometimes referred to, loosely, as **scintiscans.** It must be emphasized that scintigraphy may not always provide a positive diagnosis, and must often be used in conjunction with other diagnostic modalities. In any event, it provides a wealth of diagnostic information when used appropriately.

A summary will now include some of the more frequently used imaging studies. (See also pages 344-345 for properties of various radionuclides.)

Bone–mainly for detection of metastatic cancer, although osteomyelitis and aseptic necrosis may also be imaged. Radiopharmaceuticals include ^{99m}Tc diphosphonate, ^{99m}Tc medronate, and ^{99m}Tc oxidronate;

these compounds concentrate where osteoblasts are reacting to a destructive process. Lesions appear black, as so-called "hot" areas.

Liver-Spleen–to reveal primary or metastatic lesions, hemangiomas, and cysts in the liver, and defects such as infarcts in the spleen. The radiopharmaceutical (RPC) is usually ^{99m}Tc sulfur colloid, which deposits in the reticuloendothelial system of the liver, spleen, and bone marrow. Lesions in the liver and spleen appear light, as so-called "cold" areas.

Hepatobiliary system–for diagnosis of acute or chronic cholecystitis and biliary tree obstruction. Different segments of the biliary tree and duodenum may or may not be imaged, depending on the site of obstruction. The RPC is ^{99m}TC IDA (*imino*d*iacetic a*cid compounds).

Lung perfusion–the leading RPC ^{99m}Tc-MAA (macroaggregated albumin) consists of minute albumin particles, larger than about 5 μm which, after intravenous injection, block pulmonary capillaries. Although this occurs with only about one out of 1000 particles, there results a diffuse distribution except where an occluded artery (by infarct or embolus) prevents the radioactive particles from reaching the pulmonary zone it normally supplies. This area appears "cold" (white), but there is sufficient lack of certainty that the radiologist reports it in terms of low, intermediate, or high probability (see next paragraph).

Lung ventilation–to help differentiate between infarct or embolus, and a zone of diminished perfusion from bullae, localized emphysema, or airway obstruction. A "mismatch," that is, a high-probability perfusion study (suggesting infarction or embolus) in the face of a normal ventilation scintiphoto supports the impression of pulmonary infarct or embolus. The ventilation RPC is xenon 133, an inert, artificially radioactive gas.

Myocardium–evaluation of coronary artery perfusion of the heart muscle. The RPC is thallium 201 as thallous chloride. To reveal myocardial *infarction,* RPCs include ^{99m}Tc pyrophosphate and Cardiolite® (^{99m}Tc sestamibi).

Kidneys–detection of renal lesions and evaluation of renal perfusion. Several RPCs include ^{131}I hippuran, ^{99m}Tc DTPA, and ^{99m}Tc MAG3®.

Thyroid gland–detection and evaluation of nodules. I 123 and ^{99m}Tc pertechnetate are preferred. For uptake studies, I 123 (rarely, I 131 because of higher dose to thyroid and normal organs).

Cerebral Perfusion–determination of adequacy and pattern of blood flow to the brain, a truly dynamic study (blood in motion). Ceretec™ (^{99m}Tc exametazime) is widely used; it is a lyophilic complex, meaning that it has an affinity for brain lipids (fatty substances).

Cancer detection–especially in lungs and mediastinum, to help differentiate from infection, but abscess may also be positive. RPC is ^{62}gallium citrate.

Leukocyte and platelet labeling–for demonstration of pyogenic abscess, indium 111.

You can see that ^{99m}Tc is the most versatile and widely used radionuclide because it serves as a convenient radioactive tag for a

Table 26.2
SOME RADIONUCLIDES OF MEDICAL INTEREST

Nuclide	Physical Symbol	Half Life	Main Radiation Eγ	Eβ	Use in Medicine
			MeV	*MeV*	
Cesium 137	^{137}Cs	30 years	0.662	0.188	Teletherapy. Needles.
Chromium 51	^{51}Cr	27.8 days	0.320	0.242	Red cell volume and survival.
Cobalt 57	^{57}Co	270 days	0.122	–	Tagged vitamin B_{12} ($^{57}CoB_{12}$) in diagnosis of pernicious anemia.
Cobalt 60	^{60}Co	5.2 years	1.17 1.33	0.096	Teletherapy.
Gold 198	^{198}Au	2.7 days	0.412	0.316	Implant therapy as seeds or grains.
Iodine 123	^{123}I	13.2 hr	0.159	–	Diagnosis of thyroid disorders
Iodine 131	^{131}I	8.1 days	0.364	0.183	Diagnosis and treatment of thyroid disorders. Localization of thyroid cancer metastases. Renal function test as iodohippourate.
Phosphorus 32	^{32}P	14.3 days	–	0.700	Treatment of polycythemia rubra vera. Treatment of certain metastatic bone tumors (esp. breast and prostate). Malignant pleural and peritoneal effusions.
Strontium 90	^{90}Sr	28 years	–	0.196 0.937*	Contact therapy of superficial lesions of eye.
Technetium 99m	^{99m}Tc	6 hours	0.140	0.084	Imaging of brain, thyroid, liver, lungs, bones, and kidneys

Eγ is energy of principal gamma rays. *Eβ* is average energy of beta rays.
*Refers to yttrium 90, always associated with strontium 90.

large number of RPCs. Table 26.2 summarizes the physical date for presently available medical radionuclides.

Calibration of Radiopharmaceuticals and Gamma Camera

Radiopharmaceuticals. Before administration of an RPC, the following precautions must be taken: (1) patient must be positively identified, (2) RPC must be positively identified, (3) activity of RPC must be measured in a special calibrator to verify the dose. The accuracy of the calibrator itself must be checked daily to within ±5 percent, and the RPC to within ±10 percent.

Gamma Camera. Imaging quality must be checked periodically for:

1. *Image Resolution.* Requires a Hine-Duley or similar phantom with parallel lead (Pb) bars covering the entire field of view of the camera. The phantom is placed between the collimator and a flood-field phantom (see 2, next).

2. *Field Uniformity.* This can be determined by use of a *flood-field* phantom. It consists of a plastic (lucite or plexiglass) plate with a flat space in its center parallel to the surface; this space is filled with a radioactive liquid such as ^{99m}Tc. Alternatively, one can use a sheet of plastic that uniformly incorporates a radioactive material. Either of these phantoms should be imaged daily without a collimator, and field uniformity evaluated simply by visual inspection.

3. *Image Distortion.* This refers to the curvature of straight lines as they appear in the image. It can be evaluated by means of a phantom like the Hine-Duley, whose parallel Pb bars should normally appear straight on the scintiphoto. An image of the phantom is obtained by placing it between the flood-field phantom and the camera.

Quality assurance with a SPECT system requires greater care because small changes in quality with a conventional gamma camera are exaggerated with SPECT. Special phantoms are used and a greater number of counts (scintillations) accumulated, with the collimator in position.

QUESTIONS AND PROBLEMS

1. Define radioactivity. By whom was it discovered? Named?
2. Give an example of a decay scheme. What is meant by the statement "decay rate is constant?"
3. What makes some nuclides radioactive? Where do the ionizing radiations come from during radioactive decay?
4. Discuss radioactive series.
5. If a radioactive atom emits a beta particle, what change occurs in the atomic number? Atomic mass number?
6. How does the emission of a gamma ray affect the atomic number? Atomic mass number?
7. Briefly discuss radioactive equilibrium.
8. How does artificial radioactivity differ from natural?
9. Name and describe the three types of radiation emitted by various radionuclides. In the radium series, which members emit gamma rays that are useful in therapy?
10. Define half-life; average life.
11. Define curie; millicurie; becquerel. State the symbols for each.
12. Enumerate and explain the geometric factors in counting.
13. State three basic procedures in radionuclide diagnosis, and cite exam-

ples of each.

14. Summarize the application of three radiopharmaceuticals in diagnosis.

15. Why is ^{99m}Tc so widely used in scintigraphy?

16. Describe what happens when gamma photons enter a scintillation crystal.

17. With the aid of a simple diagram, explain how a photomultiplier works.

18. Prepare a block diagram of a gamma camera, between the patient and the CRT tube.

19. Compare the steps in image production in a gamma camera system, with those in radiographic imaging.

20. Compare the various collimators used with a gamma camera.

21. Describe the crystal-photomultiplier complex in a gamma camera.

22. Briefly outline image formation in the gamma camera.

23. Describe single-photon emission computed CT scanning.

24. How has a dual camera system enhanced SPECT (single-photon emission CT)?

25. Summarize the available diagnostic procedures in scintigraphy.

26. How are radiopharmaceuticals and the gamma camera calibrated?

27. How is gamma camera imaging quality checked?

Chapter 27

RADIOBIOLOGY

Topics Covered in This Chapter

- Definition
- History
- Physical Basis
- Cell Structure, Function, and
    Reproduction
- Deoxyribonucleic Acid (DNA)
    Genes and Chromosomes
- Radiobiologic Lesion
- Activation of Water
- Cellular Response to Irradiation
    Radiosensitivity–Cell Survival Curves

- Injurious Effects of Radiation
    Early–Limited Areas *vs* Whole Body
    Embryo and Fetus–Large Doses
- Dose-Response Curves
- Late Effects
    Somatic High Dose
    Somatic Low Dose
    Genetic–Genes and Chromosomes
        Risk Estimates
- Questions and Problems

Definition

RADIOBIOLOGY may be defined as that branch of science which deals with the modes of action and effects of ionizing radiation on living matter.

History

The biologic effects of ionizing radiation were discovered in the late 1800s when Becquerel incurred a skin reaction–reddening and irritation–induced by radium he had been carrying in his vest pocket. Later, Pierre Curie deliberately exposed his skin to radium and described the radiation effect.

We can date the beginning of experimental radiobiology to 1940, when Bergonie and Tribondeau (France) found that the radiosensitivity, or responsiveness of rabbit testes to x

rays is greatest in those cells that are actively dividing (ie, in mitosis) or have the potential to divide in the future.

In the late 1940s, D. E. A. Lea (England) did extensive research on the hydrolysis of water with the liberation of free radicals, following the initial suggestion by J. Weiss (1944). This is the mechanism of indirect action, to be described later.

Thoday and Read (England) observed in 1947 the increase in radiosensitivity of cells in the presence of oxygen, an important factor in radiotherapy.

Modern radiobiology emerged in 1956 when Puck and Marcus reported the first practical method of artificially culturing mammalian cells *in vitro* (ie, in culture media in glass containers), thereby providing material for studying their response to graded doses of radiation. Furthering their work, Elkind and Sutton-Gilbert (1959) discovered the ability of cells to recover from small doses of x rays and also explained the shape of the dose-response curve in terms of cell survival.

The Physical Basis of Radiobiology

At this point you should review the interactions of x-radiation with matter because they underlie their effects on cells. In the diagnostic region, x rays interact with matter by the electrons they release in photoelectric and Compton interactions. These *primary electrons* then initiate a *two-step process:* (1) they transfer their kinetic energy (K.E.) to tissue electrons, and (2) these high-speed secondary electrons expend their K.E. in creating ionization and excitation tracks in the tissues, and also produce bremsstrahlen which radiate away. Step (1) is called *kerma* (*k*inetic *e*nergy *r*eleased in *ma*tter), and step (2) *absorbed dose* (without brems); both are measured in gray (Gy), 1 Gy being equal to 100 red. Thus,

$$\text{absorbed dose} = \text{kerma} - \text{bremsstrahlen energy}$$

The *linear density*—crowding of the ions and excited atoms along the tracks depends on the electrons' kinetic energy: as the kinetic energy increases, ionization density decreases; and conversely, because a slower electron spends more time near an atom, increasing the probability of ionization/excitation and associated radiobiologic effects.

Note that because x rays and gamma rays deposit energy by a two-step process, they are classified as *indirectly ionizing radiation.* Charged particles such as electron beams and alpha particles (helium ions) ionize tissue by a single-step process and are therefore called *directly ionizing radiation.*

The ionization pattern along an electron or other charged particle track (see Figure 27.1) may be expressed in two ways, the second being preferred:

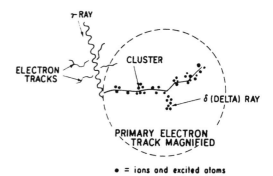

• = ions and excited atoms

Figure 27.1. Diagram of an ionization and excitation track of a primary electron that has been liberated by an x- or gamma ray interaction with matter. Clusters of ions and excited atoms along the primary and delta ray tracks produce the radiobiologic lesions in DNA, the severity of which depends on the closeness of the clusters—the closer they are, the greater the amount of energy deposited per unit length of path.

1. *Specific Ionization*—the number of ions formed per unit length of path, usually stated ion pairs per cm.

2. *Linear Energy Transfer* (LET)—the rate of energy deposit per unit length of path, expressed in *keV per μm* (micrometer). Since energy deposit increases as the ionizing particle slows, this is often stated in average LET, or LET_{av}. LET represents a measure of radiation quality and its influence on radiobiologic effectiveness in a particular tissue under specified conditions. Radiobiologic effects increase with higher LET radiation because of the denser ionization and energy deposit it engenders.

By way of comparison, x rays and electrons are rated as low-LET radiation, whereas the heavier alpha particles with their two positive charges are high-LET, directly ionizing radiation. Fast neutrons (recall, no charge) are intermediate-LET radiation, which ionizes indirectly by causing recoil of atomic nuclei, mainly protons (ie, hydrogen nuclei). Table 27.1 summarizes LET of vari-

Table 27.1
TYPICAL AVERAGE LINEAR ENERGY
TRANSFER (LET) VALUES FOR VARIOUS
KINDS OF RADIATION IN SOFT TISSUES

Radiation	Energy	LET
		keV/μ
^{60}Co γ rays	1.25 MeV (av.)	0.3
X rays	250 kVp	1.5
Electrons	1 MeV	0.25
Neutrons	20 MeV	20
Protons	10 MeV	4
Alpha particles	5 MeV	100

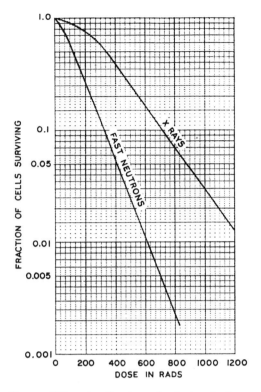

ous kinds of radiation. Again, *we* are concerned mainly with low-LET x rays.

Another criterion of radiation quality is **relative biologic effectiveness** (RBE), which compares the superiority of a test radiation over a standard type of radiation in engendering a particular biologic effect under identical conditions. The following equation applies:

$$\text{RBE} = \frac{\text{dose (cGy) of 250-kVp x rays}}{\text{dose (cGy) of test radiation}} \quad (1)$$

for same biologic effect

where 250 kVp x rays are the standard radiation. For example, in Figure 27.2 we can compare fast neutrons to x rays in their lethal effect on mammalian culture cells. Suppose we assume an endpoint of one percent survival; we select the 0.01 line on the vertical axis and follow it to both curves. The points of intersection are 600 cGy for neutrons and 1240 cGy for x rays. Substituting these values in equation (1), we

Figure 27.2. Application of relative biologic effectiveness (RBE) to cells in culture. The curves represent cell survival from increasing doses of radiation. For 1 percent (0.01) survival this line crosses the neutron curve at 600 cGy and the x-ray curve 1204 cGy. The RBE of fast neutrons at these points in the curves is 1204/600 = 2 (*approx.*)

obtain an RBE of 2.1, which means that neutrons are twice as lethal as x rays for these mammalian cells under identical conditions.

RADIOBIOLOGIC EFFECTS

We turn now to the changes induced in living matter by ionizing radiation. These will be discussed on four levels: cell structure and function, cell reproduction, radiobiologic lesion, and cell response to irradiation. But first we must review cellular structure and function.

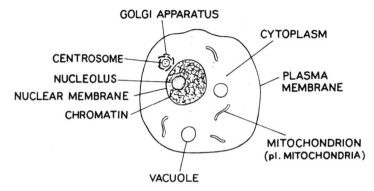

Figure 27.3. Diagram of a typical cell seen under a *light* microscope.

STRUCTURE AND FUNCTION OF CELLS

A typical animal or plant cell is a microscopic living unit composed of a central nucleus surrounded by cytoplasm (see Figure 27.3). Although these cell constituents have distinctive characteristics, they are highly interdependent so that activity or change in one has an effect on the other.

Nucleus

Not only does the nucleus serve as the cell's reproductive center, but it also regulates cellular metabolism—its life-sustaining functions. The nucleus is enclosed by a double-layered semipermeable membrane, which allows selective and nonselective passage of dissolved substances in both directions, between the nuclear contents and the cytoplasm. Within the nucleus one finds three essential components: (1) chromatin, (2) one or more nucleoli, and (3) a liquid called nuclear sap.

Chromatin, a nucleoprotein, is basophilic (blue-staining with basic dyes) and consists of two extremely important compounds, *deoxyribonucleic acid* (DNA) and *ribonucleic acid* (RNA), each of which is combined chemically with the simple protein histone.

In a nondividing cell chromatin exists in the nucleus either diffusely or in clumps, but just before cell division (mitosis), it forms thin threads. At the onset of mitosis, the DNA chromatin condenses into *chromosomes.*

Every animal and plant species has a characteristic number of chromosome *pairs* in its somatic (body) cells. But as shown in Figure 27.4A, the pairs differ from one another. Human somatic nuclei normally contain 23 pairs—46 chromosomes in all. Although one set of 23, known as the *genome* or *haploid number* called *n* is compatible with life, two sets or the *diploid number 2n* is necessary for completely normal structure

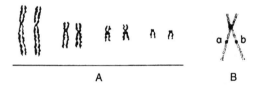

A   B

Figure 27.4. *A.* Four pairs of chromosomes from the normal human 23 pairs; the chromosomes of each pair are said to be *homologous. B.* Black dots on a chromosome pair exemplifying the *location* of a gene pair, such gene pairs being called *alleles.*

and function. Since all humans normally have the same number of chromosomes, faulty reproduction or aberrations such as abnormal number or composition may engender serious physical and/or physiologic abnormalities in the individual.

Genes. The chromosomes consist of DNA subsegments called genes, whose existence was suspected years ago as the result of G. Mendel's breeding experiments with sweet peas (1865), and T. H. Morgan's with the fruitfly *Drosophila* (1910). Such experiments revealed that specific points along each chromosome governed corresponding bodily traits or ***phenotypes.*** These chromosomal sites were later found to be genes. Each particular gene has a matching gene on the other chromosome of each homologous chromosomal pair, such matching genes being called ***alleles*** (ah-lēles) as in Figure 27.4B. This means that each gene of an allelic pair has the identical trait-producing effect. Actually, one gene may specify the trait blue eyes, while its mate specifies black eyes; the individual bearing these particular genes will have black eyes, because the trait black eyes is ***dominant*** over blue eyes. Conversely, the trait blue eyes in this example is ***recessive*** to black eyes. On the other hand, in some instances incomplete or intermediate dominance occurs, as in a type of chicken known as the Blue Andalusian Fowl. These birds have blue feathers, but their color is determined by an allelic gene pair, one for white and the other for black feathers. Note that the allelic genes (as part of the chromosomes) are derived from the parents, one from the mother and the other from the father.

Genes were originally thought to be fixed in position like beads on a string, but in the 1980s, Barbara McClintock discovered in breeding experiments with corn that they sometimes shifted in position along the chromosomes from generation to generation–so-called "jumping genes"–for which she received a Nobel Prize. Genes carry instructions for the manufacture of all the body's proteins including enzymes. The latter facilitate protein production and utilization, without themselves being used up during the process.

Cytoplasm

Surrounding the nucleus is the cytoplasm, the cell as a whole being enclosed in a triple-layered semipermeable membrane through which water and dissolved substances can flow in either direction between the cytoplasm and surrounding liquid environment.

The cytoplasm contains a number of microscopic bodies or ***organelles*** that have specific functions:

Mitochondria. These microscopic elliptical double membranes are concerned with catabolism (metabolic breakdown) of certain substances, especially carbohydrates, to provide energy for the cell.

Golgi Apparatus. The small, variously shaped bodies consist of double membranes. They are concerned with secretion, carbohydrate synthesis (manufacture), and the bonding of proteins to other organic compounds.

Endoplasmic Reticulum. A double-membrane system of tubes pervades the cytoplasm. As seen under the electron microscope, this exists in two forms: one has a *smooth* surface, while the other is *rough* owing to the presence of ribosomes on its surface. The rough type has to do with protein and enzyme synthesis and other chemical processes, but the function of the smooth type remains incompletely understood, although both types may conduct secretions to the cell surface.

Ribosomes. These bodies are composed of a nucleic acid, *ribonucleic acid (RNA),* which is intimately concerned with protein synthesis. Some ribosomes adhere to the surface of the rough endoplasmic reticulum, whereas others remain free in the cytoplasm. More will be said later about the ribosomes in the discussion of DNA and RNA function.

Lysosomes. Consisting of single-membrane microscopic sacs, the lysosomes contain enzymes that assist in the lysis (breakdown or digestion) of substances such as proteins, DNA, and certain carbohydrates. Any condition that increases the permeability of the sac membrane, or causes it to rupture, permits the enzymes to enter the cytoplasm; this may lead to actual digestion of the cell itself, a process called *autolysis.*

Those cells having the capacity to reproduce also contain, in their cytoplasm, a ***centrosome*** lying near the nucleus and containing two ***centrioles,*** which play a major role in mitosis (cell reproduction).

CELL REPRODUCTION

The mechanism by which somatic cells reproduce has been known for many years, having first been described in the early 1880s by W. Flemming; in fact, he coined the word mitosis (from Greek mitos = thread), centered in the nucleus.

In 1953, Howard and Pelc proposed a scheme for a ***complete cell cycle*** including periods of apparent cellular inactivity (see Figure 27.5). A typical cell cycle consists of two main stages, mitosis and interphase, with further subdivision of interphase into G_1, S, and G_2. G_1 is the gap after mitosis; G_2, the gap after S; and S the synthetic phase in which DNA undergoes replication (ie, duplication). Don't assume that the cell is dormant during the gaps; for example, RNA and enzymes are synthesized in G_1 and G_2 among other important activities. Some cells are truly resting in the G_0 phase.

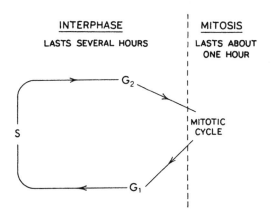

Figure 27.5. The Howard and Pelc (*Heredity,* 6:261, 1953) concept of a typical cellular generation cycle. There are two main phases–the synthetic *S* phase in which the DNA is normally doubled in amount, and the mitotic *M* phase in which the DNA is normally divided equally between the two daughter cells. The *S* and *M* phases are separated by two gaps, *G_1* and *G_2,* in which other kinds of cellular activity occur, such as RNA and-protein synthesis, *S*, *G_1*, and *G_2* constitute the interphase.

Mitosis

The process of mitosis occurs in specific phases as shown in Figure 27.6. Note again that it involves primarily the cell nucleus, with participation of the centrosome. There are four mitotic phases: prophase, metaphase, anaphase, and telophase, in which the prefixes are derived from Greek. Thus, *pro* = before, *meta* = beside, *ana* = up, and *telo* = end. Do not confuse the mitotic cycle with the cell cycle–mitosis is a ***stage*** of the cell cycle.

Prophase. The DNA, having doubled in amount during the *S* phase now aggregates

MITOSIS

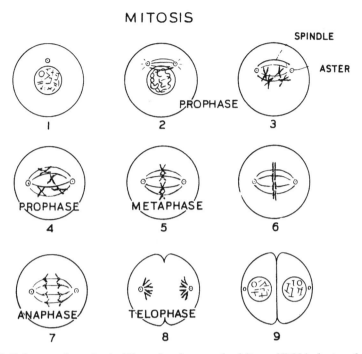

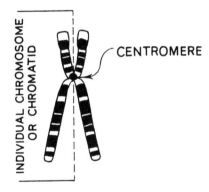

Figure 27.6. Cell division by *mitosis*. There has been a doubling of DNA during the synthetic *(S)* phase, resulting in a doubling of the normal somatic chromosome number to eight. During mitosis the normal number of somatic chromosomes is restored, that is, four (hypothetical example).

in the form of fine, paired threads. These gradually shorten, coil, and thicken to form the **chromosomes.** Every chromosome pair is joined at some point by a tiny body called a *centromere,* each member of the pair being designated a **chromatid** (see Figure 27.7). The somatic cells of each plant and animal species have a typical number of chromosome pairs, as well as unique chromosomal structure and function, and in any one species all the chromosome pairs differ from one another. During prophase the chromosome pairs *duplicate* to form twice the normal diploid number, that is, the 4n number, so there are now four chromosomes of each kind called **tetrads.** In humans, the total number of chromosomes *at this stage* is 92.

While the chromosomes are forming, the nuclear membrane and nucleolus disappear. At the same time, the two **centrioles** lying in

Figure 27.7. Schematic representation of a chromosome pair. The "banding" pattern of the chromosomes, brought out by special stains, increases the accuracy of identifying the various chromosomes.

the centrosome just outside the nucleus gradually separate and move to the opposite poles of the cell, meanwhile building a *fine threadlike spindle* between them. This part of mitosis, up to the instant when the chromosomes have assumed their definitive shape and the centrioles have reached the opposite poles, is called the *prophase*.

Metaphase. In this stage the chromosome pairs drift toward the equator of the spindle, that is, midway between the centrioles, and arrange themselves in a plane perpendicular to the spindle axis, called the *equatorial* or *metaphase plate*. The chromosomes then become attached to the spindle threads by their respective centromeres.

Anaphase. This stage starts with the separation of each chromosome pair into its respective chromatids, which move to the opposite poles of the spindle as its fibers shorten.

Telophase. Finally, during telophase, after the chromosomes have reached their respective pole of the cell, all the chromosomes begin to uncoil and lengthen, reverting to a chromatin network. The nuclear membrane reappears and the cytoplasm divides into two equal portions, each containing a full set (2n) of chromosomes characteristic of the species.

In summary, then, the normal 2n number of chromosomes, 46 in humans, doubles during prophase after the amount of DNA has doubled during the *S* phase to supply sufficient material for a quadruple complement, or 4n chromosomes. There are now 92 chromosomes in all. During anaphase the chromosome pairs separate and, at random, one chromosome of each pair ends up in a daughter cell at telophase. Thus, the original 2n number, 46, has been restored as 23 different **pairs** of chromosomes.

The mitotic process lasts about 40 minutes in humans as well as in other mammalian cells, two hours in cold-blooded animals, and up to a day or so in plants.

Meiosis

A modified form of mitosis known as **meiosis** (myo′ sis) governs the reproduction of **gametes** (sex cells, ie, ova and sperm). During fertilization of an ovum by a sperm, these two contribute all their chromosomes to the resulting embryo. If the gametes each

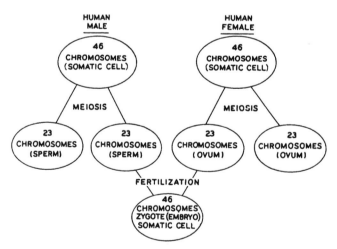

Figure 27.8. Formation of sex cells by meiosis. Each sperm and ovum receives one-half the normal somatic number of chromosomes for that species. During fertilization, the normal somatic, *2n,* number is restored in each of the daughter cells.

contained the somatic (diploid or 2n) number, the embryo would receive a total of 4n chromosomes, or twice the normal number for the species. Moreover, doubling would take place in each subsequent generation, a state of affairs incompatible with survival of the species. Instead, during meiosis the chromatids making up the homologous pairs of chromosomes separate, each going to one of the daughter gametes as shown in Figure 27.8. In humans, for example, each mature ovum or sperm contains 23 chromosomes, none being paired–the haploid or n number. During fertilization the ovum and sperm each contribute 23 chromosomes, restoring the 46 (n, or diploid) number characteristic of human somatic cells. Note that the sperm contain a Y chromosome paired with an X chromosome; the Y has few known genes, but is invaluable in tracing human male ancestry because it is transmitted only from male to male.

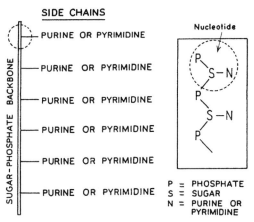

Figure 27.9. On the left is a schematic diagram of a straightened single DNA strand consisting of a sugar-phosphate "backbone" and its bonded purine and pyrimidine side chains; normally there are two such strands with complementary side chains like rungs on a ladder. The purines are A and G, and the pyrimidines are T and C; the only possible bonds are between A and T, and G and C (see Figure 27.11).

Structure of DNA

Although the chemical composition of **deoxyribonucleic acid (DNA)** had already been known in the early 1930s, its molecular structure was not discovered until the 1950s. In 1953 the American J. D. Watson, working with F. H. C. Crick in England, began an intensive study of DNA structure. At about that time, the x-ray crystallographer Rosalind Franklin, in M. H. F. Wilkin's laboratory in England, made a significant contribution to the eventual elucidation of DNA's molecular arrangement. The outcome of this story is that Crick, Watson, and Wilkins received the Nobel Prize in physiology/medicine in 1962 for discovering the structure of DNA–the **double helix.** Rosalind Franklin had died in the interim.

DNA is a **macromolecule,** the largest known today; although it is microscopic in size, it uniquely serves as the storehouse of information and instructions (not entirely unlike a computer) that direct all processes vital to cellular structure and function essential for survival–the production of enzymes and other proteins, antibodies, and secretions.

The information is embodied in a specific sequence of certain molecular subunits–the nitrogenous (nitrogen-containing) bases adenine (A), guanine (G), thymine (T), and cytosine (C)–located in each of the strand pairs (see Figure 27.9). Any change in the sequence of A, G, T, and C alters the instruction code; each of these is part of a subsegment of the DNA strand called a **nucleotide.** Every DNA strand has a phosphate-sugar "backbone," the sugar being deoxyribose (a pentose with 5 carbon atoms, oxygen, and hydrogen) bonded to the nitrogenous bases that were just mentioned. Hydrogen bonds also connect the nucleotides *between* the strand pairs, but these bonds are restricted to

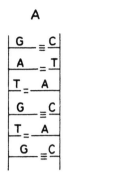
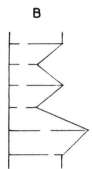

Figure 27.10. In *A* is shown the primary structure of DNA; the two strands are *parallel* and are joined by double and triple bonds between the side chains. The longer *G* (guanine) and the shorter *C* (cytosine) side chains are joined by a triple bond, whereas the longer *A* (adenine) and the shorter *T* (thymine) are joined by a double bond. In *B,* if short chains were joined, and long side chains were joined, the *phosphate backbones would not be parallel* and so would not conform to the normal DNA structure.

the A and T, and G and C components of the nucleotides (see Figure 27.10). This, then, constitutes the information code–the sequence of A-T and G-C along the DNA strands.

Just described is the ladder analogy of the DNA molecule, often called its *primary structure.* The *secondary structure* is the twist about its long axis–the **double helix**–as shown in Figure 27.11. There now follows a summary of the relation between the DNA strands:

Four Nitrogenous Bases. Only four nitrogenous bases exist in DNA: the purines adenine (A) and guanine (G); and the pyrimidines thymine (T) and cytosine (C). These connect the sugar-phosphate backbones like rungs on a ladder.

Parallelism of DNA Strands. X-ray diffraction studies of DNA show that the strands in the primary structure are absolutely parallel, being equidistant from each other at every point (see Figure 27.10). A and G

Figure 27.11. Double helix secondary structure of DNA molecule twisted around its long axis. Horizontal lines represent the weak hydrogen bonds: 2 between A and T, and 3 between G and C.

are equal in length, but are longer than C and T, which are also equal in length. This explains why bonding is restricted to A with T, and G with C (short with long in each case). Moreover, A and T are joined by a double bond, and G and C by a triple bond.

Chemical bonding of smaller molecules such as nucleotides to form chains is called **polymerization,** and the result a polymer. Reversal of this process is **depolymerization.** As will be indicated later, proteins are formed by polymerization of amino acids, which are "designed" by DNA.

RNA differs from DNA in having uracil instead of thymine as one of its pyrimidines, and in consisting of a **single strand.** Moreover, RNA may have a hairpin shape with hydrogen bonds between segments of the strand. RNA is intimately associated with DNA in its various functions, as explained in the next section.

Functions of DNA

The Crick-Watson model so clearly and correctly explains the structure of DNA that it has been universally accepted as the physicochemical basis of life and its perpetuation. We shall now turn to the functional attributes of the DNA macromolecule: (1) replication and (2) coded instructions for the synthesis of proteins as well as enzymes, which are specialized proteins.

Replication of DNA. During the synthetic *(S)* phase of the cell cycle, the amount of DNA doubles, or *replicates*. In this process, one strand serves as a model (template) for

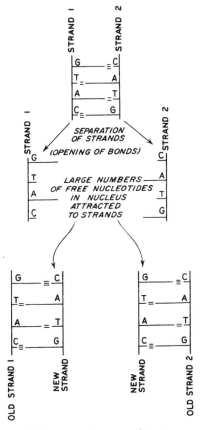

Figure 27.12. Replication (duplication) of DNA. The two strands separate from one end to the other like a zipper. Each strand then bonds with free nucleotides having complementary side chains, C to G, and A to T. The new strands are complementary to the original ones, so the new two-stranded DNA molecules have the same nucleotide sequence as original ones, which have thus been duplicated.

the production of the other strand, the two being complementary, as explained in Figure 27.12, where the double helix has been straightened out for simplicity. A specific enzyme then separates the two strands by opening the bonds "zipper" fashion, from one end to the other. Nuclear sap contains large numbers of nucleotides from which each DNA strand selects the complimentary one as shown in Figure 27.12. Thus, thymine (T) bonds to adenine (A), cytosine (C) to guanine (G): A to T, and G to C. Hydrogen bonds restore the normal two-strand state in the two daughter DNA molecules which are exact replicas of the original. This occurs during the synthetic (S) phase of the cell cycle. As a result of DNA replication all genetic material (ie, DNA) should normally be distributed equally to all daughter somatic cells during mitosis. The same applies to early gametogenesis (ie, early stages of sex cell formation), but in this case only one-half the genetic material reaches the mature gametes during meiosis.

Protein Synthesis. DNA also serves as a *model* or *template* for the manufacture of a great number of specific proteins and enzymes in the cell. Whereas, replication takes place in the nucleus, protein synthesis occurs in the cytoplasm. Furthermore, the latter process requires not only "instructions" from DNA, but also the participation of three kinds of RNA. All cells of a particular animal or plant species contain the same amount of DNA; but the quantity of RNA varies among the cells, being higher in those cells in which there is a higher level of protein synthesis. Thus, growing tissues have more RNA per cell than resting tissues. RNA exists in large concentration in the *nucleolus,* but makes up only about 10 percent of chromosomal nucleic acid.

Although RNA is produced mainly in the nucleus, it readily diffuses through the nuclear membrane into the cytoplasm. Here is a brief summary of protein manufacture:

1. Encoding of messenger RNA (mRNA) by DNA for proper sequence of amino acids in the required protein.

2. Assembly of amino acids by specific transfer RNA (tRNA), encoded in (1).

3. Bonding of tRNA to ribosomal rRNA only at complementary sidechains, by a hydrogen bonding process originally dictated by DNA in the nucleus

4. Separation of amino acids, and their polymerization–joining end-end to form protein.

To summarize, *the specificity of proteins depends on the sequence of nucleotides as initially programmed by DNA.* Even a single nucleotide out of sequence changes the nature of the protein. In the end, the genes, as assemblies of DNA, determine all bodily structures and functions. A number of missing or abnormal genes are associated with various hereditary abnormalities including, for example, breast and ovarian cancer in women harboring the BRCA1 or BRCA2 gene.

THE RADIOBIOLOGIC LESION

Passage of electrons released by x rays in tissues results primarily in damage to DNA– the *fundamental radiobiologic lesion.* Its severity depends on the *absorbed dose* and the *LET* (see page 131) of the radiation. Diagnostic x rays are low-LET radiation, as are the associated electrons. These high-speed *primary electrons* deposit energy at the cellular level by *ionization* (about 80%) and the remainder by *excitation* of molecules and atoms. Along the tracks are clusters of ions formed by interaction of the electrons with tissue atoms (see Figure 27.1), with occasional side branches–*delta rays*– produced by secondary electrons. As already mentioned, the rate of energy release (LET) along a track strongly determines the degree of radiobiologic injury.

Only a tiny amount of energy is needed to induce the radiobiologic lesion. To put this in perspective, a total body dose of about 450 cGy raises body temperature only about 0.001°C, yet this is a highly fatal dose, causing death in about 50 percent of exposed individuals. Henry Kaplan neatly summarized killing by radiation as a chain of amplified reactions:

1. *Very tiny physical energy input triggers,* in 1 microsecond

2. *Chemical changes* in DNA-protein and in water

3. *Radiobiologic lesions* which may be modified by

4. *Repair processes,* not all of which are specific.

Modes of Action of Ionizing Radiation

This section will deal with the two processes by which ionizing radiation incites the radiobiologic lesion. Recall that only absorbed radiation can be effective. Furthermore, evidence points to the fact that *multiple sensitive targets* within the cell must be "hit" to destroy its reproductive ability and/or kill it outright ("birdshot" effect). The modes of action are *direct* and *indirect* (see Figure 27.13).

1. *Direct Action.* Ionization of the sensitive "target"–DNA–specifically, breaking its chemical bonds, results in mutations, abnormal replication, or cell death. Chromosomes may undergo microscopically visible changes such as breaks, exchanges, fusions, and many others (see Figure 27.14). Finally,

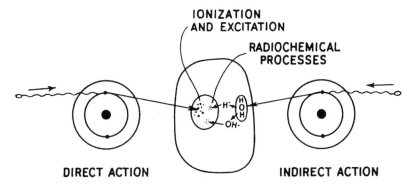

Figure 27.13. Mode of action of radiation on a cell. By direct action (shown on the left) a charged particle enters the nucleus and ionizes and excites the sensitive target (DNA). Indirect action (shown on the right) involves the interaction of charged particles with water to produce free radicals such as H· and OH·, and solvated electrons; all of these can injure a sensitive target. Actually these have a short range so if they are to reach DNA they must be released within the cell nucleus.

proteins, including enzymes, may be disrupted. Direct action is much more important with protons or heavy charged particles (eg, alpha particles), while injury by electrons is less marked due to their smaller LET, characterized by wider separation of ion clusters with a smaller chance of "hits" in multiple radiosensitive targets.

2. ***Indirect Action–Activation of Water.*** The passage of a charged particle through water, a major constituent of cells–water content ranging from about 70 percent in skin to 85 percent in muscle–initiates a process called ***radiolysis.*** This yields highly

reactive entities known as ***free radicals,*** which have an extremely short life of 10^{-10} sec. Their reactivity depends on the presence of an *unpaired* electron in their outer shell; normally, an atom or molecule has paired electrons in its outer shell, these electrons rotating on their own axes in opposite directions. Free radicals spread a short distance among cells, causing severe injury to their DNA.

Here are several examples of free radical formation during the radiolysis of water:

a. Ionization by primary electrons, released by x or γ rays, with removal of an

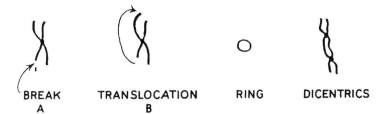

Figure 27.14. Several types of chromosome aberrations. *A.* Chromosome fragment resulting from a chromosome "break" by radiation. *B.* Fragment has become attached to another chromosome, a process called translocation. In the ring form, a single chromosome joins at its ends. A dicentric results from fusion of two chromosomes end-to-end.

electron from a water molecule:

$$H_2O \xrightarrow[\text{particle}]{\text{ionizing}} H_2O^+ + e^-$$

where H_2O^+ is a positive water ion.

b. Liberated electron e^- immediately becomes surrounded by water molecules to form an aqueous electron e^-_{aq}:

$$e^- + 4H_2O \rightarrow e^-_{aq}$$

c. Positive water-ion H_2O^+ from step (a) combines with a water molecule

$$H_2O^+ + H_2O \rightarrow H_3O^+ + OH\cdot$$
$$\text{\tiny hydroxy}$$
$$\text{\tiny free radical}$$

d. Two hydroxy free-radicals may combine to form the toxic substance hydrogen peroxide:

$$OH\cdot + OH\cdot \rightarrow H_2O_2$$
$$\text{\tiny hydrogen}$$
$$\text{\tiny peroxide}$$

e. Two hydrogen free radicals may combine to form molecular hydrogen:

$$OH\cdot + OH\cdot = H_2$$

Note that the dot in the above equations, as in $OH\cdot$ and $H\cdot$ denotes a free radical. The $OH\cdot$ radical displays a high degree of reactivity, especially with DNA from which it can remove a hydrogen atom, leaving a free radical site within DNA (Johns & Cunningham).

The aqueous (hydrated or solvated) electron mentioned above is highly reactive, especially with oxygen and carbon dioxide. It is worth mentioning again that free radicals and aqueous electrons are non-specific poisons acting on DNA.

CELLULAR RESPONSE TO RADIATION

Radiation is always injurious to living cells, with no strong evidence for a lower limit, a threshold dose, below which no effects occur. This problem is confounded by the fact that we cannot escape ionizing radiation which bombards us from outer space throughout our lives. In any case, the severity of damage depends on the absorbed dose, LET, cellular radiosensitivity, oxygenation, and phase of mitotic cycle. Observable microscopic effects include damage to the nucleus and cytoplasm.

Nuclear Damage

Through a microscope, one can see swelling of the nucleus, with severe clumping and fragmentation of chromatin (nucleoprotein) and disruption of the nuclear membrane. Chromosome damage causes aberrations such as breaks, fusions, translocations, rings, and others (see Figure 27.14). Gene damage–changes in DNA–are invisible but nonetheless important because they induce mutations, which can lead to abnormalities in cell structure and function, cell death, or cancer.

Bizarre mitoses may occur, resulting in extra or deficient chromosomes, giant cells, or abnormal distribution to daughter cells. All in all, there is a profound disturbance in the reproductive mechanism of the cell, further enhanced by delay or inhibition of mitosis, diminished production of DNA, and creation of abnormal types and amounts of DNA. Often the injury to DNA does not become manifest until the cell begins to reproduce–the cell dies during mitosis, a situation known as *mitotic death.* But some cells die during interphase, called *interphase death.*

When injury involves a single strand, the chromosome may self-repair. Sometimes radiation causes cells to go from the premitotic phase directly to the postmitotic phase, acquiring excessive amounts of chromatin and increased resistance to radiation. Finally, radiation can induce premature senility and death of cells.

Radiation changes in the nucleus may be summarized as follows:

1. Nuclear swelling
2. Nuclear membrane rupture
3. Gene mutations
4. Chromosomal aberrations
5. Delayed or abnormal mitosis
6. Giant cell formation
7. Mitotic death
8. Premature senility or death of cells

Cytoplasmic Damage

Radiation injury may also involve the cytoplasm. This occurs in the organelles– damage to the Golgi apparatus, mitochondria, and endoplasmic reticulum. The nuclear and cellular membranes swell and become more permeable, that is, more porous so that eventually the cell undergoes lysis (dissolution). It should be noted that radiation effects on the cytoplasm are non-specific, in contrast to the nucleus in which the specific target is DNA. Moreover, large doses are required to seriously injure the cytoplasm.

CELLULAR RADIOSENSITIVITY

The radiosensitivity of cells may be defined as their susceptibility to the injurious effects of ionizing radiation. This can be extended to tissues, organs, and organisms. However, this definition is not very precise. The use of tissue cultures mentioned earlier, made it possible to derive a much more scientific method of defining radiosensitivity.

Culture of mammalian cells, as applied by Puck and Marcus, was derived from the method of culturing bacteria in an artificial medium in which they can multiply and produce colonies, a colony being the outgrowth of one bacterium. Mammalian cells taken from a sample can be counted and "plated" on a suitable culture medium in a glass dish, enclosed by a removable glass cover. As they reproduce, each cell is normally expected to form a colony, although occasional ones may fail to do so.

Using the cell culture method, one spreads a counted number of identical cells on a solid growing-medium in a *series* of glass plates. One is left unirradiated as a *control,* while the others receive increasing doses of

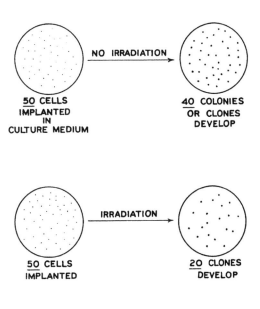

$$\frac{\text{NUMBER SURVIVING DOSE OF RADIATION}}{\text{NUMBER SURVIVING WITHOUT RADIATION}} = \frac{20}{40} \times 100 = 50\%$$

Figure 27.15. Principle of cell "survival" determination. Survival in this context means the ability of a cell to continue dividing to form a clone of at least 50 cells.

radiation. The number of cells surviving to produce colonies, or **clones,** decreases as the radiation dose increases. At **each** dosage level, the **surviving** fraction is then obtained by dividing the number of colonies in each irradiated plate by the number in the unirradiated plate. Figure 27.15 illustrates the method, greatly simplified.

Understand that survival curves do not involve direct cell killing, but rather the inability of sufficiently irradiated cells to **reproduce and form clones** of at least 50 cells. This equates to extinction of the cell lines, that is, ultimate death.

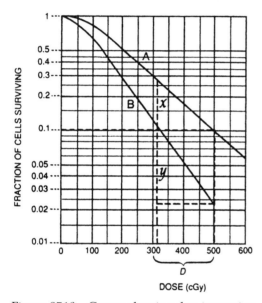

DOSE (cGy)

Figure 27.16. Curves showing that increasing radiosensitivity is indicated by a steeper straight portion of the cell survival curve. The same dose increment *D* causes a greater fractional loss in survival *y* in population *B* then the loss in fractional survival *x* in population *A*. Therefore, *B,* with a steeper straight portion, is the more radiosensitive population.

Figure 27.16 shows two typical survival curves of cells in culture exposed to x rays, representing cell survival fractions as a function of dose. Being based on response to cell populations, such curves are statistical in nature.

At *low* dose levels, there is an initial short straight-line segment which is believed to represent single-hit killing without cell recovery. This is immediately followed by a downward-curved portion, or **shoulder, typical of mammalian cell survival curves;** there is recovery from sublethal doses acting on multiple targets, as explained by Elkind. Finally, there is a declining straight-line portion, wherein cell survival (ie, number of clones) decreases by the same *fractional* amount with equal increments of dose, denoting an exponential relation between dose and response. Similar results have been obtained with cells implanted into small animals, that is, *in vivo* cell cultures.

The steepness of the final straight portion indicates the **radiosensitivity** of a cell population. In Figure 27.16 the final straight portion of curve B is steeper than that in curve A, so cell population B is more radiosensitive. This can be expressed in quantitative terms by the concept of D_o, the **mean lethal dose,** defined as the dose that reduces the surviving fraction from a particular value on the straight portion to 37 percent of that value (see Figure 27.17). D_o is inversely related to radiosensitivity: larger D_o means lower radiosensitivity. Thus, **1/D_o is a measure of radiosensitivity.**

In general, mammalian cells have a relatively narrow range of radiosensitivity in a fully-oxygenated culture–average lethal single dose is 150 cGy ± 50%, with a roughly similar range in various malignant tumors.

Modifying Factors in Radiosensitivity

A number of conditions affect the radiosensitivity of tissues (including tumors):

1. **Inherent Radiosensitivity.** Primitive cells in the red bone marrow and testes, and

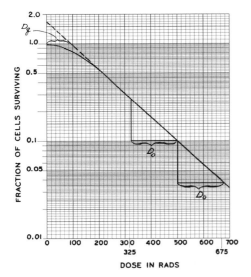

Figure 27.17. Cell survival curve showing the quantities *mean lethal dose D_o* and *quasithreshold dose D_q*. D_o is that dose which reduces the surviving fraction from a particular value (on the straight portion) to 37 percent of this value. D_o is the reciprocal of the slope (steepness), so the larger the D_o, the less the radiosensitivity. D_q measures the width of the shoulder and therefore the ability of cells to recover from small doses.

anoxic or hypoxic cells. The influence of oxygen on radiosensitivity has understandably been termed the *oxygen effect,* which requires the presence of oxygen *at the time of irradiation.* An increase in oxygen enhances radiosensitivity, but a maximum effect is reached at a pressure of 50 mm mercury. A quantitative expression for the oxygen effect is the *oxygen enhancement ratio, OER:*

$$\text{OER} = \frac{\text{dose in anoxic state}}{\text{dose in oxic state}}$$

for same radiobiologic effect

Typical OER values include 2.5 to 3 for the low-LET x rays. With high-LET radiation OER plays a much smaller role: about 1.5 for fast neutrons, and 1 (no effect) for alpha particles. Thus, OER parallels the magnitude of the oxygen effect. It seems that the poorer blood supply in the center of tumors, especially large ones, reduces the effectiveness of radiotherapy with low-LET radiation. The radiobiologic explanation for the oxygen effect on radiosensitivity is the combination of oxygen with free radicals, prolonging their lives and thereby increasing the injury to DNA.

4. *Volume of Tissue.* Normal tissues and organs are fully oxygenated so the OER does not apply. However, animal experiments have shown that an increase in the volume of tissue irradiated results in greater radiation injury. This has also been observed in radiotherapy.

5. *Dose Fractionation.* Cell survival curves, as well as experience with radiotherapy, has shown that fractionation of a given dose in a series of daily increments decreases radiosensitivity (see Figure 27.18). You can see the reappearance of the shoulders in the survival curves with a less steep resultant slope of the terminal straight portion.

6. *Age.* The age of the animal at irradiation has an important bearing on response—embryos display greater radiosensitivity than

in the crypt cells of the intestines, are more radiosensitive than muscle or connective tissue. In the *therapy* range, the spinal cord, liver, and kidneys are much more radiosensitive than the neighboring tissues

2. *Radiation Quality.* Since high-LET radiation produces greater ionization and energy deposition, it is more effective in damaging cells. In radiography, x rays liberate electrons, which are low-LET radiation and therefore less damaging than high-LET radiation, dose for dose. With fast neutrons—high LET radiation—there is a minimal shoulder and a steeper terminal straight-line portion because of direct action, a more efficient cell killer than x rays.

3. *Oxygenation.* When exposed to low-LET radiation, fully oxygenated cells are about 2.5 times more radiosensitive than

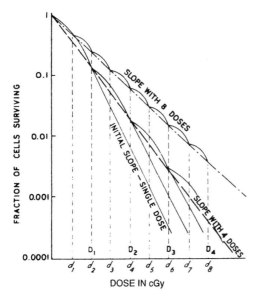

Figure 27.18. Modification of cell survival curves by splitting a single dose into multiple doses (extension of principle in Figure 10.03). Again, separation of doses by a time gap of a few hours results in reappearance of the shoulder (sometimes referred to as "return of *n*"). The larger to total number of fractions for the same total dose, the smaller the slope of the cell survival curve.

mature animals of the same species because of the presence of more, actively reproducing somatic cells (higher mitotic rate). For example, the LD_{50} dose, yielding 50 percent fatality in mice, increases almost in proportion to the $\log_{10}$ of the age up to about 13 weeks, then levels off, and finally decreases in old age.

Acute Whole-Body Radiation Syndromes

Early effects of radiation are well represented in the *acute radiation radiation syndrome (ARS)* or "radiation sickness" soon after whole body exposure to various single doses of radiation. A *syndrome* is a complex of signs and symptoms induced by a single cause. Whole-body large exposures are rare and may be incurred during industrial accidents with high-level radioactive materials, as in nuclear power plants, or during explosion of nuclear weapons. Usually involving gamma rays (low-LET, like x rays) and fast neutrons (high-LET), such exposures may entail profound and often fatal outcomes when the dose is high enough.

Before describing the acute radiation syndromes (ARSs), which are dose dependent, we shall consider the median lethal dose—LD_{50}—the whole-body dose that causes 50 percent of an exposed population to die by the end of a designated time. Small animals may survive for as long as 30 days, so for them $LD_{50/30}$ applies, but humans may survive as long as 60 days so for us it is $LD_{50/60}$. Table 27.2 indicates the LD_{50} values for various animals.

Table 27.2
MEDIAN LETHAL DOSES (LD_{50}) FOR
VARIOUS ANIMAL SPECIES*

Animal	LD_{50}
	R
Guinea pig	250
Dog	325
MAN	450 (cGy/air)**
Mouse	530
Rabbit	800
Rat	850

*Values according to the U.S. Public Health Service. The LD50 for bacteria is 20,000 to 50,000 R, and for viruses 50,000 to 1,000,000 R.
**Information quoted by Conklin and Walker.

Unfortunately, the designation of dosage units in whole-body irradiation has not been standardized. In the first place, the distinction is not always made between skin dose and deep body dose (as you know, the dose decreases as the radiation passes through the body); even if we assume a midplane dose, this will depend on the exposed individual's thickness. Secondly, specification may not be made as between low- and high-LET radiation; for example, the equivalent dose from 100 cGy of fast neutrons (100 keV) is 20 times greater than that from 100 cGy of x rays, with a correspondingly greater radiobiologic effect (see page 371)! Finally, the LD_{50} value for whole-body irradiation in humans is only an estimate, and the radiation exposures for the ARS are usually quoted in the literature in cGy (or rads), without regard to radiation quality (LET or RBE) or as to how doses were measured. It would therefore be prudent to state exposure in R (roentgens), especially because measuring instruments are calibrated in R and serve the purpose as a measure of skin exposure. The use of units such as centigray or sievert gives a false sense of precision, which is not warranted.

The four acute whole-body radiation syndromes will now be described under separate headings based on the size of the single dose: subclinical, hematopoietic, gastrointestinal, and central nervous system. There is very little agreement in the literature as to the doses responsible for these syndromes, so an estimated average will be given here. Note that the values overlap from one syndrome to the next.

1. *Subclinical Syndrome.* Whole body exposure of 50 to less than 200 R. This is a relatively benign form of radiation sickness, with prodromal complaints of nausea, vomiting, and malaise (general "feeling poorly"). The white blood count (WBC) shows moderate decline, followed by gradual recovery. Between 25 and 50 R there are no obvious symptoms, but WBC may undergo a slight temporary fall. Below 25 R, no detectable changes occur in the blood count, but special studies may reveal chromosome aberrations, mainly breaks, and slight increase in incidence of leukemia.

2. *Hematopoietic Syndrome.* Often called the bone marrow syndrome, occurs after exposures from about 200 to 800 R.

a. *Prodromal*–starts at about 2 hours and lasts 2 days. Nausea with or without vomiting, malaise, fright.

b. *Latent*–2 days to 3 weeks, when individual begins to improve, but bone marrow and lymph nodes still show loss of cells. Duration is 2 to 3 weeks. Latent period shorter with larger doses.

c. *Manifest*–full-blown syndrome. Individual now very ill with fever, malaise, sore throat, diarrhea, petechiae (tiny skin hemorrhages), and pancytopenia–decreased number of WBC, RBC, and platelets. Infection results from loss of WBC; anemia due to hemorrhage and loss of RBC. Alopecia (hair loss) in 3 weeks.

d. *Recovery*–in lower dose range (200-300 R), occurs in 5 weeks, but incomplete until about 6 months. In upper dose range (500-600 R), most likely a fatal outcome due to hemorrhage and infection. Severe leukopenia (low WBC). Need special treatment such as marrow transfusions from compatible donors, antibiotics, and ordinary blood transfusions. Note that at 400 R ($LD_{50/60}$) about one-half of exposed persons will die unless treated. When death occurs, it is usually within 2 to 3 weeks, although it may be delayed for as long as 6 months.

3. *Gastrointestinal Syndrome.* About 800-5000 R. This results in virtually 100 percent mortality in about 1 to 2 weeks (average about 6 days).

a. *Prodromal*–nausea and vomiting occurs in a few hours. If this sets in before 2 hours, it is highly probable that the whole body exposure was at least 1000 R; fatal unless treated as above.

b. *Latent*–lasts 2 to 5 days, but with a large dose, latent period may be absent.

c. *Manifest*–marked by nausea, vomiting, and step-like increase in temperature. Increasing diarrhea appears, becoming severe by the sixth day. By this time, diarrhea becomes intolerable owing to denudation of the small bowel lining epithelium, which allows fluid to leak into the bowel lumen through the bare surface. At the same time, no fluid is absorbed in the denuded colon, resulting in marked dehydration. Since bile salts cannot be reabsorbed in the distal ileum, they pass into the colon, irritating it and intensifying the diarrhea. At the same time, bacteria invade from the lumen through the gut wall into the blood stream, causing sepsis and fever. This is made worse by depletion of WBC by radiation. Finally, the individual goes into shock, dying at one to two weeks after exposure. *However,* with an exposure of about 1000 R, treatment with fluids, blood transfusions, antibiotics, and compatible bone marrow transfusions may reduce fatality. In fact, this has been used deliberately in some forms of leukemia, although chemotherapy has now replaced whole-body irradiation.

4. *Neurovascular Syndrome.* Previously called the central nervous system syndrome, this requires an exposure of at least 5000 R. Within a *few minutes,* onset of severe nausea and vomiting followed by rapid dehydration, drowsiness, ataxia (staggering), and general convulsive state. This is caused by cerebral edema (swelling), vasculitis (inflammation of blood vessels), and injury to nerve cells in the brain. Death occurs within a few days, as treatment is of no avail. Bone marrow and G.I tract are also damaged, but these two syndromes do not have time to develop.

Explanation of Acute Whole-Body Radiation Syndromes

The hematopoietic and gastrointestinal systems each have a supply of primitive (very young) precursor cells which continuously proliferate (multiply) and differentiate (mature) as the mature cell population nor-

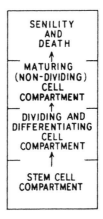

Figure 27.19. Vertical or rapid renewal population. Survival of the individual depends on prompt replacement of dying mature cells by proliferation of stem cells and eventual differentiation. Thus, a cell renewal system is at risk during the time interval between the death of mature cells and the arrival of replacement cells from the lower compartments. Examples include bone marrow cells, male gametes, and intestinal epithelium.

mally ages and dies off. We may simply regard these stages in cell development as residing in three compartments, wherein cells move progressively from the first to the last without distinct boundaries (see Figure 27.19). In the acute radiation syndromes the radiosensitive immature cells—the earliest being the *stem cells* in the red bone marrow and lymph nodes, and in the lining of the intestines—are largely destroyed. As a result, insufficient precursors remain to replenish the supply of mature functioning cells that have normally died off. Replacement of these mature cells must await sufficient recovery of precursor cells, but this may not occur quickly enough to sustain life; thus, there is a race between the recovery of the precursor cells, and the demand of the body for mature functioning cells.

In the *hematopoietic system,* the WBC, RBC, and platelets originate from a common *stem cell.* Radiation damage to these stem cells eventually results in *pancytopenia* (very low number of all circulating blood cells) because the mature elements cannot be replaced as they "wear out." If and when the stem cells recover, the mature cells reappear, giving the individual a chance to survive.

In the ***gastrointestinal tract,*** mature cells are located in and near the tips of the *villi–* microscopic projections of the small bowel lining. As these cells die off naturally, regeneration occurs by proliferation of stem cells in the bottoms of the ***crypts.*** When this process fails through injury to the stem cells, the bowel is left without a complete lining and the sequence of events described under (3) follows. Figure 27.24 (see page 378) shows a normal small intestinal villus, and one damaged by radiation.

The ***central nervous system,*** heretofore believed to be devoid of stem cells, is now known to contain them in a limited area of the brain. It has also been discovered that fetal mouse cells can multiply and differentiate into various brain-cell lines when transplanted into an adult mouse brain. As for the mature human brain, it does not have a significant pool of stem cells, so there is virtually no turnover of cell population, and brain cells that die are not replaced. Moreover, the radioresistance of mature brain cells explains why such largo doses of radiation are needed to induce the central nervous system syndrome, and why it is more properly called a ***neurovascular syndrome*** engendered by radiation injury to supporting tissues, mainly blood vessels, resulting in profound cerebral edema (swelling). A direct effect on brain cells does not have time to appear, if it were to happen at all.

DOSE-RESPONSE MODELS

The severity and frequency of a particular effect of ionizing radiation has a ***dose-effect*** relationship to the size of the single dose. Since this relationship is not always clear, several kinds of dose-response models have been proposed for specific conditions. The three most useful models include the *sigmoid, linear,* and *linear-quadratic curves.*

Sigmoid Dose-Response Curve

Characterized by an *S* shape, the sigmoid curve (see Figure 27.20) applies to high-dose effects prevailing in radiotherapy (Braestrup & Vikterlof, 1974). Such a curve has a threshold; that is, below a minimum dose under given conditions, no observable effect occurs in the exposed individual. Above this minimum, predictable effects take place. The S-shape means that a nonlinear relationship exists between effect and dose; the frequency or intensity of an effect is not pro-

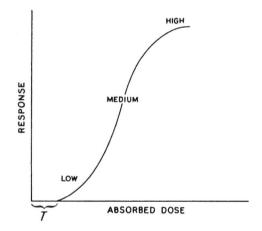

Figure 27.20. Sigmoid dose-response curve. *T* = threshold dose. As the dose increases, the response is not proportional but rather increases disproportionately for equal increments of dosage. At high doses, the effect levels off for equal increments of dosage because animals die before the specific effect can appear. Response is proportional to dose in the medium dosage range.

portional to dose. Furthermore, the shape of the curve at low doses denotes partial recovery from low doses. The curve eventually levels off and even turns downward at high doses because the animal or tissue dies before the effect appears.

The sigmoid dose-response curve has the following characteristics, according to Braestrup and Vikterlof:

1. Presence of a threshold.
2. Partial recovery from smaller doses.
3. Dose-rate effect (decreased response at low dose rate).
4. Plateau in response at upper limit of dosage; in fact, at very high dose levels curve may eventually turn downward.
5. Nonstochastic or certainty effect; predictable in the exposed individual.

We find the sigmoid-type response in radiotherapy wherein we can be reasonably certain that, for example, at a particular dosage level the skin will become red (erythema dose); or at another dosage level a severe mucosal reaction will ensue, as in the mouth; or at another dosage level a cataract will be produced. Therefore this type of response has been designated as a ***nonstochastic*** or ***certainty effect.***

Linear Dose-Response Curve

In a linear dose-effect relationship with ***no threshold*** (see Figure 27.21), any dose, no matter how small, engenders an effect. However, an effect may be present even at zero dosage if there is a *natural* incidence of the same effect. Radiation simply increases the incidence with which it occurs. The linear dose-response curve represents a direct proportion between dose and response (ie, frequency of occurrence or severity). Sometimes a threshold is present.

We may summarize the characteristics of the linear dose-response relationship as fol-

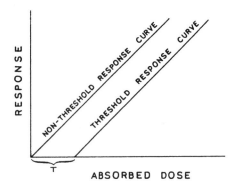

Figure 27.21. Dose-response curves for linear nonthreshold and linear threshold models. $T =$ threshold dose, that is, the minimum dose that will evoke a response. No response occurs with doses below the threshold. With a nonthreshold dose-response model, any dose, no matter how small, will evoke a response.

lows:

1. Nonthreshold or threshold.
2. Response (frequency or severity) proportional to dose.
3. No dose-rate effect (no reduced effect at small dose rates).
4. Stochastic (statistical) response; not predictable in any one exposed individual.

The nonthreshold linear dose-response curve applies to late effects that may or may not appear in a particular individual, but rather exhibit a statistically higher frequency in a population of individuals who have been exposed, in comparison with a population of like individuals who have not been exposed. In other words, such effects are ***stochastic.***

One application of the linear dose-response model has been in estimating the genetic risk of ionizing radiation. Another has been in assessing the highest possible risk of inducing leukemia or breast cancer by low doses.

Linear-Quadratic
Dose-Response Curve

In the 1980 report of the Committee on the Biologic Effects of Ionizing Radiation (BEIR), the majority proposed a *linear-quadratic* model for certain harmful effects of ionizing radiation such as cancer induction. The applicable curve (see Figure 27.22) is linear, the effect being proportional to dose x at *low dose levels,* as indicated by the straight-line curve. But at *high dose levels* the curve turns because the effect is now proportional to dose x^2, which you may remember from algebra denotes a quadratic equation. A linear-quadratic dose-response effect has no threshold and is stochastic, like the purely linear curve. However, some experts hold that the linear-quadratic model underestimates the effect at low doses. As we shall see later, when dose-effect data at high doses are extrapolated (extended) to the low-dose region, smaller (less numerous or less intense) effects are predicted than with the purely linear type response. This has led some authorities to opt for the linear model so as to err on the side of safety.

The characteristics of the linear-quadratic

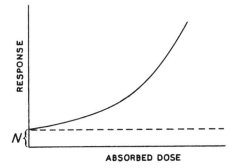

Figure 27.22. Linear-quadratic nonthreshold dose-response curve. At lower doses the response is linear, whereas at larger doses the response becomes nonlinear, that is, proportional to the square of the dose. This dose-response model has been recommended by the BEIR committee for estimating the carcinogenic risk at low doses. N indicates the natural incidence of the effect under consideration.

dose-response model may be summarized as follows:

1. No threshold.
2. Linear response at low dose levels.
3. Quadratic response at high dose levels.
4. Stochastic (statistical effect).

INJURIOUS EFFECTS OF RADIATION ON NORMAL TISSUES

This subject will be discussed under the headings of early and late effects, other than the acute whole body radiation syndromes. We must understand that no appreciable harm should accompany the diagnostic uses of radiation when all protective measures are scrupulously enforced.

EARLY EFFECTS: LIMITED AREAS OF THE BODY

The injurious effects of ionizing radiation on normal tissues depend on their cell renewal system–the way mature cells derive from primitive precursor cells. With short turnover time, mature cells are rapidly replaced by maturation of the precursor cells that display high mitotic activity and are most radiosensitive. So radiation effects appear early, within days or weeks. Such tissues include skin, gastrointestinal tract

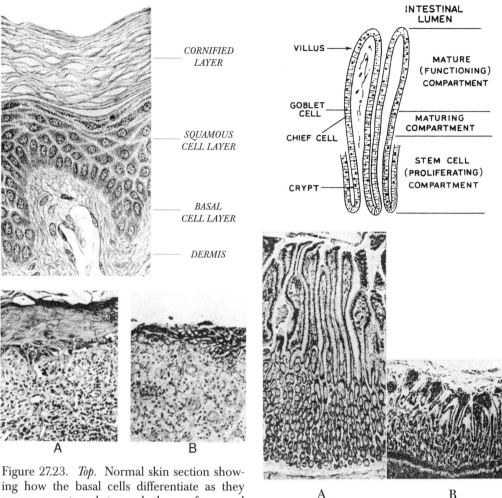

Figure 27.23. *Top*. Normal skin section showing how the basal cells differentiate as they progress outward toward the surface and become flattened squamous cells. These cornify (become horn-like) and wear off as they are replaced by underlying cells. (Adapted from Copenhaver W, et. al., (Eds.), *Bailey's Textbook of Histology*, 1978, Williams & Wilkins.) *Bottom*. Effects of radiation on skin. *A*. Early radiodermatitis: thinning of epidermis and loss of basal layer, with inflammation in underlying dermis. *B*. Reaction more severe, with ulceration, four days later. (From Lacassagne A, Gricouroff G. *Action of Radiation on Tissues*, 1958, New York, Grune & Stratton.)

Figure 27.24. *Above*. Microscopic appearance of villi (plural of villus) in the small intestine. The stem cells are located in the crypts; they proliferate and mature as they move toward the apex of the villus. As mature surface cells normally age and die off, they are replaced by proliferation and subsequent differentiation of crypt stem cells. *Below*. Effect of radiation on the small intestine of the dog. *A*. Normal intestinal lining (mucosa), with the crypts at the bottom and the villi above. *B*. Irradiated mucosa; marked damage to the crypt cells and flattening of villi, which have not been replaced by proliferation and maturation of crypt cells. (From Lacassagne A, Gricouroff G. *Action of Radiation on Tissues*, 1958, New York, Grune & Stratton.)

mucosa, red bone marrow, oral and pharyngeal mucosal lining, testes, and lymphocytes.

1. **Skin.** Renewal comes from basal cells

(see Figure 27.23), which multiply and mature toward the surface as dead cells are shed. Skin erythema (reddening) occurs just a few days after a dose of about 200 cGy with low-energy x rays (eg, 100 kVp). With increasing doses the reaction may progress to thinning of skin locally, with progression to blistering and a wet reaction. Finally, a poorly healing ulcer may ensue, becoming malignant in a matter of years. Moderate doses may cause epilation (hair loss), which may be temporary.

2. *Gastrointestinal Tract.* Mature cells on the internal surface of the gut derive from basal cells in the crypts of the lining mucosa (see Figure 27.24), the entire epithelial lining being gradually replaced daily in a continuous process. Irradiation, as in therapy, often

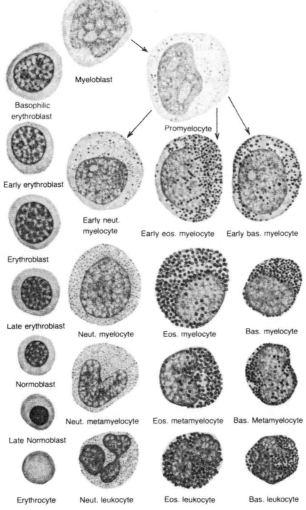

Figure 27.25. Normal development of human blood cells in the red bone marrow. They arise from a common stem cell that has not been definitely identified. Platelets have been omitted. (Adapted from Copenhaver W, et.al., (Eds.), *Bailey's Textbook of Histology,* 1978, Williams & Wilkins.)

causes diarrhea due to loss of epithelium and impaired renewal.

3. ***Red Bone Marrow.*** Mature blood cells develop in the red marrow of the skull, ribs, vertebrae, pelvis, and metaphyses of the long bones, from primitive stem cells (see Figure 27.25). Irradiation with relatively small doses suppresses renewal of the mature cell population, which begins to decline and, depending on dose, may become severe. The granulocyte count (WBC) drops first, in about two days, followed by the platelets 6 days, and the erythrocyte count (RBC) 110 days. Survival time of mature cells determines how soon suppression of precursor cell reproduction is reflected in the peripheral count of that cell line.

4. ***Oropharynx.*** Moderate to severe radi-

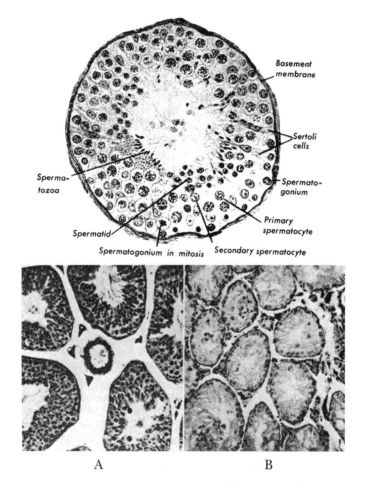

Figure 27.26. Normal development of human sperm cells (spermatozoa). The primitive stem cells are at the periphery of the testicular tubule shown in cross section. These cells move toward the center as they proliferate and mature. (From Copenhaver W, et. al., (Eds.), *Bailey's Textbook of Histology,* 1978, Williams & Wilkins.) *A.* Normal mouse testis. *B.* After a sterilizing dose of radiation, the spermatogenic cells have disappeared and mature cells have vanished as well. Only the resistant interstitial Sertoli cells remain. (From Lacassagne A, Gricouroff G. *Action of Radiation on Tissues,* 1958, New York, Grune & Stratton.)

ation reactions occur, depending on the required dose and time factors. In 1932 Coutard in France, carefully observing the sequence of radiation effects in the larynx from low-energy x rays, discovered the principle of ***fractionation therapy.*** By dividing the total dose into daily equal fractions over a period of six weeks, he noted a marked increase in curability accompanied by a significant reduction in the incidence of serious side effects. This has become the bedrock of modern radiotherapy. Typically, about two weeks after starting a course of irradiation, he noted reddening and swelling of the laryngeal lining, soon followed by a white diphtheroid membrane called an *epithelite,* and a reaction in the skin on the 26th day, an *epidermite.* Very large doses may result in slow-healing ulceration and fibrosis. Urinary bladder radiosensitivity resembles that of the oropharynx.

5. ***Testes.*** Development of sperm starts with the youngest forms, *spermatagonia,* within the periphery of the testicular tubule, then maturing into primary and secondary spermatocytes, and finally mature spermatozoa in the central region (see Figure 27.26). The spermatagonia show the greatest radiosensitivity during this maturation process. A single dose of about 500 cGy induces sterility, although this is not always permanent. Note that fractioned daily doses require a *smaller* total dose because as the radiosensitive spermatagonia die off, replacements are killed by subsequent dose fractions.

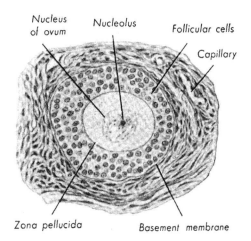

Figure 27.27. Primary follicle of a human ovum (egg cell). Each follicle produces a single ovum (290X magnification). (From Copenhaver W, et.al., (Eds.), *Bailey's Textbook of Histology,* 1978, Williams & Wilkins.)

6. ***Ovaries.*** A girl is born with a full complement of ovarian follicles, which have matured during fetal life from oogonial stem cells (see Figure 27.27). After birth, the number of follicles remains constant until the onset of puberty when a follicle then ruptures each month to release its ovum. At the time of menopause, the ovary has atrophied. The most radiosensitive stage is the ovum in the mature follicle; although 500 cGy will induce menopause in one-third of women aged 30 to 35, it will do so in 80 percent of women nearing natural menopause.

LATE EFFECTS

One should consider delayed effects (one year or more) from the standpoint of total dose: low doses of a few cGy, and high doses of at least a few hundred cGy. Common to both is a latent period before radiation effects appear, the delay ranging from about one year to many years.

Furthermore, injury may occur after a single exposure, or after repeated or chronic exposure.

Radiation injury may result from either local or whole-body exposure, the intensity of the effect in either case depending on

1. Radiation quality (LET; RBE)
2. Total absorbed dose
3. Overall exposure time (days or longer)
4. Daily dose if fractionated
5. Radiosensitivity of exposed tissue or organ
6. Volume of irradiated tissue

The magnitude of each of the above factors determines the severity of the resulting radiobiologic effects, which may be stochastic or non-stochastic.

Late Somatic Effects in High-Dose Region

We may arbitrarily assign doses of 200 cGy or more to the high dose region. Much of the available information dates back to animal experiments, as well as careless x-ray exposure of the early radiation workers. High doses engender three possible harmful effects: carcinogenesis, cataractogenesis, and life-shortening.

1. ***Carcinogenesis.*** The induction of cancer by any agent, whether radiation or chemical, is called c*arcinogenesis,* and the inciting agent a ***carcinogen.*** This can occur in experimental animals and is dose-dependent up to about 300 cGy, marked by a sigmoid dose-response curve. Experimental cancer induction in humans is out of the question, and is now strictly limited in animals by federal regulations. But there are many examples of accidental carcinogenesis in humans dating back almost to the discovery of x rays in 1895.

a. *Breast cancer* in women having received breast irradiation years earlier in (1) atomic bombing of Hiroshima and Nagasaki near the end of World War II, (2) diagnostic fluoroscopy, and (3) therapy of benign inflammatory breast disease. As indicated in NCRP Report 66, *Mammography,* there was an increased incidence with higher doses. The fluoroscopic studies involved multiple exposure of the hemithorax over months or years in following the status of pneumothorax induced for the treatment of cavitary tuberculosis; pneumothorax was in those days used to compress such cavities. There was strong correlation of breast cancer incidence on the side fluoroscoped. When the studies are combined there appears to be an increased susceptibility to breast cancer induction in the younger individuals—age 10 to 19 years.

b. *Skin cancer* in engineers, physicians, and others using x rays carelessly in the early years of radiology; and in patients years after receiving superficial x-ray therapy for acne.

c. *Bone Cancer* (sarcoma) in radium dial painters in the early l900s. Woman workers pointed fine brushes with their tongue and lips to paint clock and watch dials with radium solution to make them glow in the dark. Absorbed in the body, the radium (like calcium) was deposited preferentially in the jawbones, causing cancer after a latent period of many years.

d. *Liver Cancer* in patents who at one time received injections of thorotrast for angiography and hepatography. Thorotrast contained thorium, a natural radionuclide that accumulated in the liver.

e. *Lung cancer* in persons working in uranium mines from inhalation of radium and uranium minerals.

f. *Ankylosing spondylitis* (rheumatoid arthritis of the vertebral column). Patients who had received x-ray therapy for this painful condition many years before, showed a statistically significant increase in the incidence of leukemia.

g. *Thyroid carcinoma* in persons who, during infancy, had received x-ray therapy for so-called "enlarged thymus," actually a non-entity! Similar evidence comes from reports on patients who had been treated with x rays years earlier for acne and ringworm of the scalp, and from atom bomb fallout and nuclear power plant accidents.

h. *Leukemia* in atom bomb survivors in Japan near the end of World War II. They showed an increased incidence of chronic myelogenous leukemia after a latent period of about 14 years, with a rate of one or two cases per million population per rem (cSv*) per year averaged over the 14-year period (Brill and others). The same type of leukemia has also been reported in other accidental human exposure.

A number of possible causes have been proposed to explain the carcinogenic effect of ionizing radiation. One is *gene mutation.* The discovery of cancer-inducing **oncogenes** lends support to this theory, especially since we know that radiation induces gene mutations of which some could be oncogenic.

At least in experiments on small animals, radiation is associated with an acceleration of the *aging process.* This could explain the increased incidence of cancer, just as in the normal human aging process. But even this might result from as yet unknown spontaneous mutations.

Radiation may alter the *cellular environment* in some unidentified manner, which could then be an indirect cause of cancer induction.

After radiation injury, cells may undergo abnormal mitotic activity in attempting to recover, and incur chromosomal aberrations or gene mutations that might be carcinogenic. This has been an explanation, for example, in the appearance of skin cancer in old radiation "burns," now extremely rare. Years ago, facial acne was treated with small doses (about 50 cGy) of superficial x rays, given weekly for several weeks per course. Often, the courses were repeated several times, and even though the total dose was not expected to be carcinogenic, patients often developed numerous skin cancers in the treatment field.

2. *Cataractogenesis (induction of ocular cataract).* The normally transparent fiber cells of the lens arise from dividing precursor cells in the anterolateral portion of the lens. These fiber cells gradually migrate around toward the back of the lens, as they mature. Irradiation damages the lens by injuring the sensitive earlier dividing cells so that the mature fiber cells become opacified to form a cataract in the posterior region of the lens. As a result, vision becomes clouded by the cataract. A single x-ray dose of 250 cGy is the minimum required to produce a small, nonprogressing cataract, whereas 500 cGy is the minimum for a progressive one, with a latent period of about 8 years (Merriam, see Hall). The latent period decreases at higher doses. Note that cataractogenesis is a nonstochastic (certainty) effect, with a threshold dose and a response that increases in severity as the dose is increased. Having a high LET and RBE, *fast neutrons* are 20 times more effective than x rays in causing cataract; thus, the minimum *neutron* dose for a stationary cataract is about 12 cGy, and for a progressive one, 25 cGy.

3. *Life Shortening.* Years ago, studies on x-ray exposure of mouse colonies revealed a stochastic life-shortening effect that was dose-dependent, that is, as the dose was increased, average survival decreased. It was impossible to predict which animals would survive. With regard to Japanese atom bomb survivors, there has been no conclusive evidence of life-shortening, other than that related to cancer induction (UNSCEAR, 1977*).

Late Somatic Effects in Low-Dose Region

Low-dose effects, if any, would most likely occur during x-ray or radionuclide diagnostic procedures, and would be stochastic,

---

*1 cSv = 1 rem
*United Nations Scientific Committee on the Effects of Atomic Radiation.

occurring at random in exposed individuals. Because of even the slightest possibility of harm, we must constantly weigh benefit against risk, expressed by the benefit/risk ratio.

Unfortunately, we cannot readily estimate the benefit/risk ratio for a particular radiodiagnostic procedure, so we can assume that if a certain small dose is known to cause a certain incidence of a particular effect, a smaller dose would produce a smaller incidence of the same effect. Moreover, we assume a linear, nonthreshold response; that is, any dose, no matter how small, can induce an effect. For example, if 200 cGy were to create an excess incidence of cancer amounting to 1 percent in a particular organ, we could extrapolate (extend curve) backward to determine the excess cancer incidence at very low levels of exposure. If this errs, it most likely does so on the side of safety.

There are two major categories of *low-dose effects.* One is somatic and includes carcinogenesis in the adult and in the conceptus whereas the other involves genetic damage. They will now be described.

1. **Carcinogenesis.** Repetitive small doses of ionizing radiation do increase the incidence of cancer, as witnessed by the occurrence of skin cancer in unprotected early workers with x rays, and in patients receiving repeated small doses for facial acne. Especially important is *leukemia,* cancer of the blood-forming organs; studies of women exposed to diagnostic abdominal x rays during pregnancy indicated about a 40 percent increased incidence of leukemia in the *child* some years later (MacMahon). In fact, 1 cGy to the gravid uterus would be expected to increase the incidence in the child from the natural rate of 10/10,000 live births to 14/10,000, after a latent period of about two to four years. Note that a fetal dose of about 1 cGy would be delivered by four anteroposterior views of the abdomen

with medium speed screens, but one-fourth that dose with high-speed screens; thus, the great advance in patient protection with the introduction and widespread use of high-speed screens.

In recent years *mammography* has become a valuable tool for early detection of *breast cancer,* with attendant improvement in curability. With modern mammography using fine-grain special low-dose films and intensifying screens, and a 3:1 or 4:1 grid, the total midplane dose to the breast for two views should average less than 0.6 cGy with screen-grid technique). According to the method of Feig (see Selman, *Elements of Radiobiology,* p. 181), modified for the lower breast dose just cited, the benefit/risk ratio is about 70:1; that is, about 70 cancers detected per cancer induced. But bear in mind that this is a gross estimate.

2. **Genetic Damage.** Ionizing radiation may induce, in sperm and ova, changes that can be transmitted to future generations. We designate this the *genetic effect of radiation,* the targets of which include DNA and chromosomes. The *point* changes in the DNA macromolecules are called *gene mutations,* whereas injury to the chromosomes results in breaks and various structural changes called *chromosomal aberrations.* In 1927, H. J. Mueller conclusively demonstrated the induction of gene mutations in fruitflies (*Drosophila sp.*) by x rays; for this, he was awarded a Nobel Prize.

Not all gene mutations result from x-ray or chemical exposure—there is also a natural incidence of spontaneous mutations which, although probably due to background radiation and chemical pollution, may also be caused by errors in replication of DNA. When you stop to think of the inordinate number of times that DNA replicates during a lifetime, it's no surprise that copy errors may occur.

The following summary points out the more relevant information about gene mutations and chromosome aberrations:

a. ***Gene Mutations.*** Each gene consists of DNA segments that dictate the manufacture of specific substances–the full ranges of proteins, including enzymes–by the cell. Changes in DNA composition may consequently engender profound changes not only in cell structure and function, but also in its progeny.

1. *Mutagenic Agents.* A great variety of chemicals, besides ionizing agents, can induce gene mutations.

2. *Radiation-induced Mutations Recessive.* The vast majority of radiation mutations are recessive, that is, they become manifest only if the individual mates with another having the same mutant gene; or if the mutant gene is located in the X chromosome of a son of an irradiated *mother* (the X chromosome in the human male is paired with a Y chromosome that carries only a few genes, so that any gene in the male's X chromosome acts as a dominant and is therefore expressed in that individual).

3. *Delay in Appearance of Gene Mutation.* Because most radiation-induced mutations are recessive, their manifestation may be delayed for a number of generations, or may not appear at all. In fact, some are lethal, causing death of the embryo and therefore going undetected.

4. *Radiation-induced Mutations Not Unique.* There are no characteristic differences between radiation-induced and spontaneous mutations; radiation simply increases the natural mutation rate. Note that mutagens (mutation-inducing agents), both chemical and radiation, can have a marked impact on health and still go unrecognized. Even with a high mutation rate, mutations may be so hidden by other factors (concurrent disease or other abnormalities) that one cannot blame such effects on a specific mutagen. For example, one cannot tell whether a patient with leukemia incurred this cancer by exposure to radiation, or by spontaneous

gene mutation.

5. *High-LET Radiation More Effective.* For equal doses, high-LET radiation, like fast neutrons, is more effective in causing gene mutations and aberrations than is low-LET radiation.

6. *Linear, Non-Threshold Dose-Effect Curve.* Down to a dose of a few cGy, the dose-response curve appears linear, so a non-threshold response is assumed; that is, any dose, no matter how small, can induce a mutation.

7. *Dose-Rate Dependence.* Evidence points to the fact that low-LET radiation (x and gamma rays) delivered at low dose rates–cGy per minute–and with fractionation are less mutagenic. This undoubtedly depends on repair mechanisms occurring under these conditions.

b. ***Chromosome Aberrations.*** Sometimes referred to as mutations they are more correctly called ***aberrations*** because the radiation-induced lesions consist of breaks, translocations, rearrangement of fragments, etc (see Figure 27.14). Recall that genes are lined up on a chromosome, although they can sometimes shift positions on it. We cannot always distinguish between a mutation in a gene's DNA and a chromosome aberration. With x rays, the rate of chromosome damage is nearly proportional to dose within the medical range.

In summary, radiation is a proven mutagen for genes, and is also capable of breaking chromosomes. In either case, physical and physiological changes may occur in the individual harboring the mutant genes or aberrant chromosomes. Some effects may be so subtle that they are beyond our present capacity to detect, but we must always be vigilant to avoid unnecessary exposure to ionizing radiation. Also, note that ***single-strand breaks*** may heal themselves. No perceptible increase in incidence of anomalies has been found in children borne of parents who had been exposed to as much as *100 cSv* of radiation during the atom bombing of Japan.

Nevertheless, every effort should be made to keep gonadal exposure to a minimum during x-ray and radionuclide examinations, but still consistent with medical needs. Care should be exercised in fertile women who may knowingly or unknowingly be pregnant, especially during the first six weeks– the period of major organogenesis. Since many gene mutations are recessive, they may be hidden and not appear in the exposed conceptus after birth. As will be emphasized later (see Health Physics) various devices must be present in the radiology department, such as gonadal shields, efficient positive beam-limiting "collimators", adequate beam filtration, and optimum speed film-screen combinations with accurate technique or automatic exposure control.

The chapter on Health Physics will be devoted to the radiation sources and protective measures for patients and personnel. The units applying specifically to human radiation exposure will be included.

Risk Estimates for Genetic Damage

We have no direct proof for radiation-induced mutations in humans, but we can reasonably assume that the genetic lesions produced in small animals are basically similar, although their relationship to dose and other factors may not be the same. Still, we do have reliable evidence for the induction of *chromosome aberrations* and *gene mutations*.

For low doses and low dose rates of low-LET radiation, BEIR* (1980) extrapolates from high-dose data in mice, using a linear dose-response curve. Such data are used to estimate the genetic risk of radiation in man.

We shall now consider the increased risk of radiation mutagenesis relative to the spontaneous mutation rate in humans, but first we shall define two terms.

Genetically Significant Dose (GSD). As shown in Table 28.1, page 393, two major sources of human exposure to low level radiation are medical and dental x rays, and natural background. As for medical and dental x rays, only a *part* of the population is exposed, and so the GSD was introduced by the Bureau of Radiological Health (BRH) in 1970 as a statistical concept that estimates the genetic impact of such radiation. It is no longer widely used, but will be explained briefly for the sake of completeness and because of its theoretical validity.

The GSD regards radiation-induced mutations in the gametes (sex cells) of an individual as belonging to a genetic "pool" of the entire population of the U.S.A., so the average dose of *actually exposed* individuals during radiography is divided among this population of fertile persons, as well as their expected offspring. Although GSD is expressed as a dose, it could more logically be regarded as a *mutagenic equivalent dose* (Gaulden, see Dalrymple).

To simplify, let us suppose that four persons received an actual average gonadal dose of 1 cGy during radiography, and there were 1000 fertile individuals in the population as a whole. Then

$$GSD \times 1000 = 1 \text{ cGy} \times 4 \text{ persons}$$

$$GSD = \frac{4}{1000} = 0.004 \text{ cGy}$$

In summary, then, the gonadal doses have been prorated over the entire fertile population. For background radiation, the GSD and gonadal doses are identical because the entire population is continually receiving the same low-level exposure. More acceptable now is the concept to be

---

* Committee on the Biological Effects of Ionizing Radiation.

described next.

Mutation Rate or Frequency. The number of mutations, spontaneous or radiation-induced, occurring per generation is called the *mutation rate* or *mutation frequency.* In evaluating the genetic effects of ionizing radiation, we customarily refer to the increased mutation rate as compared with the natural mutation rate; this is, the *relative mutation risk.*

As yet we have no experimental basis for relating the incidence of complex inherited disease to gene mutation frequency in humans, but we still need at least an estimate of the genetic risk associated with ionizing radiation at low doses so as to set up guidelines for permissible exposure levels. Such estimates must be based either on experimental evidence at the *lowest possible doses and dose rates,* or on *extrapolation from data at high doses and dose rates.* Insofar as genetic changes are concerned, a purely *linear, non-threshold dose-response curve* is used for extrapolation. Furthermore, there is good evidence that the incidence of gene and chromosome effects is proportional to dose at low dose levels.

Federal guidelines recommend that the natural background level (see pages 392-394) serve as a standard, assuming that artificial radiation such as x rays in a whole body dose equal to, and in addition to, background would produce additional genetic defects similar in number and kind to those engendered by background alone. Therefore, we must aim to keep radiation exposure to a minimum relative to background, to minimize any increase in mutation rate.

A useful concept is embodied in the ***doubling dose***–that dose of ionizing radiation which causes as many mutations per generation as occurs naturally. Based on old experiments with mice and on data from Japanese survivors of the atom bomb attack at the end of World War II, the best available estimate

for the doubling dose falls within the range of 25 to 250 cSv (rem; pages 395-396). The relative mutation risk *per cSv* would then be

$1/250$ to $1/25 = 0.004$ to 0.04 mutation per cSv

What increase in the natural mutation rate would be expected from a dose of 5 cSv per generation? This has the following range:

$$5 \text{ cSv} \times 0.004 = 0.02, \text{ or } 2\%$$
$$5 \text{ cSv} \times 0.04 = 0.20, \text{ or } 20\%$$

The mutation rate would fall somewhere between 2 and 20%, a wide range, but the best estimate possible today.

We still do not know the natural mutation rate for humans, so it is difficult to appreciate the impact of an added 2 to 20% to an unknown mutational burden. We do know that natural background in the U.S.A. indoors averages about 0.8 mSv (80 mrem) per year and that it has no known deleterious effects, but which cannot be excluded. Also, it varies from place to place: 0.15 mSv (15 mrem) at sea level along the Atlantic and Gulf coasts, as high as 1.4 mSv (140 mrem) in Colorado, and even higher in Kerala, India where it is about 20 mSv (2000 mrem) per year. Still, there is no marked difference in the incidence of congenital abnormalities.

Radiation Injury to Embryo and Fetus

In humans, the *conceptus*–product of conception–includes ***embryo*** through the second month of pregnancy, and ***fetus*** thereafter until term. Obviously, radiation experiments cannot be performed in humans, so radiobiologists have had to rely on (1) experiments on rats and mice, and (2) retrospective study of radiation accidents involving pregnant women as in the atomic bomb experience in Japan, and (3) occasional instances of irradiation of pregnant women for pelvic cancer. We cannot rely on anecdotal reports of anomalies in the newborn because at least six percent of live births in humans have anomalies unrelated to known sources of radiation, other than background.

Why have small animals been useful in research on radiation injury to the conceptuses. All mammals that have been studied respond similarly during embryonic development, organ for organ, and stage for stage. Only the *timetable* of the responses differs among the species, depending on their gestation period. Whereas gestation in the mouse lasts 20 days, corresponding to 288 days in humans, the *fraction* of gestation time for various stages is not the same in humans and mice. Major organogenesis (organ formation) in mice lasts from about the sixth to the thirteenth day, or up to *two-thirds* of total gestation time. The corresponding period in humans is the eleventh to the forty-first day, or about *one-seventh* of the gestation time.

The results of irradiating mice with various doses at selected embryonic stages can be extrapolated (carried over) to humans with reasonable certainty.

In 1952 Russell and Russell, using 200-R exposures in mouse embryos, found that gestation may be divided into three stages: (1) preimplantation, (2) organogenesis, and (3) fetal development. We shall now describe these, extrapolated to the human time frame (see Figure 27.28), bearing in mind that damage to the embryo can occur with exposures as little as 10 R.

Preimplantation Stage in Humans (0 to 10 days). After fertilization of the ovum it usually takes nine days for the embryo to reach the uterus *via* the fallopian tube. This is the preimplantation period during which the embryo exhibits marked radiosensitivity. In fact, doses as small as 5 to 15 cGy may kill the embryo, but surviving embryos are normal except for genetic chromosomal damage. Irradiation during this very early stage

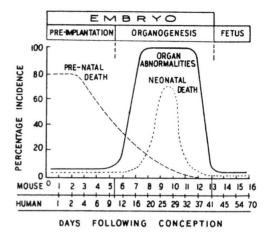

Figure 27.28. Effects of 200 R given on various days of gestation in mice, with comparative time scale for humans. The most severe organ damage occurs during period of major organogenesis, about 11 to 41 days in humans. (Adapted from Russell LB, Russell WL. J Cell Physiology (Suppl 1) 43:103, 1954.)

has an ***all-or-none effect*** insofar as survival is concerned, but genetic effects may still occur.

Organogenesis Stage in Humans (11 to 41 days). During the period of organogenesis, ***virtually every embryo surviving radiation exposure develops an anomaly.*** The organ or tissue suffering the greatest injury is the one irradiated at the time that its precursor cells are undergoing the *most active multiplication (mitosis) and differentiation (maturation).* In the mouse, brain herniation is especially prone to occur when the embryo is irradiated at eight days, nose and ear anomalies when irradiated at nine days, and microcephaly (small head) when irradiated at 11 days. The greatest *variety* of anomalies occurs in mice irradiated during the ninth to eleventh day of gestation, corresponding to the *25th to 27th day in humans.*

Murphy (1947) found in a study of women receiving pelvic radiotherapy during pregnancy that out of 28 fetuses, 16 (or about

60 percent) were microcephalic. Others had spine bifida, club feet, skull defects, hydrocephalus, or blindness, separately or in combination. Various authors using doses as small as 18 cGy have found an anomaly rate of 25 percent.

In humans, central nervous system malformations represent an especially important outcome of irradiation *in utero*. In their most primitive stage (neurectoderm) nerve cell precursors are moderately radiosensitive, requiring about 400 cGy to kill all cells (Rugh, 1965), in contrast to a dose of about 10,000 cGy for the same effect in mature neurons (nerve cells). From the 17th day of gestation to full term delivery (and a few weeks into the neonatal period) neuroblasts are present throughout the central nervous system; a dose of only 25 cGy is lethal for these extremely radiosensitive cells, showing why nervous system malformations represent the most common type of radiation anomalies in embryonic and fetal life.

Fetal Development Stage in Humans (6 weeks to term). During this stage, visible malformations are mild or absent, depending on dose. The lethal dose approaches that in adults. Functional impairment does occur but is difficult to identify, involving mainly disorders of intelligence and growth.

Dekaban (1968) has summarized the anomalies most likely to occur in human embryos and fetuses subjected to 250 cGy at various times during gestation:

Less Than 2-3 Weeks–high probability of lethal effect with absorption of embryo; low probability of anomalies.

4-11 Weeks–severe anomalies of many organs in most children.

11-16 Weeks–microcephaly, stunted growth, and genital organ anomalies.

16-20 Weeks–mild microcephaly, mental retardation, and stunting of growth.

More Than 30 Weeks–rare visible abnormalities, but functional impairment such as mental deficiencies may occur; these may be too difficult to demonstrate. Midlethal dose approaches that in adults.

There is at present no evidence of a threshold dose for radiation injury to the conceptus, but the severity of the effect is dose-dependent–the number of cells damaged increases with dose. The embryo as a whole is much more radiosensitive than the adult. Thus, Rugh (1963) found that the $LD_{50/30}$ ranges from about 16 to 60 percent of the adult dose, which comes out to 65 to 240 cGy (x rays); recall that the adult $LD_{50/30}$ is about 400 cGy.

Although no anomalies have been conclusively reported in the human conceptus from low doses, even a few cGy can induce chromosome aberrations such as *breaks*. Moreover, as already mentioned above, diagnostic x-ray exposure of the pregnant uterus entails a small but detectable increase in the incidence of childhood leukemia. In fact, Hammer-Jacobsen at one time recommended induced abortion of any pregnancy of six weeks or less if the embryo had received 10 cGy or more.

Some years ago the 10-day rule was promulgated, based on the so- called "safe" period 10 days before the onset of menstruation when pregnancy uncommonly occurs. However, due to its unreliability, this rule has been abandoned. In 1993, the ICRP* recommended the 28-day rule which states that the safe period for exposure of the possibly pregnant uterus is 28 days *after* menstruation.

This rule has also been discarded and a policy adopted that if the examination is necessary it should be done with minimal exposure to the lower abdomen; and if the examination is not really necessary, it should be omitted or postponed, or reordered at the discretion of the referring physician after consultation with the radiologist.

QUESTIONS AND PROBLEMS

1. Summarize the early history of radiobiology.
2. Who initiated modern radiobiology in 1956?
3. State the two-step physical process in radiobiology.
4. How is ion track density related to the ionizing electron's energy.
5. Define directly and indirectly ionizing radiation.
6. Define linear energy transfer (LET) and give examples of low- and high-LET radiation.
7. State the equation for relative biologic effectiveness (RBE).
8. Describe with a diagram a typical mammalian cell, labeling the important constituents.
9. Explain the nature of chromosomes, and describe their function.
10. Define haploid and diploid number.
11. Describe and label a diagram of normal mitosis.
12. What is the Howard-Pelc cell cycle?
13. Explain the chromosome numbers in meiosis.
14. Discuss the structure of DNA.
15. What are genes—their composition and location?
16. What are the two functions of DNA?
17. Explain the parallelism of the DNA strands.
18. What is the fundamental radiobiologic target of ionizing radiation?
19. Differentiate direct action from indirect action. How do they differ from directly and indirectly ionizing radiation?
20. What are free radicals ? Why are they important?
21. Describe radiation damage to nucleus and cytoplasm, as seen under a microscope.
22. How are cell survival curves obtained? Explain their three main segments with the aid of a typical curve.
23. What does a steeper curve indicate?
24. Define oxygen effect; oxygen enhancement ratio.
25. What is meant by an acute whole body radiation syndrome?
26. Name the four acute radiation syndromes, and explain their basis.
27. Draw the shapes of three types of dose-response curves with the graph axes, but without numerical values.
28. Define threshold dose; nonthreshold dose.
29. List the early effects of x rays on limited areas of skin, GI tract, bone marrow, and gonads.
30. How does a dose-response curve differ from a cell-survival curve?
31. In what time-frame in human pregnancy is the embryo most sensitive to ionizing radiation in terms of major organ damage? Why?
32. What is your opinion of the 10-day rule?
33. Name the three main late effects of high-dose radiation.
34. List the major late effects in the low-dose region.
35. What is the difference between gene mutation and chromosome aberration?
36. Would you say the risk of injury to a radiologic technologist in a modern supervised radiology department is (a) extremely low, (b) moderate, (c) high?

Chapter 28

PROTECTION IN RADIOLOGY–HEALTH PHYSICS

Topics Covered in This Chapter

- Introduction
- Background Radiation
- Dose Equivalent Limits
    Derivation of Units
    Occupational Limits
    Nonoccupational Limits
    ALARA Concept
- Personnel Protection
    Monitor Badges

    Protective Measures
- Acceptance Testing
- Protection Surveys
- Protection of Patient
    Dose Reduction in Radiography
    Dose Reduction in Fluoroscopy
- Protection in Nuclear Medicine
- Questions and Problems

Introduction

THE HARMFUL EFFECTS of exposure to ionizing radiation were largely unsuspected at the time of the discovery of x rays and radium. However, the pioneer workers in this field soon recognized the injurious potentiality of such radiation. It is therefore surprising that so many of the early radiologists and technicians carelessly exposed themselves to x rays and radium, incurring serious local and general radiation injuries, which all too often resulted in death.

In the intervening years we have learned a great deal about radiation hazards and their prevention. As a result, protective measures have become ever more stringent, especially with the extensive use of radionuclides in medicine and industry, and nuclear reactors in the generation of electric power.

Every radiologic technologist should have access to pertinent *National Council on Radiation Protection and Measurements (NCRP)* *Reports; Bureau of Radiologic Health (BRH) publications;* and *Code of Federal Regulations 21.* These are listed in the Bibliography, together with information about their availability.

What is the "safe" limit of exposure to radiation? It is doubtful whether any amount of radiation is really safe. However, an attempt will be made to answer this question on the basis of acceptable exposure limits, as well as to point out what steps can be taken to eliminate the risk of overexposure. The subject of protection or *health physics* will be taken up principally as it affects the *radiologic technologist* and the *patient.* Attention will be directed to the use of artificial radionuclides, as well as x rays.

It is generally assumed that no threshold dose exists for radiation effects, that is, there is no minimal dose below which the risk of radiation injury is zero. Therefore, with any exposure to artificial radiation one must weigh benefit against risk. Thus, if an x-ray or nuclear medicine procedure will contribute nothing to the diagnosis, then the matter has to be decided between the radiologist and referring physician. The risk-benefit ratio, R/B, should be as low as possible, short of harming the patient by omitting the examination.

Related to the R/B ratio are the concepts of *ALARA* and *NIRL* (NCR Report 91). ALARA is the acronym for *A*s *L*ow *A*s *R*easonably *A*chievable, which assumes a linear, nonthreshold dose-effect response. ALARA is an excellent goal in radiation protection, as it errs on the side of "safety." However, this may be limited by excessive cost related to vanishingly small benefit. Therefore one may follow NIRL, the acronym for *N*egative *I*ndividual *R*isk *L*evel, defined as the level of exposure below which the risk of fatality is trivial compared to other risks in daily life (eg, occupational and recreational). Thus, NIRL implies the lower limit of ALARA except that the latter pertains to risks other than fatal outcome. According to NCRP, an annual effective NIRL dose of 0.01 mSv (1 mrem) is a good estimate, as it is trivial when compared to other risks in daily life (fatality risk ranges from $1/10^8$ to $1/10^6$).

BACKGROUND RADIATION

A sensitive radiation detector, such as a GM counter or ion chamber survey meter should indicate the presence of environmental ionizing radiation even in the absence of all known sources, such as x rays and radioactive nuclides. As it turns out, man has virtually no control over environmental radiation, which we call *natural background radiation,* and to which all matter, living and nonliving, is continuously exposed.

Natural background includes not only external sources of radiation, but also radioactive materials located within the body and even the detector itself. Let us survey the sources of natural background radiation, which may be classified as *external* and *internal.*

1. *External sources* comprise three categories. First are the *cosmic rays,* which include two types. The *primary cosmic rays,* arising in the sun and other stars, consist mainly of high energy protons (more than 2.5 billion electron volts), but also include alpha particles, atomic nuclei, and high energy electrons and photons. *Secondary cosmic rays* are produced by interaction of primary cosmic rays with nuclei in the earth's atmos-phere and consist mainly of mesons (a type of nuclear particle), electrons, a small number of neutrons (absorbed mainly at high altitudes), and gamma rays. Most cosmic rays detected on earth are secondary and are extremely penetrating. However, human exposure indoors is about *80 percent* of that outdoors due to residential shielding.

A *second* external source of background radiation comprises the *naturally radioactive minerals* within the earth. These occur in minute amounts almost everywhere in the earth's crust, including building materials, and in larger amounts where the minerals of uranium, thorium, and actinium have been deposited. It must be emphasized that terrestrial (earthly) background radiation varies from place to place; for example, in Kerala, India, background averages about 500 – 600 mrem/yr, with a maximum of about 3000 mrem/yr (Sunta & others, quoted in Eisenbud) .

A *third* external source is the radionuclides produced by interaction of cosmic rays with nuclides in the earth's atmosphere. The main products are carbon 14 and hydrogen 3 (tritium) resulting from capture of slow cosmic-ray neutrons by stable atmospheric nitrogen, sodium 22, and beryllium 7.

2. **Internal sources** comprise the naturally radioactive nuclides incorporated in the tissues of the body, and in the materials of which the detector itself is constructed. The main naturally-occurring radioisotopes of stable nuclei are potassium 40, carbon 14, and lead 210 (polonium 210).

To the natural background radiation must be added the **artificial radiation** arising from the following sources: medical and dental x rays; occupational, such as x-rays and radionuclide technology; environs of atomic energy plants; and residues of atomic bomb fallout. In contrast to natural background, man can exercise some measure of control over exposure to artificial sources.

Table 28.1 summarizes the **average** exposures of the general population of the United States on an annual basis as to the site of exposure, dose per exposed person, and the

Table 28.1
ANNUAL DOSE RATES FROM IMPORTANT SIGNIFICANT SOURCES
OF RADIATION EXPOSURE IN THE U.S. (BEIR 1980).*
Multiply each value by 0.01 to convert to mSv.

Source	Exposed Group	Body Portion Exposed	Average Dose (exposed persons)	Prorated over Population
			mrem/yr	mrem/yr
NATURAL BACKGROUND				
cosmic radiation	total population	whole body	28	28
terrestrial radiation	total population	whole body	26	26
internal sources	total population	gonads	28	28
		bone marrow	24	24
		Total (rounded)	80	80**
MEDICAL X RAYS				
Patients				
medical diagnosis	adults	bone marrow	100	77
dental diagnosis	adults	bone marrow	3	1.4
		Total (rounded)	103	78
Personnel (occupational)				
medical	adults	whole body	320	0.3
dental	adults	whole body	80	0.05
RADIOPHARMACEUTICALS				
*Patients (diagnosis)****	adults	bone marrow	300	13.6
Personnel	adults	whole body	300	0.1
ATMOSPHERIC WEAPONS TESTING	total population	whole body	4-5	4-5

*NCRP REPORT No. 105 (1989) gives natural background values that differ slightly from BEIR.

**Based on shielding by dwellings and other buildings; otherwise, average 100 mrem/yr. Range on Atlantic and Gulf coasts 15 to 35 mrem/yr, and Colorado Plateau 75 to 140 mrem/yr.

***The Environmental Protection Agency (Report ORP/CDS 72-1) estimates the exposure of patients during nuclear medicine procedures to be about 20 percent of the total radiation exposure in medical diagnosis.

prorated exposure to the population as a whole (based on *BEIR Report, 1980).* Obviously, the concern with these small exposures is with genetic effects appearing in future generations from damage to the reproductive cells, and with injury to the bone marrow. Bear in mind that these data are approximate and do not necessarily apply to a particular individual.

DOSE EQUIVALENT LIMIT

Our knowledge about the deleterious effects of ionizing radiation on living cells demands that an upper limit of absorbed dose be established for personnel and for the population at large. Even when radiation exposure is required for medical diagnosis, it must be kept as low as possible–ALARA (see page 392).

The concept of dose limitation implies *risk limitation.* The dose equivalent limit will be detailed later, but at this point we shall consider it on the basis of stochastic and nonstochastic effects:

A *stochastic effect* pertains only to the *statistical chance* or probability that a particular effect will occur, usually from a small absorbed dose. This can be plotted graphically as incidence of effect as a function of dose (see Figure 27.22, page 377). Note that the severity of a radiation injury is not included. A stochastic effect is assumed *not* to have a threshold dose (ie, minimal dose below which there is no effect). Examples of stochastic effects include cancer induction (solid cancer and leukemia) and genetic effects.

A *nonstochastic effect* is one which increases in *severity* with increasing absorbed dose by way of injury to increasing numbers of normal cells. It may be regarded as a *certainty effect* in that there are threshold doses for particular clinically significant radiation injuries. Such thresholds are high enough that they may be avoided. Examples include cataracts (opacities of the ocular lens), suppression of bone marrow, and impaired spermatogenesis.

DERIVATION OF UNIT FOR RISK ASSESSMENT

Before taking up the special dose unit for radiation protection, we must define two underlying concepts: (1) linear energy transfer (LET), and (2) relative biologic effectiveness (RBE).

1. *Linear energy transfer* is the amount of energy deposited per unit length of path by an ionizing particle, expressed in $keV/\mu m$ (kiloelectron volts per micrometer of path). The higher the average LET, the more severe will be the expected biologic effect. For example, neutrons produce a stronger reaction than x rays for an equal absorbed dose (cGy or reds).

2. *Relative biologic effectiveness* is used primarily in radiobiology to express the ratio of the absorbed dose of a standard radiation–200-kV x rays–to that of a given kind of radiation, to produce the same intensity of biologic effect, expressed by

$$\text{RBE} = \frac{\text{dose of standard radiation (200-kV x rays)}}{\text{dose of radiation in question}}$$

for a given effect

We may therefore conclude that equal absorbed doses of radiations having different

LETs would pose different degrees of risk. Hence, in assessing risk, it would be misleading to consider only the absorbed dose of single or mixed radiations when one or more has an RBE exceeding unity.

Dose Equivalent

To express on a common scale for protection purposes, the risk of radiation injury from all kinds of ionizing radiation, the quantity **dose equivalent H** was introduced by the ICRU many years ago and adopted by the NCRP. The dose equivalent is obtained by simply multiplying the absorbed dose of a given type of radiation by its relevant **quality factor $\overline{Q}$.** Derived experimentally from LET in water, $\overline{Q}$ for a given kind of radiation is then related to an arbitrary value of unity for low-LET radiation (x, beta, gamma) (see Table 28.2). The **SI unit** of dose equivalent **H** is the sievert (Sv) = 100 cSv (centisievert).

Table 28.2
RECOMMENDED VALUES OF $\overline{Q}$ FOR
RADIATION OF VARIOUS KINDS*

Radiation	Approximate Value of $\overline{Q}$
X-rays, gamma rays, beta particles, and electrons	1
Thermal neutrons	5
Neutrons (other than thermal), protons, alpha particles, and multiple-charged particles	20

*NCRP Report No. 91

The following equation states how dose equivalent H^* is derived:

$$H = \textbf{absorbed dose} \times \overline{Q} \quad (1)$$
$$rem = rad \times \overline{Q}$$
$$cSv = cGy \times \overline{Q}$$

---

*NCRP Report No. 91

The appropriate value of $\overline{Q}$ converts cGy to cSv, and rad to rem; the units are related as follows:

$$1 \text{ sievert (Sv)} = 100 \text{ rem}$$
$$\text{and } 1 \text{ gray (Gy)} = 100 \text{ rad}$$

Therefore,

$$1 \text{ centisievert (cSv)} = 1 \text{ rem}$$
$$\text{and } 1 \text{ centigray (cGy)} = 1 \text{ rad}$$

As already noted, *x, gamma, and beta (electron)* radiation have been arbitrarily assigned a $\overline{Q}$ value of 1, so

$$H = \text{absorbed dose} \times 1$$
$$= 1 \text{ cGy (rad)} \times 1 = 1 \text{ cSv (rem)}$$

for these specific kinds of radiation. In protection surveys for low-LET radiations, we may use the roentgen (R) in place of cSv or rem because they are numerically about the same; moreover, survey instruments are still calibrated in R. But organ doses should still be expressed in *H* units–cSv or rem.

For fast neutrons, $\overline{Q} = 20$, that is, 1 cGy of fast neutrons is 20 times as effective as 1 cGy of x rays for induction of ocular cataract; conversely, the cataractogenic dose of fast neutrons is $1/20$ that of x rays. Since the x-ray dose is 250 cGy for stationary cataract induction,

$$H = \text{absorbed dose} \times 1/20$$
$$= 250 \text{ cGy} \times 1/20 = 12.5 \text{ cSv}$$

for fast-neutron cataract induction

You can see clearly how the dose equivalent expresses the relative risk from radiations having different $\overline{Q}$ values.

The dose equivalents of various kinds of radiation (ie, different values of $\overline{Q}$) can be added, whereas absorbed doses cannot. For example, if a person were exposed to a mixture of radiations, receiving 1 cGy of x rays and 1 cGy of fast neutrons, the total dose would be 2 cGy. But this would seriously underestimate the radiation hazard. Instead, conversion to *H* would give

x rays	1 cGy =	1 cSv
fast neutrons	1 cGy =	20 cSv
	Total	21 cSv (rems)

Thus, the value of *H* in cSv gives a more realistic measure of radiobiologic hazard when $\overline{Q}$ differs from unity. In medical radiology and radiotherapy, however, by far the most commonly used radiation comprises x and γ rays whose $\overline{Q} = 1$.

Effective Dose Equivalent. With uniform irradiation of the whole body, stochastic effects (cancer induction and genetic injuries) may be expressed by the dose equivalent to the whole body, symbolized by H_{wb}. However, truly total body irradiation, such as one might expect in a radiology department, does not entail uniform tissue dosage because of variations in absorption of the low-energy x rays, as well as uncertainties introduced by shielding with lead aprons and other protective devices.

Because of this inherent nonuniformity of a single exposure to ionizing radiation, and the different stochastic limits for various organs, the ICRP introduced, and the *NCRP* Report No. 91 pursued, a new concept–*effective dose equivalent H_E*–to provide a more realistic assessment of stochastic risk by assigning *weighting factors* to individual organs. This attempts to correct for the appreciable differences in dose equivalent *H* to individual organs, especially when a mixture of radiations is involved.

At present, there is some question about the relevance of H_E in diagnostic radiology where the main source of exposure to personnel is in fluoroscopy and special procedures, and in diagnostic nuclear medicine–these all deal with low-LET radiation. Recall that we are concerned with stochastic limits. The attempt to relate exposure monitoring with film badge, Luxel® and TLD, to H_E, is complicated by uncertainties in shielding and in absorbed dose to various organs, so the readings obtained with these monitors worn on the collar outside the lead apron remain the most reliable index of exposure. It turns out that *NCRP Report No. 107* (p. 9) assumes the monitor readings adequately represent H_E when reported for low-LET radiation (x and gamma rays) in units of either dose equivalent *H* (mSv or rem), absorbed dose (cGy or rad), or exposure (R).

NUMERICAL DOSE EQUIVALENT LIMITS

Occupational

As explained in the preceding section, effective dose equivalent may be replaced by the simple *dose equivalent H* for setting exposure limits for radiation workers in diagnostic imaging. According to *NCRP Report No. 91,* a uniform dose equivalent of 1 rem (10 mSv) to the whole body is estimated to entail a lifetime fatality risk of 10^{-4} (ie, 1 in 10,000) as a *stochastic* effect for adults. But data from the Environmental Protection Agency (quoted in *NCRP Report No. 91)* indicate an average of only about 2.3 mSv (0.23 rem) per year for monitored radiation workers with measurable exposure; that is, $1/20$ the earlier NCRP prospective annual whole body exposure limit of 50 mSv (5 rem). Thus, a reduction in occupational exposure limits is achievable on the basis of current practice, so the NCRP makes the following recommendations (see Table 28.3 for summary):

1. For *Stochastic Effects:*

a. *Abandon the previous dose limit 5 (age – 18) rem.*

Table 28.3
DOSE LIMIT RECOMMENDATIONS

OCCUPATIONAL EXPOSURE *(ANNUAL)*		
Effective Dose Equivalent Limit*		
(stochastic effects)	50 mSv	5 rem
Dose Equivalent Limits for Tissues and Organs		
(nonstochastic effects)		
Lens of eye	150 mSv	15 rem
All other organs (eg, red marrow, breast,		
lung, gonads, skin, and limbs)	500 mSv	50 rem
Guidance: cumulative exposure**	10 mSv × age	1 rem × age in years
PUBLIC EXPOSURE *(ANNUAL)*		
Effective Dose Equivalent Limit		
continuous or infrequent exposure***	1 mSv	0.1 rem
infrequent exposure	5 mSv	0.5 rem
Dose Equivalent Limits:		
lens of eye, skin, and limbs	50 mSv	5 rem
EMBRYO-FETUS (CONCEPTUS) EXPOSURES*		
Total dose equivalent limit	5 mSv	0.5 rem
Dose equivalent limit in any month	0.5 mSv	0.05 rem (50 mrem)

* After NCRP Report No. 105 (1989)
** Not mandatory at present
*** Sum of external and internal doses

b. Set *annual* whole body dose equivalent limit of *50 mSv (5 rem)* * *for routine procedures* (occupational). This is a limiting condition— one that may be approached only when the cost (financial or otherwise) becomes prohibitive.

c. Set *lifetime* dose equivalent limit to the whole body at tens of mSv, that is, in rem, not to exceed the radiation worker's age in years. For example, at age 25 years, the cumulated dose shall be no greater than 25 × 10 mSv (25 × 1 rem = 25 rem). This does not mean that it is acceptable for younger individuals to receive larger exposures than older ones. Note that this limit is less than the old MPD at age 25. Moreover, it has been estimated to result in about the same lifetime fatality rate as in so-called "safe" industries and in auto travel to and from

work (ICRP, 1977a).

d. *During pregnancy* the specific dose equivalent limit applies to the **conceptus** (embryo and fetus) and **not** to the mother. Therefore, all pregnant women are subject to the dose equivalent limit for the conceptus. Naturally, radiation workers are continually monitored while the general public is not. Monitor badges overestimate the dose to the conceptus because they do not take into account attenuation of x rays in the lead apron and abdominal tissues. This can be estimated in abdominal radiography of fertile women as described on page 410. The recommended dose equivalent limit is *5 mSv (0.5 rem) for the entire gestational (pregnant) period,* but at a rate not exceeding *0.5 mSv (0.05 rem) in any one month.*

2. For *Nonstochastic Effects.* The annual

---

* *International Commission on Radiologic Protection* (ICRP) in 1990 reduced this to 2 rem per year, but it has not yet been implemented in the U.S.

dose equivalent limits include:

a. *Ocular lens 150 mSv (15 rem),* and

b. All *other organs* such as bone red marrow, breast, lung, gonads, skin, thyroid gland *500 mSv (50 rem).* However, experience with personnel monitoring by film badge or TLD shows that radiation workers receive only about one-tenth (or less) of the annual dose equivalent limit for stochastic effects in diagnostic imaging. Therefore, the lifetime dose equivalent to these organs should be expected to be **no greater than [age in years × 10 mSv (1 rem)]** (see paragraph 1c above).

Radiation workers should know that exposure to x rays for their own medical diagnosis or treatment does not count as an occupational exposure, but during pregnancy the conceptus should be protected as described below.

Fertile or Pregnant Radiation Workers

As already stated, the **conceptus** (embryo or fetus) is the primary object of protection in the pregnant radiation worker, generally the radiologic technologist. The conceptus is an **embryo** from the time of implantation in the uterine wall to the end of the second month, and a **fetus** thereafter until delivery.

The monitored surface dose to the skin of the abdomen will be higher than the dose to the conceptus; the technologist should be wearing an acceptable lead apron (0.5 mm Pb), which reduces the exposure rate at the skin to approximately 10 percent of that at the outside surface of the apron in the usual fluoroscopic kVp range. Further attenuation occurs in the intervening soft tissues of the abdomen and amniotic fluid.

According to the recommendations in *NCRP Report No. 105,* all fertile, premenopausal radiation workers **shall*** be informed by a responsible person, in advance, about the risks to which a conceptus may be exposed. If the technologist is uncertain, counseling **should*** be available. Personnel records, exposure records, departmental radiation surveys, sources of radiation, and availability of protective devices *should* be reviewed to aid in evaluating risk. Once a technologist learns she is pregnant, she should notify her employer or supervisor.

NCRP Report No. 105 does not advocate restriction of the pregnant technologist in fluoroscopy, mobile radiography, and special procedures, other than **not** being allowed to hold patients during these or any other x-ray examinations. However, she must consistently wear a lead protective apron and two personnel monitor (film, Luxel®, or TLD), one on the collar for the face and neck, and the other on the anterior abdomen beneath the apron for the conceptus. She should wear a wrap-around lead apron in C-arm fluoroscopy and radiography, and in special procedures.

A U.S. Supreme Court ruling in March 1991 emphasizes the necessity of radiology departments to have carefully written policies regarding the employment of pregnant women and restrictions in their duties. A conflict exist between the exposure of a conceptus and discrimination against the mother. Moreover, the ICRP in 1990 recommended lowering the annual limit for all radiation workers to 20 mSv (2 rem). It is advisable for every radiology department to remain current on future regulations. The American College of Radiology discusses this in its *Bulletin* of September 1992.

Nonoccupational (General Public) Limit

It is extremely difficult to estimate accurately the radiation risk to the general public

---

* *Shall* means mandatory. *Should* means advisable.

because of the confounding effect of the great variety of risks from nonradiation sources. Radiation sources include: (1) natural background (see pages 392-394) and (2) man-made radiation. For the general public, *NCRP Report No. 91* recommends:

1. An ***annual dose equivalent limit of 1 mSv (0.1 rem) for continuous or frequent exposure,*** to limit the annual fatality risk to the level of common nonradiologic risk of about 10^{-5} (1 per 100,000) per year.

2. A ***maximum annual dose equivalent limit of 5 mSv (0.5 rem) for infrequent exposures.***

Dose equivalent applies here instead of effective dose equivalent because the exposures involve diagnostic x rays, other than those used for medical procedures performed on the individual. Moreover, they do not include background radiation. For other kinds of radiation such as high-energy x or gamma rays, or artificial radionuclides, *effective* dose equivalent applies to the above

ALARA Concept

The *NCRP in Report No. 107,* and the *ICRP** in *Publication 26,* have both addressed the ***ALARA*** (*A*s *L*ow *A*s *R*easonably *A*chievable) concept, which deals with the problem of reducing exposure limits further than presently recommended. They maintain that the "reasonably achievable" aspect of ALARA implies a balancing of benefit (risk reduction) *vs* cost (financial and social). In other words, further reduction should yield a definite net benefit after the resulting economic and social factors have been carefully evaluated.

It is difficult to determine ALARA quantitatively, but we can approach the problem in two ways: (1) planning radiology departmental protection in advance of construc-

tion, and (2) promoting awareness within the department of the need to carry out all radiologic procedures utilizing all appropriate protective measures, with *rare,* deliberate exceptions. This resembles the constant attention of operating room personnel to aseptic technique, which requires a certain "frame of mind."

The day-to-day procedures in diagnostic radiology demand risk-limitation not only in the department itself but also in mobile radiography, mammography, angiography, cardiac catheterization, and interventional procedures wherever radiography or fluoroscopy is done. Technologists and other personnel must keep "patient holding" to a bare minimum; protective lead gloves, aprons, and shields, and other appropriate devices should be readily available and used properly.

It is especially important in ***C-arm fluoroscopy and radiography*** that the technologist wear a lead protective apron and gloves, and particularly to avoid exposure to the direct beam. A qualified radiation safety officer should establish guidelines and regularly check on their implementation for these special procedures.

ALARA does not require expensive modifications of rooms, equipment, or devices (provided department has adequate built-in protection), if the monitored personnel exposures are within the recommended limits. This depends on continuous vigilance in using the necessary protective devices. Whenever there is an appreciable rise in the monitor readings, a search must be made promptly for "breaks" in technique, which, if found, should be corrected without delay.

The principles of ALARA for other radiologic personnel–nuclear medicine, radiation oncology, and dentistry–appear in detail in *NCRP Report No. 107.*

---

* *International Commission on Radiological Protection.*

PERSONNEL PROTECTION FROM EXPOSURE TO X RAYS

We turn now to the protection of personnel against exposure to x rays, under two main headings:

1. ***Determination of personnel exposure*** in the radiology department and other areas where radiation is used. This is called ***radiation monitoring.***

2. ***Protective shielding*** of radiographic and fluoroscopic equipment, as well as the personnel.

Personnel Monitors. Serving to record whole body or local exposure to ionizing radiation by personnel, these are placed on appropriate sites on the body. Several kinds of personnel monitors are available including pocket dosimeter, film badge, thermoluminescent dosimeter (TLD), and Luxel® Landauer.

1. ***Pocket Dosimeter***–outwardly resembles a fountain pen but contains a thimble ionization chamber at one end. The exposure may be read by means of a separate electrometer; or, in the self-reading type, by means of a built-in electrometer. A dosimeter of this type, while sensitive to exposures up to 0.2 R (200 milliroentgens), is easily damaged and is unreliable in inexperienced hands. Besides, it does not provide a permanent record. However, it does give an immediate reading when necessary.

2. ***Film Badge***–offers a most convenient method of personnel monitoring under average conditions. Commercial laboratories specialize in furnishing and servicing the film badges. The x-ray film badge consists of a dental film, covered by a copper and plastic filter to show the quality of the radiation, and allow conversion to tissue dose. The film is backed by a sheet of lead foil to absorb scattered radiation from behind the badge. It is usually worn on the clothing over the front of the hip or chest, but during fluoroscopy it should be worn at the neck, clipped to top of the front of the apron, to monitor exposure of the thyroid gland and eyes. Pregnant personnel should wear a second badge over the abdomen beneath the leaded apron during fluoroscopy and mobile radiography.

The laboratory regularly supplies a fresh badge, which, after being worn for the prescribed interval (usually one month), is returned to the laboratory for processing under standard conditions. Dosimetric comparison is then made with standard films exposed to known amounts of radiation. The resulting film exposure in roentgens is reported to the Radiology Department for permanent filing. Badges should never be exchanged among personnel and should be clearly marked for identification. With exposure levels well below the recommended dose equivalent limit, the badge should be worn and returned ***monthly,*** especially if the work load is reasonably constant. The one-month period is preferred because of convenience and greater accuracy of calibration.

Advantages of Film Badge
a. Simple to use.
b. Inexpensive.
c. Readily processed by commercial laboratory
d. Permanent record by laboratory and in radiology department.

Disadvantages of Film Badge
a. Danger of damage by accidental laundering.
b. Not reusable.
c. Lower limit of sensitivity about 0.1 mSv (10 mrem).
d. Error about ±10 to 20%.
e. Can be fogged by heat or, rarely, light leak.

3. ***Thermoluminescent Dosimeter (TLD)***– depends on the ability of certain crystalline materials to store energy on exposure to ionizing radiation because of the trapping of

valence electrons in crystal lattice defects. When the crystals are heated under strictly controlled conditions, the electrons return to their normal state, the stored energy being released in the form of light. Measurement of the light by a photomultiplier device gives a measure of the initial radiation exposure, since the two are very nearly proportional. The dosimeters should be returned to a commercial laboratory for readout, unless this can be done accurately on the premises. Lithium fluoride (LiF) is at present the most prevalent material for TLD.

Advantages of TLD

a. Can be made very small.
b. Sealed in Teflon®, minimizes chance of damage.
c. Low exposure limit, 0.05 mSv (5 mrem).
d. Response to x rays proportional up-to about 400 R.
e. Response almost independent of x-ray energy from about 50 kV to 50 mV.
f. Accuracy about ±5 percent.
g. Response very similar to tissues.
h. Less sensitive to environmental heat, than is film badge.
i. Can be fashioned into rings (e.g., finger badge) and other jewelry.
j. Can be worn three months and are reusable

Disadvantages of TLD

a. Cannot be stored as a permanent record.
b. More expensive than film badge, but reusable.

Luxel®. A new personnel monitor, the Luxel® badge (see Figure 28.1) uses optically stimulated luminescence (OSL) technology. The detector is aluminum oxide (Al_2O_3) which, when exposed to certain frequencies of laser light, becomes luminescent in proportion to the previous exposure to ionizing radiation. A thin layer of the detector is sealed within a packet containing a three-element filter to identify static, contaminant, and dynamic exposure to low-energy x rays (also, beta particles). A static image shows a fine grid filter clearly, indicating that the exposure took place under stationary conditions; for example, the badge was not being worn when exposed. The packet is sealed in a hard plastic badge with a spring clip on the back for attachment to the leaded apron or other garment. Landauer developed this new system.

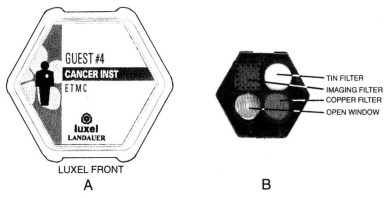

Figure 28.1. Features of Luxel® personnel dosimeter. *A.* Front, showing icon with position of badge on chest, name, and place. *B.* An insert with tin and copper filters, an imaging grid to show motion, and an open window. Other inserts include an additional filter and the layer of aluminum oxide detector.

Advantages of LUXEL® system

1. Dose measurement range very wide: 1 mrem to 1000 rem (0.01 mSv to 10 Sv). (Film 0.1 mSv and TLD 0.05 mSv, minimums.)
2. Accuracy ±15% for shallow and deep exposures.
3. Precision within ±1.0 mrem (0.01 mSv).
4. Energy range 5 keV to over 40 MeV.
5. Complete reanalysis available: Al_2O_3 can be restimulated many times.
6. Bar code and other useful information available.
7. Bimonthly readout offered (monthly service still available).
8. Tamper-proof sealed badge.
9. Services include badges for whole body, collar, waist, wrist, and ring exposure to x rays, gamma rays, and beta particles.
10. Reports available in a great variety of forms.

On termination of employment all technologists must receive an appropriate summary of occupational exposure, regardless of the monitoring system. If reemployed, the technologist should present this summary report to the new employer.

Harmful Effects of Radiation. Why has health physics become so important in radiology? The answer lies in the fact that radiation is always harmful to tissues, the amount and nature of any damage being dependent on radiation quality (LET), daily dose, total dose, and overall exposure time. The chapter on radiobiology deals with radiation effects in some detail. However, as stated at the outset, extreme precautions are taken in diagnostic radiology to keep exposure of personnel patients well below the permissible limits so that the risk of radiation injury is vanishingly small. *The objective of a program in health physics is to keep the exposure of personnel as low as reasonably achievable so as to avoid the slightest radiation injury.*

Protective Measures

Protection in the diagnostic radiology department involves three basic principles: (1) exposure time, (2) distance, and (3) lead barriers.

1. *Exposure Time.* The total dose equivalent H to a person for a particular *H* rate is directly proportional to the exposure time:

$$H_{total} \; cSv = H \; cSv/s \times time \; (s) \qquad (3)$$

Note that the unit s appears in both the numerator and denominator of the *right* side of the equation and therefore cancel out, leaving the answer in cSv (rem).

2. *Distance.* The inverse square law applies closely enough to be valid for protection purposes (see pages 214-217). Whenever possible, the technologist should stand at least 2 m (6 ft) from the patient, x-ray tube, and x-ray beam; *preferably,* in the control booth or behind a fixed, full-length lead (Pb) shield. During fluoroscopy, if the technologist stands 1 m (3 ft) from the x-ray tube, moving back only one good step more (about 1 m) reduces the exposure rate to about one-fourth. Thus, distance is a powerful factor in radiation protection.

3. *Lead (Pb) Barrier.* Lead is an efficient absorber of x rays in the kV region in diagnostic radiology because of the low photon energy. Therefore, placing a relatively small but appropriate thickness of Pb between the x-ray source and the personnel contributes greatly to exposure reduction. The thickness of protective barrier is usually stated in *half-value layer (HVL)* for kilovoltage x rays (tenth-value layer or *TVL* for megavoltage photons). Recall that HVL applies to any material and is that *thickness* which reduces the exposure rate by one-half; for kV x rays, this is specified in mm or cm Pb. For any protective barrier, *1 TVL = 3.3 HVL.*

Application of the HVL concept means that if the initial dose equivalent *(H)* is 10 cSv (rem), and the HVL is 1 mm Pb for the given kVp, then 1 mm Pb will reduce *H* to 5 cSv (rem). A second HVL of 1 mm Pb will reduce *H* to another one-half, that is, to 2.5 cSv (rem), or one-fourth the initial value. Note that adding 2 HVLs has the same effect as doubling the distance.

Protective Barriers in Radiography and Fluoroscopy

Radiation Sources. All x-ray tubes must be ray proof, that is, the manufacturer must enclose the tube in a metal housing that reduces leakage radiation to a prescribed acceptable level (leakage radiation is that radiation which penetrates the protective housing). The specifications for housing appear in *NCRP Report No. 102:* leakage radiation from a diagnostic tube, when operated at its maximum kV and mA, shall not exceed 0.1 R/hr at a distance of 1 meter in all directions from the tube housing, except the tube aperture.

Wall Protection. This requires the use of built-in *protective barriers* of suitable radiation-absorbing material(s) to restrict exposure to the recommended limit. Wall protection should be planned in advance by a *health physicist* qualified in this field, to avoid expensive alteration after completion of the building; poor planning may result in excessive wall protection, which becomes unnecessarily expensive. However, it costs relatively little to reduce the x-ray exposure rate from 100 mR/week to 10 mR/week, a policy consistent with the ALARA concept*. A basic principle in permanent barriers is that joints and holes must be covered by the same or equivalent protective barrier as the wall. Proper wall protection varies with the *energy of the radiation* modified by certain factors described in *NCRP Report No. 49.*

Terms applied to specific areas include:

Controlled or restricted area is one that is under the supervision of the Radiation Safety Officer and in which the occupational limit is 1 cSv (1 rem) per year.

Uncontrolled or unrestricted area is one that is not under supervision and in which the annual exposure limit to the general public prevails, that is, 0.5 cSv (0.5 rem) for infrequent or occasional exposures.

There are four main types of radiation to consider in radiation protection:

Useful beam is the radiation passing through the tube aperture and the beam-limiting device, previously called the primary beam.

Leakage radiation includes all the radiation passing through the tube housing, other than the useful beam.

Scattered radiation is that which has undergone a change in direction during passage through matter.

Stray radiation is the sum of scattered and leakage radiation.

There are two kinds of wall barriers. A *primary protective barrier* is one whose atomic number and thickness suffice to reduce the exposure rate of the useful beam to the occupational exposure limit. A *secondary protective barrier* is one whose atomic number and thickness suffice to reduce the exposure rate of the stray radiation to the occupational exposure limit.

Although, as stated above, wall barriers should be planned by a health physicist before construction, typical *approximate* values are as follows: the *primary protective barrier,* that is, where the useful beam can hit the wall *directly,* in radiography up to 140 kVp is about 1/16 in. lead (Pb) extending 7 ft up from the floor when the x-ray tube is 5 to 7 ft from the wall. This barrier also takes care of leakage radiation, that is, radiation coming through the tube housing. A *secondary protective barrier* (covering areas exposed only to scattered and leakage radia-

---

* At present, survey meters read exposure rates, that is, mR/hr, with options for higher outputs.

tion) is about $1/32$ in. Pb under the same operating conditions. This secondary barrier extends from the top of the primary barrier to the ceiling, but should overlap the primary barrier at least $1/2$ in. at the seam. Ordinary plaster often suffices as a secondary barrier without added lead, especially in large radiographic rooms. Control booths should have the same protection as walls, and radiation should scatter at least *twice* before reaching the opening of the booth.

The leaded glass observation port in the control booth should have the same lead equivalent as the adjacent wall and should be overlapped about $1/2$ in. by the lead in the wall. Leaded glass must be about four times as thick as sheet lead for equivalent protection; for example, $1/4$ in. leaded glass is equivalent to $1/16$ in. sheet lead. Other, more sophisticated observation systems are available.

Working Conditions

Certain general rules govern personnel safety. These should be impressed repeatedly on all Radiology Department personnel, since their cooperation is mandatory.

1. Never expose a human for demonstration purposes alone.

2. Never remain in a radiographic or radiotherapy room while an exposure is in progress. In general, approximately 0.1 percent of the useful beam is scattered perpendicular to the beam at a distance of 1 meter from the patient.

3. Patients should *rarely* be held for radiography, since many efficient devices are available, in addition to the technologist's ingenuity. However, if a patient must be held, this should preferably be done by a person not habitually exposed to ionizing radiation; in any case, protection by a lead apron and lead gloves is mandatory, and

exposure to the direct beam must be avoided.

4. Give yourself the same protection as a loaded cassette!

5. In *mobile radiography,* always wear an approved lead apron and stand at least 180 cm (6 ft) from the x-ray tube, with the useful beam directed away from all personnel. Protective measures are extremely important in *C-arm radiography.*

6. In *fluoroscopy* personnel protection is extremely important and you should be sure that the following items are available and in use:

a. A *protective drape* having a lead (Pb) equivalent of at least 0.25 mm, or preferably 0.5 mm, situated between the fluoroscopic assembly and the fluoroscopist to block scattered radiation, mainly from the patient.

b. A *protective apron* to be worn by personnel in the fluoroscopy room. This apron must have a Pb equivalent of 0.5 mm. When not actually assisting the radiologist, stand either in the control booth or behind the radiologist.

c. A *lead shield* at least 0.5 mm thick covering the *Bucky slot* to give additional protection to the personnel's gonads from scattered radiation.

d. Special precautions against the increased scattered radiation engendered by large format image intensifiers (12 in. × 12 in. or larger).

7. Check *lead-protective gloves* (same Pb requirements as for aprons) periodically for cracks by means of a radiographic test using medium speed intensifying screens and exposure factors of 100 kV and 2.5 mAs at a 40-in. (100 cm) distance, with a 400-speed film-screen combination.

8. Always wear an approved personnel monitor when on duty.

9. If fertile or pregnant, you must wear two personnel monitors; one at the neck, the other on the abdomen under the apron.

10. On leaving employment, always

obtain a complete record of your x-ray and nuclear medicine exposure, as well as the total time of employment. This should be given to your new employer if radiologic technology is involved.

ACCEPTANCE (COMPLIANCE) TESTING

Before new x-ray equipment is put into service, a Health Physicist must test its compliance with requirements of the State Regulatory Agency. These tests will be briefly stated for radiographic and fluoroscopic units.

1. *Timer Accuracy*–shall be within ±10% of the indicated time, the test being performed at 0.5 sec.

2. *Exposure Reproducibility*–four exposures are averaged for the same kVp and mA settings, and the following formula applied: $E_{av} \geq 5 (E_{max} - E_{min})$; E_{max} is the highest exposure and E_{min} the lowest.

3. *kVp Test*–indicated kVp shall be accurate to within ±5% of the nominal kVp setting at no less than three different settings.

4. *Linearity of mAs*–constant x-ray output in R/mAs at a fixed distance when mA is changed and time adjusted to maintain constant nominal mAs. Outputs measured at 2 successive mA stations shall not differ by more than ±10%.

5. *Tube Stability*–for any selected fixed position, the tube shall not move during an exposure. When specially designed for movement (tomography), free movement of the tube shall occur.

6. *Beam-limiting device ("collimator")*

a. *Numerical indicators of field size*–measured field size must coincide with selected field size to within 2% of SID for manual device, and 3% for positive beam-limiting device (PBLD) or semiautomatic device.

b. *Light field vs x-ray field congruence*–within 2% of SID.

c. *Center of x-ray field alignment with center of light field*–within 2% of SID.

Special regulations apply to fluoroscopic equipment as follows:

1. *Timer*–special device provided to preset cumulative "on" time of fluoroscopic unit. Timer shall be set for 5 min or less. An audible signal shall indicate that either this time has expired, or has been terminated.

2. *Primary Barrier*–consists of image intensifier assembly itself. No exposure shall be possible until this is firmly in place. Transmitted radiation shall not exceed 2 mR/hr for each R/min exposure (eg, 2 mR/hr = 0.033 mR/min, so for a 5-min cumulative exposure, patient would receive $5 \times 0.033 = 0.165$ mR).

3. *kVp and mA Indication*–shall be continuous during fluoroscopy, at the control panel and/or fluoroscopist's position.

4. *Deadman Switch*–x-ray production shall occur only when pressure applied to foot- or handswitch.

5. *High Levels Control*–requires special means of activation, continuous manual activation, and continuous audible signal.

6. *Exposure Rate Limits*–exposure rate to patient shall not exceed 10 R/min with ordinary exposure control. With high-level control, exposure rate shall not exceed 5 R/min.

7. *Source-Skin Distance (SSD)*–shall be no less than 38 cm (15 in.) for stationary unit, and 30 cm (12 in.) for mobile unit. For special surgical procedures, may be 20 cm (8 in.).

8. *X-Ray Field Sizes*–stepless (continuous) field size selection shall be provided, with a minimum 5×5 cm² at longest SID. For input phosphor field size, neither the length nor the width shall exceed 3% of the

SID; nor shall the length + width, 4% of the SID. This applies also to spot film size, but for equipment with only manual beam limitation ("collimation"), the x-ray field shall be limited to the area of the spot film cassette at 16 in. above the table top.

It is recommended that compliance tests for radiography and fluoroscopy be performed every two years.

PROTECTION SURVEYS

Every State in the Union has a Radiation Safety Agency within its Health Department, charged with setting up regulations governing the medical use of ionizing radiation–x rays and radionuclides. A hospital or clinic **Radiation Safety Officer (RSO)** is responsible for assuring compliance with the State regulations. An RSO must have at least the training and experience mandated by the regulations; he or she does not necessarily have to be a physicist, **but the latter is preferred.**

The RSO sets up the mandated procedures assuring that personnel and general public do not receive radiation exposure exceeding the prescribed limits. The success of such a program requires constant vigilance by the RSO, who meets quarterly with the Radiation Safety Committee to report on above-limit personnel monitor readings, licensing of new equipment, spills or misadministration of radiopharmaceuticals, corrective actions, special communications with the State Agency following its inspections, and other pertinent matters. The minutes of these meetings, kept on permanent file, are reviewed periodically by an inspector from the State Agency.

To verify compliance with protective measures, the RSO conducts periodic **surveys** according to regulations, which are extremely detailed and will only be touched upon here:

1. **Initial Survey**–mandatory for new, or newly modernized radiology facilities. This includes testing of protective barriers in walls and control booths for leakage, reliability of x-ray equipment (collimation, proportionality of mAs over useful range, kVp accuracy, beam filtration, focal spot size, fluoroscopic x-ray output, adequacy of protective gear).

2. **Re-surveys**–more limited than initial survey and are indicated as follows:

a. When there has been a change or addition of x-ray equipment or radiopharmaceuticals, or a change in procedures that require additional protective barriers or change in beam direction.

b. To confirm whether previous deficiencies have been corrected.

c. To verify adequate cleanup of a spill of a liquid radiopharmaceutical.

d. When RSO discovers or suspects that personnel have received exposures exceeding limits in controlled or uncontrolled area, one or more of the following steps should be taken:

1. Determine the cumulative dose in the area in question.

2. Use personnel monitoring in the area in question.

3. Add barrier material to comply with authoritative recommendations in *NCRP Report No. 105.*

4. Impose restrictions on the use of the equipment, or on the direction of the beam.

5. Impose restrictions on the occupancy of the area if this is controlled.

A tissue-equivalent phantom such as stacked acrylic plates must be used as a source of scattered radiation to check the adequacy of the secondary barrier.

All "on-off" control mechanisms (control panel, entrance door, emergency cutoff switch) should be checked routinely every 6 months, or whenever they malfunction, and necessary repairs made.

Official yellow warning signs reading RADIATION AREA (see Figure 28.2) in black letters should be mounted on the outside of all doors having access to radiographic/fluoroscopic rooms. This would cover prevailing exposure rates of 5 to 100 mR per hour. Other signs are available for areas in which the exposure rate could exceed 100 mR per hour, but such levels are highly unlikely in diagnostic radiology departments.

Whenever any restriction has been placed on the equipment, such as limitation of beam angulation to prevent penetration of a secondary barrier by the useful beam, the specified restriction must be enforced.

All reports of calibrations and surveys should be made in writing, signed by the RSO, and filed *permanently*. The report should indicate whether a resurvey is needed, and when.

Instructions must be posted conspicuously in the control area describing the steps to be taken in the event equipment cannot be turned off.

The name, address, and phone number of the Radiation Safety Officer should be posted in a conspicuous place in the control area for prompt service in the event of an emergency. A responsible substitute should be available, on call, if the Radiation Safety Officer cannot be reached.

Figure 28.2. Yellow radiation warning labels. A badge must be posted on the entrance door to a room, the badge shown on the left when prevailing exposure rates are 5 to 100 mR/hr, and the one on the right when exposure rates are higher, but less than 500 mR/hr.

PROTECTION OF THE PATIENT IN DIAGNOSTIC RADIOLOGY

All x-ray exposures, regardless of how small they may be, entail some degree of risk to the patient. In other words, there is no threshold dose for stochastic effects. However, this should not deter us from using x rays for diagnostic purposes when medically indicated; you must keep in mind the benefit/risk ratio in each case, although as a technologist you have no control over the ordering of x-ray examinations. At the same time, you should continually strive to minimize patient exposure; this is especially important in fertile women who may unknowingly be pregnant.

Because of the urgency of protecting the general public from unnecessary radiation, the U.S. Congress in 1968 passed the Radiation Control for Health and Safety Act which set up the Bureau of Radiological Health (BRH) under the Food and Drug Administration (FDA). The BRH has the responsibility, through its five divisions, of (1) establishing manufacturing standards for all x-ray-producing electronic equipment, (2) following up on compliance with these standards, (3) studying the biologic effects of radiation, (4) developing programs for training and medical applications, and (5) overseeing the field of radioactive materials and nuclear medicine.

Table 28.4
SELECTED RESULTS OF THE X-RAY EXPOSURE STUDIES
OF THE BUREAU OF RADIOLOGICAL HEALTH

	1964	1970	Change
# having 1 or more examinations	108×10^6	130×10^6	+20%
# radiographs per examination	2.2	2.4	+10%
mean ratio of beam area to film area	1.9	1.2	–30%
av. skin exposure per film (AP or PA abdomen)	480 mR	620 mR	+25%
av. skin exposure per film (PA chest)	28 mR	27 mR	0
genetically significant dose per year*	16 mrads	20 mrads	+25%

*According to NCRP Report No. 93 (1987) the annual GSD is 22-30 mrem (0.22 to 0.030 mSv) from diagnostic x rays, especially of hips and pelvis in men, and lumbar spine and barium enema in women. This may be overestimated because of gonadal shielding.

In 1964, and again in 1970, the BRH surveyed a number of installations to find possible trends in the exposure of patients during diagnostic radiologic procedures. The results of these surveys are summarized in Table 28.4. Note the 20 percent increase in the utilization of diagnostic x rays during the six-year interval. Also, while the mean ratio of the beam area to the film area has shown a significant decrease of 30 percent, the average skin exposure for a plain radiograph of the abdomen has increased by 25 percent! According to the later survey, the average skin exposure for a posteroanterior radiograph of the chest is about 30 mR (0.03 R), and for an anteroposterior radiograph of the abdomen about 600 mR (0.6 R).

Because of possible genetic damage, gonadal exposure must be minimized by appropriate beam limitation ("collimation"), filtration, protective Pb shielding, and high speed film-screen imaging systems. Rare earth screens have contributed much to the reduction of gonadal as well as whole-body exposure of patients.

The last line in Table 28.4 shows that the genetically significant dose (GSD) has increased about 25 percent, to 20 mrads per year, between the two surveys. GSD is explained on pages 386-387.

In 1987, *NCRP Report No. 93* estimated the annual GSD from diagnostic x rays to be 22 to 30 mrem (0.22 to 0.30 mSv) for women (mainly from lumbar spine and barium enema examinations), and for men (mainly from hip and pelvis examinations). However, this is probably overestimated because of the prevalence of gonadal shielding in men.

Table 28.5 has been adapted from *NCRP Report No. 91* (1987) to give the dose equivalent H_E for a number of typical x-ray examinations. Recall that H_E is the weighted (effective) dose equivalent to an organ, which entails the same detriment or risk as though the same dose equivalent were given to the whole body. For example, in Table 28.5, the average H_E for a lumbar spine examination is 1.3 mSv (130 mrem); this is equivalent to 1.3 mSv to the whole body. Thus, H_E estimates the level of risk to the whole body from H_E mrem to the lumbar spine. However, as noted on page 396, *NCRP Report No. 107* states that monitor readings in terms of exposure, absorbed dose, or dose equivalent in diagnostic radiology may serve, instead of effective dose equivalent, as a measure of risk on account of uncertainties in determination of weighting factors and the low LET of x rays and medical radionuclides.

The values of H_E from nuclear medicine tests, included in Table 28.6, are adapted from *NCRP Report No. 91.*

Table 28.5
AVERAGE EFFECTIVE DOSE EQUIVALENT H_E FOR THE U.S. IN 1980
FROM DIAGNOSTIC MEDICAL EXAMINATIONS*

	Annual number of examinations (in thousands)	Average effective dose equivalent per examination
		mrem**
Computed tomography (head and body)	8,300	110
Chest	64,000	6
Skull and other head and neck	8,200	20
Cervical spine	5,100	20
Biliary tract	3,400	190
Lumbar spine	12,900	130
Upper gastrointestinal tract	7,600	245
Kidney, ureters, bladder	7,900	55
Barium enema	4,900	405
Intravenous pyelogram	4,200	160
Pelvis and hip	4,700	65
Extremities	45,000	1
Other***	8,400	50

* Adapted from *NCRP Report No. 91* (1987).

** 1 mrem = 0.01 mSv.

*** *Other* is estimated from mean value of examinations such as thoracic spine, full spine, mammography, etc.

Under the Radiation Control for Health and Safety Act, the standards for medical x-ray equipment went into effect on August 1, 1974. Of major importance is the recommendation that the components of a diagnostic x-ray unit must make up an integrated system. Furthermore, the manufacturer must test and adjust the equipment at the factory as the final step in the production line. Finally, the installers of the equipment must follow closely the manufacturers' instructions. Some of the more significant of the numerous specifications include:

1. Ability of the equipment to reproduce an exposure for any permissible combination of kVp, mA, and time.

2. Proportionality between exposure and time.

3. Positive beam limitation for each film size in stationary equipment.

4. Automatic beam limitation within one inch of the margin of the fluoroscopic screen.

5. Automatic limitation of the x-ray beam in fluoroscopy at the patient's entrance port for spotfilming.

6. Built-in filtration (2.5 mm Al) to

Table 28.6
AVERAGE EFFECTIVE DOSE EQUIVALENT H_E FOR THE U.S. IN 1980 FROM DIAGNOSTIC NUCLEAR MEDICINE TESTS*

Examination	Annual number of examinations (in thousands)	Average effective dose equivalent per examination
		mrem**
Brain	810	650
Hepatobiliary	180	370
Liver	1,400	240
Bone	1,800	440
Lung	1,200	150
Thyroid	680	590
Kidney	240	310
Tumor	120	1,200
Cardiovascular	950	710

* Adapted from *NCRP Report No. 93* (1987).

** 1 mrem = 0.01 mSv

Table 28.7
ABDOMINAL-ORGAN DOSE (mrad) FOR 1,000 mR ENTRANCE SKIN
EXPOSURE IN AIR AP ABDOMEN, 40-IN SID, 14 in. × 17 IN. FILM

	Organ Dose in mrads				
Beam Quality HVL(mm Al)	1.5	2.0	2.5	3.0	3.5
Ovaries	97	149	203	258	313
Embryo (uterus)	133	199	265	330	392
Testes	<0.01	<0.01	<0.01	<0.01	<0.01

Data abbreviated from Table 7 in *Handbook of Selected Organ Doses for Projections Common in Diagnostic Radiology,* by M. Rosenstein (HEW/FDA 76-8081), 1976.

remove low-energy radiation that is radiographically useless.

Every effort should be directed toward minimizing patient exposure during radiologic examinations, especially of the gonads and the pregnant or potentially pregnant uterus, because of the hazard to the genetic material or the conceptus. Protection requires lead shielding and tight collimation of the x-ray beam whenever these do not hamper the examination. At the same time, unnecessary exposure to other parts of the body should be avoided because of possible somatic injury by radiation, especially to the blood-forming bone marrow.

The absorbed dose to the ovaries or conceptus during radiography of the **abdomen/pelvis** can be found by using x-ray output of the radiographic unit to derive the gonadal absorbed dose from the appropriate table in *Handbook of Selected Organ Doses for Projections Common in Diagnostic Radiology* by M. Rosenstein (HEW/FDA 76-8031). Table 28.7 has been adapted from Table 7 in that publication to show in particular the absorbed doses in the ovaries, uterus (no larger than 2-month's pregnant), and testes, resulting from an anteroposterior view of the abdomen.

Note that the basic data include the organ dose per 1000 mR (1 R) exposure at the skin, for an SID of 40 in. (100 cm), film size 14 in. × 17 in. (35.6 cm × 43.2 cm), and a variety of x-ray beam HVLs (mm Al). X-ray output in mR/mAs having been measured locally at 40 in. using the selected kVp, this value is recorded and kept in an accessible place.

To find the ovarian or uterine (normal size) dose for an anteroposterior view of the abdomen in a particular patient, suppose that with exposure factors 90 kVp, 100 mAs, 40-in. SID, and 2.5 mm Al filter, the output is 5 mR/mAs, then for 100 mAs the exposure at the skin (entrance exposure) would be

$$5 \text{ mR/mAs} \times 100 \text{ mAs} = 500 \text{ mR}$$

In Table 28.7 we see that for a radiographic beam having an HVL of 2.5 mm Al, the uterine dose is 265 mrad for an entrance exposure of 1000 mR. Since we have just found that the required skin exposure in this case is 500 mR, the uterine dose for a KUB would be

$$500 \text{ mR/1000 mR} \times 265 \text{ mrad} = 123 \text{ mrad}$$
$$\text{or 1.23 mGy}$$

On the assumption that this dose results from a medium-speed film-screen system, we can state that with a 400-speed system, the uterine dose can be reduced to about

$$100/400 \times 123 \text{ mrad} = 31 \text{ mrad}$$
$$\text{or 0.31 mGy}$$

Dose Reduction in Radiography

We have at least ten ways to minimize patient exposure during radiography. They are extremely important and should be learned and used by every radiologic technologist. We shall now discuss them individually, and then summarize them.

1. ***Beam Filtration.*** Patient exposure can be greatly reduced by the insertion of an aluminum (Al) filter in the beam to remove the lower energy photons which would otherwise be absorbed by the patient and not contribute to the radiographic image. Recommendations for filter thickness have been expressed either as (a) minimum total thickness in mm aluminum equivalent (Al eq), including inherent filtration by the glass window of the x-ray tube and the cooling oil in the tube housing, or as (b) half-value layer (HVL) of the x-ray beam in mm Al.

Table 28.8 shows the minimum Al eq filtration needed for each of three useful kVp ranges according to *NCRP Report No. 33* (1968). You can see that above 70 kVp, the ***total*** filtration must be no less than ***2.5 mm Al equivalent.***

It turns out that *21 CFR* (see Table 28.8) also recommends a ***half-value layer*** of nearly 2.5 mm Al for x-ray beams with energy above 70 kVp. Since there may be uncertainty as to the inherent filtration of a particular x-ray tube, it is much simpler for the physicist to add sufficient aluminum filtration until the correct HVL has been attained. Thereafter, the filter is permanently fixed in position and used for the entire range of radiographic and fluoroscopic kVp. *(The U.S. Code of Federal Regulations–21 CFR 800.10, April 1983*–specifies minimum HVLs for various kVp ranges on all new equipment manufactured after December 1, 1983.)

How much dose reduction can we expect from beam filtration? Table 28.9 shows that a 3 mm Al filter (added) reduces exposure to the testes to about 1/3 to 1/2 that with an unfiltered beam at 65 to 68 kVp (Ardran and Crooks). Another study with a pelvic phantom showed that a 3 mm Al filter reduces skin exposure to about 1/5 that without filtration at 60 kVp, other exposure factors (mAs) being adjusted to maintain equal radiographic density (Trout and others).

Table 28.8
REQUIREMENTS FOR FILTRATION IN
RADIOGRAPHY*

Operating kVp	Minimum Total Filter* (inherent plus added) in mm Al eq	Minimum HVL** in mm Al
Below 50	0.5	0.3
50-70	1.5	1.2
Above 70	2.5	2.3

* From NCRP Report No. 33 (1968).
** *From U.S. Code of Federal Regulations–21 CFR 800.10, April 1983.*

Table 28.9
REDUCTION IN EXPOSURE OF MALE GONADS (TESTES)
WHEN 3 mm Al FILTER IS ADDED*

Examination	Added Filtration	kVp	Testes Uncovered	Testes Covered 1 mm Sheet Lead
			mR	mR
Pelvis, AP	0	65	2,000	42
15 × 12 in. film	3 mm Al	65	670	24
Lumbar Spine, AP	0	68	24	5
10 × 12 in. film	8 mm Al	68	6	2

*Data of Ardran, G. M., and Crooks, H. E. (Courtesy of authors and *British Journal of Radiology.*) (Higher kVp and faster screens have been in use for a number of years, but the effect of beam filtration still holds.)

Furthermore, in this latter study it was found that at 60 kVp the mAs had to be increased to compensate for the removal of x-ray photons by the filter, but this did not increase the actual exposure of the patient. At 130 kVp, filters up to 3 mm thick did not require an increase in mAs. (With modern higher kVp technique and rare earth screens, and prevailing gonadal shielding, testicular doses have been markedly reduced.)

2. *Beam Limitation ("Collimation").* As already explained (see pages 247-253), a decrease in the cross-sectional area of the beam not only avoids unnecessary exposure of tissues outside the area of interest, but it also reduces the amount of scattered radiation generated within the patient. We achieve this most efficiently by means of beam-limiting devices that are loosely called collimators. The variable shutter type is preferred. Federal regulations require that all new equipment must have *positive variable beam-limiting devices* interlocked with the bucky tray so that the beam size is automatically restricted to the particular cassette size. A manual override may be provided for use, with certain limitations, in the event of system failure. Neither the length nor the width of the x-ray field in the plane of the film (or other image receptor) may differ from the corresponding film dimension by more than 3 percent of the SID. Thus, at a 40-in. (100-cm) SID the x-ray field dimension at the film must be *within* 3 cm of the corresponding film dimension.

3. *Consistent mAs.* The product of mA and time—mAs—should be consistent throughout the range of mA and time settings; for example, at any given kVp, 100 mA × 1 sec should match an exposure at 200 mA × 1/2 sec, within ±10 percent.

4. *Light Beam Intensity.* Illumination of the x-ray field by the collimator light must be bright enough to be easily visible, with a contrast ratio of at least four at the field edge.

5. *Gonadal Shielding.* Although the beam should be so restricted that direct exposure of the gonads does not occur unnecessarily, additional precautions should be taken to use specific gonadal shields when these do not obscure the area of interest. In the majority of radiographic examinations, specific shielding of the testes should be used when they lie within the beam, or within 5 cm beyond the edge of the beam. The shield must have a lead equivalent of at least 0.5 mm and should be cup-shaped for AP, and flat for PA projections. "Shadow" shields mounted on tube housing are also available. Whenever possible, when not interfering with the examination, the ovaries should be shielded with at least 0.5 mm Pb, although this is limited by uncertainty as to their exact location. Table 28.10 gives the attenuation in various thicknesses of lead with diagnostic x rays.

6. *Modified Radiographic Projection.* In radiography of girls for scoliosis one should

Table 28.10
THICKNESS OF LEAD REQUIRED TO REDUCE THE INCIDENT
RADIATION TO 0.5 PERCENT OF THE INITIAL EXPOSURE RATE,
WITH RADIOGRAPHIC BEAMS*

kVp	Half Value Layer	Thickness of Lead for 0.5% Transmission		Usual Thickness of Lead Available
		mm.	in.	
60	0.5 mm Al	0.21	0.008	1/100 in.
100	1.0 mm Al	0.68	0.026	3/100 in.

*From data of Trout ED, Geiger RM. (Courtesy of *American Journal of Roentgenology.*)

use a posteroanterior projection. This reduces the breast dose by at least 98 percent without appreciable loss of radiographic quality.

7. ***High-speed Image Receptors.*** A great reduction in patient exposure results from the use of the highest speed intensifying screens and films consistent with satisfactory image quality. Rare earth screens having a speed of 400, that is, about four times that of medium speed screens, decrease patient exposure to about one-fourth relative to medium speed without significantly impairing recorded detail. Therefore, 400-speed film-screen systems should prevail in routine radiography, although slower systems still play an important role in special situations where fine recorded detail is needed.

8. ***Optimum Film Processing.*** Automatic film processing has become universal in modern radiology departments. Processors require careful maintenance for optimal function. This assures consistently high radiographic quality in properly exposed films and helps avoid repeat examinations.

9. ***High Kilovoltage.*** As we pointed out earlier (see pages 115, 145-146) higher kVp entails more efficient tube operation. Also, for equal radiographic density, higher kVp with lower mAs delivers a smaller absorbed dose to the patient. Therefore, we should use the highest kVp consistent with optimum image quality. With modern equipment capable of a wide range of mAs values, an optimum kVp can be selected for a particular anatomic part. The most modern radiographic equipment is provided with automatic exposure control (phototiming) which selects optimum kVp and mAs for various anatomic regions.

10. ***Careful Technique Selection*** to minimize repeat examinations. As reported in the literature, they range from about 5 to 15 percent. The ratio of exposure/positioning errors is about 5:3 (Dowd). Often overlooked is the extreme importance of SID

measurement in mobile radiography. Applying the inverse square law, you would find that at an estimated distance of 36 in. an error of six in. on the short end would give a 70 percent overexposure, and on the long end an underexposure of 36 percent! That is why SID should be measured by a retractable tape securely attached to the tube housing, or by a 36 inch ruler. More advanced mobile units have a laser system to ensure the correct SID. Automatic exposure control is available on stationary x-ray equipment, reducing errors in technique selection.

Based on the preceding discussion, we may summarize as follows the important factors in reducing patient exposure during radiography—so-called ***minimum dose radiography:***

Beam filtration to provide an ***HVL of 2.5 mm Al*** (see also pages 411-412).

Beam limitation ("collimation") to the smallest possible dimensions by a variable beam-limiting device. Equipment of the most modern type is provided with positive (automatic) variable beam-limiting devices with manual override.

Lead shielding of the gonads by at least 0.5 mm lead equivalent (see Table 28.10 for attenuation of diagnostic x rays in lead). Omit the shield if it hides the area of interest.

PA projection for scoliosis examination of girls to minimize breast dose.

Highest speed screen-film systems with good image quality; 400-speed systems using rare earth screens and appropriate films in general radiography.

Optimum film processing for consistent results.

Highest practicable kVp that still yields good image quality.

Careful technique to minimize repeat examinations.

In any procedure using ionizing radiation, whether radiography, fluoroscopy, or nuclear medicine, the patient should be questioned about previous exposures. The total exposure, especially to gonads, embryo, and bone marrow, should be kept at the lowest level consistent with medical needs.

Radiographic examinations that expose the gonads or uterus in fertile women should be performed only when absolutely necessary, since the patient may not be aware that she is pregnant. It is during the first few months that the embryo is most susceptible to fatal injury or the induction of serious congenital anomalies. Authorities used to recommend the ***10 day rule,*** which argued that such examinations should be limited to the first 10 days following a menstrual period (ie, the "safe" period) on the assumption that ovulation and pregnancy are unlikely during this time. However, this is no longer widely accepted. If a radiographic examination is not urgently necessary, it should be postponed. But if urgent, the examination should be performed with strict adherence to all available protective measures and careful technique selection. There is also a ***28-day rule,*** quoted by Dowd, which states that abdominopelvic examinations should be done 28 days after the onset of a menstrual period because pregnancy at that time would fall within the "all-or-none" period in which the conceptus would not survive. This rule is not often applied.

Protection in Mammography

As stated earlier on pages 289-291, skillful technique minimizes breast dose while producing mammograms of superb quality. To achieve these goals requires dedicated mammographic machines with special mammographic tubes having molybdenum targets and filters; low-dose mammographic screens and films, with or without grids having ratios of 3:1 or 4:1; and an efficient breast-compression device.

Superb ***quality control,*** as well as careful positioning and technique selection, are all necessary to avoid repeats. The subject of quality control occupies a separate section (see pages 289-294).

Computed Tomography Scanning

In general, the tight collimation for each slice with relatively small side scatter to adjacent slices engenders about the same dose as a conventional radiographic examination. For example, a complete CT scan of the head delivers about the same dose as a full radiographic skull series.

One can determine the dose in CT scanning by measuring the absorbed dose at the ***center*** of one "slice" with a small dosimeter in a water phantom, while scanning this slice and three adjoining slices on both its sides (see Figure 28.3). Under these conditions,

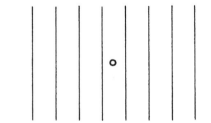

Figure 28.3. Method of measuring absorbed dose in CT scanning. A small dosimeter is placed in the center of the middle "slice" in a water phantom, remaining there while all seven slices are scanned. The dosimeter measures the absorbed dose of the primary beam through the middle slice, as well as that of the scattered radiation from the adjoining slices.

the dosimeter will record the doses from the direct beam through the center slice, as well as the scattered radiation from the adjoining slices (Curry and others).

Noise, which degrades image quality, can be reduced by increasing the exposure, but this obviously raises patient dose. Thus, techniques are set to use the smallest exposure that will produce optimal contrast resolution. The collimator should also be checked periodically to assure its proper function; if too wide, scattered radiation increases and acts as another noise factor.

Patient Protection in Fluoroscopy

Although the general principles of dose reduction in radiography apply to fluoroscopy, special precautions must be taken with the latter:

1. *Intermittent Fluoroscopy.* It is good practice to use intermittent activation of the fluoroscopic tube to decrease patient exposure and prolong tube life.

2. *Restriction of Field Size.* The size of the fluoroscopic field must be limited by suitably collimating lead shutters placed between the tube and the patient. As we have seen, the skin and depth doses, as well as the scattered radiation into the surroundings, decrease as the area of the radiation field is decreased. Therefore, the fluoroscopic beam should be restricted to the smallest field that includes the area of interest, to reduce exposure of both the patient and personnel. According to the Federal Register Code *21 CFR 800.10,* neither the length nor the width of the useful beam at the image receptor shall exceed that of the visible image area by more than three percent of the source-image receptor distance. In this case, the image receptor is the input screen of the image intensifier.

3. *Correct Operating Factors.* An increase in kVp produces an increase in fluoroscopic image brightness. If this is accompanied by an increase in filtration and a compensating decrease in mA, exposure rate to the patient is actually decreased with no change in image brightness. In other words, for a given image brightness, x-ray exposure of

the patient decreases as kVp is raised and mA is lowered. Since further improvement becomes insignificant above 100 kVp, the recommended operating factors are 90 to 100 kVp, 2 to 3 mA for bright (image intensified) fluoroscopy. Image resolution can be improved by increasing the exposure rate, but this is strictly limited by maximum permissible exposure of the patient. For children the kVp should be reduced to a value depending on body thickness, just as in radiography. The source-skin distance must be at least 15 in. (38 cm) with stationary and 12 in. (30 cm) with mobile fluoroscopic equipment according to the federal code cited above. Distances longer than these, with compensating increase in exposure factors, do not significantly reduce patient exposure.

4. *Filtration.* We can increase the hardness of an x-ray beam by using a suitable filter. This removes relatively more soft than hard x rays, thereby removing a large portion of non-image-producing radiation which would otherwise be absorbed in the patient's skin. In fluoroscopy, the same filtration is required as in radiography: total of 2.5 mm Al equivalent, or HVL of 2.6 mm Al.

5. *Exposure Limits.* Although there is the hazard of cumulative small doses of radiation during one's lifetime, there is as yet no official limit to x-ray exposure of patients in diagnostic radiology. Obviously, this should be reduced to a minimum consistent with medical needs. According to the federal code cited above, the exposure rate at the table top (ie, patient's entrance exposure rate) must be no greater than 10 R/min for fluoroscopic equipment with automatic exposure rate control and no greater than 5 R/min without automatic exposure rate control. This does not apply to the exposure rate in the radiographic spotfilm mode. In addition, a timer must be provided to preset the on-time of the fluoroscopic tube for a cumulative total of no more than 5 min. At

the end of this time, an audible signal must indicate that the time limit has been reached and must continue to sound as long as fluoroscopy continues until the timer is reset.

Some fluoroscopic units have an optional high-level control which temporarily increases x-ray output above the 5 R/min limit at the patient's skin entrance surface. This control shall be subject to continuous manual operation only, and be accompanied by an audible signal. Extreme caution should be exercised in the use of the high-level control, as radiation injuries have occurred from ignoring the audible signal and continuing fluoroscopy.

In recent years fluoroscopy has been used increasingly for special procedures such as interventional cardiology (eg, cardiac catheterization with angioplasty, and C-arm fluoroscopy), with reports of significant skin reactions in the patient's radiation field. This must be avoided insofar as possible by adequate history of previous fluoroscopic exposure and more efficient use of fluoroscopic equipment. In-service training sessions with cardiologists and other nonradiologist users of fluoroscopic equipment should be conducted by the health physicist or radiation safety officer.

7. *Primary Protective Barrier.* This should be equivalent to 2 mm lead (Pb), which is provided by the image intensifier assembly. Note that design of the system prevents activation of the fluoroscopic tube when the image intensifier is in the parked position.

PROTECTION FROM ELECTRIC SHOCK

On initial installation, all radiographic equipment must be properly **grounded** to prevent electric shock. Ordinarily, secure connection by a suitable conductor to an underground metal water pipe should be adequate. In some instances, with very high energy x-ray units, grounding may have to be made through a metal rod that has been driven into the ground until it reaches permanently wet ground; this may have to be evaluated by actually measuring the resistance of the grounding circuit.

The building codes of most cities require all electric wall outlets wherever located, to be grounded in all newly constructed buildings. It is advisable to ground outlets in old buildings.

Personnel should always observe the **one-hand** rule: never touch an electric device with one hand while the other hand is submerged in the processing tanks or is touching a conductor.

High-voltage, low-amperage shock tends to "throw," whereas low-voltage, high-amperage shocks tends to "hold" the victim. In the latter instance, do not grasp the victim directly, but first either open the main switch or remove him/her by means of a dry board, a dry wad of newspaper, or a dry rope. Immediately call for medical aid, in the meanwhile applying cardiopulmonary resuscitation (CPR).

PROTECTION IN NUCLEAR MEDICINE

The handling of radionuclides requires certain precautions because of the harmful effects of chronic exposure to ionizing radiation, as already described. But the problems associated with radionuclides differ in some respects from those prevailing in the diagnostic x-ray department; hence, the need of a separate section dealing with health physics in nuclear medicine. Medical compounds containing radionuclides are called **radiopharmaceuticals.** In this section we consider protection as it relates to the diagnostic use of radionuclides.

Types of Radiation. Before discussing the harmful effects of radiation emitted by radionuclides and the methods of protection, let us review the types of radiation, involved. Since radiopharmaceuticals emit only gamma rays and beta particles, our discussion will be limited to them. Not only do gamma rays have much greater penetrating ability than beta particles, but they also differ from them in character: gamma rays are electromagnetic waves or photons, whereas beta particles consist of high-speed electrons. Consequently, protection must differ for these two kinds of radiation.

Harmful Effects. These have already been described in detail on pages 377-389. The same effects are produced by beta and gamma radiation as by x radiation for equal absorbed doses; but betas are much less penetrating, their effects being limited to the skin from external sources, or in their immediate vicinity from internal sources. However, recall that fast-moving electrons give rise to bremsstrahlung as they slow down when they interact with atoms. The effects of x and gamma rays occur within the body, even from external sources, depending on their energy and dose.

Dose Equivalent Limit. Since the quality factor H for gamma rays is the same as that for x rays, the dose equivalent limit is also the same, especially since exposure is low, in the diagnostic range.

Principles of Radiation Protection. There are basically four factors, suitable combinations of which are used to reduce or eliminate radiation hazards in nuclear medicine. These include: (1) **distance,** (2) **shielding,** (3) **time of exposure,** and (4) **limitation of activity** of source to smallest required amount. In this section, we shall assume that the radioactive sources are sealed in capsules or needles. Liquid preparations of radionuclides, except for the hazard of spill, ingestion, or inhalation, may also be regarded as sealed sources insofar as external protection is concerned.

1. **Distance.** This is often the easiest and least expensive method of reducing the exposure rate to the permissible level. In the case of gamma rays, the inverse square law applies with reasonable accuracy for protection in storage and handling. Long forceps should be used for picking up encapsulated radiopharmaceuticals; **capsules should be used whenever possible*** to avoid spillage of liquids. Storage containers that emit gamma rays should be as remote as possible from areas habitually occupied by personnel and general public.

2. **Shielding.** Protective materials placed between a radioactive source and its surroundings to reduce the exposure rate are called **shields** or **barriers.** Proper shielding contributes materially to the reduction of **gamma-ray exposure rate,** especially when combined with distance. The shield may be incorporated in the container itself, or it may be placed as a barrier around the source in the storage or working area. **Lead** of proper thickness serves as the most satisfactory barrier material for storage of radiopharmaceuticals because of its efficiency as an absorber of gamma rays. As a general rule, gamma-ray emitting radiopharmaceuticals should be stored in their shipping container and surrounded with lead bricks 5 cm (2 in.) in thickness, on all sides that might constitute radiation hazard. Routine monitor surveys

---

* There is a strong trend in the U. S. for delivery of ordered doses in capsules from local radiopharmaceutical services. Not only is this very convenient, but it also obviates contamination by spill of liquid radiopharmaceuticals in the hospital or clinic.

gist has been conscientious in observing all the rules, can we be certain that radiation hazards have been eliminated? Insofar as we know, there have been no reported harmful effects from exposures approximating the recommended dose equivalent limit. Periodic blood counts, while they may detect early radiation injury, are a poor substitute for preventative monitoring by badges: film, TLD, or Luxel®. If the dose equivalent limit is exceeded, prompt corrective steps must be taken.

Upon accepting a new job, the technolo-

gist must bring records of previous exposure and undergo a complete blood count and physical examination. Upon terminating employment, the technologist must be given a complete summary of the incurred radiation exposure.

The danger from ionizing radiation in the well-controlled Radiology Department is virtually nil, provided all personnel make full use of all protective devices and equipment and constantly bear in mind the ALARA principle.

QUESTIONS AND PROBLEMS

1. State the causes of background radiation. What is the average individual background exposure per year from natural and from artificial sources?

2. State how dose equivalent (H) and effective dose equivalent (H_E) are derived. Why may we use H instead of H_E in diagnostic radiology?

3. Describe two methods by which one may determine the amount of whole-body radiation being received in the Radiology Department.

4. Discuss the effects of excessive whole-body exposure to ionizing radiation. What are the locally damaging effects of overexposure to ionizing radiation?

5. Name the materials widely used in protective barriers. What material (and thickness) is required for the walls of ordinary radiographic rooms?

6. State the annual occupational dose equivalent limits for radiologic technologists. Why may this be used instead of effective dose equivalent?

7. What are the dose equivalent limits before and after pregnancy is known? For the general public annually?

8. What special precautions should

nuclear medicine technologists take when injecting radiopharmaceuticals?

9. Where should the personnel monitor (Luxel®, film badge, or TLD) be worn during fluoroscopy?

10. What is ALARA? Should it be enforced regardless of cost?

11. Discuss minimum exposure radiography in (1) fluoroscopy and (2) radiography, including the important factors in each.

12. Assume that the conditions of your x-ray department are "safe," from the standpoint of ionizing radiation. State three rules of conduct on your part that will help keep your exposure below the occupational dose equivalent limit.

13. Name and justify the four main protective measures against whole body exposure in nuclear medicine.

14. A radionuclide is stored in a container that is transmitting twice the acceptable limit of gamma radiation. How many HVLs of lead must be added to the shielding?

15. Calculate the dose equivalent in cSv (rem) from a mixture of radiation sources yielding these doses: x rays 5 cGy, gamma rays 2 cGy, and fast neu-

trons 12 cGy.

16. Why must x-ray equipment be grounded? Is a gas pipe satisfactory? A water pipe?

17. What is the difference between acceptance testing and radiation safety surveys? Who performs these tests and what do these entail?

Chapter 29

NONRADIOLOGIC IMAGING

Topics Covered in This Chapter

- Magnetic Resonance Imaging (MRI)
    Physical Principles
    Nuclear Magnetic Resonance (NMR)
    Signals
    From NMR to MRI
    T_1 and T_2 Relaxation
    Spin Echo
    T_1, T_2, and Proton Density Contrast
    Gradient Coils
    Noise in MRI
    External Field Magnet

    Surface Coils
- Ultrasound Imaging
    Nature and Properties
    Transducer
    Beam
        Resolution
        Reception
    Image Display
    Doppler
- Questions and Problems

I N THIS CHAPTER we shall deal with two important diagnostic devices that do not employ x rays. These include magnetic resonance and ultrasound imaging systems.

MAGNETIC RESONANCE IMAGING

To a casual observer, a magnetic resonance imaging (MRI) unit bears a resemblance to a computed tomography (CT) scanner. However, they operate on radically different physical principles. Whereas CT requires a rotating x-ray beam, MRI depends on the behavior of positively charged particles–atomic nuclei–in an external magnetic field that undergoes modification in the process of creating an MR image. Note again that MRI is a nonradiologic process in that the radiation it uses, while electromagnetic in nature, has a much lower frequency (longer wavelength) than x rays and depends on continually-spinning atomic nuclei within the body, rather than on x rays generated by electron interactions with an x-ray tube target. A physical phenomenon called *nuclear magnetic resonance (NMR)* underlies MRI.

Nuclear Magnetic Resonance

Although relatively new in its application to the imaging of atomic structures, nuclear magnetic resonance (NMR) has been in use for about 50 years in spectrography to identify atoms and molecules and to study their properties. When applied to imaging (MRI), it provides, an extremely wide range of diagnostic information.

In particular, MRI achieves **superb low-contrast resolution** without bone or air artefacts, and does not employ ionizing radiation. Low contrast resolution with MRI is many times greater than with x rays.

The physical basis of NMR is extremely complex, but it fortunately lends itself to some degree of simplification for our purpose. A review of magnetism should be helpful at this point.

Nuclear Spin. A magnet consists ultimately of atomic-size magnets, each with a north and south pole, induced by the intrinsic continual spinning of atoms, nuclei, protons, neutrons, and electrons, on their individual axes. These spinning particles are called **magnetic dipoles,** the strength of which is unique to each kind of chemical element. An ordinary magnet contains innumerable iron atomic dipoles, arranged in domains and lined up in the form of bulk magnetization, with a north pole at one end and a south pole at the other (see pages 60-61).

Since MRI is based on NMR, which involves spinning nuclei, and since hydrogen nuclei–protons–are the most abundant in the body and have strong magnetic dipoles, we shall base our description of NMR primarily on protons. The fields of the spinning proton dipoles point in a particular direction and are therefore vector quantities indicated by the **right hand rule.** If you allow your right fingers to curl in the direction of proton spin (see Figure 29.1), your thumb will point in the direction of the magnetic field, that is, toward the dipole's north pole. The direction and intensity of this field define the proton's **magnetic dipole moment.** The term moment refers to the tendency of a dipole to rotate in an external magnetic field due to the torque (twisting force) gener-

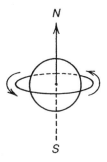

SPINNING PROTON

Figure 29.1. Inherent spinning of protons, which behave as miniature magnets. They therefore possess magnetic moment, the tendency to turn in an external magnetic field due to the force between them.

ated by interaction between the two fields–dipole and external. Since forces have direction, they are **vector** quantities.

Alignment of Magnetic Dipoles. In the body, the hydrogen nuclear (proton) dipoles normally arrange themselves in random directions (helter skelter) like the iron atoms in a piece of soft iron. This randomness results from thermal agitation of the particles, in the absence of an external direc-

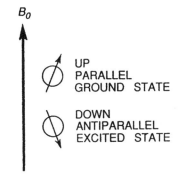

Figure 29.2. Precession of spinning protons around their axes, when situated in an external magnetic field. The "up" position represents a lower energy state and the "down" position a higher energy state.

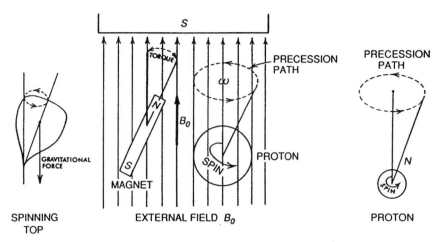

Figure 29.3. Precession of protons compared with a small bar magnet in a magnetic field, and with a spinning top which precesses on account of the force of gravity.

tional force. Therefore, the algebraic sum of all the magnetic moments is zero and there is no net magnetization of the body. The earth's magnetic force is too weak to align the dipoles to any appreciable degree. But when a person is placed in a strong external magnetic field, B_0, the resulting torque causes some of the proton magnetic moments to line up with B_0 (ie, "parallel"), while an almost equal number remain directed opposite to the field ("antiparallel"). It so happens that only a tiny *excess* (about 1 per million) align themselves parallel with the field. By convention, the parallel orientation is designated the "up", and the antiparallel the "down" position (see Figure 29.2). The up proton dipoles are in the ground (lowest) energy state, with their north poles in the direction of the south pole of the external field. As shown on Figure 29.3, a small, freely-moving magnet like a compass needle lines up similarly when placed in an external magnetic field.

Nuclear Precession. At the same time the protons spin on their axes, these axes also rotate (see Figure 29.3), an effect known as *precession.* This is caused by the torque generated between the proton dipole

moment and the external field. Although the precessing protons are said to line up parallel with external field B_0, they actually have their axes inclined at a small angle with B_0. An analogy is the precession or wobble of a toy top under the influence of gravitational force. In summary, then, we have proton magnetic dipole moments induced by spontaneous, continual spin of protons on their axes and, at the same time, precession of the axes due to the torque of the external field.

Precession has a specific *frequency,* that is, number of rotations per second, symbolized by ω (Greek *omega*) called the ***Larmor frequency;*** this is proportional to the intensity of the external field B_0 and related to it by a constant γ (Greek *gamma)* that is unique to the specific particle. Gamma is known as the ***gyromagnetic*** or ***magnetogyric ratio.*** Table 29.1 gives values of for a few typical nuclei that lend themselves to NMR; it so happens that only nuclei with odd numbers of protons *and* neutrons are subject to NMR (Curry and others). Note the high value of γ for ordinary hydrogen–43 MHz/tesla (Hz is the unit of frequency, which is 1 cycle per s, and MHz is 1 million Hz). Tesla (T) is the SI

Table 29.1
APPROXIMATE GYROMAGNETIC RATIOS*

Nuclide	Gyromagnetic Ratio
	MHz/T
Hydrogen 1 (^{1}H)	42.6
Hydrogen 2 (^{2}H) (deuterium)	6.5
Carbon 18 (^{13}C)	11
Sodium 28 (^{23}Na)	11
Phosphorus 31 (^{31}P)	17

*Modified from Curry TS III, Dowdey, JE, Murry RC Jr: *Christensen's Physics of Diagnostic Radiology*, 4th ed., Philadelphia, Lea & Febiger, 1990. By permission.

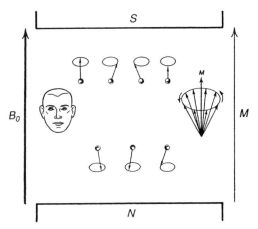

Figure 29.4. Basic relations in magnetic resonance imaging. On the left is a frontal view of a patient's head in the supine position within the external, main magnetic field B_o directed according to the thick arrow. The field also contains precessing protons, in the lower energy state with their N-poles directed toward S-pole of the magnet; and precessing protons in the higher energy state with their N-poles directed toward the N-pole. Actually, the lower-state protons are in extremely slight excess. On the right is a figure (taken from Figure 24.5) representing the summation of the protons in their lower energy state, with a net magnetization M parallel to B_o. Note that whole figure is oriented as it would appear to one looking vertically downward at the MRI couch.

unit of magnetic field strength, in this case, of B_o.

Larmor frequency ω is expressed by

$$\omega = \gamma B_o \quad MHz \qquad (1)$$

You can see that γ is in MHz/T and B_o is in T, so

$$\frac{MHz}{T} \times T = MHz$$

Also, the precessional frequency or speed of axis rotation of the nuclear dipole can be increased by using a stronger magnet B_o. The field strengths of clinical MRI imagers range from 0.15 to 1.5 T and are constant for each imager.

Considering only the precessing protons in the parallel (up or ground state) position (lowest energy level), we find that their small excess over those in the down position creates a resultant net magnetization M superimposed on B_o (see Figure 29.4). In other words, M represents the algebraic sum of the individual precessing nuclear magnetic moments (ie, difference between up and down protons), and precesses with the same frequency ω. As will be shown later, M gives rise to an NMR signal under certain conditions.

Resonance. By mechanical analogy, resonance is exemplified by the sympathetic vibration of one object when its natural vibrational frequency matches that of another vibrating object. For example, striking the middle A key on a piano will transfer energy to the air in the form of sound waves of the same frequency, 435 Hz, which will then set the A string of a nearby violin into vibration with the same frequency–the violin A string is said to be tuned to the piano A, or the two strings are in ***resonance.***

Similarly, a condition of resonance is created by application of a radiofrequency (RF) electromagnetic pulse with the same Larmor frequency as the precessing net magnetiza-

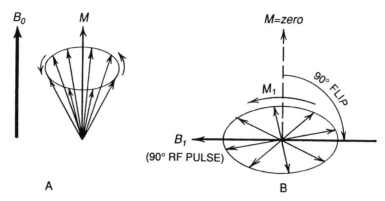

A   B

Figure 29.5. *A.* Group of precessing protons creating a net magnetization vector *M;* the diagram is oriented as explained in Figure 29.4, with the net magnetization vector being the summation of the precessing protons. *M* rotates with the same Larmor frequency as the precessing protons. *B.* 90° radiofrequency (RF) magnetic pulse B_1 has been applied at 90° to *M* (ie, transversely across the MRI couch), causing the net magnetization to "flip" (spiralling) down to the patient's *transverse* plane, where *M* continues to rotate, initially at Larmor frequency. Now *M* is zero and M_1 at maximum.

tion M, if the RF pulse is directed perpendicular to M or B_o. In this process, energy is transferred to some of the precessing protons (and M), which abruptly turn down to a higher energy state. If the RF field designated as B_1 lasts one-fourth cycle (90°), these protons "flip" down 90° together with their net magnetization onto the transverse plane (ie, perpendicular to B_o) where they continue to precess, as shown in Figure 29.5. In the transverse plane, net magnetization is labeled M_1. Note that the rotation of M_1 in the transverse plane has the Larmor frequency of the protons. When the RF was applied, it caused the precessing "up" protons to get into step so as to assume identical point-for-point positions in their rotational paths; the precessing protons are said to be *in phase* in the transverse plane. While rotating in the transverse plane, the net magnetization energy sets up an alternating voltage (AC), whose radiofrequency matches the Larmor ω, in a nearby receiver coil; this *signal* in the receiver coil eventually contributes to the MR image.

When the RF turns off at the end of

one-fourth cycle, the RF signal fades and disappears. This is called *free induction decay (FID)* of the signal because it occurs spontaneously in the absence of the RF field (see

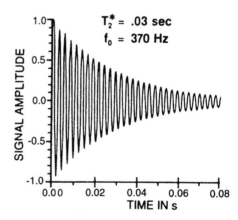

Figure 29.6. Free-induction decay (FID) wave form due to a single radiofrequency of $f_o = 370$ Hz. The amplitude of this signal decays exponentially with time constant T_2^*. * represents the effects of inhomogeneities in the external magnet. Without *, the arbitrary time constant T_2 here is 0.03 s. *(From Wehrli FW, by permission of the author and GE Medical Systems.)*

Figure 29.6). You can see that FID is a damped oscillation wave with exponentially declining amplitude, lasting only a fraction of a second. FID represents so-called T_2 relaxation, to be explained below.

Factors Governing NMR Signal. Three conditions determine the intensity of an NMR signal, and the brightness of the resulting MR image: (1) proton density, (2) T_1 relaxation, and (3) T_2 relaxation. These will now be explained.

1. ***Proton density.*** This is simply the number of mobile hydrogen nuclei–protons–per unit volume of tissue in the NMR field. This translates into the number of protons in a given voxel (corresponds to voxel or unit tissue element in CT). Obviously, the greater the proton density, the greater will be the number of precessing protons contributing to net magnetization and the resulting signal. Note that proton density differs from tissue to tissue.

2. ***T_2 Relaxation.*** On delivery of the 90° RF pulse perpendicular to B_0, the protons in the net magnetization in the transverse plane precess in phase (in step) as noted above. Then cessation of the RF pulse causes the precessing protons to ***dephase*** (get out of step) due to three factors: thermal molecular agitation, "spin-spin" interactions with

neighboring protons, and inhomogeneity (nonuniformity) of the external field B_0. As a result, transverse magnetization decays exponentially (FID) with time, according to the time constant T_2^* (like exponential decay of a radionuclide, mathematically) (see Figure 29.7). T_2^* may be defined as the time it takes for transverse magnetization to decay to 37 percent of its maximum (initial) value. The asterisk in T_2^* indicates the inclusion of B_0 inhomogeneity. T_2 denotes ***spin-spin relaxation*** because it results only from interactions among spinning protons; it is also called ***transverse relaxation*** because it occurs perpendicular to B_0. As T_2 relaxation or demagnetization occurs, the transverse net magnetization M_1 spirals upward along a path with decreasing diameter (like the upper part of a beehive) until M_1 reverts to M parallel to B_0. This corresponds to FID mentioned earlier. It must be emphasized that RF signals are emitted *only* during transverse relaxation.

3. ***T_1 Relaxation.*** This is longitudinal remagnetization as the protons return to the lower energy state, with their net magnetization M parallel to B_0. While M_1 is spiraling upward as described under T_2 relaxation, it also has a vertical component which ends up as M. T_1 relaxation occurs exponentially with time, but this is a ***growth*** of magnetization from M = 0, to M = M (see Figure 29.8); when M turned down 90° due to the RF pulse, its value declined to zero along its initial axis; and during relaxation it returned to its maximum value along its initial longitudinal axis.

T_1 is the time constant and is unique to various tissues. The protons absorbed energy from the RF pulse during excitation at the resonant frequency, to precess in the transverse plane. They return this energy to the tissues during relaxation, so T_1 relaxation is also called ***spin-lattice relaxation*** (tissues are a lattice or network of structures). Exponential buildup of net magnetization M

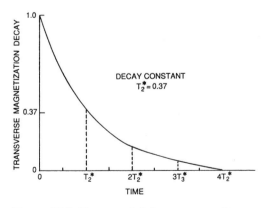

Figure 29.7. Exponential decay curve of transverse relaxation T_2.

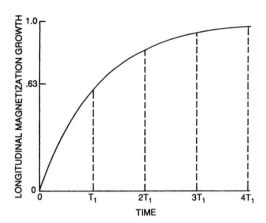

Figure 29.8. Exponential growth curve of longitudinal relaxation T_1.

means that the time constant T_1 represents the time it takes for M to return to 63 percent of its original equilibrium value parallel to B_o (see Figure 29.8). T_1 is also called *longitudinal relaxation* because it occurs along the axis of the longitudinal field B_o. Again note that T_1 relaxation *does not emit an RF signal;* as will be shown later, it must first be converted to a T_2 signal.

Relaxation Times. Relaxation times and proton densities have a strong effect on the MR images produced by various normal and pathologic tissues. In water, $T_1 = T_2$, but in soft or solid tissues T_1 is much longer than T_2; T_1 relaxation therefore occurs more slowly than T_2. As it turns out, T_2 can be changed to a larger extent than T_1 by various tissue abnormalities such as cancer, resulting in a greater likelihood of detection.

Several factors influence relaxation times, especially T_2. For example, the rate at which complex molecules "tumble": the slower the tumble, the faster the relaxation. Thus, fatty molecules such as the lipids in the white matter of the brain tumble relatively slowly, so their protons relax faster than water molecules.

Spin-Echo. This is a technical procedure in which a sequence of 90° and/or 180° RF pulses at selected time intervals are used for one or more of the following reasons: (1) minimizing the effects of B_o inhomogeneity on T_2 relaxation, (2) changing T_1 relaxation to T_2 signals, and (3) preferentially increasing the intensity of signals from T_2 relaxation and processed T_1 relaxation, and (4) proton density. Spin-echo technique involves the production of so-called "echoes" which are simply enhanced selected signals. An example of this process is the one that intensifies the signal from transverse relaxation–T_2–to increase brightness and improve contrast in the resulting MR image. The following steps are involved (see Figure 29.9):

1. 90° RF pulse generates free induction decay according to time constant T_2^* which includes B_o inhomogeneity, followed by a

2. 180° RF pulse after a selected interval TR (about 400 to 6000 ms), followed by the spin-echo signal at time TE–this is the time interval between the first RF pulse of 90° and the middle of the spin-echo, also called *echo time.*

3. Additional pulses may be applied.

4. Echoes decay with time constant T_2, since B_o inhomogeneity has been counteracted by the spin echo process. Image brightness depends on the intensity of the echo signal.

Spin-echo may be explained by the effect of B_o inhomogeneity on protons as they precess in the transverse plane after a 90° pulse. Owing to different local values (nonuniformity) of B_o, precession of the protons will occur at different frequencies (recall that Larmor frequency ω is proportional to B_o) As shown in Figure 29.9, if one thinks about the simple case of two different precessional frequencies in the transverse plane, one must be faster than the other. After the first 90° RF pulse, dephasing occurs as the faster precessions get farther and farther ahead of the slower ones. Following the 180° pulse, transverse magnetization moves down and around 180° and comes to lie on the oppo-

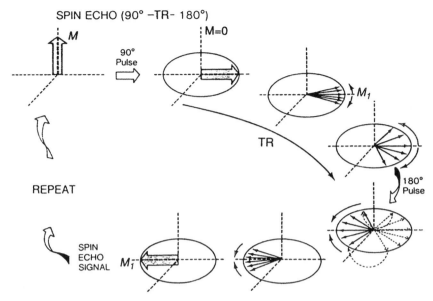

Figure 29.9. A typical spin-echo sequence. First a transverse 90° RF pulse flips net magnetization down 90° to rotate in the transverse plane, where spinning protons get out of step, some rotating behind the others. After a time period TR (repetition time) a 180° RF pulse is applied transversely, which now causes net magnetization M_1 to rotate 180° down and around into the horizontal plane on the mirror image side of the 90° plane (arrows pointing to the left). Now the lagging protons catch up with those ahead, and magnetization becomes focused (thick arrow toward the left), during which a spin-echo signal is emitted. Note that this is how T_1 relaxation, which by itself does not emit a signal, has been converted to transverse relaxation to emit a signal. Other sequences may be used to vary the kind of MRI weighting. *(Courtesy of Harms SE, et al, and Radiographics.)*

site (mirror image) side of the transverse plane, with the faster precession now ***behind*** the slower; they are still dephased. When the faster precession catches up with the slower one, they rephase and generate an echo–a relatively intense signal. The TR value (time between the pulses) has a pronounced effect on the signal intensity and the resulting MRI brightness and contrast.

From NMR to MRI

What has been said thus far describes the signals generated by the basic NMR processes. In recent years, the principles of NMR have been developed into magnetic resonance imaging (MRI) to produce exquisitely detailed images of anatomic structures, both normal and pathologic.

Although MR images bear a superficial resemblance to CT images, MRI: (1) requires no ionizing radiation; (2) provides much better low-contrast image resolution; (3) yields images directly in multiple sections; (4) is not subject to artefacts from high and low density materials such as bone and air, although magnetic materials such as iron, cobalt, and nickel introduce artefacts and may cause injury due the strong external magnetic field; (5) does not require injection of organic iodides for image enhancement, although gadolinium injection improves contrast with much less hazard than organic

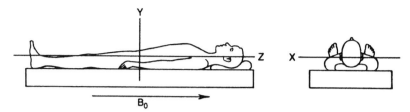

Figure 29.10. Orientation of standard axes relative to the patient's position and external field B_o.

iodides; and (6) serves as an excellent method of measuring blood flow, but this will not be included here.

Let us now apply and extend the principles of NMR to MRI—the actual imaging of the body in multiple slices. But first we must clarify some basic terminology related to the patient's and MRI axes. Figure 29.10 shows the position of a patient on an MRI couch. The **longitudinal axis** corresponds to the **z-axis**, as well as the axis of the external field B_o and net longitudinal magnetization M. The **x-axis** runs from side to side and is the transverse axis, so it lies in the transverse plane. Finally, the **y-axis** is anteroposterior and, with the x-axis, defines the **transverse plane** (recall that application of a 90° RF pulse generating a field perpendicular to the longitudinal or z-axis flips net magnetization onto the transverse or *xy* plane).

The next three sections deal with MR image brightness and contrast and how they are affected by T_1 and T_2 relaxation, and by proton density.

Image Brightness. The brightest images are produced for tissues with a short T_1—fast remagnetization after a 90° RF pulse. As already explained, tissues with a high fat or lipid content, such as the white matter in the brain, have a low tumbling rate and therefore a short T_1 and T_2. Aqueous molecules such as water in cysts and cerebrospinal fluid have a rapid tumbling rate and therefore a long T_1 and T_2 (see Table 29.2).

T_1-weighting preferentially intensifies the T_1 contribution to the MR image, but T_1

Table 29.2
APPROXIMATE T_1 AND T_2 RELAXATION
TIMES FOR VARIOUS TISSUES

Tissue	T_1	T_2
	msec	*msec*
Cerebrospinal fluid	2000	150
Fat	220	80
Brain		
White matter	650	90
Gray matter	750	100
Muscle	600	40
Liver	480	50
Kidney		
Medulla	680	140
cortex	860	70
Blood	800	180

*Averaged from sources in the literature. Values differ among various authors. Data here are for a 1-T magnet.

relaxation does not generate an RF signal, so it must be changed to transverse magnetization and T_2 relaxation by sequential pulsing based on spin-echo technique. A short TR (time between pulses) stops T_1 remagnetization early, along the *lower* part of the longitudinal magnetization growth curve (see Figure 29.8). A 90° RF pulse now would convert the relatively small longitudinal magnetization to a correspondingly small transverse magnification giving rise to a weak signal. But a long TR allows more time for the growth of longitudinal magnetization, say in the upper part of the growth curve, resulting in a stronger transverse magnetization after a 90° pulse and yielding a stronger signal and image brightness. Thus,

SHORT TR and WEAK MAGNETIZATION
→ WEAK TE SIGNAL

LONG TR and STRONG MAGNETIZATION
→ STRONG TE SIGNAL

where TE equals time between echoes.

Optimal brightness of image occurs when TR (time between pulses) exceeds a $3T_1$ interval for a particular tissue. Note that in the process just described, T_1 (longitudinal) relaxation has been changed to T_2 (transverse) magnetization to acquire an RF signal. However, this must be modified to achieve optimum contrast resolution with spin-echo technique, as described below.

With T_2 *relaxation,* which also follows an exponential decay curve, a T_2-weighted RF pulse sequence causes tissues with a longer T_2 to give a stronger echo and a ***brighter image.***

Proton density is an inherent property of various tissues and has a strong effect on signal intensity and eventual image brightness. As indicated above, proton density is simply the number of protons per unit volume of tissue—in practice, per ***voxel.*** (It is assumed that the protons are mobile so that they can participate in magnetization and demagnetization.) In and of itself, proton density has a major influence on image brightness because more protons are available to contribute to signal intensity.

Image Contrast. Besides image brightness, image contrast has a major impact on image quality. As in radiography, contrast in an MR image, as recorded on film, denotes variation of image density from tissue to tissue. Pulse sequences TR and TE present white matter as white, and gray matter as dark, but manipulation of these factors can alter the shades of contrast, and even invert the images so that white becomes dark and dark becomes white. In any event, we must be able to distinguish easily between various tissues by producing optimum contrast and brightness.

Contrast is the resultant of three basic NMR factors: T_1 relaxation, T_2 relaxation, and proton density. These can be manipulated to optimize contrast.

1. ***T_1 Contrast.*** Growth of longitudinal magnetization after an RF pulse occurs at different rates for different tissues. To obtain optimum contrast we must find a TR (time after an RF signal) which yields an echo at the instant that the ratio between the magnetization of both tissues is at maximum (see Figure 29.11). However, this may occur at

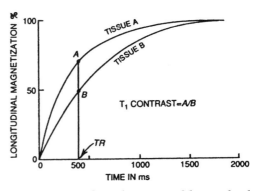

Figure 29.11. Growth curves of longitudinal magnetization for tissues having different relaxation constants T_1. Contrast, as derived from the curves, reaches a maximum at A/B. Spin-echo repetition time TR here is 500 ms. *(Adapted from Keller PJ, by permission of the author and GE Medical Systems.)*

points on the curves where signal intensity is too low. In general, a $TR = 0.63T_1$ will provide a satisfactory combination of brightness and contrast (analogous to density and contrast in a radiograph). For T_1-weighting, one should select an intermediate TR and short TE (time between echoes) because a short TE decreases T_2 contrast more than T_1 contrast. A T_1-weighted MR image displays reasonably long scale contrast, with good overall image resolution of various anatomic structures.

2. ***T_2 Contrast.*** T_2 relaxation can be used to enhance image contrast because different

tissues have different relaxation times according to their T_2 constants. Comparison of the decay curves of transverse demagnetization for two hypothetical tissues (see Figure 29.12) shows how contrast can be maximized. Contrast is the ratio of the residual transverse magnetization for the two tissues at a particular time, for example, A/B.

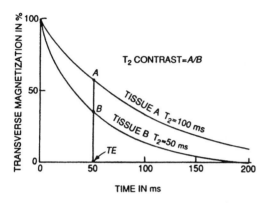

Figure 29.12. Decay curves of transverse magnetization for tissues having different relaxation constants T_2. Contrast reaches a maximum at A/B, spin-echo TE (time to echo) is here 50 ms. *(Adapted from Keller PJ, by permission of the author and GE Medical Systems.)*

For maximum contrast an intermediate TE should be selected (TE is spin-echo time, between the first RF pulse and the center of the following spin-echo). Recall that spin-echo (see pages 428-429) is used to intensify RF signals from transverse magnetization, to produce a bright image. But if the selected TE turns out to be too long, image brightness suffers on account of the low residual transverse magnetization. A long TR interval (time between successive pulses) minimizes T_1 contrast, thereby favoring T_2 contrast. Thus, by adjusting the spin-echo sequence to a long TR and intermediate TE, we obtain a T_2-weighted image, which reveals excellent contrast—bright gray matter and cerebrospinal fluid, and dark white matter.

3. ***Proton-Density Contrast.*** Proton density is the number of mobile protons (ie, capable of producing a signal), per unit volume of tissue—in practice, per voxel. A high proton density creates a bright image (pixel) of the voxel. Proton density contributes to contrast, which depends on the ratio of proton density in one voxel to that in another.

Figure 29.13 shows how proton density contributes to image contrast. Just as with T_1, growth of longitudinal magnetization after a 90° RF pulse occurs at different rates for different tissues, but when based on different proton densities, the magnetization curves have a different shape. In Figure 29.13 you see the longitudinal magnetization curves for two voxels that have the same T_1 values but different proton densities; the upper curve represents the voxel with higher proton density. Contrast due to proton density is the ratio of the longitudinal magnetization of the two voxels at a particular instant. The ratio A/B in the figure is the contrast at time TR. Note that as you follow the curves upward, contrast progressively increases to a maximum at TR, 1500ms,

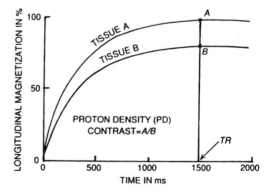

Figure 29.13. Growth curves of longitudinal magnetization resulting from tissues having different PD (proton density) relative values. Maximum contrast is reached at A/B; repetition time (TR) here is 1500 ms although it may be much longer. *(Adapted from Keller PJ, by permission of the author and GE Medical Systems.)*

Table 29.3
MR IMAGE CONTRAST PRODUCED BY WEIGHTING

Weighting	TR	TE	Brain Matter White	Gray
Proton Density	long (min* T_1)	short (min T_2)	dark	light
T_1	intermediate (enh** T_1)	short (min T_2)	gray	gray
T_2	long (min T_1)	intermediate (enh T_2)	dark	light

* minimizes
** enhances

Note: Cerebrospinal fluid appears gray-to-black in T_1-weighted and PD images, and white in T_2-weighted image (*see* Figure 29.17).

after which it remains constant up to full magnetization M (equilibrium values). Since maximum contrast is reached late in the magnetization process (high on the curves) as compared to T_1, a long TR is desirable; this also minimizes T_1 contrast. At the same time, a short TE minimizes T_2 contrast. Therefore, proton-density weighting requires a long TR (as in Figure 29.13) and a short TE.

Table 29.3 presents the weighting factors, rationale, and effect on white and gray matter of the brain as they appear on the MR image.

Conversion of RF Signals to MR Images. This requires a procedure called *signal encoding* which makes it possible to create images from RF signals. A signal's amplitude (strength) is proportional to the degree of transverse magnetization elicited by the resonant RF pulse which, in turn, is proportional to the number of excited protons in the voxel. Signals can be enhanced by weighting of T_1, T_2, and proton density to produce spin-echoes. These signal differences may be small from tissue to tissue, but by MRI manipulation, they may become distinguishable—*low contrast resolution.* In fact, MR image contrast resolution exceeds that of CT by a factor of 30 or so.

Signal-to-image conversion is highly complex, requiring the use of *auxiliary coils* to generate RF field gradients across the main field B_o. A *field gradient* causes a change of field strength along a particular axis. Wehrli uses a simple device to demonstrate the basic principle (see Figure 29.14). Two cylindrical holes are drilled in a teflon block and filled with water; they have been aligned with the *z-axis,* parallel to B_o, at the same distance *along* the *y-axis,* but at different dis-

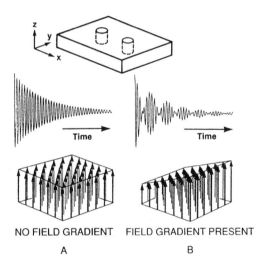

Figure 29.14. *A.* In the absence of a field gradient ($G_x = G_y = G_z = 0$) both samples experience the same field, resulting in a simple free-induction decay (FID) wave form. *B.* In the presence of a field gradient G_x, the two samples experience different fields, resulting in a complex FID wave form. (Figure 29.14A *from Wehrli FW, by permission of the author and GE Medical Systems.*)

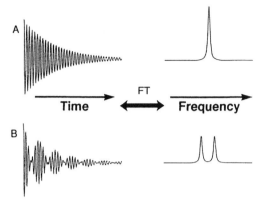

Figure 29.15. By a mathematical process called Fourier transformation (FT) applied to the FID wave forms in Figure 29.14, the time line is changed to a frequency line showing a single frequency signal in *A*, and two signals in *B*. *(From Wehrli FW, by permission of the author and GE Medical Systems.)*

tances *along* the x-axis. Without a field gradient along the *x-axis,* both samples experience the same field B_0, so free induction decay (FID) displays a single frequency, as in Figure 29.14A. But with a field gradient applied along the *x-axis,* the samples experi-

ence different fields and the resulting FID consists of two superimposed signals (see Figure 29.14B). By a mathematical process called Fourier (foory-ay) transformation (FT), the plot of amplitude *vs* time is changed to amplitude *vs* frequency, which now shows a single frequency spike without the field gradient, and two spikes with the field gradient (see Figure 29.15A and B). Thus, with this simplified example, application of a field gradient and Fourier transform make it possible to recognize the two samples.

In practice, three sets of field gradient coils are required. They cooperate with the main field B_0 to produce the electromagnetic image by defining (1) slice thickness and (2) signal encoding, the location and intensity of the various spin-echoes emitted by the excited voxels in the selected slice. The position and function of these field gradient coils will receive a brief description (see Figure 29.16).

1. ***Slice Thickness.*** RF field gradient coils surround the patient, with the gradient field along the *z-axis* (parallel to B_0). Then a 90° pulse tuned to the Larmor frequency of the precessing protons within the selected slice

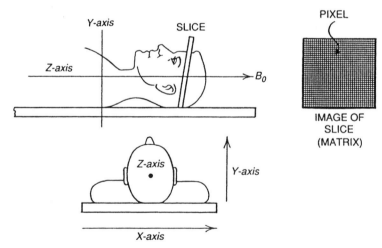

Figure 29.16. Matrix obtained by computer processing of innumerable bits of position and signal-intensity data accumulates during MR imaging.

will establish its position, and its band width (range of frequencies in the pulse) will determine its thickness (usually 5 to 10 mm). The margins of the slices are not sharply defined—some of the RF energy spills over to adjoining tissue to produce "noise." A gap of about 30 percent between slices reduces this effect (Curry and others). Note that slices can be selected for other orientations—oblique, longitudinal, etc, depending on the orientation of the z coils.

2. ***Signal Encoding.*** Signal intensities and their precise location must be identified so that each voxel will have a distinctive pattern that is sent to the MRI computer network for processing. This requires two kinds of encoding of the information: ***phase*** and ***frequency.***

a. *Phase encoding* involves the use of ***field gradient coils*** which cause a shift in RF wave phase along the *y-axis* while the signals are being generated within the slice (described in [1] above). In essence, phase encoding divides the slice into horizontal rows of voxels, each now being identified by a particular wave phase.

b. *Frequency encoding* employs field gradient coils that generate an RF field gradient along the *x-axis* to give each voxel in the slice a unique frequency, thereby dividing the slice into vertical columns.

The intersections of the rows and columns characterize the voxels, each with its distinctive information. This is digitized by the computer, processed, and converted to the final MR image.

Figure 29.17 demonstrates a normal brain MRI, axial view, single "cut".

Noise in MRI

In assessing image quality, we must take into account the amount of ***noise*** present. As noted before, noise in any system has a random distribution in both time and space, and impairs image quality by covering up information. Any factor that improves the signal-to-noise (S/N) ratio (ie, signal intensity relative to noise intensity) will improve image quality. The important elements that enhance the S/N ratio include:

1. ***Voxel volume.*** As this is increased (smaller matrix), the S/N ratio improves and the time needed to acquire an image decreases, but this occurs at the cost of some

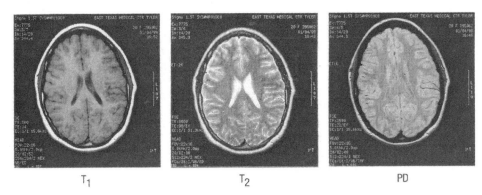

T_1　　　　　　　T_2　　　　　　　PD

Figure 29.17. Magnetic resonance images of a normal brain. T_1-weighted image shows mild contrast between gray and white matter—peripheral and central regions, respectively; cerebrospinal and ventricular fluid dark. T_2-weighted image shows bright gray matter and ventricular fluid, and dark white matter. Proton density (PD)-weighted image has moderate contrast between gray and white matter, but CSF is dark. Images represent axial projection.

decrease in image resolution.

2. *TE Value (time-to-echo).* A short TE results in a stronger signal, increasing the numerator of the S/N ratio without contributing to noise. Thus, a short TE improves the S/N ratio.

3. *TR Value (repetition time).* Signal intensity increases with a longer TR interval because it allows more time for longitudinal remagnetization, producing a stronger signal. Therefore, a longer TR improves the S/N ratio, but it also increases the time needed to create a slice.

4. *Signal Averaging.* As with digital fluoroscopy, smoothing of signal and noise intensities decreases the effect of noise. In MRI, this is accomplished by computer averaging of the signal and noise intensities of each voxel several times. The signal intensities are more consistent over time than the noise intensities because the latter are random events and have greater fluctuations. Thus, the average noise intensity is lower than that of the signals, resulting in a higher S/N ratio. At the same time, averaging increases the time needed to acquire an image.

External Field Magnets for MRI

There are three types of magnets available to produce the external B_o field: (1) superconducting, (2) resistive, and (3) permanent. These will be briefly described.

1. *Superconducting Magnets.* This type is the one most widely used at present. Its strong, constant magnetic field is generated by coils within the gantry surrounding the patient; the magnetic field (lines of force) parallel the patient's cephalocaudal axis. *Superconductivity* refers to the ability of certain metals to have virtually no electrical resistance at extremely low temperatures of about –270°C! A magnetic field intensity of 2 T can be obtained, although most clinical models operate at 1.5 T.

The magnet is actually a *solenoid,* composed of a copper coil containing particles of a superconducting metal such as niobium and titanium and operating at a temperature of –270°C. This temperature is maintained by concentric chambers, the inner one containing liquid helium and the outer one liquid nitrogen. Between and around them are vacuum chambers to serve as insulators from heat. The liquid helium and nitrogen must be replenished periodically because of gradual vaporization.

Superconducting magnets provide the strongest and most homogeneous fields available. In operation, a current is supplied to the magnet and then turned off. The current, however, flows steadily because of the superconductivity of the coil, and generates a strong magnetic field by electromagnetic induction.

2. *Resistive Magnets.* A resistive magnet consists of a series of four solenoids within the gantry surrounding the patient. The name *resistive* means that the solenoids must be continuously supplied by electric current during MRI. Field intensity and homogeneity are reasonably good. Power consumption is extremely high, but purchase and maintenance are much less than with superconducting magnets. Resistive magnets generate a field intensity of about 0.5 T.

3. *Permanent Magnets.* These resemble ordinary permanent bar magnets, but are fabricated from special ceramics that can be magnetized. In the form of building blocks, they can be stacked to form magnets large enough to surround a patient. Although inexpensive in purchase and maintenance, field intensity is low—about 0.1 to 0.15 T—and homogeneity less than that of the other types of magnets.

Surface Coils

When especially good recorded image detail is required for a small structure, such

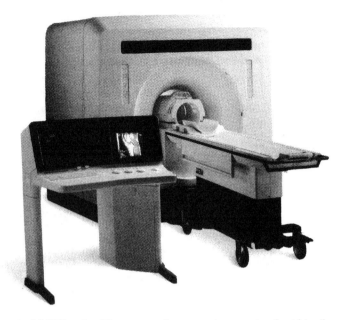

Figure 29.18. A typical MRI unit. The external magnet is contained within the massive gantry. The patient is slid into the gantry opening head-first. Note that the entire body is not contained within the opening at any one time. (*By courtesy GE Medical Systems.*)

as an arm or leg, ***surface*** receiving coils can be placed directly in contact with the zone of interest. They provide a limited field of view with smaller pixels than ordinary MRI, with resulting enhancement of detail. Moreover, the S/N ratio is improved. Surface coils must be very carefully positioned with relation to the main magnet to avoid loss of signal intensity.

A Typical MRI Unit

At the outset, it was stated that an MRI unit resembles externally a CT scanner. As shown in Figure 29.18, the *gantry* (outer rectangular shell) contains the field magnet. Within the circular opening lies a *shim coil,* which improves the uniformity of the main magnetic field B_o. To prepare for the examination, the technologist slides the patient head-first into the gantry opening by means

of the movable couch top. To the left of the gantry you can see the control system.

Hazards of MRI

The strong magnetic stationary field can be hazardous because many magnetic objects can become severely damaging projectiles. This includes such objects as iron tools, surgical instruments, and even metal carts and oxygen tanks. Internally, any iron-containing implants may malfunction (eg, pacemaker) or be displaced (metallic foreign bodies). Especially dangerous are intracranial surgical clips which could tear brain tissue, although surgical clips elsewhere in the body pose no significant hazard.

When a patient requires resuscitation during MRI, he or she must be moved elsewhere, unless special, nonmagnetic resuscitation equipment is available.

Damage can occur outside the body to calculators, watches, credit cards, and electronic equipment, among others, unless warranteed to be immune to strong magnetic fields.

So-called "fringe" fields around the magnet may extend for some distance—about 11 m from a 0.5-T field—although proper shielding will minimize it.

RF gradient fields may cause mild muscle contractions and cardiac arrhythmias, or light flashes during orbital MRI, but these have been temporary and without serious consequences. Examination of pregnant women with MR has not been proven safe.

ULTRASOUND IMAGING

This section will concentrate on the use of ultrasound waves in medical diagnosis. But first we must acquire an understanding of the physical nature of sound in general, and then turn to the special properties and application of ultrasound.

Nature of Sound

As a form of mechanical energy with wave characteristics, sound differs from electromagnetic radiation in a number of ways. Electromagnetic radiation, as described earlier on pages 109-111, consists of transverse waves that can be transmitted in a vacuum as well as in various media, depending on the energy of the radiation. Sound waves, on the other hand, are longitudinal in nature and require a material medium (solid, liquid, or gas) for their transmission; they will *not* pass through a vacuum.

Sound must be generated mechanically by an oscillating (vibrating) body of matter. The bell is a familiar example (see Figure 29.19); when struck by the clapper, the entire bell is set into vibration, a rapid, periodic back-and-forth motion of its wall. This transfers mechanical energy to the surrounding air in all directions (isotropic). To simplify matters, let us look at only the right side of the bell: the outward component of its vibration compresses the adjacent air on the right (molecules move closer together). The right wall of the bell instantly moves to the left (inward component of vibration), now rarefying the air on the right (molecules move farther apart). This rapid, periodic, alternating compression and rarefaction of the air creates a wave form, which moves outward as a *longitudinal wave.* It is important to note that the air molecules do not move continuously in the direction of the wave, but rather in a to-and-fro motion in the alternating zones of compression and rarefaction; in fact, they move over a distance of only a few μm (millionths of a meter). The to- and-fro motion has the same frequency as the vibrating bell, that is, the same number of cycles per second, so the resulting *wave form,* representing the summation of the alternate compressions and rarefactions, also has the same frequency. When it reaches an ear, the eardrum is set into vibration, again with the same frequency, now perceived as sound.

The above description of sound waves also applies to the transverse waves in water relative to a fishing float: throwing a pebble into the water generates visible waves that pass outward in all directions, but the float only bobs up and down indicating that the water molecules also bob up and down; it is only the wave form that moves out over the water.

Properties of Ultrasound

Like sound generally, ultrasound obeys the ***wave equation,*** which also applies to electromagnetic radiation:

$$\mathbf{v} = f\lambda \qquad (2)$$

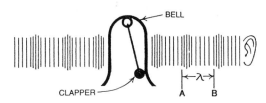

Figure 29.19. An example of audible sound-wave production by a bell. When struck by the clapper, the bell rings, sending out sound waves in all directions. Shown here are only a portion of the laterally transmitted waves in air. The waves proceed in trains of alternating zones of compression and rarefaction of air, symbolized by the closeness of the vertical lines. Wavelength λ is the distance between the beginning of two successive compressions *(A* to *B),* which pass outward ahead of zones of rarefaction. Upon entering the ear, the waves set up vibrations in the tympanic membrane (eardrum) with the same frequency. These vibrations set up electrical impulses in the auditory nerve, which carries them to the auditory center in the brain for processing.

where v is the velocity of sound in m/s; (meters per second), *f* the frequency in Hz (hertz = 1 cycle/sec), and λ the wavelength in m. In Figure 29.18, λ is the distance between two corresponding points in successive cycles, such as A and B. Simply put, one cycle equals the distance between the start of one compression and the start of the next compression. Frequency *f* is the number of compression-rarefactions per s measured in Hz. The relation of units in the wave equation (2) is revealed by:

$$\frac{m}{s} = \frac{1}{s} \times m$$

The components of equation (2) will now be explained under separate headings, with the addition of amplitude and direction.

1. ***Frequency.*** Sound waves with frequencies ranging from about 15 Hz to 20,000 Hz are audible to the normal human ear (dogs can hear higher frequencies than humans). The mechanism for hearing is the transfer of sound wave energy to the eardrum, which goes into vibration with the same frequency as the sound. Thus, if one strikes an A note (frequency 435 Hz), this excites a frequency of 435 Hz in the air in the form of sound waves which, in turn, produce the same frequency in the ear. Medically useful ultrasound involves frequencies of 1 to 10 MHz (1 MHz [megahertz] = 1 million Hz).

2. ***Velocity.*** The term velocity refers to speed in a given direction. If direction is along a straight line, velocity and speed are the same. Velocity of any object is defined by

$$\textbf{velocity = distance/time} \qquad (3)$$

For example, in ordinary units, we drive to Dallas over a distance of 100 miles in 2 hours; what is the velocity or speed? Substituting these values in equation (3),

$$\text{velocity} = 100 \text{ mi/2 hr} = 50 \text{ mi/hr}$$

The same equation applies to sound waves: the velocity of sound waves equals the distance between any two points such as A and B in Figure 29.18 divided by the time it takes the wave front to move through this distance (like the front end of our car going from Tyler to Dallas).

The velocity of sound, and ultrasound in particular, depends on the density (g/cm^3) and the compressibility of the conducting medium. Since various soft tissues are essentially liquids and have about the same density and compressibility, they conduct ultrasound with about the same velocity–1540 m/s. The most rapid conduction occurs in bone (very dense and rigid), the slowest in gas (low density and high compressibility); if gas were not highly compressible, how could it be stored under pressure and used to inflate a tire?

3. *Wavelength.* Since velocity changes little between various soft tissues, you can see from equation (2) that a higher frequency ultrasound must be accompanied by a proportional decrease in wavelength (doubling f goes with halving λ). According to the wave equation (2), an ultrasound frequency of 2 MHz and a velocity of 1540 m/sec gives rise to a wavelength found by

$$1540 \text{ m/s} = 2 \text{ MHz} \times \lambda$$

Rearranging, and changing 2 MHz to $2 \times (10)^6$ or 2 million Hz,

$$\lambda = 1540/2,000,000 = 0.0008 \text{ m, or 0.8 mm}$$

With a higher f of 5 MHz, λ would be shorter by inverse proportion:

$$2/5 \times 0.8 \text{ mm} = 0.32 \text{mm}$$

4. *Amplitude or Intensity.* Although all forms of sound can be measured by physical methods, ordinary sound is unique in that its intensity can be perceived as *loudness.* Amplitude increases with more rapid back-and-forth movement and greater excursion of the molecules producing increased compression, resulting in increased intensity. The more vibrational energy available for transfer to the medium, the greater will be the intensity or amplitude of the ultrasound waves. In fact, the dimension of amplitude is *power/unit area,* expressed in *watts/cm².* *

5. *Direction.* At the outset, it was stated that ordinary sound waves are emitted in all directions from a vibrating bell; this also applies to other sources such as vocal cords, musical instruments, and so on. But ultrasound waves are *directional,* that is, they move straight forward or backward from the source, at least in their initial path.

Production of Ultrasound

A special device called a *transducer* generates ultrasound waves for medical imag-

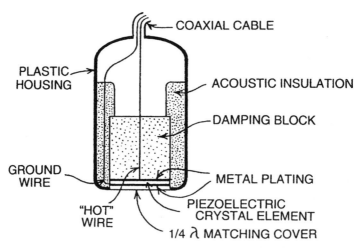

Figure 29.20. Idealized drawing of an ultrasound transducer, which is capable of both sending ultrasound waves and receiving their reflected waves. (Modified from Curry TS III, Dowdey JE, Murry RC Jr: *Christensen's Physics of Diagnostic Radiology,* 4th ed., Philadelphia, Lea & Febiger, 1990. By permission.)

---

* This is a mixed cgs and SI unit, employed by acousticians.

ing. At its heart is a type of material known as a ***piezoelectric crystal.*** Piezoelectricity (studied intensively by Pierre Curie in the 1890s) is a property peculiar to certain natural and artificial crystals which, when deformed or compressed, generate a voltage; and, conversely, when subjected to a voltage, becomes stressed or deformed. Application of a high-frequency AC voltage causes the crystal to vibrate with its resonant frequency, thereby converting electrical energy to mechanical energy. A sufficiently high frequency will generate ultrasound waves. Figure 29.20 shows a typical transducer.

The piezoelectric crystal now widely used consists of *lead zirconate titanate* or ***PZT***® of which there are at least two varieties. (Natural quartz has piezoelectric properties, but is not so efficient as PZT.) The molecules of a piezoelectric crystal are polarized, one end positive and the other end negative. When the high frequency voltage pulse is applied across the crystal, it alternately thickens and thins along its short axis, at its resonant frequency; and generates ultrasound waves as a beam in the air in front of the crystal face, and also back toward the damping block. Note that both faces of the crystal vibrate synchronously, thereby increasing the intensity of the ultrasound. In the usual B-mode operation (to be described later), the damping block must stop the crystal vibration within a microsecond or so because the transducer must be ready immediately to receive reflected waves (echoes) from tissue interfaces within the body.

The ***damping block*** usually consists of powdered rubber and tungsten blended with an epoxy resin. The proportion of these ingredients depends on the transducer's frequency and is chosen to achieve optimum suppression of the backward-directed ultrasound.

Crystal ***resonant frequency*** depends on its thickness, which should be $1/2\ \lambda$ of the ultrasound waves being produced. At this frequency, both faces of the crystal vibrate in synchrony and generate ultrasound waves with maximum intensity. It turns out that a 2-MHz crystal element composed of PZT-4® requires a thickness of 1 mm, and a 1 MHz crystal a thickness of 2 mm (Curry and others). A trick called ***quarter-wave matching*** improves the efficiency of ultrasound energy transfer from the crystal to the skin–this requires a layer of material of appropriate thickness and acoustic impedance (see below) to be equivalent to $1/4\ \lambda$.

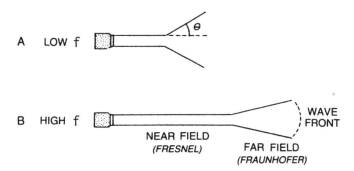

Figure 29.21. Transducer crystal diameter is the same in *A* and *B*. At higher-frequency ultrasound, the near field is longer (ie, reaches deeper into the body), while the far field spreads less (ie, has a smaller divergence angle *θ)*. The zone of maximum lateral resolution lies at the junction of the near and far fields.

Ultrasound Beam Characteristics

Although an ultrasound beam is directional (anisotropic), it is not entirely focused without certain modifications.

1. *Unfocused Beam.* An ultrasound beam leaving a flat crystal element has an initial cylindrical segment, followed by a diverging conical portion (see Figure 29.21). The length of the former and the divergence of the latter depend on ultrasound frequency and crystal diameter, as follows:

a. *Frequency.* The higher the frequency, the longer will be the cylindrical segment or *near field* (Fresnel [frā-nel'] zone). At the same time, the *far field* (Fraunhofer zone) becomes less divergent at higher frequencies (compare Figures 29.21A and B). The best *lateral resolution* exists at the junction of the near and far fields (ie, ability to display two closely spaced points in the same plane, as two separate images—see also pages 205-206. *Depth resolution* improves at higher frequencies (points closely spaced in depth, displayed as two separate images). But note that as frequency is increased, greater absorption of sound energy occurs in the tissues, weakening beam intensity.

b. *Crystal Diameter.* Increasing crystal diameter increases near zone length, but worsens lateral and depth resolution (see

Figure 29.22). The effect of crystal diameter on near-field length and far-field beam spread can be obtained from the following equations (Sprawls):

$$NFL = D^2/4\lambda \quad cm \qquad (4)$$

where NFL is near-field length, and λ wavelength, both in cm.

The divergence angle DA may be derived from

$$\mathbf{DA = 70\lambda/D} \qquad (5)$$

where DA is defined in Figure 29.21, and D is the crystal diameter. Beam intensity in power/cm^2 decreases as the divergence angle increases, so the far field in A is less intense than in B.

2. *Focused Beam.* By focusing the ultrasound beam, one can narrow it at a particular depth, thereby improving both lateral and depth resolution. This can be effected in two ways:

a. *Concave crystal face.* The crystal element must be fabricated with a *concave face;* the greater the concavity (crystal curvature inward), the nearer the beam will focus relative to the crystal. Stated technically, a shorter focal length goes with a more highly concave crystal; the focal length approximates the diameter of curvature or the concave face of the crystal according to Curry and others (see Figure 29.23).

b. *Acoustic lenses.* Plastic *acoustic lenses* having a flat surface on one side and a concave surface on the other are placed with the flat side against the crystal face (see Figure 29.24). The shorter the focal length (greater curvature), the nearer the lens will focus the beam relative to the lens face.

The *quality* or *Q-factor* of an ultrasound beam comprises its bandwidth (total range of frequencies in the beam) and its pulse duration. Thus, the narrower the bandwidth, the purer will be the sound because of the fewer frequencies present. Also, the shorter the duration of the pulse, the more time the

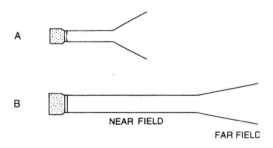

NEAR FIELD

FAR FIELD

Figure 29.22. Crystal *A* has a smaller diameter than crystal *B*. With the larger crystal, the near field is longer and the far field has a smaller divergence angle.

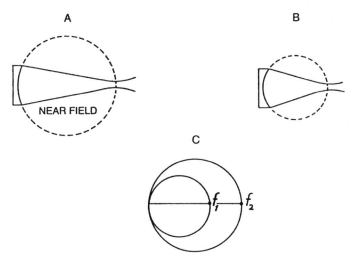

Figure 29.23. On the left in *A* and *B,* in contact with the circles, are concave-faced scintillation crystals. The crystal face in *B* has a greater degree of curvature (smaller radius) than in *A,* so its near field is shorter, that is, its focal zone is closer to the skin and is narrower than in *A.* Lateral resolution is maximum at the focal zone and is obviously better than with an an unfocused beam. In *C* is shown how the degree of curvature affects the focal distance f_1 and f_2. Recall that a circle of smaller diameter has a greater curvature. The focal length approximates the circle of curvature. (Curry & others.)

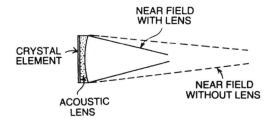

Figure 29.24. A concave acoustic (plastic) lens in front of a flat crystal focuses the ultrasound beam in the same manner as a concave crystal.

transducer can be in the "receive" mode for reflected (echo) signals, since the transducer serves alternately as *both* a sender and receiver of ultrasound. For optimum depth resolution, the pulse should contain only a few wavelengths; the shorter the distance they occupy, the closer two points can be in

depth and still be imaged as two separate points.

Echo Reception of Ultrasound

As already stated, the transducer has two functions: not only does it generate an ultrasound beam when stimulated by a high-frequency electric pulse with the selected frequency, but it also receives the beam echo after its reflection from the boundary between dissimilar tissues within the body. Ordinarily, the applied pulse lasts for about 1 μs (microsecond), then abruptly goes into the "receive" mode for about 999 μs. Thus, the total duration of the send-receive sequence is about 1000 μs = 1 msec. However, a transducer cannot, with a single crystal element, simultaneously send and receive.

Behavior of Ultrasound in Matter

Upon entering the body, an ultrasound beam may interact with tissues by one or more of the following processes: (1) reflection, (2) refraction, and (3) absorption (attenuation). Each of these will be explained.

1. *Reflection.* All of us have seen the reflection of light from a mirror or other smooth surface. Although this process involves transverse (electromagnetic) waves, the longitudinal sound waves also undergo reflection–with ultrasound, from interfaces or boundaries between different tissues.

Both ultrasound and light obey the *law of reflection,* namely, the angle of incidence equals the angle of reflection (see Figure 29.25). Thus, sound and light "bounce off"

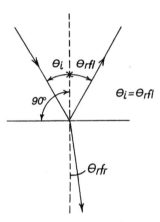

Figure 29.25. Above the horizontal line (reflecting tissue interface), the left arrow shows the direction of an ultrasound beam making an angle of incidence θ_i with the perpendicular (dashed line). The beam is *reflected* at the same angle θ_{rfl} as shown by the arrow upward and to the right. Thus, the angle of incidence equals the angle of reflection. The line directed downward below the interface indicates the direction of the *refracted* beam; which bends toward the perpendicular at an angle θ_{rfr}, the angle of refraction. Refracted ultrasound waves serve no useful purpose and create noise.

an appropriate surface in a particular direction.

The factor that determines the percent of the incident ultrasound beam undergoing reflection is a property, peculiar to various tissues, called *acoustic impedance.* The greater the difference in acoustic impedance, the greater will be the percentage reflection. A simple relation for acoustic impedance is given by:

$$Z = \rho v \ \ rayls \qquad (6)$$

where, for a given tissue, Z is its acoustic impedance, ρ its density, and v the speed of sound in it, in cm/s. The unit of acoustic impedance is the *rayl,* named after the British physicist Lord Rayleigh. As the velocity of sound in all soft tissues is virtually the same–1540 m/s–Z is nearly proportional to the density of the particular tissue. The velocity of sound in dense bone is high, so it has a high Z ($= 7.8$ rayls) and a high percentage of reflection. Air has an extremely small Z, 0.0004 rayl, with virtually *100 percent reflection.* Note that it is the *difference* in Z between two tissues, not their position relative to the boundary, that determines the degree of reflection; the same percent reflection occurs whether the ultrasound beam goes from soft tissue to air, or from air to soft tissue.

A simple and convenient way to calculate the percent reflection at an interface employs the following equation (Curry and others):

$$R = \left(\frac{Z_1 - Z_2}{Z_1 + Z_2} \right)^2 \times 100 \qquad (7)$$

where R is the percent of the incident beam reflected when the incident beam is perpendicular (90°) to the surface, and Z_1 and Z_2 are the acoustic impedances of the two tissues (see Table 29.4). For example, at the interface between fat and muscle, Z_1 for fat = 1.38, and Z_2 for muscle = 1.70. What is the percent reflection at the interface?

Table 29.4
APPROXIMATE VALUES OF ACOUSTIC
IMPEDANCE FOR AIR, WATER AND
VARIOUS ORGANS*

Material	Acoustic Impedance
	rayls
Air	0.0004
Fat	1.38
Water	1.54
Brain	1.58
Kidney	1.62
Liver	1.65
Muscle	1.70
Eye lens	1.84
Skull	7.8
PZT-4 (piezoelectric crystal)	30.0

* Modified from Curry TS III, Dowdey JE, Murry RC Jr: *Christensen's Physics of Diagnostic Radiology,* 4th ed., Philadelphia, Lea & Febiger, 1990. By permission.

$$R = \left(\frac{1.70 - 1.38}{1.70 + 1.38}\right)^2 \times 100$$

$$= \left(\frac{0.32}{3.08}\right)^2 \times 100 = 1\% \text{ (approx)}$$

Note the very small percent reflection. Yet, it suffices to produce an echo that can be amplified by the system to produce a satisfactory image. It also leaves a large percent of transmitted beam for reflection (echo production) deeper in the body.

For an interface between the transducer crystal ($Z_1 = 30$) and air ($Z_2 = 0.0004$),

$$R = \left(\frac{0.0004 - 30}{0.0004 + 30}\right)^2 \times 100$$

The value 0.0004 is so tiny relative to 30, that it may be ignored, leaving

$$\left(\frac{-30}{30}\right)^2 \times 100 = -1^2 \times 100 = 100\%$$

This means that the entire ultrasound beam has been reflected back to the transducer at the crystal-air interface! Incidentally, the fact that the right side of equation (7) is squared means that the value of R is always

positive, and independent of the direction of the perpendicular beam. It also confirms the need for *coupling oil* between the transducer and skin to eliminate the chance of air being trapped between them. Table 29.4 gives the acoustic impedance of air, water, and a number of tissues of interest in medical ultrasound studies.

The ***intensity*** of the reflected beam relative to the incident beam is important in diagnostic ultrasound. Relative amplitude is expressed as the ***logarithmic ratio*** of the reflected intensity to the incident intensity, by analogy to the sensitivity of the human ear to sounds of various degrees of intensity. When one sound is 100 times as intense as another, it is perceived by the normal ear as twice as loud; if 1000 times intense, it is perceived as 3 times louder. It turns out that 2 is the logarithm of 100, and 3 is the logarithm of 1000. At this point you must review pages 14-15 for an explanation of logarithms, to enable you to follow the discussion of relative amplitude in ultrasound.

In evaluating the intensity or power of a reflected beam relative to an incident ultrasound beam, one obtains an extremely small ratio, requiring the use of negative powers of ten. Since the unit for the ratio of audible sound intensity—the decibel—was already in use, it was only natural for it to be adopted for ultrasound. Logarithms, being small numbers for ultrasound, are much more convenient to use.

The following equation defines ***relative amplitude (RA):***

$$RA = log \; \frac{A_r \; watts/cm^2}{A_i \; watts/cm^2} \quad \text{bels (B)} \quad (8)$$

where A_r is the amplitude of the reflected ultrasound beam, and A_i that of the incident beam. To convert bels to decibels: 1 B = 10 dB. Suppose the numerator in equation (8) is 0.001 watts/cm^2 and the denominator 100 watts/cm^2. Substituting these values in the equation,

$$\log \frac{0.001}{100} = \log \frac{10^{-3}}{10^2} = \log 10^{-5} = -5 \text{ B, or } -50 \text{ dB}$$

The negative answer indicates that the reflected intensity is much weaker than the incident intensity. (If we invert the fraction, it gives the reciprocal value–how much greater the incident amplitude is than the reflected amplitude; in this case, 50 dB.)

As a general rule, a 50 percent loss of intensity of the reflected beam relative to the incident beam represents a loss of 3 dB (ie, the amplitude ratio –3 dB).

The *angle of incidence* of an ultrasound beam has an important bearing on imaging. As the transducer serves both to send and receive ultrasound energy, you can see from Figure 29.26A that when the transducer is aimed toward the skin at a large angle of

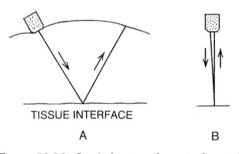

TISSUE INTERFACE

A   B

Figure 29.26. In *A*, the transducer is directed at a large angle to the vertical, so it misses the returning ultrasound echo. In *B*, the transducer is directed at a small angle, so it detects the ultrasound echo.

incidence, the reflected beam (echo) will return at some distance from the transducer and be missed. With a small angle of incidence as in Figure 29.26B, typically less than 3 degrees, the echo will enter the transducer during its reception mode. At a particularly large angle of incidence known as the *critical angle,* total reflection occurs at the skin.

2. *Refraction.* This occurs when an ultrasound beam passes, at an angle other than

90 degrees, from one tissue into another with a change in velocity. The frequency remains constant, so the wavelength has to change according to the wave equation (2). Refraction involves only a fraction of the beam, increasing with the angle of incidence, and is manifested by a change in direction (see earlier Figure 29.25). The refracted ultrasound contributes nothing useful to the image, but passes deeper into the body where it gives rise to *artifacts.* In practice, however, with the angle of incidence of the beam relatively small, less than 3 degrees, very little ultrasound is refracted.

3. *Absorption.* Due to friction among molecules in their back-and-forth motion, reduction in intensity of the ultrasound beam occurs as it traverses matter. Friction results in degradation of a part of the molecules' kinetic energy to heat. Analogous to x rays, *attenuation* of ultrasound beam intensity is exponential with depth in the tissue, governed by the *attenuation constant;* therefore we can apply the concept of *half-value layer (HVL)* to ultrasound. HVL in this case depends on the frequency of the ultrasound waves: the greater the frequency, the greater the attenuation coefficient. This means that a *high-frequency beam shows less penetration than a low-frequency beam,* the attenuation of ultrasound per cm of tissue being approximately proportional to frequency so that doubling the frequency doubles attenuation, and halving frequency halves attenuation (both per cm). Bushong states that attenuation in soft tissue is about 1 dB/cm/MHz.

Ultrasound Image Displays

There are four main categories of ultrasound image display: (1) *A* mode, (2) *T* or *TM* mode, (3) *B* mode, and (4) Doppler. These will be considered under separate headings.

1. *A-Mode Display.* At one time the only method of displaying echoes received by the

Figure 29.27. An ultrasound A-mode display used some years ago to determine the displacement of a "midplane complex" in the brain. It is here identified as a spike along the horizontal depth-line. Measuring the distance of the spike from the actual midsagittal plane of the skull indicated pressure from a subdural hematoma or other mass on the side opposite the shift. A-mode is still used for depth measurement in other conditions, but with more sophisticated equipment.

Figure 29.28. TM-mode ultrasound display. This shows the mobility of an internal structure such as a heart valve leaflet, recorded on a moving strip.

transducer and converted to electric signals, the A mode shows the signals as spikes along a horizontal axis (see Figure 29.27). The height of a spike represents the amplitude of the reflected signal; A mode stands for *a*mplitude mode. The position of a spike along the horizontal axis is proportional to the return-time of the echo, which depends on the depth of the relevant tissue interface. Because the signal loses amplitude by virtue of the attenuation of its reflected beam along its path (dB/cm), a specially designed amplifier automatically corrects for this lost amplitude, taking into account the distance the echo has traveled. Early on, the A mode was used to determine the possible lateral shift of the midplane "complex" of the brain by a contralateral subdural hematoma (see Figure 29.27). A good deal of subjective input by the operator impaired the reliability of this procedure. Modern A-mode equipment has greatly improved the accuracy of *depth* measurement of structures within the body, but signal amplitude is largely ignored.

2. *TM or M Mode.* The TM (time-motion) mode has its main application in echocardiography to show the motility of the cardiac valve leaflets and muscle wall. In this mode, the spikes representing ultra-

sound echoes are electronically converted to dots on a single B-mode display. The scanner incorporates a paper recording strip, the vertical axis of which records time, and the horizontal axis depth of the structure under study (see Figure 29.28). The motion of a valve leaflet or heart wall proceeds back and forth along this depth axis. If the dots (signal "blips") are moved electronically to the top of the recording strip and allowed to drop to the bottom, they will describe a series of transverse waves, which denote the amplitude of motion of the structure. Thickened leaflets or scarred muscle will show diminished motion. A radionuclide added to the process with appropriate imaging greatly enhances the visibility of heart wall motion and improves diagnostic power.

3. *B-Mode Display.* We use ultrasound most frequently in the *brightness* or *B mode* because of its extensive diagnostic application, especially in the abdomen and pelvis. The B-mode display involves the *pulse echo sequence* with the buildup of an image from innumerable echo blips. The dot elements in the image correspond to the pixels in CT and MR images, and to dots in a half-tone print. You can see the latter by viewing a newspaper photo with a magnifying glass. In the B mode, the brightness of the dots is proportional to the intensity of the echoes they represent.

The imaging process in B mode requires that the technologist move the transducer

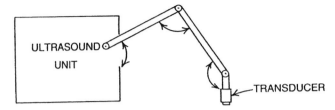

Figure 29.29. The triple-jointed arm of an ultrasound unit. These joints contain electrical resistance sensors that "tell" the computer the position and depth of the tissue interface, as the operator scans the region with the transducer.

over the skin in a scanning fashion, the transducer crystal being "coupled" to the skin by a layer of oil (see page 445). This is why it is called "contact scanning." The transducer may be moved straight across the skin as in a *linear scan,* but much more frequently as a combination of linear and rocking motion in a *compound scan* in which the operator must time transducer motion to coincide with the image on the screen. The innumerable dots have to be assembled and processed by the computer to build up an intelligible image. To accomplish this, the computer is linked to a jointed arm, which consists of three segments connected by two joints (see Figure 29.29). Thus, the assembly has a total of three joints, each of which contains a sensor. The three sensors contain variable resistors (potentiometers) whose resistance varies with the position of the arm and its attached transducer. This continually "informs" the computer about the sizes of the angles at the joints as the transducer scans the area of interest. From this information, the computer determines the position and depth of the echoes. In addition, motion of the linked arm is limited so as to image a single slice at a time: the slice is made in one direction, so the computer "turns" the image, allowing the observer to see the slice in face view.

Until 1972, B-mode scanners produced images directly on storage cathode ray tubes, giving extremely high-contrast images (short scale) with no shades of gray (similar to high-contrast radiographs). Such images were difficult to interpret. Marked improvement, notably in long-scale contrast, resulted from introduction of the *gray-scale imaging* that is in general use today. This was achieved by the use of a *scan converter,* which discards superfluous echoes from the same point and allows only the strongest echo blips to be stored for imaging. The final image is displayed on a TV monitor or film.

4. *Doppler Scanning.* We have all experienced the Doppler Effect–the increasing frequency (higher pitch) of a fire engine siren as it approaches, and the decreasing frequency

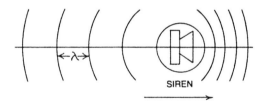

Figure 29.30. Simplified diagram to show the Doppler effect. Motion of a sound source like a siren causes the sound waves to be pushed together in front and pulled apart behind. Therefore, in the direction of the siren's motion, the wavelength shortens and the frequency increases, so the sound pitch increases (like high notes compared to low notes). At the same time, the wavelength behind the siren increases and the frequency (pitch) becomes lower.

(lower pitch) as it recedes, caused in the first instance by compression of sound waves (higher frequency) ahead of the siren and, in the second instance, by rarefaction of sound waves (lower frequency) behind the siren (see Figure 29.30). This is called the the *Doppler Shift,* in honor of the Austrian physicist who discovered it about 150 years ago both for light and sound.

Doppler scanning employs *two transducers* side-by-side in a single unit, one to send and the other to receive ultrasound waves, either in pulsed or continuous mode. *Pulsed Doppler* uses individual pulses to determine the depth and velocity of moving internal structures by controlling the space between the pulses according to their depth.

The *continuous* mode allows us to evaluate the velocity of blood flow in an artery or vein by reflecting ultrasound waves off blood cells. A Doppler shift occurs first when the moving cells receive the incident beam, and again when the echo returns from the moving cells. For this procedure, the transducer unit is hand-held at a relatively wide angle to the skin surface because of the double-transducer design, in which the receiving transducer is separate from the sending one. If blood motion is slow enough, the Doppler shift engenders audible frequencies–the operator can assess the approximate velocity of blood flow by listening to the pitch of the sound: the lower the pitch, the slower the flow. In addition, the shift signal can be processed by an appropriate electronic device to yield a frequency distribution which can be recorded. This gives a more accurate estimate of blood flow through the vessel.

Continuous Doppler scanning has contributed greatly to noninvasive assessment of heart valve mobility, and adult and fetal cardiac wall motion, as well as the patency of blood vessels on the basis of blood flow pattern.

Question of Biohazard

Although diagnostic ultrasound has been in use for a number of years, no biologic injury has thus far been reported at the prevailing dose levels. Still, there is ongoing investigation of possible harmful effects, especially with regard to embryonic and fetal injury, since ultrasound is widely and frequently used to diagnose fetal abnormalities.

High-intensity ultrasound, in the range of about 10 or more watts/cm^2 in experimental animals, has revealed tissue changes such as release of free radicals, tissue disruption at interfaces, and production of tiny cavities. However, no harmful effects have been reported from diagnostic ultrasound.

Although opinion may change in the future, it is generally held that the benefit/risk ratio is so high that the major contribution of ultrasound to medical diagnosis warrants its continued use.

QUESTIONS AND PROBLEMS

1. What are magnetic dipoles? Which one is used in MRI, and why?
2. Describe the effect on the magnetic dipoles when a person is placed in a strong external magnetic field.
3. Define and explain the precession of magnetic dipoles. What is meant by Larmor frequency and its relation to resonance?
4. Give an example of resonance, in your experience.
5. State and describe the factors contributing to the MRI signal.
6. Define "free-induction decay" and

what it represents.

7. How is T1 relaxation converted to an MR signal?

8. State the factors contributing to MRI contrast.

9. What are gradient coils? State their purpose.

10. Briefly discuss the spin-echo process.

11. Explain the purpose of signal encoding and its relation to voxel creation.

12. List the sources of noise in MRI. How can it be suppressed?

13. Describe the types of external field magnets in MRI. Which one is the most advanced? What is meant by tesla (T)?

14. What are some of the hazards of MRI?

15. Describe sound waves. How do they differ from, and resemble, electromagnetic waves?

16. How are ordinary sound and ultrasound waves generated?

17. State the wave equation and explain the terms.

18. What is a piezoelectric crystal? On what does its resonant frequency depend?

19. Describe a transducer with the aid of a simple diagram. What are its functions?

20. Why does a transducer crystal have to be coupled to the skin by oil?

21. Discuss the characteristics of an unfocused ultrasound beam. How can such a beam be focused? Compare the shapes of a focused and unfocused beam.

22. Describe the fate of an ultrasound beam in the body. How are ultrasound waves reflected and received?

23. What happens to the depth of the "near field" as ultrasound frequency is increased?

24. What percentage of an ultrasound beam is reflected from the boundary between soft tissue and air? Explain acoustic impedance.

25. Discuss relative amplitude and its applicable units.

26. Describe three modes of ultrasound image display.

27. Explain the principle of Doppler imaging.

28. Briefly outline the biologic hazards of ultrasound.

Appendix

ANSWERS TO PROBLEMS

Chapter 1.

1. (a) $^1/_2$ (b) $^3/_4$ (c) $^1/_{15}$
   (d) $^4/_5$
2. (a) $1^4/_5$ (b) $1^{71}/_{84}$ (c) $1^{29}/_{30}$
3. 25 bu
4. (a) $1^1/_3$ (b) $1^1/_{63}$ (c) $1^{13}/_{27}$
5. b = ac   c = b/a
6. (a) 8   (b) –3   (c) $2^1/_3$
   (d) $4^4/_9$ (e) 44
7. (a) $^2/_3$ (b) 5   (c) $10^1/_2$
   (d) 15
8. 20 cm
9. tripled
10. 12.56 cm^2
11. 600 sq ft
12. 3,600,000
13. 4.24×10^5
14. 5
15. 10^4, or 10,000
16. a. 10^{11} b. 11
17. a. 10^3 b. 3
18. 10^9
19. 7

Chapter 2.

10. 1 yard; 1/28 oz; 1 quart
11. 10^6; 10^9
12. 100,000 V, or 10^5 V
13. 0.001g, or 10^{-3} g
14. 20°C
15. 104°F
16. –40°F = –40°C

Chapter 3.

7. 1000 energy units (eg, Calories)
10. 1960 joules
11. 10,000 joules

Chapter 4.

7. 12
8. Z 6; M 8

Chapter 6.

8. 2 A
9. 150 V
11. 1.8Ω; 4, 2, and $^2/_3$ A in branches; $6^2/_3$
    A in main lines
13. 24 watts (W)

Chapter 9.

8. 56.6 kV
9. 42.3 A
10. 18,000 watts, or 18 kilowatts (kW)

Chapter 10.

3. 110,000 V, or 110 kV
10. 100/240 = 5/12   (Hint: first find primary voltage needed on transformer)
16. 85 V

Chapter 12.

21. 0.67%
22. 100 R
23. 12.5 R/min
24. 100 keV
25. 1.7 times
26. 30 R-cm^2 (Hint: first change mR to R.)

Chapter 13.

10. $1/10$ s
18. 80,000 HU remain. 120,000 HU could be added back.

Chapter 14.

19. Underexposed
20. Correctly exposed

(Hint: $\dfrac{mAs}{mA} = s$

Chapter 15.

23. Reduce mAs by $1/2$, or kVp by 15%

Chapter 18.

6. 0.03
14. 80 R/min; 5 R/min

15. 300 mA
17. 20 mAs
18. 86 kVp
30. Image diameter 10 in. (25.4 cm). Magnification factor 1.11.

Chapter 19.

2. 15:1
4. 125 mAs
19. 16 in. (40.6 cm)
20. 7 mm
22. 8 in. × 10 in. (20.3 cm × 25.4 cm)
23. 0.03 mm

Chapter 20.

4. 36
8. 2.5 times
9. 5000

Chapter 23.

19. 11100111
20. 45

Chapter 28.

14. 1 HVL
15. 247 cSv (rem)

BIBLIOGRAPHY

Batson OV, Carpentier VE. Stereoscopic depth perception. *Am J Roentgenology* 51:202, 1944.

Bergeron RT. Manufacturers' designation of diagnostic x-ray tube focal spots. *Radiology* 111:421, 1974.

Bernstein H, Bergeron RT, Klein DJ. Routine evaluation of focal spots. *Radiology* 111:427, 1974.

Bitter F. *Magnets: The Education of a Physicist.* Garden City, NY: Anchor Books, 1959.

Bloom WL Jr, Hollenbach JL, Morgan JA. *Medical Radiographic Technic,* 3rd ed. Springfield, Thomas, 1969.

Bookstein JJ, Voegeli E. A critical analysis of magnification radiography. *Radiology* 98:23, 1971.

Bucky GP. A grating-diaphragm to cut off secondary rays from the object. *Archives of the Roentgen Ray,* June 1913.

Bureau of Radiological Health (BRH), U.S. Department of Health, Education, and Welfare. Rockville, MD 20852.
> *Population Exposure to X rays U.S.,* 1970.
> *Pre-release Report: X-ray Exposure Study (XES) Revised Estimates of 1964 and 1970 Genetically Significant Dose,* 1975.
> *Gonadal Shielding in Diagnostic Radiology,* 1974.
> *Handbook of Selected Organ Doses for Projections Common in Diagnostic Radiology,* 1976.

Bushong SC. *Radiologic science for technologists. Physics, Biology, and Protection,* 4th ed. St. Louis, Mosby, 1988.

Chamberlain WE. Fluoroscopes and fluoroscopy. *Radiology* 38:383, 1942.

Code of Federal Regulations (21 CFR 800 to 1299). Federal Register, U.S. Government Printing Office. April 1, 1983.

Conklin JJ, Walker RI. *Military Radiobiology.* Boston, Academic Press, 1987.

Copenhaver W, Kelly DE, Wood RL. *Textbook of Histology.* Baltimore, Williams & Wilkins, 1978.

Cullinan JE, Cullinan AM. *Illustrated Guide to X-Ray Technics.* 2nd ed. Philadelphia, J. B. Lippincott Company, 1980.

Curry TS III, Dowdey JE, Murry RC. Jr. *Christensen's Physics of Diagnostic Radiology,* 4th ed. Philadelphia, Lea & Febiger, 1990.

Dawkins R. *The Selfish Gene.* New York, Oxford University Press, 1989.

Doi K, Rossman K. Effect of focal spot distribution on blood vessel imaging in magnification radiography. *Radiology* 114:435, 1975.

Dowd SB. *Practical Radiation and Applied Radiobiology.* Philadelphia, W.B. Saunders, 1994.

Drew PG. The NMR phenomenon: Putting spinning nuclei to work. *Diagnostic Imaging,* August 1983.

Eastman Kodak Company, Rochester, N.Y.:
> *Introduction to Medical Radiographic Imaging,* 1993.
> *Sensitometric Properties of X-ray Films.*

Eisenbud M. *Environmental Radioactivity.* New York, Academic Press, Inc., 1987.

Feig SA. Can breast cancer be radiation induced? In *Breast Carcinoma, the Radiologist's Expanded Role.* Logan WW (Ed.) New York, John Wiley, 1977.

Gaulden ME. Possible effects of diagnostic X-rays on the human embryo and fetus. *J Arkansas Med Soc,* 70:424, 1974.

Gould RG, Hale J. Control of scattered radiation by air gap techniques: application to chest radiography. *Am J Roentgenology* 122:109, 1974

Grigg ERN. *The Trail of the Invisible Light.* Springfield, Thomas, 1965.

Harms SE, Morgan TJ, Yamanashi WS, Harle TS, Dodd GD. Principles of nuclear magnetic imaging. *RadioGraphics* 4:26, 1984 (Special Edition).

Haus AG ed. *Film Processing in Medical Imaging.* Madison, WI, Medical Physics Publishing, 1993.

Hendee WR, Ritenour ER. Medical Imaging Physics. St. Louis, Mosby-Yearbook, Inc., 1992.

Humphries RE. Personal communication.

Jaffe C, Webster EW. Radiographic contrast improvement by means of slit radiography. *Radiology* 116:631, 1975.

Jans RG, Butler PF, McCrohan JL, Thompson WE. The status of film-screen mammography: Results of the BENT study. *Radiology* 132:197, 1979.

Johns HE, Cunningham JR. *The Physics of Radiology,* 4th ed. Springfield, Thomas, 1983.

Keller PJ. *Basic Principles of Magnetic Resonance Imaging.* General Electric Medical Systems, 1991.

Liebel-Flarsheim Company. *Characteristics and Applications of X-ray Grids.* Cincinnati, 1968.

Littleton JT, Rumbaugh CL, Winter ES. Polydirectional body section roentgenography: A new diagnostic method. *Am J Roentgenology* 89:1179, 1963.

Mammography Quality Control: Radiologic Technologist's Manual. American College of Radiology, 1999.

Mettler FA, Guiberteau MJ. *Essentials of Nuclear Medicine Imaging.* 3rd ed. New York, Grune & Stratton, 1990.

Milne ENC. Characterizing focal spot performance. *Radiology* 111:483, 1974.

National Council on Radiation Protection and Measurements (NCRP) Reports: NCRP, P.O. 7910 Woodmont Ave., Suite 800, Bethesda, MD 20814.

66   *Mammography* (1980)

68   *Radiation Protection in Pediatric Radiology* (1981)

71   *Operational Radiation Safety–Training* (1983)

74   *Biological Effects of Ultrasound: Mechanisms and Clinical Implications* (1983)

82   *SI Units in Radiation Protection and Measurements* (1985)

85   *Mammography–A User's Guide* (1986)

86   *Biological Effects and Exposure Criteria for Radiofrequency Electromagnetic Fields* (1986)

91   *Recommendations on Limits for Exposure to Ionizing Radiation* (1987)

93   *Ionizing Radiation Exposure of the Population of the United States* (1987)

94   *Exposure of the Population in the United States and Canada from Natural Background Radiation* (1987)

95   *Radiation Exposure of the U.S. Population from Consumer Products and Miscellaneous Sources* (1987)

99   *Quality Assurance for Diagnostic Imaging* (1988)

102   *Medical X-Ray, Electron Beam and Gamma-Ray Protection For Energies Up To 50 MeV (Equipment Design, Performance and Use)* (1989)

103   *Control of Radon in Houses* (1989)

104   *The Relative Biological Effectiveness of Radiations of Different Qualities* (1990)

105   *Radiation Protection for Medical and Allied Health Personnel* (1989)

107   *Implementation of the Principle of As Low As Reasonably Achievable (ALARA) For Medical and Dental Personnel* (1990)

Panichello JJ. *X-Ray Repair: A Comprehensive Guide to the Installation and Servicing of Radiographic Equipment.* Springfield, Charles C Thomas, 1998.

Pfalzner EM. Attenuation coefficient: Finite or infinitesimal? *Physics in Med & Biol* 11:132, 1966.

Pizzutiello RJ, Cullinan JE. *Introduction to X-ray Imaging.* Rochester, NY, Eastman Kodak Company, 1993.

Potter HE. The Bucky diaphragm principle applied to roentgenology. *Am J Roentgenology* 7:292, 1920.

Potter HE. History of diaphragming roentgen rays by the use of the Bucky principle. *Am J Roentgenology* 25:396, 1931.

Quimby EH, Feitelberg S, Silver S. *Radioactive Isotopes in Clinical Practice.* Philadelphia, Lea & Febiger, 1968.

Ridway A, Thumm W. *The Physics of Medical Radiography.* Reading, PA, Addison-Wesley, 1968.

Rosenstein M. *Handbook of Selected Organ Doses for Projections Common in Radiology.* HEW/FDA 76-8031, 1976.

Rossmann K, Lubberts G. Some characteristics of line spread-function and modulation transfer function of radiographic films and screen-film systems. *Radiology* 86:235, 1966.

Rossman K. Point spread-function, line spread-function, and modulation transfer function. *Radiology* 93:257, 1969.

Seemann HE. *Physical and Photographic Principles of Medical Radiography.* New York, John Wiley, 1968.

Selman J. *The Basic Physics of Radiation Therapy,* 3rd ed. Springfield, Thomas, 1990.

Selman J. *Elements of Radiobiology.* Springfield, Thomas, 1983.

Spiegler P, Breckenridge WC. Imaging of focal spots by means of the star test pattern. *Radiology* 102:679, 1972.

Sprawls P Jr. *Physical Principles of Medical Imaging.* 2nd ed. Rockville MD, Aspen, 1993.

Stent GS ed. *The Double Helix.* New York, W.W. Norton & Co., 1980.

Steves A. *Review of Nuclear Medicine Technology.* The Society of Nuclear Medicine, 1992.

Ter-Pogossian M. *The Physical Aspects of Diagnostic Radiology.* New York, Hoeber, 1967.

Theron WO, Moore R, Amplatz K. The evaluation of high-speed screen-film combinations in angiography. *Radiology* 114:449, 1975.

Trout ED, Gager RM. Protective materials for field definition in radiation therapy. *Am J Roentgenology* 63:396, 1950.

Trout ED, Kelley JP, Larson VL. A comparison of an air gap and a grid in roentgenography of the chest. *Am J Roentgenology* 124:404, 1975.

INDEX

A

Absorbed dose, gray (Gy), 131, 356
Absorption (*see* Attenuation)
Absorption "edge effect," 132
Acoustic lens, ultrasound, 442-443 (Figure 29.24)
Active layer, screen, 177 (Figure 15.3)
    thickness, 152-153
Activity, radionuclide, 333, 341-342
    minimal detectable, 342
    specific, 333
    units, 337
Acutance, 205
Acute whole body radiation syndromes, 372-375
    explanation of, 374-375
    gastrointestinal, 373-374
    hematopoietic (bone marrow), 373
    neuromuscular, 374
    subclinical, 373
Aerial image, 179-180, 194-195 (Figure 17.4)
Afterglow, 178, 179-180 (Figure 15.7)
Air gap, 226-227, 246-247 (Figure 19.6)
ALARA, 392, 394
Algebra, 6-9
Algorithm, 320
All-or-none effect on embryo, 388
Alnico, 59
Alpha particles, 325
Alternating current (AC), 71, 73-74, 89-90
    advantages over DC, 77-78
    capacitor, 76
    curve, 74 (Figure 9.3), 75 (Figure 9.4)
    cycle, 74
    dynamo, 72-75 (Figure 9.2)
    effective value, 75
    frequency, 74
    generator, 72-75 (Figure 9.2)
    heating effect, 77
    impedance, 76
    maximum, 75
    measurement, 81-82
    motor, 78-81 (Figures 9.9-9.11)
    Ohm's Law, 76
    peak value, 75
    power loss, 77
    production of, 73-75 (Figure 9.2)
    root mean square value (RMS), 75
    transmission, 77-78
    voltage, 74-75 (Figures 9.3, 9.4)
Aluminum (Al) filter, 117, 122-123 (Figure 12.11)
    protective, 411-412 (Table 28.8, Figure 28.9)
Ammeter, 52 (Figure 6.8)
    filament, 156
Amperage, 49, 81
Ampere (A), unit, 49
Analog *vs* digital image, 310 (Figure 24.5)
Anger camera, 348-350 (Figure 26.20) (*see also* Gamma camera)
Angular distribution of x rays (anode cutoff, heel effect) 253-254 (Figures 19.25, 19.26)
Annihilation reaction, 128
Anode, electric cell, 50 (Figure 6.4)
Anode/target, x-ray tube, 138-147
    angle, 138-139 (Figures 13.4-13.6)
    atomic number, 138
    connection, 159 (Figure 14.12)
    cutoff ("heel" effect), 253-254 (Figures 19.25, 19.26)
    damage, 143-146 (Figure 13.10)
    effective focal spot, 139 (Figures 13.4-13.6), 141 Figure 13.8)
    failure, 143, 145
    focal spot, 138
        measurement, 208-210 (Figures 18.5, 18.7)
    focal track, 140, 141 (Figure 13.8)
    fractional focal spot, 141
    heat dissipation, 146-147
    heat units, 145
    heat storage (thermal capacity), 143-146 (Figure 13.11)
    "heel" effect (*see* cutoff, above)
    housing cooling, 147
    life extension, 143-148
    line-focus effect, 138-139 (Figures 13.4-13.6)

mammography, 286-287
material, 115, 138-139, 140
microfocus, 141
molybdenum ("moly") target, 286
pitting, 143-144 (Figure 13.10)
rating chart, 145 (Figure 13.11)
rhenium target, 146
rhodium target, 287
rotating, 139-141
tungsten target, 138
vanadium target, 287
Aperture diaphragm, 248 (Figures 19.17, 19.19), 251 (Figure 19.22)
Applied kV (kVp), 117, 123
Apron, lead protective, 404
Area, 10-11 (Figures 1.3-1.6)
exposed in radiography, 247-248 (Figure 19.17)
Arithmetic, 3-6
Armature
generator, 72
motor, 79
Artifacts
radiographic, 196-197
screen, 185-186
Artificial radioactivity, 329-335
Atom, atomic, 26-38, 102, 103, 125
acceptor atoms, 103
atomic mass, 30-31
atomic number *(Z)*, 30
atomic weight, 31
binding energy, electronic, 125
building blocks of, 29-30
Bohr model, 28-29 (Figure 4.3)
bonds, 26, 34-36
compounds, 28
definition of, 28
donor atoms, 103
electrical nature, 28
electrons, 29-30
energy levels, 29
excited state, 126
ground state, 126
ionization, 36-38, 130
ions, 34
isotopes, 30
mass number *(A)*, 30
model, 28-29 (Figure 4.3)
molecule, 26-27
neutrons, 29
nucleons, 30

nucleus, 29-30
nuclides, 31
octet rule, 102
orbits, 28-29 (Figure 4.3)
periodic table, 31-34 (Table 4.1)
protons, 29, 30
radionuclides, 323-324, 332, 335-336, 344-347 (Table 26.2)
shells, 29-30
structure, 28-30
subdivisions, 26-28
substance, 27
valence, 34-35, 102 (Figure 11.4)
weight, 31
Attenuation of x rays, 122-123 (Figure 12.11), 132-133, 176
Automatic exposure control, 153-154 (Figure 14.7)
Automatic processing, 197-200 (Figure 17.7) (*see also* Film processing, automatic)
Automatic processor care, 202-203
Autotransformer, 88-89, 150
construction, 88-89 (Figure 10.6)
law, 89
principle 89
Average gradient (film contrast), 221 (Figure 18.11)
Average life, radionuclide, 327-328

B

Back emf, 75-76, 79, 89-90 (Figure 10.7)
Background radiation, 392-394 (Table 28.1)
counting, 343-344
external sources, 392
internal sources, 393-394
Backup timer, 153-154
Bands
energy, intensifying screens, 178-180 (Figure 15.9)
valence, atomic, 101 (Figure 11.13)
Barium lubrication (rotating anode), 140
Barriers, protective, 402, 403-404, 416
Barrier voltage (solid state), 104
Base (film), 174
Batteries, electric, 50
Battery-powered mobile unit, 168
Beam limitation ("collimation"), 247-253
aperture diaphragm, 248 (Figures 19.7, 19.18)
centering field light, 252

"cones," 248-250 (Figures 19.19, 19.20)
coverage, film, 251-252
fluoroscopy, 415
illumination of field, 251
moving slit, 252-253 (Figure 19.24)
positive beam-limiting device (PBLD), 251
scatter limitation, 247-248
unwanted radiation ("penumbra"), 249-250 (Figures 19.18-19.21)
variable beam-limiting device (VBLD), 250-251 (Figure 19.21B)
Becquerel (Bq)
    physicist, 323
    unit, 333
Beta decay, 327
Beta particles, 325, 332
    shielding, 418
Bias
    definition,95
    forward, 95, 104
    reverse, 95, 104
Binding energy, atomic, 125
Biologic effects of radiation, 377-389
Blooming, focal spot, 209 (Figure 18.6)
Blur ("penumbra," edge gradient)
    focal spot, 206-207
        evaluation, 208-210 (Figures 18.5, 18.7)
    geometric factors, 206-207
    intensifying screen) 183-184 (Figures 15.10, 15.11), 185 (Figure 15.12)
    magnification 222, 227
    motion, 210
    mottle, screen, 211
    object-image distance (OID), 207-208 (Figure 18.4)
    shape of object, 212 (Figure 18.8)
    source-image distance (SID), 206 (Figure 18.3)
Bonds, 26, 34-36
Bone marrow (red), radiation injury, 380-381
    in acute radiation syndrome, 373
Breaker, circuit, 155 (Figure 14.8)
Breast (mammography), benefit/risk ratio, 381
Bremsstrahlung (braking radiation), 113 (Figure 12.5)
Bright fluoroscopy (*see* Fluoroscopy)
Brushes, generator, 73 (Figure 9.2)
Bucky grid (*see* Grid, radiographic)
Bucky grid factors, exposure, 242-243 (Table 19.2)

C

Cables (shockproof), 157 (Figure 14.11)
Calcium tungstate, screens, 177-178 (Figure 14.11)
Capacitance, 57
Capacitive reactance, 76
Capacitor, 56-57 (Figure 6.14), 168-169
    in AC circuit, 76
    in DC circuit, 76
Capacitor discharge unit (mobile), 168-171 (Figures 14.15, 14.20)
Carbon anode, dry cell, 50 (Figure 6.4)
Carcinogenesis, radiation-induced, 382-383
Cardboard film exposure holders, 176-177
Carrier-free radionuclide, 332
Cassettes, 177 (Figure 15.3), 184-185 (Figure 15.12)
Cataractogenesis, radiation-induced, 383, 384
Cathode, x-ray tube (*see* Filament/cathode, x-ray tube)
CCD, 267
Cell (biologic) structure, 358, 360 (Figure 27.3)
    cytoplasm, 359-360 (Figure 27.3)
    nucleus, 358-359 (Figure 27.3)
    reproduction, 360-363
Cell cycle, biologic, 360-362 (Figure 27.5, 27.6)
Cell generation cycle (Howard and Pelc), 360 (Figure 27.5)
Cell, electric, 50-51
Cellular radiosensitivity, 369-372
    cell cultures and, 369 (Figure 27.15)
    cell survival curves, 370 (Figures 27.16, 27.17)
    modifying factors
        age, 371-372
        dose fractionation, 371 (Figure 27.18)
        inherent, 370-371
        oxygen effect, 371
        radiation quality, 371
        tissue volume, 371
Cellular response to irradiation, 368-369
    cytoplasmic damage, 369
    interphase death, 368
    mitotic death, 368
    nuclear damage, 368-369
Celsius (temperature), 19
    conversion to Fahrenheit, 19
Centering lights, beam-limiting devices, 251
Centigrade, 19
    conversion to Fahrenheit, 19
Cerebral perfusion imaging, 352

Chain reaction (nuclear), 331
Chamber, ionization, 116
  cutie pie, 418-419 (Figure 28.5)
Changeover switch, 161
Characteristic curves
  film sensitometric, 220-222 (Figure 18.11, 18.12)
  GM tube, 337 (Figure 26.11)
  radiographic tubes, 142 (Figure 13.9)
Characteristic radiation, 114, 115, 126
Charge-coupled device, 267
Charge, electric (*see* Electric charge)
Chemical behavior of atoms, 34-36
  bonds, 34-36
Chemistry of radiography, 192-195
  aerial image, 194-195 (Figure 17.4)
  automatic processing, 197-203 (Figure 17.7)
  crystal development, 193 (Figure 17.2)
  development process, 194
  fixation, 194
  Gurney-Mott hypothesis, 193
  hypo, 194
  latent image, 192
  manifest image, 194 (Figure 17.3)
  manual processing, 195-197 (Figure 17.5)
  metallic silver, 193-194
  nucleation, 193 (Figure 17.2)
  radiograph, defined, 194
  radiographic chemistry, 193-195
  radiographic photography, 192-193
  sensitivity specks, 192-193 (Figure 17.1)
  silver sulfide, 192
Choke coil, 89-90 (Figure 10.7)
Chromosomes, 358-359 (Figure 27.4)
  aberrations, 366-367 (Figure 27.14), 385-386
  banding, 361 (Figure 27.7)
  diploid, 358
  haploid, 358
  single-strand breaks, 385
Circle, 16-17 (Figure 1.5)
Circuit breaker, 155 (Figure 14.8)
Circuit, electric
  amperage, 49, 81, 156
  breaker, 56
  capacitor, 56-57 (Figure 6.14)
  components, 46-47 (Figure 6.2)
  "condenser" (capacitor), 56-57 (Figure 6.14)
  current, 46-47 (Figure 6.1)
  definition of, 47
  elementary, 47 (Figure 6.2)

factors in, 47-50
filament, x-ray tube, 111-112, 135, 155-156 (Figure 14.9)
fuse, 56, 150 (Figure 14.2, 14.3)
ground, 157 (Figure 14.10)
high voltage, 156-158
main, 237 (Figure 14.1)
meters, 51-52
Ohm's Law, 50
parallel, 54-60 (Figures 6.11-6.13)
polarity, 51
primary, 150-156
rectifier, 157, 162-163
resistance, 49-50
secondary, 156-158
series, 53-54 (Figures 6.9, 6.10)
short, 56
simple, 47 (Figure 6.2)
three-phase, 162-164
x-ray machine, 149-172 (*see also* x-ray circuits)
x-ray tube, 85, 89, 111-113, 136-137
Clear-Pb™ compensating filter, 255
Coherent (unmodified) scatter, 127
Collimation (*see* Beam limitation)
  gamma camera, 348-350
Commutator (DC generator), 76 (Figure 9.6)
Compass, magnetic, 59-60 (Figure 7.2)
Compensating filters, 254-255 (Figure 19.27)
Compensator
  space charge, 156
  voltage, 156
Compliance testing (equipment), 405-406
Compound, 27
Compton interaction, 127-128 (Figure 12.14)
Computed tomography (CT), 314-322
  back projection, 318-319 (Figure 25.6, 25.7)
  components, 314-315
  computer, 315
  contrast enhancement, 317
  CT number, 318
  first and second generation, 315-316 (Figure 25.2)
  fourth generation, 317 (Figure 25.4)
  gantry, 315
  image reconstruction, 317-319 (Figure 25.6)
  matrix, 318 319 (Figure 25.6)
  medical applications, 321-322
  pixels, 317-318 (Figure 25.5)
  principle, 314-315 (Figure 25.1)
  scanning motion, 315-317

rotation and translation, 315-316 (Figure 25.2)
rotation only, 316-317 (Figure 25.3)
spiral CT, 319-320
advantages, 321
algorithm, 320
image reconstruction, 320-321
interpolation, 320
pitch, 321 (Figure 25.8)
principle, 320-321 (Figure 25.8)
slip-ring technology, 320
third generation, 316-317 (Figure 25.3)
voxels, 318 (Figure 25.5)
x-ray tube, 314-315, 320
Computer science, 295-305
binary numbers, 303-304
CD-ROM, 301
central processing unit (CPU), 300
definition of, 295
floppy disk, 299-300
hard disk, 301
hard drive, 300
hardware *vs* software, 298
history of, 295-296
input devices, 298-299
language, 303
layout, 298 (Figure 23.2)
microprocessor, 300
operating system, 297
operations, 297-298
output devices, 301-303
primary memory (RAM), 300
secondary memory, 300
summary for radiology, 304-305
Condenser (capacitor), 56-57 (Figure 6.14), 168-169
Conduction band
intensifying screens, 178-180 (Figure 15.5, 15.6)
semiconductors, 225-226 (Figure 11.18)
Conductors, electric, 46-47
Cones, radiographic (*see* Beam-limiting devices)
Cones, retinal, 258
Console (control panel), 158-161 (Figure 14.13)
Contrast (radiographic), 217-222
film, 220-222
average gradient, 221 (Figure 18.11)
characteristic (H&D) curves, 220-222 (Figures 18.11, 18.12)
inherent 220-222
kVp, 218-219 (Figure 18.10)

scale, 217
screens, 220
step wedge, 219 (Figure 18.10)
subject, 218-220
Controlled area, 403
Coolidge tube, 112 (Figure 12.4)
Copper losses, transformer, 87
Corpuscular nature of x rays, 124
Counters, radionuclide
efficiency of, 341
GM, 336-338 (Figures 26.9-26.11)
scintillation, 338-339 (Figure 26.12)
sensitivity, 341-342 .
well (scintillation), 338-339 (Figure 26.13)
Counting, radionuclides
absolute, 343-344
comparative, 344
focused collimator, 342-343 (Figure 26.17)
geometric factors, 342
methods, 343-344
Covalent bonds, 35
Crossover (screens), 185 (Figure 15.13)
CT (computed tomography)
CT numbers, 318
Curie, Marie, 323
unit, 323, 325
Curie, Pierre, 441
Current, electric, 46-47
alternating (AC), 89-90
ammeter, 52 (Figure 6.8)
amperage, 49, 81
battery, 47
cell, 47
circuit, definition of, 47 (*see also* Electric circuit)
factors in, 47-50
conduction, 46-47
coulomb (C), unit, 49
current, definition of, 49
direct (DC), 51, 76-77 (Figure 9.6)
direction (polarity), 51 (Figure 9.6)
dynamo, 72 (Figure 9.1)
dynamo rule (left hand), 69 (Figure 8.6)
effective current (RMS value), 75
electromagnetic induction, 68-69
electromotive force, 48-49
factors in circuit, 47-50
filament (x-ray tube), 111-112, 136-137, 155-156 (*see also* Filament/cathode)
frequency (Hz), 74
generator, 72-75

heating effect, 57-58
induced, 68-71
in gas, 46
in vacuum, 46
ionic solution, 37-38 (Figure 4.14)
IR drop, 53
left hand rule, 69 (Figure 8.6)
measuring devices, 51-52, 156
motor principle, 78-79 (Figures 9.9, 9.10)
    right hand rule, 78
multiphase DC, 114-115 (Figure 9.8)
nature of, 46-47
ohm, unit, 49
Ohm's Law, 50
overload, 56
parallel, 54-56
polarity, 51 (Figure 9.6)
potential difference (voltage), 48-49
power loss, 57
resistance, 49-50
RMS value, 75
saturation, 142 (Figure 13.9)
self-induced, 69-70
semiconductors, 49, 101-105
series circuit, 53-54 (Figures 6.9, 6.10)
simple circuit, 47-52 (Figure 6.2)
solid state diode, 100-106, 162
sources, 47
speed, 46-47
transmission (AC), 77-78
velocity, 46-47
voltage, 48-49
work, 57-58
x-ray tube current (mA), 117
Cutie pie, 418-419 (Figure 28.5)
Cutoff
    anode, 253-254 (Figures 19.25, 19.26)
    grid, 235, 238, 239
    wavetail, 169 (Figure 19.23)
Cylinder "cone," 249, 252 (Figure 19.23)
Cytoplasm, 359-360
Cytoplasmic injury, radiation, 364

D

Darkroom, 188-191
    automatic processor, 197-202 (Figure 17.7)
    benches, 190 (Figure 16.2)
    building essentials, 189
    entrance, 189-190 (Figures 16.1-16.3)

film bin, 191
    illumination, 190-191
    lighting, 190-191
    location, 188-189
    manual processing tanks, 195-196 (Figure 17.5)
    passboxes, 189
    safelight, 190-191
    size, 190
    ventilation, 190
    walls
        interior finish, 189
        shielding, 188
Dead time, counters, 340-341
Decay constant, 327
Decay curves
    iodine 131, 334 (Figures 26.7, 26.8)
    radium, 326 (Figure 26.2)
    radon, 328 (Figure 26.3)
Decay, radioactive, 324-328, 333-335
Decay scheme (Ra), 326 (Table 26.1, Figure 26.2)
Decimals, 5
Decubitus grid, 235
Definite proportions, law of, 36
Definition (recorded detail, image sharpness), 205
Densitometer, 293 (Figure 22.5)
Density gradient, 212
Density (optical; radiographic), 213-217
    anatomic part, 216-217
    base (film), 174
    defined, 213
    distance (inverse square law), 214-217 (Figure 18.9)
    fifteen percent rule, 213-214
    intensifying screens, 279-280
    kVp, 213-214
    mA, 214
    mAs, 214
    mass density, 19
    object, 216-217
    scattered radiation, 214
    thickness of part, 216-217
    time, 214
Deoxyribonucleic acid (*see* DNA)
Dependent variable, 12
Derived units, 18-21
Detail, recorded, 205-230
    factors in
        blur ("penumbra"), 206-208
        contrast, 217-222
        distortion, 222-226

macroradiography, 226-228
modulation transfer function (MTF), 228-230
resolution, 205
visibility of, 205
Detection of ionizing radiation, 130
Deuterium, 30-31 (Figure 4.5)
Development, 163 (Figure 17.2), 194, 198-200
automatic, 197-200
Devices to enhance quality (*see* Radiographic quality, devices)
Diamagnetism, 64
Dielectric, 57
Digital fluoroscopy, 307-312 (Figure 24.2)
Digital radiography, 312
Digital x-ray imaging, 306-313
analog *vs* digital, 310 (Figure 24.5)
fluoroscopy, 307-312 (Figure 24.2)
introduction, 306
matrix, 308 (Figure 24.3)
noise, 311
photostimulable phosphor, 312
pixel width, 310
radiography, 312
signal/noise ratio, 311-312
subtraction, conventional, 306-307 (Figure 24.1)
subtraction, digital, 308-312 (Figure 24.4)
x-ray generator, 312
Diodes, solid state, 100-106, 162
Dipoles magnetic, 61
Direct current, 46-57, 76-77
battery, 51
capacitor in circuit, 57
current, 49
curves, 77 (Figure 9.7)
generator, 76-77 (Figure 9.6)
multiple coil, 77 (Figure 9.8)
potential difference, 47-49
power, 57
pulsating, 76-77 (Figure 9.7)
resistance, 49-51
steady, 57
voltage, 47-49
Direct magnification (macroradiography), 226-228
Direct proportion, 9
Directly ionizing radiation, 356
Discharge, electric, 44-45
Distance effect (inverse square law) on x rays, 117, 214-216 (Figure 18.9)

compensation in radiography, 216
density effect, 214-216 (Figure 18.9)
exposure rate, 214-216 (Figure 18.9)
fluoroscopy, 415
inverse square law, 214-216 (Figure 18.9)
object-image distance (OID), 207-208 (Figure 18.4)
protection, 402
recorded detail, 206-207
source-image distance (SID), 206-207 (Figures 18.2, 18.3)
Distortion, 222-226
shape, 225-226
size (magnification), 223-225 (Figures 18.13-18.15)
DNA (deoxyribonucleic acid)
functions, 364-366
protein synthesis, 365-366
replication, 365 (Figure 27.12)
structure, 363-364 (Figures 27.10, 27.11)
Domains, magnetic, 61
Donor atom, 103
Dose reduction (health physics)
fluoroscopy, 415-416
radiography, 410-414
Dose-response models, 375-377
linear, 376 (Figure 27.21)
linear-quadratic, 377 (Figure 27.22)
sigmoid, 375-376 (Figure 27.20)
threshold dose, 375-376 (Figures 27.21, 27.22)
Double filament x-ray tube, 137 (Figure 13.3)
Doubling dose, 387
dual nature of x rays, 124
Duplicating film, 175
Dynamo (left hand) rule, 69 (Figure 8.6)
Dynode, 338 (Figure 26.12)

E

Earth's magnetism, 64-65
Eddy current losses, transformer, 87-88 (Figure 10.5)
Edge band effect (focal spot), 209 (Figure 18.6)
Edge effect, 132, 182
Edge gradient, 206
Effective dose equivalent (H_E), 396
Effective focal spot, 139 (Figures 13.4-13.6), 206
Effective voltage (RMS), 75
Efficiency
intensifying screens, 179-181

transformers, 86-87
x-ray production, 115
Einstein, Albert
mass-energy equation, 24-25
photoelectric equation, 126
Electric capacitor, 56-57 (Figure 6.14), 76, 168-169
Electric charge
electron, 29-30, 39-43, 44
ionization, 36-38
magnetic effect of moving charge, 66-67
moving electrons (current), 46-47
nuclear, 29-30
Electric circuit, simple (*see also* Electric current)
components, 51-52
definition of, 47
factors in, 49-50
fuse, 56
meters, 51-52
overload, 56
parallel, 52, 54-56
polarity, 51
series, 52-53
Electric generator (dynamo), 72-78
armature, 72
essential features, 72-73
frequency (Hz), 74
simple, 73-75
turbine, 72
Electric motor, 78-81
induction, 80-87
principle, 79, 81 (Figure 9.9)
rotating anode, 81
simple, 79 (Figure 9.10)
slip ring, 79
split single-phase, 81
synchronous, 79-80
types, 79-81
Electrical potential energy, 44
Electrification, 39-41
contact, 39
friction, 40
induction, 40
Electrode, 37-38 (Figure 4.14)
Electrodynamics, 46-58
Electrodynamometer, 82
Electrolysis, 37 (Figure 4.14), 38
Electromagnet, 67-68, 72
Electromagnetic induction, 68-71
direction of induced current, 69

emf, factors, 68-69
left hand rule, 66
magnitude of emf, 68-69
mutual induction, 71
self-induction, 69-70
Electromagnetic radiation, 109-111
energy, 118, 121 (Figure 12.9), 122, 125-126 (Figure 12.12)
frequency (Hz), 110
photon, 111
quantum, 111
spectrum, 110 (Figure 12.3)
speed or velocity, 110
wavelength, 110
Electromagnetism, 66-71
definition of, 66
discovery of, 66
electromagnet, 67-68
left hand thumb rule, 66 (Figure 8.1)
magnetic flux, 67 (Figures 8.2-84)
solenoid, 67 (Figure 8.3)
Electromotive force (emf), 48
electromagnetic induction, 68-69
Electron(s), 29-30, 43-45
absorption, 126-127
beta particles, 325, 327
binding energy (atomic), 125
bound, 125, 127
bremsstrahlung, 113-114 (Figure 12.5)
cathode rays, 115
charge, 30
cloud, 95, 112
Compton, 127-128 (Figure 12.14)
current, 45-47 (Figure 6.1), 51 (Figure 6.5)
energy conversion to heat and x rays, 113
energy levels, atomic, 29
free (loosely bound), 125, 127
kinetic energy of, 13
kVp and electron velocity, 112-113, 121-122
negatron, 128-129 (Figure 12.15)
orbital, 28-29 (Figure 4.3)
photoelectrons, 126
space charge (x-ray tube), 95, 112, 141
speed in x-ray tube, 112-113, 121-122
thermionic emission, 112, 135
transition in atoms, 126
traps, in crystals, 178-179 (Figures 15.5-15.7), 192-193
Electronic timer, 153
Electron volt (eV), 119

Electroscope, 43-44
Electrostatics, 39-45
  Coulomb's Law, 42
  laws of, 41-43
  lightning, 44-45
Elements, 27, 31-33 (Table 4.1)
Emf, 48, 69, 79
  electromagnetic induction, 68-69
Embryo and fetus (conceptus) radiation injury to, 387-389
Emissivity, 140
Emulsion (film), 174-175 (Figures 15.1, 15.2)
Energy, 22-28
  bands, 90-101 (Figure 11.13)
  binding (atomic), 125
  Einstein's equation, 24-25
  electron volt, 119
  forms of, 24
  kinetic, 22-23
  law of conservation, 23-24
  mass equivalence, 24-25
  maximum electron energy *vs* kVp, 119
  photon, 118, 122
  potential energy, 23-24 (Figures 3.1, 3.2)
  relation to kVp, 119-120
  states, semiconductor, 100 (Figure 11.12)
Energy bands, intensifying screens, 178-180
Envelope, x-ray tube, 111-112 (Figure 12.4), 136
Excretion test, ^{57}Co, 347
Exit (remnant) radiation, 179-180, 194-195 (Figure 17.4)
Exponents, 13-14
Exposure latitude, 175
Exposure rate (output), 117
  definition of, 116-117
  distance effect, 117, 128, 214-217 (Figure 18.9)
  factors in, 117
  filtration, 117, 132
  half-value layer, 122-123 (Figure 12.11)
  intensity, 117
  inverse square law, 214-217 (Figure 18.9)
  kilovoltage and, 117
  milliamperage and, 117
  unit (R/min), 116-117
Exposure (roentgen, R)
  conversion to absorbed dose (gray, Gy), 131-132, 356
  definition of, 116
  measurement, 117, 356

  timers, 151-155
  unit (R), 116
Exposure switches, 151-152

F

Fahrenheit (°F), 19
Falling load generator, 167-168
Fast neutrons, 311
Fat, radiolucency, 219
Ferromagnetism, 64
Fertile workers in radiology, dose equivalent limit, 398
Field magnet, 72
Fields
  electric, 41-42 (Figures 5.3, 5.4)
  magnetic, 61-62 (Figure 7.7)
Filament/cathode, 136-138 (Figures 13.1-13.3)
  ammeter, 156
  booster circuit, 143
  circuit, 111-112, 135, 155-156, (Figure 14.9), 159 (Figure 14.12)
  composition, 135
  current, 136-137 (Figures 13.1, 13.2)
  current control, 89-92 (Figures 10.8, 10.10)
  double, 137 (Figure 13.3)
  evaporation, 137
  failure, 148
  heating, 135-136
  space charge, 141-142
    compensator, 141-142 (Figure 13.9)
  sparkover, 140, 143
  stabilizer, 156
  thinning, 137-138
  transformer (step-down), 156 (Figure 14.9)
  voltage, 136
Film badge (monitor), 400
Film contrast, 220-222 (Figures 18.11, 18.12; Table 18.2)
  inherent, 220
  intensifying screens, 220
  sensitometric (H&D) curves, 220-222 (Figures 18.11, 18.12)
Film coverage (radiographic), 249-252 (Figures 19.20, 19.23)
Film holders, 176-177 (Figure 15.3)
Film processing, automatic, 195-197 (Figure 17.5)
  agitation, 198
  care of, 203

chemicals, 200
comparison with manual method, 200, 202
    (Table 17.1)
emulsion thickness, 200, 201 (Figure 17.8)
filters, 198
gel swell, 198-200, 202 (Figure 17.8)
replenishment, 200
sensitometry, 202
solutions, 199 (Figure 17.1)
squeegee action, rollers, 198
temperature control, 200
timing, 198
transport system, 198
typical processor, 199 (Figure 17.7)
Film processing, manual ("time-temperature"),
    195-197 (Figure 17.5)
Films (*see also* X-ray films), 173-176
Film-screen contact, 185 (Figure 15.12), 211
Filters, filtration
added, 133
action of, 117-118 (Figure 12.7)
aluminum, 117-118 (Figure 12.7), 411-412
    (Tables 28.8, 28.9)
atomic number, 117-118
clear Pb™, 255
compensating, 254-255
edge effect, 132
exposure rate, with, 132-133
fluoroscopic, 415
inherent, 133
mammography, 282-287
primary, 133
protective, 410-412 (Tables 28.8, 28.9)
quality effect, 132-133
radiographic, 410-412 (Figures 28.8, 28.9)
secondary, 33
total, 133
trough, 255 (Figure 19.27)
wedge, 254-255
Fission, nuclear
products, 332
reactor, 330-332
Fixation, film processing, 194, 200
Fluorescence, 108, 116, 178-179 (Figures 15.5,
    15.6)
Fluoroscopy, 258-272
human eye, 258-259 (Figure 20.1)
image intensification, 259-264 (Figures
    20.2-20.5)

image intensifier, 259-264 (Figure 20.2)
brightness gain, 261
contrast, 261
conversion factor, 261
exposure rate, 262
flux gain, 261
image magnification (Figure 20.4)
image sharpness, 160
input screen, 259
minification gain, 261
multiple field, 263-264 (Figure 20.5)
output screen, 260
pincushion effect, 262
viewing options, 262 (Figure 20.3), 264-268
vignetting, 262
spotfilming modalities, 268-269
cinefluorography, 269
cut film camera, 269
film strip camera, 269
television viewing, 264
video image recording, 269-271
laser disc, 270-271 (Figure 20.11)
magnetic tape and disc, 269-270 (Figures
    20.9, 20.10)
video viewing, 264-268
brightness stabilization, 268
charge-coupled device (CCD), 267
interlacing, 266-267 (Figure 20.8)
television image quality, 267-268
television monitor, 266-267 (Figure 20.7)
video cameras, 264-266 (Figure 20.6)
Flux
electromagnetic, 68-69
magnetic, 61, 67
Focal spot, 136, 138
blooming, 209 (Figure 18.6)
edge band effect, 209 (Figure 18.6)
effective focal spot size, 139 (Figures 13.4-13.6)
focal track, 140-141 (Figure 13.8)
image quality, effect of size, 206-207 (Figure
    18.2)
line-focus effect, 138-139 (Figure 13.4)
measurement, 208-210 (Figures 18.5, 18.7)
resolution test patterns, 209-210 (Figure 18.7)
size, 208-210 (Figures 18.5, 18.7)
Focal track, 140-141 (Figure 13.8)
Focusing distance, grids, 239-240 (Figure 19.9)
Fog, film, 188, 194
Forbidden band, 101 (Figure 11.13)

Force
    electromotive (emf), 48
    mechanical, 22
Forward bias (voltage), 95, 104 (Figure 11.20A)
Fractional focal spot, 141, 226-228
Fractions, 3-6
Frequency (f),
    alternating current (AC), 74
    photon, 110
Full-wave rectification, 97-100 (Figures 11.5, 11.11)
Fundamental units, 17-18

G

Galvanometer, 81-82 (Figure 9.12)
    motor principle, 81
Gametes, 362
Gamma camera, 348-350 (Figure 26.20)
    calibration, 353
    collimators, 348-350 (Figure 26.21)
    crystal-photomultiplier complex, 350
    dual camera, 351
    image quality, 353
    medical applications, 351-353 (Table 26.2)
Gamma rays, 109 (Figures 12.1, 12.2), 325
    background radiation, 392-393
    camera, scintillation, 348
    emission, 327
    nature of, 109 (Figures 12.1, 12.2), 110 (Figure 12.3)
    radium, 325
    source, 325-326, 332-333
Gamma rays, health physics, 416-420
Gantry
    computed tomography (CT), 315 (Figure 25.1)
    magnetic resonance imaging (MRI), 437 (Figure 28.18)
Gas, 27
Gastrointestinal tract, radiation injury to, 379-380 (Figure 27.24)
    acute whole body effects, 373-374
Gel swell, 200-201 (Figure 17.8)
General ("white") radiation, 113-115 (Figure 12.5)
Generator, electric, 72-75 (Figure 9.2)
    AC 72-75 (Figure 9.2)
    DC, 76-77 (Figure 9.6)
    left hand rule, 66
    principle, 72
    turbine, 72

Genes, 359
    composition (DNA), 363-364
    dominant, 359
    "jumping," 359
    mutation rate, 387
    mutations, 384
    phenotype, 359
    recessive, 359
Genetically significant dose (GSD), 386-387
Genetic damage, radiation-induced, 384-386
Genetic effects of x rays, 366-368
    mutations, 385-386
Genome, 358
Geometry, 10-12
Glass envelope, x-ray tube, 136
Gloves, lead protective, 404
G-M tube, 336-338 (Figures 26.9-26.11)
Grain size, intensifying screens, 182-184
Graphs, 12-13
Gray (Gy) unit, 131
Gray-scale image, 308
Grenz rays, 130 (Table 12.4)
Grid-controlled x-ray tube, 170-171 (Figure 14.23)
Grids, radiographic, 234-246
    absorption of scattered x rays, 234
    angulation cutoff, 240-241
    Bucky's original grid, 234 (Figure 19.2)
    bucky or grid factor, 242-244 (Table 19.2)
    centering tube, 241-242
    "cleanup," 236
    construction, 234
    contrast improvement factor, 234, 237-238
    convergence line, 239-240 (Figure 19.9)
    crossed, 238 (Figure 19.7)
    cutoff, 235, 238, 239
        definition of, 235
    decentering
        angulation, 239
        distance, 238-239 (Figure 19.8)
        off-axis, 241-242 (Figure 19.12)
    decubitus grid, 235
    exposure factors, 242-244 (Table 19.2)
    focused, 239-240 (Figure 19.9)
    frequency (grid), 237
    functional factors, 237-238
    grid factors, 242-243
    grid-front cassette, 238
    grid lines, 382-384
    grid ratio, 236 (Figure 19.5)

high ratio, 243-244
kVp *vs* grid ratio, 242-244 (Tables 19.2-19.4)
lead strips/cm or lines/in., 237
lead content, 237
moving grid, 239, 244-246
noise, 233
patient dose increase, 244 (Table 19.5)
physical factors, 236-237
Potter-Bucky mechanisms, 244-246
practical aspects, 242-244
precautions in use, 240-242
principle, 232-234
ratio, grid, 236-237 (Figures 19.4-19.6)
reciprocating, 245
reciptomatic, 245
relation of scattered to primary photons, 233
     (Table 19.1)
scattered radiation, 232-233 (Figure 19.1)
selectivity, 237
single-stroke bucky, 244-245
specifications, 244
stationary grid, 234-235 (Figure 19.3), 246
synchronism, 245 (Figure 19.15)
trill mechanism, 245 (Figure 19.14)
tube side *vs* film side, 242 (Figure 19.13)
types, 238-240 (Figures 19.8, 19.9)
Ground(ing), 41
cables, 157-158 (Figure 14.11)
milliammeter, 157
transformer, 157 (Figure 14.10)
Ground state, atoms, 126
Gurney-Mott hypothesis, 143
Gyromagnetic ratio, 424 (Table 29.1)

H

Handbooks (health physics), 391
H&D curves, 220-222 (Figures 18.11, 18.12)
Half-life, 327, 333-335
biologic, 335
effective, 334-335
physical, 333-334
radon, 328
radium, 327
Half-value layer (HVL), 122-123 (Figure 12.11)
in ultrasound, 446
Health physics (radiation protection), 391-420
ALARA concept, 392, 394
apron, lead protective, 404

background, 392-394 (Table 28.1)
barriers, protective (walls), 403-404
beam limitation, 247-253
beta particles, 418
blood-forming organs, 394
chromosome aberrations, 516
computed tomography, 414-415 (Figure 28.3)
"collimation," 247-253
controlled areas, 402
distance, 402
dose equivalent (H), 395
dose equivalent limits, 396-399 (Table 28.3)
dose reduction, patients, 410-416
dose reduction, personnel, 400-405
effective dose equivalent (H_E), 396
electrical protection, 416
film badge, 400
filtration, x-ray beam
     fluoroscopy, 415
     radiography, 411-412
gamma rays, 417
gloves, lead protective, 404
gonadal shielding, 412
handbooks, 391
lead barriers
     radionuclides, 417
     wall, 403-404
leakage radiation, 403
Luxel® 401-402 (Figure 28.1)
mammography, 414-415
monitoring, 400-402 (Figure 26.1)
nuclear medicine (*see* radionuclides, under
     Health physics)
patient protection, 407-417
     fluoroscopy, 415-417
     radiography, 410-414
personnel protection, 396-402
     general principles, 402
     monitoring, 400-402, 418-419
     protection surveys, 406-407
     working conditions, 404-405
positive beam-limiting device (PBLD), 251
pregnancy (and fertility) in workers, 307, 308
protection surveys, 406-407
protective barriers, 403-404, 417
quality factor ($\overline{Q}$), 395
radionuclides (nuclear medicine), 416-418
     beta radiation, 418
     distance, 417

lead barriers, 417
lead-shielded syringes, 418
limitation of stored radionuclides, 418
monitoring, 418-419
principles, 417-418
shielding, 417
unsafe practices, 418-419
waste disposal, 419-420
rem, 395
sievert (Sv), 395-396
stray radiation, 403
survey meters, 418-420
uncontrolled areas, 403
useful beam, 403
wall barriers, 403-404
warning signs, 407
whole body exposure limits, 396-399
working conditions, 404
Heat, x-ray tube, 135-148
anode, 143-147
filament, 135, 137, 143, 148
housing, 147-148
Heating effect, electric current, 57-58
Heat units, 145-146
Heavy hydrogen, 30-31 (Figure 4.5)
Heel effect (anode), 253-254, (Figures 19.25, 19.26)
Helium, 31-34 (Table 4.1), 325
Helix, 67
Hertz (Hz), 74, 91, 110, 118
High-frequency generators, 91-92, 165-166 (Figure 14.19)
High voltage, 156-158
control, 88-89, 91-92
high frequency, 91-92, 165-166
production, 83-88, 162-164
three-phase, 162-164
Holes, in solid state diodes, 103
Horsepower, 57
Hounsfield numbers, 318
Hydrogen, 31 (Figure 4.5)
Hypo, 194, 200
Hysteresis losses, 88

I

Image intensifier, 259-264 (Figures 20.2-20.5) (*see also* Fluoroscopy)
Image quality (*see* Quality, radiographic)
Impedance (AC), 76

Independent variable, 12
Indirectly ionizing radiation, 356
Induced magnetism, 63
Induced emf (*see* Electromagnetic induction)
Induction
electrification, 40
electromagnetic, 68-69
electrostatic, 40-41
magnetic, 63
motor, 80-81
mutual, 71
self, 69-70, 89-90 (Figure 10.7)
Inductive reactance, 76
Inert elements, 36 (Figure 4.13)
Inertia, 22, 26
in mAs meter, 157
Information *vs* noise, 212
Inherent filtration, 132
Injury (radiation) (*see* Radiation effects on)
Insulation
autotransformer, 89
cable, 157
radiographic tube, 136
transformer, 85-86
Insulators (nonconductors), 40
Intensifying screens, 177-186 (Figure 15.4)
active layer, 177, 182-183 (Figure 15.9)
aerial image, 194-195 (Figure 17.4)
afterglow, 179
artifacts, 185-186
blur, 182-183 (Figure 15.9), 211-212
calcium tungstate (phosphor), 177-178
care of, 185-186
composition, 177 (Figures 15.4, 15.5)
conduction band (phosphor), 178-189 (Figures 15.5-15.7)
contact (film), 184-185 (Figures 15.10, 15.11)
contrast, 220
conversion efficiency, 179, 181
crossover, 185 (Figure 15.13)
crystal size, 182-184, 211
detail (effect on), 185
edge effect, 182
efficiency, 179, 181
energy bands (phosphor), 178-180 (Figure 15.9)
extrinsic factors, speed, 183-185
film-screen contact, 184-185 (Figures 15.11, 15.12)
fluorescence, phosphor, 178-179 (Figure 15.6)

grain size, 182-184, 211
image receptor, 181
intensifying factor, 180-182
intrinsic factors, speed, 181-183
kVp effect on speed, 183
lag, 179
luminescence, phosphor, 178
macroradiography, 228
mottle, 211
noise, 211
packing density, 182
phosphorescence, phosphor, 178, 179-180
    (Figure 15.7)
photographic effect, 180
principle of, 178-180
punch-through, 185 (Figure 15.13)
rare earth phosphors, 181-182 (Table 15.1)
recorded detail with, 185
reflectance, 182
remnant radiation and, 194-195 (Figure 17.4)
speed, 180-182
summary, 189
temperature, effect on speed, 183
thickness, active layer, 182-183 (Figure 15.9)
valence band, phosphor, 178-180 (Figures
    15.5-15.7)
Intensity (x-ray output), 117
    R/min/cm^2, 117
Interactions of ionizing radiation with matter,
    124-130
    coherent (unmodified), 127 (Figure 12.13)
    Compton, 127-128 (Figure 12.14)
    pair production, 128-129 (Figure 12.13)
    photoelectric, 125-127 (Figure 12.12)
    relative importance, 129-130
    scattering (x-ray), 127-128, 129
    secondary radiation, 129
Interphase death, 369
Interpolation, 320
Inverse proportion, 9-10
Inverse square law, 117, 124, 214-217 (Figure 18.9)
Inverse voltage, 75-76, 79, 89-90 (Figure 10.7), 95,
    97, 104 (Figure 11.20B)
Iodine, radioactive (*see* Radioactive iodine 123
    and 131)
Ionic bonds, 35
Ionization, 36, 38, 130
    alpha particles, 325
    beta particles, 325
    chamber, 116, 418-419 (Figure 28.5)

chemical dissociation, 37-38 (Figure 4.14)
    photoelectrons, 37, 153-154 (Figure 14.7)
    x rays, 37
    thermionic effect, 38, 135
Ionization chamber, 116, 153-154 (Figure 14.7),
    418-419 (Figures 28.4, 28.5)
Ionization in air by x rays, 44
Ions, 34
    IR drop (voltage), 53-54
Iron, magnetization of, 63-64
Iron filings and magnetism, 61-62, 66
Isotopes, 30-31, 329-335

J

Junction, n-p, 104 (Figure 11.19)

K

Kerma, 356
Kidney imaging (radionuclide), 352
Kilogram (kg), 18
Kilovoltage
    and electron speed, 112-113, 121-122
    applied, 117, 123
    contrast, radiographic, 218-219 (Figure 18.10)
    control, 88-89, 91-92
    density, 213-214
    efficiency of x-ray production, 115
    heat loading of anode, 145-146
    maximum photon energy, 122
    minimum wavelength, 122
    peak kV (kVp), 96 (Figure 11.2), 100 (Figure
        11.11)
    penetrating ability, 117-118 (Figure 12.7)
    quality, radiographic, 205-230 (*see also* Quality,
        radiographic)
    radiographic (optical) density, 330-337
    ripple
        high-frequency, 106 (Figure 14.20)
        single-phase, 94, (Figure 11.11)
        three-phase, 163-164 (Figure 14.17)
    scattered radiation, 185, 233 (Figure 19.1)
    selector, autotransformer, 88-89 (Figure 10.6),
        150, 159 (Figure 14.12)
        speed of electrons and, 119
        tube loading, anode, 145-146
        x-ray "hardness" (quality), 123-124, 118-120
            (Figure 12.7)
Kilovolt meter (prereading), 151, 160-161

Kilovolt selector (autotransformer), 88-89 (Figure 10.6), 150, 159 (Figure 14.12)
K-shell, 29

L

Lag
    intensifying screens, 179
    video camera, 264
Large and small numbers, 13-14
Latent image, 192
Latitude (film), 175
    contrast, 175
    exposure, 175
Lattice, crystal, 192-193 (Figure 17.1)
Law(s)
    autotransformer, 89
    conservation of energy, 23-24
    definition of, 17
    electrostatics, 41-43
    inverse square, 214-217 (Figure 18.9)
    magnetism, 60
    magnification, 223-225
    Ohm's, 50
    reciprocity, 214
    transformer, 83-85
Lead barriers, 402-404, 416
Leaded aprons, 404
    gloves, 404
    observation window, control booth, 404
    protective drape, 404
    shield cover for bucky slot, 404
Lead (Pb), radium series, 326
Lead-shielded syringe, 418
Leakage radiation, 403
Left hand rule, 66
Length, 18
Lenz's Law, 70, 90
LET, 131, 366-367, 394
Life shortening, radiation-induced, 383
Light, velocity, 110
Lightning, 44-45
Linear energy transfer (LET), 131, 356-357, 333-367, 394
Line-focus effect, 138-139 (Figure 13.4)
Lines of force (magnetic), 62-63 (Figures 7.7, 7.8)
Line spread function, 229 (Figure 18.21)
Line voltage, 149
Line voltage compensator, 151

Lithium, 31, 34 (Figure 4.8)
Liver imaging, 351
Lodestone, 59
Logarithms, 14-15
    density (radiographic), definition of, 213
    in ultrasound, 445-446
Low-dose mammography, 285-289
L-shell, 29
Lubrication, rotating anode, 140
Luminescence, 178-179 (Figure 15.6)
Lung imaging, 351
Luxel® 401-402 (Figure 28.1)

M

Macromolecule, 363
Macroradiography, 226-228
Magnetic compass, 59 (Figure 7.2)
Magnetic dipole moment, 423
Magnetic resonance imaging (MRI), 422-438
    auxiliary field coils, 433
    axes, patient, 430 (Figure 29.10)
    external field magnets, 436
    field gradient, 433 (Figure 29.14)
    free induction decay, 426 (Figure 29.6)
    gyromagnetic ratio, 423 424 (Table 29.1)
    hazards, 437-438
    image, brain, 435 (Figure 29.17)
    image brightness, 430-433 (Table 29.2)
    image contrast, 431-433 (Table 29.3)
    Larmor frequency, 424
    matrix, 434 (Figure 29.16)
    noise, 435-436
    nuclear magnetic resonance, 424-429
    nuclear precession, 424-425 (Figures 29.2, 29.3)
    radiofrequency signal (MRI conversion), 433-435 (Figure 29.14)
    signal averaging, 436
    signal encoding, 435
    slice thickness, 434-435 (Figure 29.14)
    typical MRI unit, 437 (Figure 29.18)
Magnetism, 59-65
    alnico, 59
    atomic, 61
    classification of matter, 64
    cobalt, 59
    compass, 59 (Figure 7.2)
    definition of, 59

detection, 59 (Figure 7.2)
diamagnetism, 64
dipoles, 61
    dipole moments, 423
domains, 61
earth's, 64-65
ferromagnetism, 64
fields, 61
flux, 61, 67
force, 60
induction of, 63-64 (Figure 7.10)
iron, 59
laws of, 60
lines of force, 62-63 (Figure 7.7)
magnetization, 63
nature of, 60-61
nickel, 59
paramagnetism, 64
permeability (magnetic), 64
poles, 60
retentivity, 64J
Magnetogyric ratio, 424 (Table 29.1)
Magnets (*see* Magnetism)
Magnification, radiographic, 223-225 (Figures 18.13-18.15)
Main switch, 150
Mammography, 285-294
    benefit/risk ratio, 286, 384
    compression, breast, 289
    high-frequency generator, 287-289
    low-dose film screen, 289
    mammography unit, dedicated, 288 (Figure 22.4)
    mobile mammography, 294
    patient followup, 290
    quality standards, 289-291
        equipment, 291
        qualifications, personnel, 291
        quality assurance (QA), 291
        quality control (QC), 291
        quality control tests, scheduled, 292-294
    soft tissue radiography, 285-287
    x-ray tube, mammographic, 286-287
Manifest image, 194 (Figure 17.3)
Manual (time-temperature) processing, 195-197
Marks (artifacts), 185-186, 196-197
mAs, 214
    and quantum mottle, 211
    and signal/noise ratio, 212

meter, 157
Mass, 18
    atomic mass number, 30
    inertia, 22
    unit of mass, 18
Mass-energy equation, 24
Mass number (atomic), 30
Matrix
    computed tomography, 318-319 (Figure 25.6)
    digital imaging, 308 (Figure 24.3)
    magnetic resonance imaging, 434-435 (Figure 29.16)
Matter, 26-38
    analysis of, 26-28
    atomic structure, 28-31
    definition of, 26
    inertia, 22
    states of, 26
    subdivisions, 26-28
Maximum kV (kVp) *vs* electron velocity, 121 122
Maximum photon energy, 122
Mean lethal dose (d_o), 370-371 (Figure 17.17)
Median lethal dose (LD_{50}), 372-373 (Table 27.2)
Meiosis, 362-363 (Figure 27.8)
Metallic silver (manifest image), 193-194, 301 (Figure 17.2)
Meters, electric
    AC, 52 (Figure 6.8)
    connections in circuit, 51-52
    DC, 81 (Figure 9.12)
    electrodynamometer, 82
    mAs, 157
Microfocus tube, 141, 226
Milliammeter, 157-158, 159 (Figure 14.12)
Milliamperage (mA), 117
    average, 95
    control, 89-92
    exposure rate, 117
    filament current, effect on, 136-137 (Figure 13.2)
    peak value, 95-96
    selector, 161
    stabilizer (filament), 156
Milliampere-second (mAs) meter, 153-157
Milliampere-seconds (mAs), 214
Minimum detectable activity, 342
Minimum exposure time, 154-155
Minimum wavelength, 122
Mitosis, 360-362 (Figure 27.6)

Mitotic death, 369
Mixture, 26
MKS system, 18
Mobile radiography units, 168-171
Modes of action of ionizing radiation, 366-368
    activation of water, 367-368
    direct action, 366-367 (Figure 27.13)
    free radicals, 367-368
    hydrolysis, 367-368
    indirect action, 367-368
Modulation transfer function (MTF), 228-230
    (Figure 18.21)
Molecule, 26
Molybdenum ("moly") target, 286
Monitors
    personnel, 400-402 (Figure 28.1)
    television, 266-267
Monoenergetic radiation (characteristic), 114-115,
    118
    in mammography, 286
Motor, electric, 78-82 (Figures 9.9-9.11) (*see also*
    Electric motor)
Mottle, screen, 211
Moving grids, 244-246 (Figure 19.14)
MRI (*see* Magnetic resonance imaging)
M-shell, 29
Mutual induction, 71, 83-84 (Figure 10.1)
Mutation(s), 384
    rate, 387

N

Natural background, 392-394 (Table 28.1)
Natural radioactivity, 323-328
Natural science, 16 (Figure 2.1)
Negatron, 128
Neutron, 29-30, 330-332
    capture, 331
    chain reaction, 331
    fast, 331, 395 (Table 28.2)
    flux, 330-331
    quality factor ($\overline{Q}$), 395
    radiative capture, 332
    slow (thermal), 331, 395 (Table 28.2)
Neutrons and RBE, 357 (Figure 27.2)
Newton's second law, 22
NMR (*see* Nuclear magnetic resonance)
Noise
    CT scanning and, 414-415
    digital imaging, 311

fog
    light, 128
    x rays, 232-234
information *vs* noise, 212
macroradiography, 228
MRI and, 435-436
scattered x rays, 127-129, 218, 233
screen, intensifying, 211, 282
video camera, 268
Nonconductors, 40
Nonstochastic effect, 376, 394
n-p junction, 104 (Figure 11.19)
N-type silicon, 102-103 (Figure 11.16)
Nuclear (cell) radiation injury, 368-369
Nuclear magnetic resonance (NMR), 422-429 (*see
    also* MRI)
    factors in, 427-428
    free-induction decay, 426 (Figure 29.6)
    gyromagnetic ratio, 424
    magnetic dipole moment, 422
    magnetogyric ratio 424
    nuclear spin, 423 (Figure 29.1)
    precession, 424 425 (Figures 29.2, 29.3)
    relaxation, 427-428
    resonance, 425
    spin echo, 428-429 (Figure 29.9)
Nuclear medicine, 335-353 (*see* Radionuclides)
Nuclear reactor, 330-331 (Figure 26.4)
Nuclear transformation, 332-333
Nucleation (in film development), 193
Nucleon, 29
Nucleotide, 363 (Figure 27.7)
Nucleus
    atomic, 28-31 (Figure 4.3)
    cell, 358-359 (Figure 27.3)
Nuclide, 30-31

O

Object blur, 211 (Figure 18.8)
Occupational dose equivalent limits, 396-398
    (Table 28.3)
Octet rule, 102
Off-focus radiation, 120, 248
Ohm, Georg S., 50
Ohm's Law, 50-56, 75-76
    alternating current (AC), 75-76
    direct and alternating currents
        parallel circuit, 54-56 (Figures 6.11-6.13)
        rheostat, 90-91

series circuit
    simple DC circuit, 51
Ohm, unit Ω, definition, 49
One-hand rule, 416
Open core transformer, 85-86 (Figure 10.2)
Optical (radiographic) density, 213
Ovaries, radiation injury to, 381
Overload, circuit, 56
Oxygen effect, 371
Oxygen enhancement ratio (OER), 371

P

Packing density (screens), 182, 261
Pair production, 129-130 (Figure 12.15)
Parallel circuit, 52, 54-56
Paramagnetism, 64
Passbox, 189
Patient protection, 410-416
Penetrating ability (x rays), 117-124 (Figure 12.7)
    kVp and, 218-219 (Figure 18.10)
Penumbra, 206-208, (Figures 18.2-18.4)
Percent, 5
Percent standard deviation, 339-340
Periodic system of elements, 31-33 (Table 4.1)
Phenotype, 359
Phosphorescence, 178-180 (Figure 15.7)
Phosphors, screen, 177, 181-182 (Table 15.1)
Phosphorus, radioactive (^{32}P), 332 (Figure 26.6)
Photoelectric cell, 37, 153-154 (Figure 14.7)
Photoelectric interaction, 125-130
Photomultiplier tube (PMT), 317, 338 (Figure 26.12), 344 (Figure 26.17)
    in CT scanner, 317
    in nuclear medicine, 338, 344
Photons, 110-128
    characteristic, 125-127 (Figure 12.12)
    energy, 118, 121-126 (Figures 12.9, 12.12)
    frequency, 110
    pair production, 125-129 (Figure 12.15)
    scattered, 124-128 (Figures 12.13, 12.14), 129-130
Photostimulable phosphor, 312
Phototimer, 153-154 (Figure 14.7)
Piezoelectric crystal (PZT), 440-441 (Figure 29.20)
Pinhole camera, 208-209 (Figure 18.5)
Pitting, target, 143-144 (Figure 13.10)
Pixels, 267, 308 (Figure 24.3), 310 (Figure 24.5)
    charge coupled device (CCD), 267

CT scanner, 317-318 (Figure 25.5)
Planck's constant, 118
Plane geometry, 10-12 (Figure 1.7-1.9)
Pocket dosimeter, 400
Polarity, electric current, 51 (Figure 6.5)
Pole transformer, 149
Polonium, 323
Polyenergetic radiation (brems), 114, 119-120
Positron, 128, 332
Potential difference, 47-49
    mechanical model, 23 (Figure 3.1)
Potential energy
    electrical, 44
    mechanical, 23-24 (Figure 3.1)
Potter-Bucky (*see* Grids, radiographic)
Power
    electric, 57, 77-78
    loss in electric circuit, 57
    loss in transformer, 86-88 (Figure 10.5)
    surge, 162
    unit, watt (W), 57
    x-ray generator, 166-167
Powers-of-ten, 13-14
Prefixes to units, 21 (Table 2.1)
Pregnant/fertile workers, 398-399
Prereading kVp meter, 151, 159 (Figure 14.12), 160-161
Primary electrons, 131, 365
Primary x-ray beam, 114, 247
Primary x-ray circuit, 150-156, 165-166
    autotransformer, 88-89 (Figure 10.6)
    transformer, 83-88 (Figures 10.1-10.4)
    x-ray machine
    high-frequency, 91-92, 165-166 (Figure 14.19)
    three-phase, 162-163 (Figures 14.15, 14.16)
Processor care, 202-203
Projection, radiographic, 11 (Figure 1.9), 206 (Figure 18.1)
Proportion, 9-10
    direct, 9
    inverse, 9-10
    inverse square, 117, 124, 214-217 (Figure 18.9)
    similar triangles, 11-12 (Figures 1.7-1.9)
Protection in radiology, 391-420 (*see* Health physics)
Protection, patients, 410-416
Protection, personnel, 400-409
Protection surveys, 406-407
Protective wall barriers, 403-404
    primary, 403

secondary, 403-404
Protein synthesis, 365-366
Proton, 29-30
p-type silicon, 103 (Figure 11.17)
Punch-through (screens), 185 (Figure 15.13)
Pushbutton switch, 151 (Figure 14.4)

Q

Quality factor, $\overline{Q}$, 395 (Table 28.2)
Quality (HVL), 117-124 (Figure 12.15)
Quality, radiographic (*see* Radiographs and radiography)
Quality, x rays,
  defined, 117-124
  distortion, 222-226
  energy, 118-123
    electron volt, 119
    relation to kVp, 119-120
  filtration, 117
  frequency (photon), 110
  half-value layer (HVL), 122-123 (Figure 12.11)
  hard, 123-124
  kVp and, 119-120
  minimum wavelength, 122
  maximum photon energy, 122
  penetrating ability, 117-124 (Figure 12.7)
  polyenergetic, 114, 119-120
  scattered, 127-128 (Figure 12.14), 232-234
  secondary, 126, 128
  soft, 124
  specifications of beams, 116-124
  spectral distribution curves, 120-122
Quantity (R), 116-117 (*see also* Exposure)
Quantum, 111
Quantum mottle, 211
  mAs effect, 211
Quantum theory, 111, 118-119 (Table 12.1)

R

R (unit), 116-117, 131
Rad (unit), 131-132
Radiation, electromagnetic, 109-111 (Figures 12.1, 12.2)
Radiation injury, normal tissues, 377-389
  early, limited areas, 377-381
  embryo and fetus, 387-389
  genetic, 386-389
  late, limited areas, 381-386

high dose, 382-383
  low dose, 383-386
Radiation safety officer, 406
Radiative capture, 322
Radioactive (*see also* Radioactivity)
  average life, 327-328
  cobalt 59 (^{59}Co), 332, 347
  cobalt 60 (^{60}Co), 332
  decay, 326-328 (Figure 26.2), 333-335 (Figures 26.7, 26.8)
  decay constant, 327, 334
  decay curves, 328 (Figure 26.3), 334 (Figures 26.7, 26.8)
  equilibrium, 328
  half-life, 327-328 (Figure 26.3), 333-335 (Figures 26.7, 26.8)
  radiations, 325-326
  series, 324
Radioactive iodine (^{123}I), 344-345
Radioactive iodine (^{131}I)
  decay, 333-335 (Figures 26.7, 26.8)
  imaging (thyroid gland), 348-350
  source, 332
  uptake (thyroid gland), 344-345, 346-347 (Figure 26.19)
Radioactive isotopes (*see* Radionuclides)
Radioactive phosphorus (^{32}P), 332 (Figure 26.6)
Radioactivity, 323-354 (*see also* Radionuclides)
  artificial, 329-335
  average life, 327-328
  decay constant, 327
  definition of, 324
  discovery, 323
  half-life, 327-328 (Figure 26.3)
  natural, 323-328
  radioactive decay, 326-327
  radioactive equilibrium, 328
  radioactive series, 324
  radium, 325-328 (Figures 26.1, 26.2)
  units, 325
  types of radiation, 325-326
Radiobiologic lesion, 366-368
Radiobiology, 355-390
  acute whole-body radiation syndromes, 372-375
    explanation, 374-375
  cell reproduction, 360-362
    cell generation cycle, 360 (Figure 27.5)
    meiosis, 362-363 (Figure 27.8)
    mitosis, 360-362 (Figure 27.6)

cell structure, 358-360 (Figure 27.3)
    cytoplasm, 359-360
    nucleus, 358-359
cellular radiosensitivity, 369-371
    cell survival curves, 369-370 (Figures 27.15,
        27.16)
    modifying factors, 370-371
cellular response to radiation, 368-369
chromatin, 358
definition of radiobiology, 355
deoxyribonucleic acid (DNA), 358-359,
    363-366 (Figures 27.9-27.12)
direct action, 366-367
directly ionizing radiation, 356
dose-response models, 375-377 (Figures
    27.20-27.22)
fractionation, dose, 371-372 (Figure 27.18)
genes, 359
history, 355-356
injuries, radiation, 377-385
    early, 377-381 (Figures 27.23-27.27)
    embryo/fetus, 387-389 (Figure 27.28)
    late, 382-385
oxygen enhancement ratio (OER), 371-372
physical basis of radiobiology, 356-357 (Figure
    27.1)
    linear energy transfer (LET), 356-357
    specific ionization, 356
radiobiologic effects, 357-358
radiobiologic lesion, 366-368
    free radicals, 367-368
    modes of action, 366-368 (Figure 27.13)
relative biologic effectiveness (RBE), 357
Radiograph, definition of, 194
Radiographic chemistry, 193-195
Radiographic exposure (summary), 256 (Figure
    19.28)
Radiographic photography, 192-193
    aerial image, 179-180, 194-195 (Figure 17.4),
        218
    image formation, 194-195 (Figure 17.4)
    latent image, 192
    lattice, crystal, 192-193 (Figure 17.1)
    manifest image, 301 (Figure 17.2)
Radiographic quality, devices to improve,
    232-256
    air gap, 246-247
    anode "heel" effect, 253 (Figure 19.25), 254
    beam-limiting devices, 247-252

compensating filters, 254-255
    grids, 234-246
Radiographs and radiography, 205-231
    acutance, 205
    aerial image, 179-180, 194-195 (Figure 17.4)
    air gap, 226-227, 246-247 (Figure 19.16)
    blur, image, 206-212 (*see also* Blur, factors in)
    contrast, 217-222 (*see also* Contrast, radiograph-
        ic)
    defined, 217
    definition (recorded detail), 205
    densitometer, 213, 293 (Figure 22.5)
    density (optical), 213-216
    density gradient, 221
    detail, 205-213
    devices for improving, 232-255
    direct magnification, 226-228
    distance (SID), 214-216 (Figure 18.9)
    distortion, 222-226
    fifteen percent rule, 213
    film contrast, 220-222
    focal spot size, 206
        evaluation, 208-210
    focus-film distance, 214-216 (Figure 18.9)
    geometric factors, 206-209
    grids, 234-246
    inverse square law, 214-216 (Figure 18.9)
    kVp, 218
    macroradiography, 226-228 (Figures 18.18-
        18.20)
    magnification, 223-225 (Figures 18.13-18.15)
    milliamperage, 214
    modulation transfer function (MTF), 228-230
    motion blur, 210
    mottle, screen, 211
    noise, 127-129, 211, 218, 222, 238
    object blur, 212 (Figure 18.8)
    object-film distance, 207-208 (Figure 18.4)
    object-image receptor distance (OID), 207-208
        (Figure 18.4)
    penumbra (geometric), 206-208 (Figures
        18.2-18.4)
    quantum mottle, 211
    recorded detail, 205
    resolution, 205
        test pattern, 209-210 (Figure 18.7)
    scattered radiation, 127-129, 232-234 (Figure
        l9.1)
    screen blur, 210-212

shape distortion, 225-226 (Figure 18.16)
source-image receptor distance (SID), 206 (Figure 18.13)
unsharpness (*see* Blur)
visibility of detail, 205
Radiology, 101
Radiolucency of tissues, relative, 216-217
Radiolysis of water, 367
Radionuclides, 323-354
activity, 333
artificial, 320-326
average life, 327-328
calibration, 353
counting errors, 339-341
counting methods, 343-345 (Figures 26.17, 26.18)
decay constant, 327
efficiency/sensitivity, counters, 341-343
half-lives, 328 (Figure 26.3), 333-335 (Figures 26.7, 26.8)
imaging (gamma camera), 348-351 (Figures 26.20, 26.21)
available examinations, 351-352 (Table 26.2)
instrumentation, 336-339 (Figures 26.9-26.12)
isotopes, 352 (Table 26.2)
medical applications, 335-336, 344-346
natural radioactivity, 323-328 (Figures 26.1, 26.2)
properties of, 332-333
radiations, 325-336
radiation counters, 336-339
GM tube, 336-338 (Figures 26.9-26.11)
scintillation counter, 338 (Figure 26.12)
well (scintillation) counter, 338-339 (Figure 26.13)
radiations, 325-336
tracer studies, 345-347
units of activity, 333
Radiosensitivity of cells (*see* Cellular radiosensitivity)
Radium, 325-328
Radon, 328
Rare earth phosphors, 181-182 (Table 15.1)
Ratio and proportion, 9-10
Ratio, grid, 236-237 (Figures 19.4-19.6)
Rayl (unit), 446
RBE, 359 (Figure 27.2)
Reactance
capacitive, 76
inductive, 76
Reactor, nuclear, 330-331 (Figure 26.4)
Reciprocating bucky, 245
Recorded detail (*see* Radiographs and radiography)
Recovery time, counters, 340-341
Rectangle, 10 (Figure 1.3)
Rectification, 94-100
definition of, 94
forward bias, 95, 104 (Figure 11.20A)
four-diode, 97-100
full-wave, 97-100 (Figures 11.5-11.11)
half-wave, 85-97
inverse voltage, 95, 97, 104
location in circuit, 159 (Figure 14.12)
methods of, 95-100
rectifiers, 100-106
reverse bias, 95, 97, 100
self, 95 (Figure 11.2)
single diode, 96-97 (Figure 11.3)
single pulse, 95-97 (Figure 11.2)
three-phase, 162-164
tube rating, 94, 144
twelve-pulse, 106, 162-164 (Figure 14.17)
two-diode, 97 (Figure 11.4)
two-pulse, 97-100 (Figure 11.11)
Rectifiers, solid state, 100-106
failure, 105-106
oscilloscope, 106
principle, 101-105
spinning top, 105
tests, 105
Reducing agent (developer), 194, 200
Relative biologic effectiveness (RBE), 357 (Figure 27.2), 394-395
REM, 395
Remnant radiation, 179-180, 194-195 (Figure 17.4)
Remote control switch, 152 (Figure 14.5)
Replenishment, 200
Replication of DNA, 365 (Figure 27.12)
Resistance, electrical, 49
definition of, 49
factors in, 49
parallel circuit, 54-55 (Figure 6.12)
series circuit, 52-54 (Figure 6.9)
Resistivity, 49-50
Resolution
longitudinal, 227 (Figure 18.20)
radiographic, 205, 209-213 (Figure 18.7)
test pattern objects, 209-210 (Figure 18.7)

transverse, 227 (Figure 18.20)
Restricted area, 403
Reverse bias, 75-76, 89-90 (Figure 10.7), 95, 97, 104 (Figure 11.20B)
Rhenium target, 140
Rheostat, 90-91 (Figure 10.8), 155
Rhodium target, 287 (Figure 22.3)
Ribonucleic acid (RNA), 364, 365-366
Right hand (motor) rule, 79
Ripple, in
    high-frequency circuit, 166 (Figure 14.20)
    single phase circuit, 94 (Figure 11.11)
    three-phase circuit, 163-164 (Figure 14.17)
R-meter, 418-419 (Figures 28.4, 28.5)
RNA, 364, 365-366
Rod vision, 258, 259
Roentgen (R), unit, 116-117, 131
Roentgenology, 109
Roentgen, Wilhelm C., 108-109
Root mean square (RMS), value of
    current, 75
    voltage, 75
Rotating anode, 139-141 (*see* Anode)
Rotor, induction motor, 80-81 (Figure 9.11)
    rotating anode, 139-140 (Figure 13.7)

S

Safelight, darkroom, 190-191
Saturation current, 142 (Figure 13.9)
Scanning, radionuclide, 348-353 (Figure 26.20)
Scattered radiation
    air gap, 226-227, 246-247 (Figure 19.16)
    area exposed, 247
    beam limitation and, 247-248
    coherent scattering, 127 (Figure 12.13)
    "collimation," 247-253
    Compton interaction, 127-128 (Figure 12.14)
    contrast and, 218, 232-234 (Figure 19.1)
    kilovoltage effect, 233 (Table 19.1)
    photon energy, 233 (Table 19.1)
    positive beam-limiting device (PBLD), 250-251 (Figure 19.21)
    primary *vs* scattered, 233 (Table 19.1)
    radiography, effect on, 232-233
    ratio to primary, 233 (Table 19.1)
    reduction by beam restriction, 248-252
    removal by a grid, 234-246
    unmodified, 127 (Figure 12.13)

variable aperture beam-limiting device, 250-251
    volume of tissue, effect on, 233
Schilling test, 347
Science, definition of, 16
Scintillation counter, 338-339 (Figure 26.12)
Scintiscanner, 348-350 (*see also* Gamma camera)
Screen-film contact, 183-186 (Figures 15.10-15.12)
Screens (*see* Intensifying screens)
Secondary circuit, x-ray unit, 156-159 (Figure 14.12)
    ballistic mAs meter, 157
    cables, shockproof, 157-158 (Figure 14.11)
    ground connection, 157, 159 (Figure 14.12)
    milliammeter (grounded), 157
    rectifier, 157
    transformer, 156-157
    x-ray tube, 158
Secondary x rays, 126-128
Secular equilibrium, 328
Self-induction, 69-70 (Figures 8.7, 8.8)
    autotransformer, 88-89 (Figure 10.6)
    choke coil, 89-90 (Figure 10.7)
Self-rectification, 95-96 (Figure 11.2)
Semiconductors, 40, 101-105
    rectifying, 104 (Figures 11.20, 11.22)
Sensitivity specks, 192-193 (Figures 17.1, 17.2)
Sensitometric (H&D) curves, 220-222 (Figures 18.11, 18.12)
Sensitometry, 202, 292-293 (Figure 22.5)
Series circuit, 52-53
Shells, atomic, 29, 34, 31-36 (Figures 4.3-4.13)
Shielding in nuclear medicine, 417-418
    lead-shielded syringe, 418
    plastic shield (beta particles), 418
    stored radiopharmaceuticals, 417
Short circuit, 56
Sievert (Sv), unit, 395-396
Significant figures, 5-6
Signal/noise (S/N) ratio, 212, 311-312
Silicon rectifier, 101-105
Silver halides, 173-174
Silver lubricant, rotating anode, 140
Silver recovery from fixer, 203
Silver sulfide sensitivity specks, 192-193 (Figures 17.1, 17.2)
Sine curve (AC), 74-75 (Figures 9.3, 9.4)
Single-strand breaks, chromosomes, 385
SI units, 21 (Table 2.2)

Size distortion, 223-225 (Figures 18.13-18.15)

Skin, radiation injury to, 378-379 (Figure 27.3)

Slip rings, 73 (Figure 9.2), 320
  AC generator, 73 (Figure 9.2)
  spiral CT, 320

Small numbers, 13-14

Soft-tissue radiography
  mammography, 285
  xeroradiography, 282

Soft x rays, 124

Solenoid, 67 (Figure 8.3)

Solid state rectifiers, 100-106, 162

failure, 105-106

principle 101-105

tests, 105

Sound, nature of, 438-440
  amplitude (intensity), 440
  direction, 438
  frequency, 439
  properties of, 439-440 (Figure 29.29)
  wavelength, 440
  wave nature of, 438-439 (Figure 29.19)

Source-image receptor distance (SID), 206-207
    (Figures 18.2, 18.3)
  film exposure (inverse square law), 214-216
      (Figure 18.9)
  recorded detail, 206-207 (Figure 18.3)

Source of x rays, 111

Space charge, 95, 112, 141
  effect, 141-142
  compensator, 141-142 (Figure 13.9)

Space-charge limited region, 142 (Figure 13.9)

Spark, 64

Spark gap, 151

Sparkover, x-ray tube, 143

Specific activity, 333

Specific gravity, 19

Specific ionization, 356-357

SPECT, 350-351

Spectral distribution curves, 120-121 (Figure
    12.9), 286 (Figure 22.1)

Spectral sensitivity matching (film-screen), 174

Spectral window, mammography tube, 286

Spectrum
  electromagnetic radiation, 110 (Figure 12.3)
  x-ray, 120-122

Speed or sensitivity
  film, 175
  intensifying screens, 180-182

Speed, x rays, 110

Spinning top, 105-106 (Figures 11.23, 11.24)

Spotfilm changeover switch, 161

Standard deviation, 339-340 (Figure 26.14)

Standard units, 17

Star pattern (resolution test object), 209-210
    (Figure 18.7)

Static discharge, 44-45
  electric, 39-45
  marks on films, 197

Statistical counting errors, 339-340

Step-down transformer (filament), 155-156
    (Figure 14.9), 159 (Figure 14.12)

Step-up transformer (x-ray generator), 156-157
    (Figure 14.12)

Stereoradiography, 273-276 (Figures 21.1, 21.2)
  method, 274 (Figures 21.2, 21.3)
  principle, 273-274 (Figures 21.1, 21.2)
  stereoscope, 275-276 (Figure 21.4)

"Stick" rectifier, 108

Stochastic effects, 376, 394

Storage (radionuclides), 417-418

Stray radiation, 403

Subject contrast, 218-220 (Table 18.1)
  noise, 220
  radiation quality, 218-219 (Figure 18.10)
  radiographic subject, 218-220
  scattered radiation, 220

Substance, 27

Subtraction
  conventional, 306-307 (Figure 24.1)
  digital, 308-312 (Figures 24.2-24.5)

Survey meters, 418-420 (Figures 28.4, 28.5)

Surveys, protection, 406-407

Switches
  changeover, 161
  main, 150 (Figure 14.2)
  pushbutton, 151 (Figure 14.5)
  remote control, 152 (Figure 14.5)
  timers, 151-153

Synthesis, 26

Systems
  English, 17-18
  exponential, 13-14
  metric, 17-18
  periodic, atomic, 31-34 (Table 4.1)
  SI (système internationale), 18, 20 (Table 2.1)
  three-wire, 149-150 (Figure 14.1)

T

Tabular grains, 174-175
Tantalum 182 (^{182}Ta), 336
Target angle, anode, 140
Target, x-ray tube, 111, 138-141 (*see also* Anode/target)
Technetium (^{99m}Tc), 351-353 (Table 26.2)
Television monitor, 266-267
Temperature, 19
    effect on film development, 196 (Figure 17.6), 200
    effect on electrical resistance, 49
    effect on screen speed, 183
Temperature control, processor, 200
Temperature-limited region, x-ray tube, 142
Ten-day rule, 389
Tenth-value layer, 402
Testes, radiation injury to, 381 (Figure 27.26)
    Bergonie and Tribondeau, 355
Theory, definition of, 17
Thermionic emission, 38, 112
Thermoluminescent dosimeter (TLD), 400-401
Thorium, 324
Three-phase generator, 162-164
Three-wire system, 149-150 (Figure 14.1)
Thumb rule (left hand), 66 (Figure 8.1)
Thyratron, 152 (Figure 14.6), 153
Thyristor, 153
Thyroid gland uptake, 346-347 (Figure 26.19, Table 26.1)
Timers, 151-155
    automatic exposure control, 153
    backup, 153-154
    electronic, 153
    mAs meter, 153
    mechanical, 152
    minimum exposure, 154
Time, unit, 18
Time-temperature development, 153 (Figure 17.2), 194, 198-200
TLD, 400-401
Tomography, computed (*see* Computed tomography)
Tomography, conventional, 276-282 (Figures 21.5-21.13)
    applications, 280-281 (Figure 21.12)
    exposure angle, 278 279 (Figures 21.7-21.9)
    image, 279-280

    linear tomograph, 280-281 (Figures 21.11, 21.12)
    multitomography, 281-282 (Figure 21.13)
    objective plane, 277, 280
    patient exposure, 282
    present status, 282
    principles, 277 (Figures 21.5, 21.6)
    section thickness, 278-279
    zonography, 282
Transducer, ultrasound, 440-441 (Figure 29.20)
Tracer studies, nuclear medicine, 335
Transformer (x-ray generator), 83-88 (Figures 10.1-10.4)
    coils (windings), 85-86
    construction, 85-87
    copper losses, 87-88
    effect on current, 85
    effect on voltage, 84-85
    efficiency, 86-87
    insulation, 85
    law, 83-85
    power output *vs* power input, 85
    principle, 83
    single phase, 83-88
    three-phase, 162-166
Transmission, electric power, 77-78, 83
Transmutation, nuclear, 332
Transport system, automatic processor, 198
Traps, electron, 178-179 (Figures 15.5-15.7), 192-193
Triangles, 10-12 (Figures 1.5-1.9)
Triode tube, 170 (Figure 14.23)
Tritium, 30-31 (Figure 4.5)
True absorption, x rays, 125-127
Tube current control 89-91
Tube rating chart, 144-145 (Figure 13.11)
Tube (x-ray) life, 143-148
Tubes, x-ray (*see* X-ray tubes)
Tungsten
    filament, 111, 136-138
    target, 138
    x-ray spectrum, 120-122 (Figure 12.9)
Turbine, 72 (Figure 9.1)
Twenty-eight day rule, 389

U

Ultrasound, 440-450
    absorption, 446

acoustic impedance, 444-445(Table 29.4)
acoustic lens, 442-443 (Figure 29.24)
A-mode display, 446 (Figure 19.27)
amplitude (intensity), 440
angle of incidence (effect on image), 446
artifacts (noise), 446
attenuation, 446
behavior in matter, 443-446
   focused beam, 442-443 (Figures 29.3, 29.4)
   half-value layer, 446
   unfocused beam, 442-443 (Figures 29.3, 29.4)
biohazard, 449
B-mode display, 447-448
damping block, 441
Doppler scan, 448-449
echo reception, 443-446 (Figures 29.25, 29.26)
frequency, ultrasound, 446
M- or TM-mode display, 447 (Figure 29.28)
piezoelectric crystal, 440-441 (Figure 29.20)
production of ultrasound, 440-441
properties, 439-440
PZT® 441
Q factor, 442-443
quarter-wave matching, 441
rayl (unit), 444
reception (echo), 443
reflection, 443-446 (Figures 29.25, 29.26)
refraction, 444 (Figure 29.25), 446
relative amplitude, 444-445
resolution, 442
resonant frequency, 441
sound, nature of, 438-440 (Figure 29.19)
transducer, 440-441 (Figure 29.20)
velocity, 430-431
wavelength, 440
Uncontrolled area, 403
Units
   absorbed dose (gray, Gy), 131
   exposure (R), 131
   manipulation of units, 19
   radioactivity (curie, Ci), 323, 325
   standard, 17
   table, 20 (Table 2.1)
Unmodified scatter, 127 (Figure 12.13)
Unrestricted area, 403
Unsharpness (*see* Blur)
Uptake (radioiodine), 346-347
Uranium-235, 324-325 (Figure 26.5)

Uranium-237, 324
Useful beam, 113, 403
Useful voltage (forward bias), 95, 104 (Figure 11.20A)

V

Valence, 34-35, 102 (Figure 11.4)
   band, 101 (Figure 11.13), 178 (Figure 15.5)
   electrons, 34, 101-103 (Figures 11.14-11.17)
Vanishing equipment, 273-284
   stereoscopic radiography, 273-276 (Figures 21.1-21.4)
   tomography, 276-282 (Figures 21.5-21.13)
   xeroradiography, 282-284
Variables, 12
Vector quantity, 22, 42
Velocity, 19, 28 (Table 2.1)
Video
   camera, 264-266
   monitor, 266-267
Vidicon, 264-266
Volt, 48-49
Voltage (potential difference), 47-48
   back emf, 75-76, 79, 89-90 (Figure 10.7)
   barrier, 104
   compensator, 151
   concept, 48-49 (Figure 6.3)
   drop (IR), 53
   effective (RMS), 75
   forward bias, 95, 104
   inverse, 95, 97-104
   line, 149
   measurement of, 51-52
      kilovoltage, 151, 160-161
   measurement of, 51-52
   Ohm's Law, 50
   parallel circuit, 54-56
   radiographic tube, 94, 111, 112-113
   reverse bias, 95, 97-104
   ripple, 94, 106, 163-164
   series circuit, 52-53
   unit, 49
   useful voltage, 95, 104
Voltage control (kVp)
   transformer, 88-89 (Figure 10.6)
   high frequency, 91-92 (Figure 10.10)
Voltage drop, 53
Voltmeter, 51-52 (Figure 6.8)

compensator, 151
connection, 51-52 (Figure 6.6)
prereading kilovoltmeter, 151, 159 (Figure 14.12)
Volume, 19
Voxel, 318 (Figure 25.5)

W

Wall barriers (protective), 403-404
Washing films, 196, 201 (Figure 17.8), 202 (Table 17.1)
Waste disposal, radionuclide, 420
Water (structure), 35-36
Watt, unit W, 57
Wave aspect of electromagnetic radiation, 109-111
Wavelength (λ) (Figures 12.1-12.2)
defined, 109-110 (Figure 12.2)
minimum, 122
quality and, 119 (Table 12.1)
range, useful, 110 (Figure 12.3)
ultrasound, 440
x rays, 110, 122
Wave aspect of x rays, 109-110
Wave-tail cutoff, 169
Wedge filters, 254-255
Weight, 18
atomic, 31
Well counter, 338-339 (Figure 26.13)
"White" radiation, 114
Whole-body radiation syndromes (*see* Acute whole body. . .)
Wiring diagrams
high-frequency, 165 (Figure 14.19)
single phase, 158-159 (Figure 14.12)
three-phase, 162-164 (Figures 14.15, 14.16)
Work and energy, 22
electric, 57-58
kinetic, 22-23
potential, 23-24 (Figures 3.1, 3.2)

X

Xeroradiography
advantages, 283-284
disadvantages, 284
electrostatic latent image, 283
present status, 284

principle, 282-283
processing of image, 283
X-ray beam restriction (*see* Beam limitation)
X-ray beam classification, 130 (Table 12.4)
X-ray beam specification, 116-124
X-ray circuits, 149-172
alternating current (AC), 71, 73-74, 89-90
automatic exposure control (AEC), 153-154 (Figure 14.7)
autotransformer, 88-89, 150 (Figure 10.6)
backup timer, 153-154
battery-powered, 168
booster, filament, 143
cables, 157 (Figure 14.15)
capacitor-discharge unit, 168-171 (Figures 14.24, 14.25)
changeover switch, 161
circuit breaker, 155 (Figure 14.8)
compensator voltmeter, 151
console, 158-161 (Figure 14.13)
digital fluoroscopy, 307-312 (Figures 24.2-24.4)
exposure switches, 151, 152
falling load generator, 167-168 (Figure 14.21)
filament ammeter, 156
filament circuit, 111-112, 136-137, 155-156 (Figure 14.9)
filament stabilizer, 156
filament transformer, 155-156
fuses, 150 (Figures 14.2, 14.3)
ground connection, 157 (Figure 14.10)
high-frequency, 91-92, 165-166 (Figures 14.19, 14.20)
kVp selector (autotransformer), 88-89 (Figure 10.6)
kVp meter (prereading), 151, 160-161
line voltage, 149, 151
compensator, 151
low voltage, 150-156
main switch, 150
milliammeter, 157-158, 159 (Figure 14.12)
milliampere-second (mAs) meter, 153-157
milliamperage selector, 161
minimum exposure time, 154-155
pole transformer, 149
power rating (transformer), 166-167
power supply, 149
primary (circuit), 150-156,
rectifier, 157, 162-163
remote control switch, 152 (Figure 14.5)

rheostat, 90, 155

ripple, 94, 163-164, 166

secondary circuit, 156-158

space charge compensator, 141-142

stabilizer, filament, 156

three-phase circuit, 162-164 (Figure 14.16)

three-wire system, 149-150 (Figure 14.1)

thyratron, 152-153 (Figure 14.6)

thyristor, 153

timers, 151-155

transformer, single phase, 83-88, 157 (Figure 14.10)

transformer, three-phase, 162-163 (Figure 14.14)

x-ray tube, connection in circuits, 159 (Figure 14.12)

X-ray films, 173-176 (Figure 15.1)

    base, 174

    characteristic curves (H&D), 220-222 (Figures 18.11, 18.12)

    composition, 173-175 (Figures 15.1, 15.2)

    contrast latitude, 175

    defects, 196-197

    direct exposure, 176

    duplicating, 175

    emulsion, 174-175 (Figures 15.1, 15.2)

        T-grain, 174-175

        ultraviolet-sensitive, 182 (Table 15.1)

    exposure holders

        cardboard, 176-177

        cassettes, 177, 184-185 (Figure 15.2)

    fog, 188, 194

    gelatin, 174

    graininess, 174-175

    image formation, 173, 194-195 (Figure 17.4)

    latent image, 192

    latitude, 175

    mammographic, 175

    sensitometric (H&D) curves, 220-222 (Figures 18.11, 18.12)

    spectral sensitivity matching, 174

    speed (sensitivity), 175

    supercoat, 174

    types, 175-176

    xerographic (paper), 283

X-ray generators

    battery-powered, 168

    capacitor discharge, 168-171 (Figures 14.2, 14.3)

digital fluoroscopy, 307 (Figures 24.2-24.4)

falling load, 167-168 (Figure 14.21)

high-frequency, 165-166 (Figures 14.19, 14.20)

single phase, 150-158 (Figure 14.12)

three-phase, 162-167 (Figure 14.16)

X-ray production, 111-113, 135-136

X-ray production, details, 112-113 (Figure 12.4)

    electron interactions in target, 113-115 (Figures 12.5, 12.6)

X-ray quality

    defined, 117-124

    energy, 118-123

    electron volt (eV), 119

        relation to kVp, 119-120

    filtration and, 117

    frequency (photon), 110

    half-value layer (HVL), 122-123 (Figure 12.11)

    hard, 123-124

    kVp and, 119-120

    maximum photon energy, 122

    penetrating ability, 117-124 (Figure 12.7)

    polyenergetic, 114, 119-120

    scattered, 127, 127-128 (Figure 12.14)

    specifications of beams, 116-124

    spectral distribution curves

        molybdenum, 286 (Figure 22.1)

        tungsten, 120-122 (Figure 12.9)

X rays

    absorbed dose (Gy), 131-132

    applied kV (kVp), 117-123

    attenuation, 218 (*see also* Filtration)

    beam classification, 123-124, 130 (Table 12.4)

    beam specification, 116-120

    biologic effects, 368-389

    bremsstrahlung, 113-114

    characteristic, 114-115, 121 (Figure 12.9)

    chemical effect of, 130

    classification, 130 (Table 12.4)

    corpuscular theory, 124

    detection of, 130

    discovery of, 108-109

    dose, 187-188

    dual nature, 124

    efficiency of production, 115

    electromagnetic nature, 109-111

    energy, 118-123

        electron volt (eV), 119

        relation to kVp, 119-120

    excitation by, 125

exposure ("quantity"), R, 116-117, 131
exposure rate (output), 117
filtration, 117
fluorescent, 114-115, 126
frequency (f), in Hz, 110
genetic effects, 366-368
half-value layer (HVL), 122-123 (Figure 12.11)
"hard" x rays, 123-124
harmful effects (*see* Radiobiology)
intensity (output), 117
interactions with matter, 124-130
inverse square law, 117, 124, 214-217 (Figure 18.9)
ionization, 3, 130
ionization chamber, 116, 418-419 (Figure 28.5)
linear energy transfer (LET), 131, 356-357, 366-367, 394
luminescence, 178
maximum photon energy, 122
measurement of, 116-117
monoenergetic, 119
nature of, 109-110
off-focus, 120
origin, 111
output, 117
penetrating ability, 117-124
photographic effect, 130
photons, 110-111, 118, 122
Planck's constant, 118
polyenergetic, 114, 119-120
primary, 114, 247
production of, 112-115, 135-136
properties, 115-116
protection (*see* Health physics)
quality (HVL), 117-124 (Figure 12.11)
quantity, 116-117 (*see* Exposure)
quantum (hμ), 118-120

Roentgen, W.C., 108
scattering, 127-129 (*see also* Scattered radiation)
secondary, 126, 128
"soft," 124
source of, 111
specification (quality), 122-123
spectrum, 120-121 (Figure 12.9), 386 (Figure 22.1)
speed, 110
true absorption (photoelectric), 125-127 (Figure 12.12)
unit of exposure (R), 116-117
useful beam, 113
wavelength (λ), 109-110 (Figure 12.2)
    minimum, 122
wave aspect, 109-110
X-ray tubes, 111-112 (Figure 12.4), 135-148 (Figures 13.1-13.7)
    anode, 111, 138-147 (*see also* Anode/target, x-ray tube)
    cathode, 111, 136-138 (*see also* Cathode/filament, x-ray tube)
    circuits, 89, 95, 111-113, 136-137
    Coolidge, 112 (Figure 12.4)
    electron interactions in target, 113-115 (Figures 12.5, 12.6)
    failure, 138
    kilovoltage, 117, 123, 135-136
    mammography tubes, 286-287 (Figures 22.2, 22.3)
    milliamperage (tube current), 111, 112-113
    saturation current, 142 (Figure 13.9)
    space charge compensation, 141-142

Z

Zonography, 282